Rule Number N	Left — Right Transformation $T^\dagger[N]$	Global Complementation $\bar{T}[N]$	Left — Right Complementation $T^*[N]$
64	8	253	239
65	9	125	111
66	24	189	231
67	25	61	103
68	12	221	207
69	13	93	79
70	28	157	199
71	29	29	71
72	72	237	237
73	73	109	109
74	88	173	229
75	89	45	101
76	76	205	205
77	77	77	77
78	92	141	197
79	93	13	69
80	10	245	175
81	11	117	47
82	26	181	167
83	27	53	39
84	14	213	143
85	15	85	15
86	30	149	135
87	31	21	7
88	74	229	173
89	75	101	45
90	90	165	165
91	91	37	37
92	78	197	141
93	79	69	13
94	94	133	133
95	95	5	5

Rule Number N	Left — Right Transformation $T^\dagger[N]$	Global Complementation $\bar{T}[N]$	Left — Right Complementation $T^*[N]$
96	40	249	235
97	41	121	107
98	56	185	227
99	57	57	99
100	44	217	203
101	45	89	75
102	60	153	195
103	61	25	67
104	104	233	233
105	105	105	105
106	120	169	225
107	121	41	97
108	108	201	201
109	109	73	73
110	124	137	193
111	125	9	65
112	42	241	171
113	43	113	43
114	58	177	163
115	59	49	35
116	46	209	139
117	47	81	11
118	62	145	131
119	63	17	3
120	106	225	169
121	107	97	41
122	122	161	161
123	123	33	33
124	110	193	137
125	111	65	9
126	126	129	129
127	127	1	1

A NONLINEAR DYNAMICS PERSPECTIVE OF WOLFRAM'S NEW KIND OF SCIENCE

Volume III

WORLD SCIENTIFIC SERIES ON NONLINEAR SCIENCE

Editor: Leon O. Chua
University of California, Berkeley

Series A.   MONOGRAPHS AND TREATISES*

*To view the complete list of the published volumes in the series, please visit:
http://www.worldscibooks.com/series/wssnsa_series.shtml

WORLD SCIENTIFIC SERIES ON
NONLINEAR SCIENCE

Series A   Vol. 68

Series Editor: Leon O. Chua

A NONLINEAR DYNAMICS PERSPECTIVE OF WOLFRAM'S NEW KIND OF SCIENCE

Volume III

Leon O Chua

University of California at Berkeley, USA

World Scientific

NEW JERSEY • LONDON • SINGAPORE • BEIJING • SHANGHAI • HONG KONG • TAIPEI • CHENNAI

Published by

World Scientific Publishing Co. Pte. Ltd.

5 Toh Tuck Link, Singapore 596224

USA office: 27 Warren Street, Suite 401-402, Hackensack, NJ 07601

UK office: 57 Shelton Street, Covent Garden, London WC2H 9HE

British Library Cataloguing-in-Publication Data

A catalogue record for this book is available from the British Library.

A NONLINEAR DYNAMICS PERSPECTIVE OF WOLFRAM'S NEW KIND OF SCIENCE
Volume III

ISBN-13 978-981-283-793-6
ISBN-10 981-283-793-0

Printed by FuIsland Offset Printing (S) Pte Ltd, Singapore

To

Katrin

For her sublime courage and steely composure in battling, against all odds,
a dreadful disease, on which she herself has been conducting basic research
at her Stanford laboratory.

CONTENTS

Volume III

Volume I

Volume II

AN ODE TO JOY

Barely a year has passed since declaring in our *Epilogue*, that Volume II would herald the *"end of the beginning"* of an *analytical* approach to cellular automata and complexity theory, and we have already found ourselves in the midst of a surrealistic odyssey that will plunge us into the uncharted and deeper analytical seas of cellular automata. As luck would have it, our seafaring explorations discovered vast colonies of *Isles of Eden*, hidden among almost 90% of the 256 elementary CA local rules. Some are rare, isolated, picturesque gems; others form *continuum* of exquisite island chains (metaphorically speaking), occasionally interrupted only by nature's numerical nuances when the array length is divisible by 2, as in rules $\boxed{45}$ and $\boxed{154}$, or by 3, as in rules $\boxed{105}$ and $\boxed{150}$.

The *analytical* tools invoked to prove the existence of these Isles of Eden are both novel and illuminating, as befit of the pedagogical, entertaining, yet rigorous standards that have been the hallmark of this series of *tutorials*. A colorful *ballad* of its highlights first published in [Chua *et al.*, 2007a] and [Chua *et al.*, 2007b] has been reprinted in this multi-volume set, in edited form.

Volume III continues our tradition, aiming at the *non-experts* who would otherwise be intimidated by the fallacy that Cellular Automata is the exclusive province of gifted physicists and mathematicians. In this respect, we cannot overemphasize to the *experts* that our purpose is not to seek recognition for originality, nor to establish the all too- mundane vanity of identifying who might have first thought of the ideas presented herein, but rather to enlighten interested readers. Consequently, only relevant *expositions* are listed in the reference section.

Chapter 1

ISLES OF EDEN

This paper continues our quest to develop a rigorous *analytical* theory of 1-D cellular automata via a nonlinear dynamics perspective. The 18 yet uncharacterized local rules are henceforth partitioned into ten *complex Bernoulli σ_τ-shift* rules and eight *hyper Bernoulli σ_τ-shift* rules, the latter including such famous rules $\boxed{30}$ and $\boxed{110}$. All exhibit a bizarre *composite wave dynamics* with arbitrarily large Bernoulli *velocity* σ and Bernoulli *return time* τ as the length $L \to \infty$.

Basin tree diagrams of all ten complex Bernoulli σ_τ-shift rules are exhibited for lengths $L = 3, 4, \ldots, 8$. Superficial as it may seem, these basin tree diagrams suggest general qualitative properties which have since been proved to be true in general. Two such properties form the main results of this paper; namely,

> - Rule $\boxed{90}$ has *no Isles of Eden*.
> - Rules $\boxed{105}$ and $\boxed{150}$ are composed of nothing but *Isles of Eden* for all string lengths L not divisible by 3.

Explicit global state transition formulas are given for local rules $\boxed{90}$, $\boxed{105}$ and $\boxed{150}$. Such formulas led to the rigorous proof of several surprising *periodicity constraints* for rule $\boxed{90}$, and to the discovery of a new global, *quasi-equivalence* class, defined via an *alternating transformation*. In particular, local rules $\boxed{105}$ and $\boxed{150}$ *are globally quasi-equivalent* where corresponding space-time patterns can be derived from each other by simply complementing every other row.

Another important result of this paper is the discovery of a *scale-free phenomenon* exhibited by the local rules $\boxed{90}$, $\boxed{105}$ and $\boxed{150}$. In particular, the *period "T"* of all *attractors* of rules $\boxed{90}$, $\boxed{105}$ and $\boxed{150}$, as well as of all *isles of Eden* of rules $\boxed{105}$ and $\boxed{150}$, increases *linearly with unit slope*, in logarithmic scale, with the length L.

Keywords: Cellular automata; nonlinear dynamics; attractors; Isles of Eden; Bernoulli shift; shift maps; basin tree diagram; Bernoulli velocity; Bernoulli return time; complex Bernoulli shifts; hyper Bernoulli shifts; rule 90; rule 105; rule 150; binomial series; scale-free phenomena.

1. Recap of Main Results from Parts I to VI

A rigorous *analytical* theory of one-dimensional cellular automata composed of $L \stackrel{\Delta}{=} I + 1$ identical cells, as shown in Fig. 1, has been studied in the following series of papers from a *nonlinear dynamics* perspective[1]:

Part I: Threshold of Complexity [Chua *et al.*, 2002]
Part II: Universal Neuron [Chua *et al.*, 2003]
Part III: Predicting the Unpredictable [Chua *et al.*, 2004]
Part IV: From Bernoulli shift to $1/f$ spectrum [Chua *et al.*, 2005a]
Part V: Fractals everywhere [Chua *et al.*, 2005b]
Part VI: From Time-reversible attractors to the arrow of time [Chua *et al.*, 2006]

1.1. *Local rules and Boolean cubes*

Observe that the "zeros" and "ones" in Wolfram's truth tables [Wolfram, 2002] are symbolic variables denoting a logic "Yes" or "No" state, or a "high" or "low" state in digital electronic circuit implementations. In order to exploit powerful mathematical tools from nonlinear dynamics, it is necessary to work with *real* numbers. Consequently, in the papers cited above, the *symbolic* truth table shown in Fig. 1(c) is converted into the *numeric* truth table shown in Fig. 1(d).

One could also redefine the "0" and "1" in the symbolic truth tables as *real* numbers, instead of changing "0" to "−1". There are two reasons why we opted for the latter choice. First, each of the 256 local rules can be implemented on a cellular neural network (CNN) chip [Chua & Roska, 2002] with at least three orders of magnitude faster speed than computing on standard digital computers. Such CNN implementations require that the truth tables be formulated in terms of "1" and "−1" [Chang & Muthuswamy, 2007]. The second reason is that the numeric truth table shown in Fig. 1(d) can be conveniently represented by merely coloring the eight vertices of a "unit Boolean cube" whose center is

located at the origin of the (u_{i-1}, u_i, u_{i+1}) — input space, as shown in Fig. 1(e). Such a representation in turn leads to simple visualizations of many rotational symmetrical transformations [Chua *et al.*, 2003]. Each of the 256 local rules corresponds to exactly one Boolean cube in Table 1 (extracted from [Chua *et al.*, 2003]). Observe that the number N printed under each cube corresponds to the local rule number in [Wolfram, 2002]. This number is easily obtained by adding the "vertex weights" of all *red* vertices in the Boolean cube, where the vertex weight for vertex $\textcircled{k}$ is equal to 2^k, as specified in Fig. 1(e), as well as in the lower part of Table 1.

1.2. *Threshold of complexity*

Observe also that the identification number N of each Boolean cube is colored in *red, blue* or *green*, depending on whether the red vertices can be segregated and separated from each other by $\kappa = 1, 2$, or 3 parallel planes, where κ is called *the index of complexity* of the local rule N [Chua *et al.*, 2002]. Table 2 lists all 256 local rules along with their index of complexity.

The *index of complexity* κ is *not* a *definition* of complexity. Rather it measures the relative number of electronic devices needed to implement each local rule. A $\kappa = 1$ local rule requires the smallest number of transistors. More transistors *must* be added to realize a $\kappa = 2$ local rule. Still more transistors are required to implement a $\kappa = 3$ local rule. In other words, the index of complexity κ measures the relative "cost" of hardware (Chip) implementations.

While the *asymptotic* qualitative *behaviors* of all $\kappa = 1$ local rules, and all $\kappa = 3$ local rules, have been completely understood and characterized in [Chua *et al.*, 2006], and in this paper (for Rules $\boxed{105}$, and $\boxed{150}$), there are some $\kappa = 2$ local rules that have not yet been characterized, including rules $\boxed{110}$, $\boxed{124}$, $\boxed{137}$ and $\boxed{193}$ [Chua *et al.*, 2004]. Since these four rules are universal Turing machines, they can never be completely characterized. In other words, it seems that $\kappa = 2$ can be considered as the *threshold of complexity*, in the sense articulated in [Wolfram, 2002].

---

[1]These 6-part papers have been republished, with errors corrected, in two recent edited books [Chua, 2006] and [Chua, 2007].

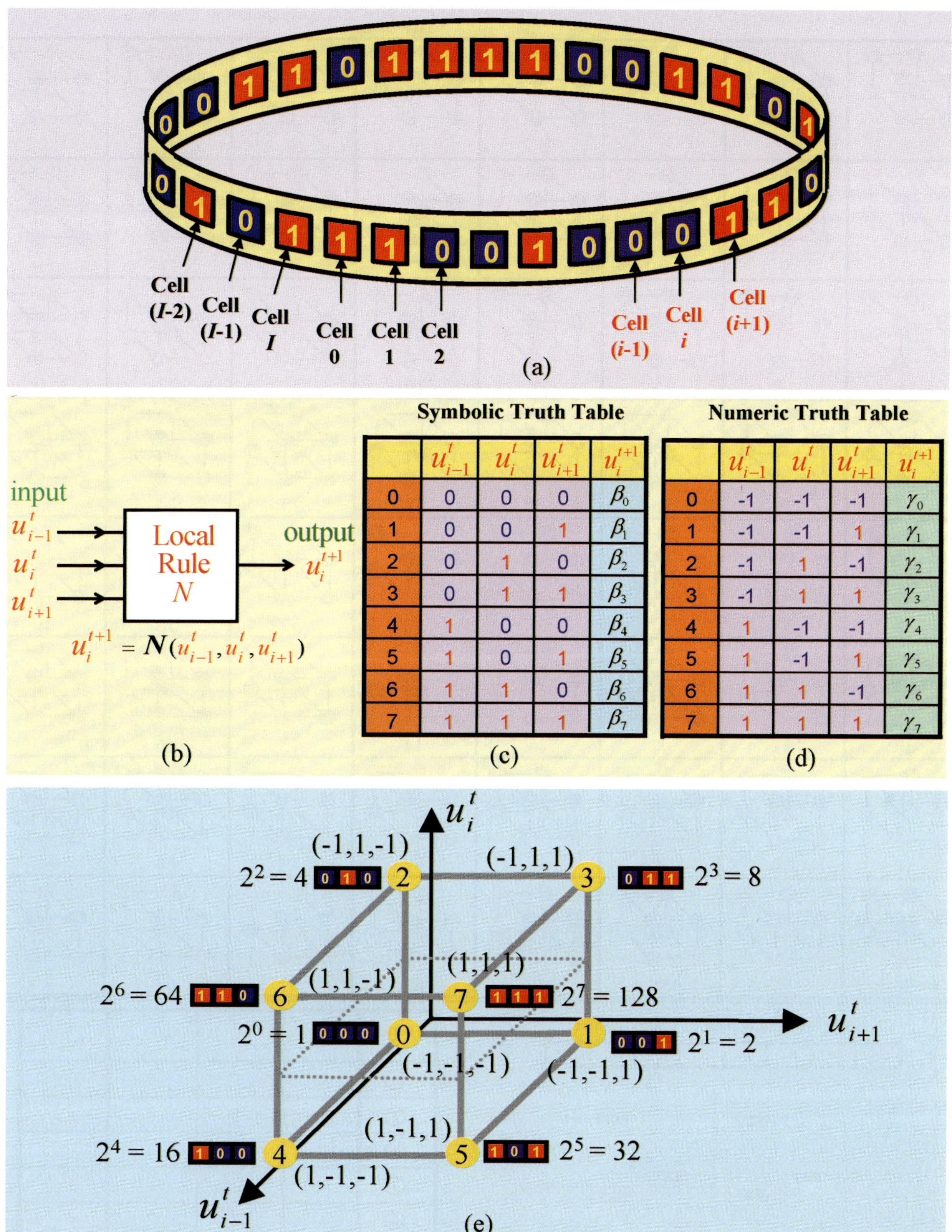

Fig. 1.   (a) A one-dimensional Cellular Automata (CA) made of $L = I + 1$ identical cells with a periodic boundary condition. Each cell "i" is coupled only to its left neighbor cell $(i-1)$ and right neighbor cell $(i+1)$. (b) Each cell "i" is described by a local rule N, where N is a decimal number specified by a binary string $\{\beta_0, \beta_1, \ldots, \beta_7\}$, $\beta_i \in \{0, 1\}$. (c) The symbolic truth table specifying each local rule N, $N = 0, 1, 2, \ldots, 255$. (d) By recoding "0" to "-1", each row of the symbolic truth table in (c) can be recast into a numeric truth table, where $\gamma_k \in \{-1, 1\}$. (e) Each row of the numeric truth table in (d) can be represented as a *vertex* of a Boolean Cube whose color is red if $\gamma_k = 1$, and blue if $\gamma_k = -1$.

Table 1.   Encoding 256 local rules defining a binary 1D CA onto 256 corresponding "Boolean Cubes".

vertex ⓚ	u_{i-1}^t	u_i^t	u_{i+1}^t	u_i^{t+1}
⓪	-1	-1	-1	γ_0
①	-1	-1	1	γ_1
②	-1	1	-1	γ_2
③	-1	1	1	γ_3
④	1	-1	-1	γ_4
⑤	1	-1	1	γ_5
⑥	1	1	-1	γ_6
⑦	1	1	1	γ_7

Table 1.   (*Continued*)

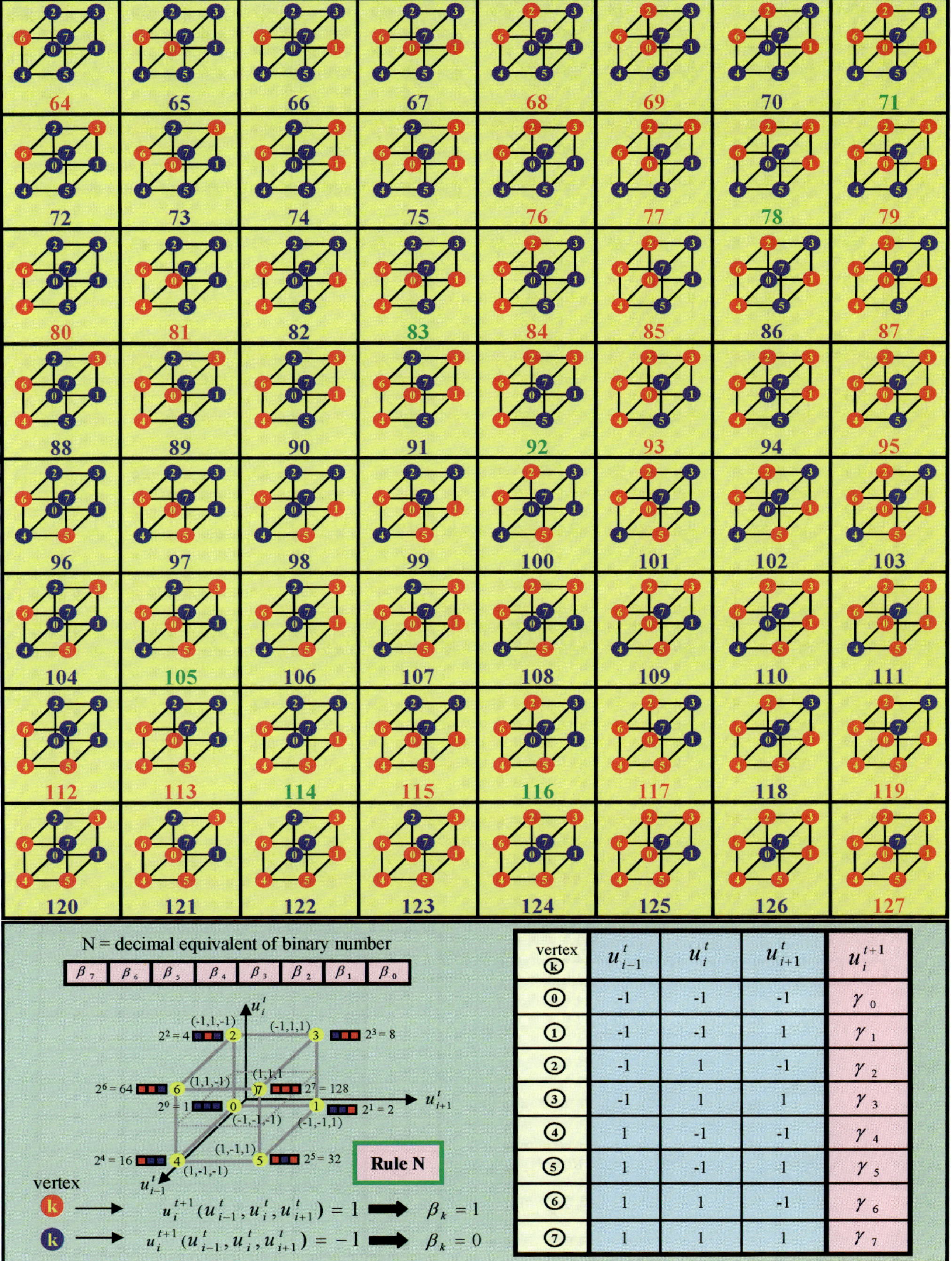

Table 1.   (*Continued*)

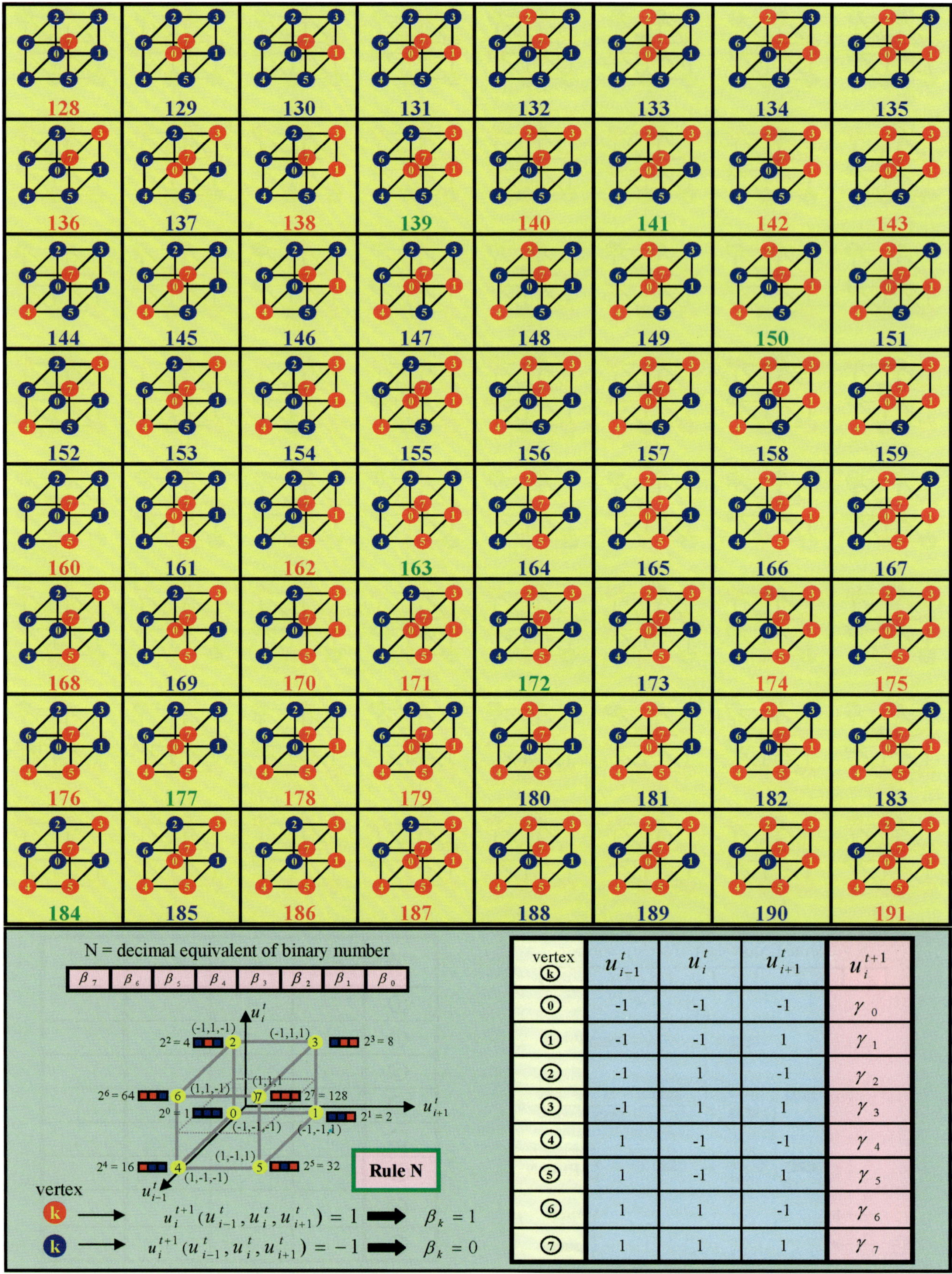

vertex k	u_{i-1}^t	u_i^t	u_{i+1}^t	u_i^{t+1}
0	-1	-1	-1	γ_0
1	-1	-1	1	γ_1
2	-1	1	-1	γ_2
3	-1	1	1	γ_3
4	1	-1	-1	γ_4
5	1	-1	1	γ_5
6	1	1	-1	γ_6
7	1	1	1	γ_7

$$u_i^{t+1}(u_{i-1}^t, u_i^t, u_{i+1}^t) = 1 \implies \beta_k = 1$$
$$u_i^{t+1}(u_{i-1}^t, u_i^t, u_{i+1}^t) = -1 \implies \beta_k = 0$$

Table 1. (*Continued*)

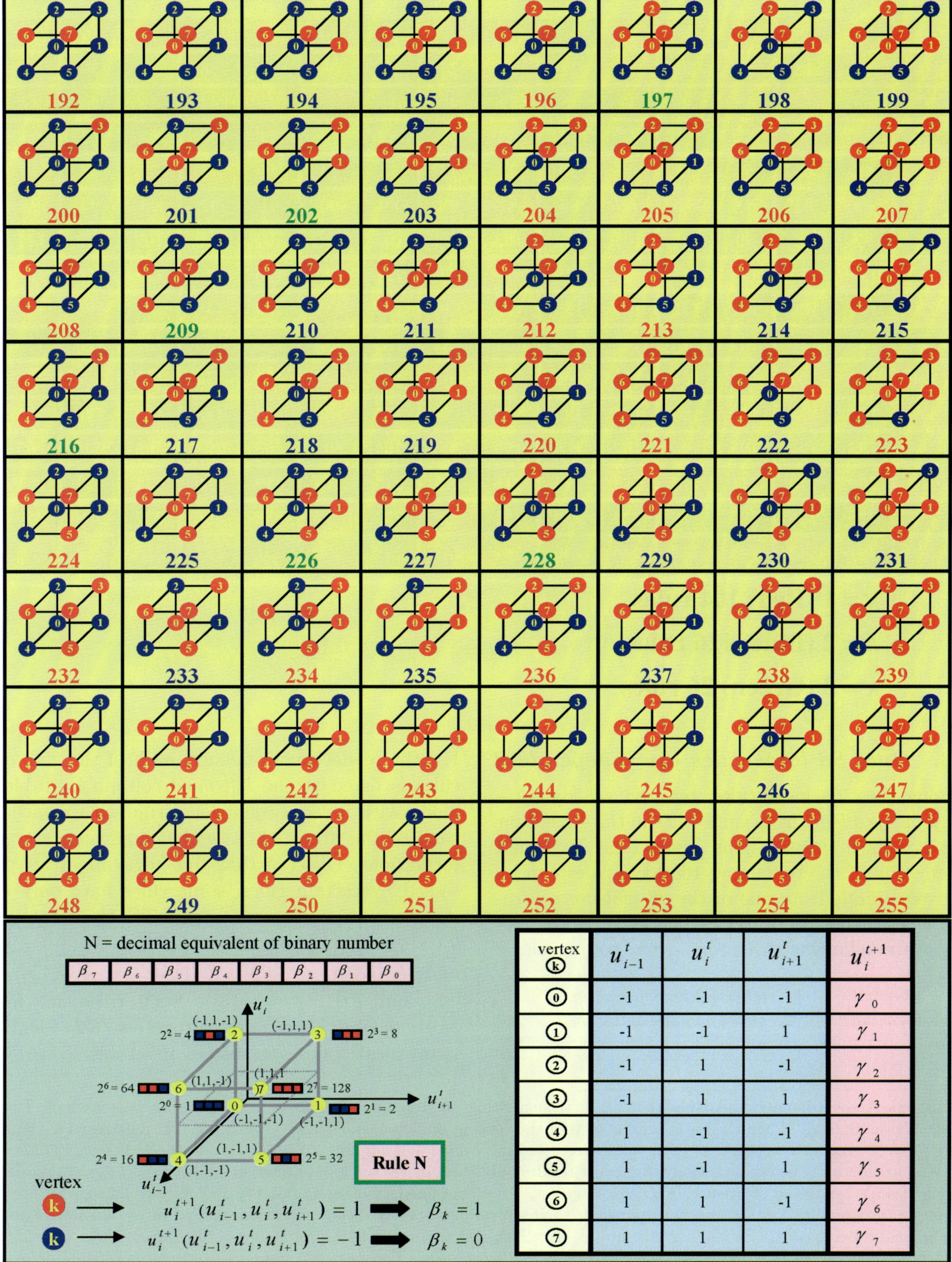

Table 2. List of 256 local rules with their complexity index coded in red ($\kappa = 1$), blue ($\kappa = 2$) and green ($\kappa = 3$), respectively.

0	1	2	3	4	5	6	7	8	9	10	11	12	13	14	15
16	17	18	19	20	21	22	23	24	25	26	27	28	29	30	31
32	33	34	35	36	37	38	39	40	41	42	43	44	45	46	47
48	49	50	51	52	53	54	55	56	57	58	59	60	61	62	63
64	65	66	67	68	69	70	71	72	73	74	75	76	77	78	79
80	81	82	83	84	85	86	87	88	89	90	91	92	93	94	95
96	97	98	99	100	101	102	103	104	105	106	107	108	109	110	111
112	113	114	115	116	117	118	119	120	121	122	123	124	125	126	127
128	129	130	131	132	133	134	135	136	137	138	139	140	141	142	143
144	145	146	147	148	149	150	151	152	153	154	155	156	157	158	159
160	161	162	163	164	165	166	167	168	169	170	171	172	173	174	175
176	177	178	179	180	181	182	183	184	185	186	187	188	189	190	191
192	193	194	195	196	197	198	199	200	201	202	203	204	205	206	207
208	209	210	211	212	213	214	215	216	217	218	219	220	221	222	223
224	225	226	227	228	229	230	231	232	233	234	235	236	237	238	239
240	241	242	243	244	245	246	247	248	249	250	251	252	253	254	255

$\kappa = 1$ (Red) 104 rules

$\kappa = 2$ (Blue) 126 rules

$\kappa = 3$ (Green) 26 rules

1.3. *Only 88 local rules are independent*

Among the 256 local rules, only 88 are dynamically *independent*[2] from each other in the sense that the dynamics and solutions (space-time diagrams) of any one of the remaining 168 local rules can be derived *exactly* from one of the 88 *globally equivalent* rules, listed in Table 3 [Chua *et al.*, 2004], via one of the following three topological conjugacies:

3 Global Equivalence Transformations	1. *left-right transformation* $T^\dagger$ 2. *global complementation* $\overline{T}$ 3. *left-right complementation* T^*

For the reader's convenience, each of the 256 local rules is listed in the *left-most* column in Table 4, along with its equivalent local rule with respect to each of the above three global equivalence transformations. Observe that due to symmetries possessed by certain rules, some rules have only two *distinct* equivalent rules (e.g. $T^\dagger(\boxed{1}) = \boxed{1}$ and $\overline{T}(\boxed{1}) = T^*(\boxed{1}) = \boxed{127}$; $T^\dagger(\boxed{29}) = \overline{T}(\boxed{29}) = \boxed{71}$ and $T^*(\boxed{29}) = \boxed{29}$; $T^\dagger(\boxed{15}) = T^*(\boxed{15}) = \boxed{85}$ and $\overline{T}(\boxed{15}) = \boxed{15}$). Such rules are *identical twins*. There are altogether 72 identical twin local rules, as listed in Table 5. A few

[2]We thank Andy Adamatzky [Adamatzky, 2009] for suggesting possible intersections of our work with [Wuensche & Lesser, 1992]. We thank Andy Wuensche for informing us that the concept of global equivalence classes was first mentioned in [Walker, 1971]. The 88 equivalence classes of local rules were listed in [Walker & Aadryan, 1971] and [Wuensche & Lesser, 1992], using differing numbering schemes. It is likely that other results published, or yet to be published, in our series of *tutorial expositions* on "Wolfram's New Kind of Science" may also intersect, if not contained, in other works. We apologize to all such authors for not citing their publications, and we will appreciate their informing us of any such intersections so that future acknowledgments can be made. Being novice on the mature subject of *cellular automata*, the high probability of such inadvertent omissions is what prompted the authors to publish their papers as *expositions* for a nonspecialist audience, and not as original papers, in the *Tutorial-Review* section of this journal.

Table 3. The first 88 *globally-independent* local rules among the 256 listed in Table 2.

88 Global Equivalence Classes

0	1	2	3	4	5	6	7
8	9	10	11	12	13	14	15
18	19	22	23	24	25	26	27
28	29	30	32	33	34	35	36
37	38	40	41	42	43	44	45
46	50	51	54	56	57	58	60
62	72	73	74	76	77	78	90
94	104	105	106	108	110	122	126
128	130	132	134	136	138	140	142
146	150	152	154	156	160	162	164
168	170	172	178	184	200	204	232

local rules are endowed with additional symmetries such that $T^{\dagger}(\boxed{N}) = \overline{T}(\boxed{N}) = T^{*}(\boxed{N}) = \boxed{N}$. Such rules are *identical quadruplets*. There are only eight identical quadruplet rules, as listed in Table 6.

1.4. Robust characterization of 70 independent local rules

By virtue of the three *global equivalence* transformations derived in [Chua *et al.*, 2004] it suffices to conduct an in-depth analysis of *only* the 88 local rules listed in Table 3, out of 256, a saving of nearly 70% of otherwise wasted man hours! By using *random* bit strings (with at least $L = 400$ bits) as *testing* signals, we have found via extensive computer simulations, and supplemented by analytical studies [Chua *et al.*, 2006], that the *robust time asymptotic* dynamics of 70, out of 88, local rules can be characterized by only one of four *steady-state* behaviors.

1.4.1. Steady-state behavior 1: Period-1 attractors or period-1 isles of Eden

Table 7 lists 26 local rules from Table 3 which exhibit a robust *period-1* steady-state behavior

corresponding to *fixed points* of the *time-1 characteristic function* $\chi^{1}_{\boxed{N}}$ of local rule $\boxed{N}$ [Chua *et al.*, 2004]. Except for rule $\boxed{204}$ where all orbits are period-1 *isles of Eden*, the *generic* steady-state behavior of the other 24 rules in Table 7 are all *period-1 attractors*. This asymptotic behavior holds for *almost all* initial random bit strings, and for *arbitrary length* $L \triangleq I + 1$.

1.4.2. Steady-state behavior 2: Period-2 attractors or period-2 isles of Eden

Table 8 lists 13 local rules from Table 3 which exhibit a robust *period-2* steady-state behavior corresponding to *fixed points* of the *time-2 characteristic* function $\chi^{2}_{\boxed{N}}$ of local rule $\boxed{N}$ [Chua *et al.*, 2006]. Except for rule $\boxed{51}$ where all orbits are period-2 *isles of Eden*, the *generic* steady-state behavior of the other 12 rules in Table 8 are all *period-2 attractors*. This asymptotic behavior holds for almost all initial random bit strings, and for arbitrary L.

1.4.3. Steady-state behavior 3: Period-3 attractors

There is only one rule from Table 3 which exhibits a robust *period-3 attractor*, namely, rule $\boxed{62}$. As demonstrated in, Figs. 5–14 of [Chua *et al.*, 2006], almost all initial bit strings of $\boxed{62}$ converge to a *period-3 orbit* corresponding to *fixed points* of the *time-3* characteristic function $\chi^{3}_{\boxed{62}}$ of local rule $\boxed{62}$ [Chua *et al.*, 2006]. The other attractors of $\boxed{62}$ have a relatively small *basin of attraction*. The period-3 isles of Eden of $\boxed{62}$ have no basins of attraction and therefore require an initial bit string falling exactly on one of the three bit strings forming an isle of Eden.

1.4.4. Steady-state behavior 4: Bernoulli σ_τ-shift attractors or isles of Eden

Table 9 lists 30 local rules from Table 3 which exhibit a robust Bernoulli σ_τ-shift steady-state behavior corresponding to a *period-T attractor* or a *period-T isle of Eden*, where $T \leq \tau L$. The three parameters (σ, τ, β) characterizing each Bernoulli rules are listed in Table 10 for each of the 30 robust Bernoulli rules listed in Table 9.[3] We will henceforth call "σ" the *Bernoulli Shift Velocity*, "τ" the *Bernoulli Return Time* and "β" the *Bernoulli Complementation sign*, or simply Bernoulli

---

[3]Table 10 is constructed from Table 16 of [Chua *et al.*, 2005, pp. 1159–1162].

Table 4.   Table of globally equivalent local rules. All local rules in each row are globally equivalent to each other. Rows with red, blue, or green background colors denote local rules with a *complexity index* $\kappa = 1, 2$, or 3, respectively.

Rule Number N	Left — Right Transformation $T^{\dagger}[N]$	Global Complementation $\bar{T}[N]$	Left — Right Complementation $T^{*}[N]$	Rule Number N	Left — Right Transformation $T^{\dagger}[N]$	Global Complementation $\bar{T}[N]$	Left — Right Complementation $T^{*}[N]$
0	0	255	255	32	32	251	251
1	1	127	127	33	33	123	123
2	16	191	247	34	48	187	243
3	17	63	119	35	49	59	115
4	4	223	223	36	36	219	219
5	5	95	95	37	37	91	91
6	20	159	215	38	52	155	211
7	21	31	87	39	53	27	83
8	64	239	253	40	96	235	249
9	65	111	125	41	97	107	121
10	80	175	245	42	112	171	241
11	81	47	117	43	113	43	113
12	68	207	221	44	100	203	217
13	69	79	93	45	101	75	89
14	84	143	213	46	116	139	209
15	85	15	85	47	117	11	81
16	2	247	191	48	34	243	187
17	3	119	63	49	35	115	59
18	18	183	183	50	50	179	179
19	19	55	55	51	51	51	51
20	6	215	159	52	38	211	155
21	7	87	31	53	39	83	27
22	22	151	151	54	54	147	147
23	23	23	23	55	55	19	19
24	66	231	189	56	98	227	185
25	67	103	61	57	99	99	57
26	82	167	181	58	114	163	177
27	83	39	53	59	115	35	49
28	70	199	157	60	102	195	153
29	71	71	29	61	103	67	25
30	86	135	149	62	118	131	145
31	87	7	21	63	119	3	17

Table 4.   (*Continued*)

Rule Number	Left — Right Transformation	Global Complementation	Left — Right Complementation	Rule Number	Left — Right Transformation	Global Complementation	Left — Right Complementation
N	$T^{\dagger}[N]$	$\overline{T}[N]$	$T^{*}[N]$	N	$T^{\dagger}[N]$	$\overline{T}[N]$	$T^{*}[N]$
64	8	253	239	96	40	249	235
65	9	125	111	97	41	121	107
66	24	189	231	98	56	185	227
67	25	61	103	99	57	57	99
68	12	221	207	100	44	217	203
69	13	93	79	101	45	89	75
70	28	157	199	102	60	153	195
71	29	29	71	103	61	25	67
72	72	237	237	104	104	233	233
73	73	109	109	105	105	105	105
74	88	173	229	106	120	169	225
75	89	45	101	107	121	41	97
76	76	205	205	108	108	201	201
77	77	77	77	109	109	73	73
78	92	141	197	110	124	137	193
79	93	13	69	111	125	9	65
80	10	245	175	112	42	241	171
81	11	117	47	113	43	113	43
82	26	181	167	114	58	177	163
83	27	53	39	115	59	49	35
84	14	213	143	116	46	209	139
85	15	85	15	117	47	81	11
86	30	149	135	118	62	145	131
87	31	21	7	119	63	17	3
88	74	229	173	120	106	225	169
89	75	101	45	121	107	97	41
90	90	165	165	122	122	161	161
91	91	37	37	123	123	33	33
92	78	197	141	124	110	193	137
93	79	69	13	125	111	65	9
94	94	133	133	126	126	129	129
95	95	5	5	127	127	1	1

Table 4.   (*Continued*)

Rule Number	Left — Right Transformation	Global Complementation	Left — Right Complementation	Rule Number	Left — Right Transformation	Global Complementation	Left — Right Complementation
N	$T^{\dagger}[N]$	$\bar{T}[N]$	$T^{*}[N]$	N	$T^{\dagger}[N]$	$\bar{T}[N]$	$T^{*}[N]$
128	128	254	254	160	160	250	250
129	129	126	126	161	161	122	122
130	144	190	246	162	176	186	242
131	145	62	118	163	177	58	114
132	132	222	222	164	164	218	218
133	133	94	94	165	165	90	90
134	148	158	214	166	180	154	210
135	149	30	86	167	181	26	82
136	192	238	252	168	224	234	248
137	193	110	124	169	225	106	120
138	208	174	244	170	240	170	240
139	209	46	116	171	241	42	112
140	196	206	220	172	228	202	216
141	197	78	92	173	229	74	88
142	212	142	212	174	244	138	208
143	213	14	84	175	245	10	80
144	130	246	190	176	162	242	186
145	131	118	62	177	163	114	58
146	146	182	182	178	178	178	178
147	147	54	54	179	179	50	50
148	134	214	158	180	166	210	154
149	135	86	30	181	167	82	26
150	150	150	150	182	182	146	146
151	151	22	22	183	183	18	18
152	194	230	188	184	226	226	184
153	195	102	60	185	227	98	56
154	210	166	180	186	242	162	176
155	211	38	52	187	243	34	48
156	198	198	156	188	230	194	152
157	199	70	28	189	231	66	24
158	214	134	148	190	246	130	144
159	215	6	20	191	247	2	16

Table 4.   (*Continued*)

Rule Number	Left — Right Transformation	Global Complementation	Left — Right Complementation	Rule Number	Left — Right Transformation	Global Complementation	Left — Right Complementation
N	$T^{\dagger}[N]$	$\overline{T}[N]$	$T^{*}[N]$	N	$T^{\dagger}[N]$	$\overline{T}[N]$	$T^{*}[N]$
192	136	252	238	224	168	248	234
193	137	124	110	225	169	120	106
194	152	188	230	226	184	184	226
195	153	60	102	227	185	56	98
196	140	220	206	228	172	216	202
197	141	92	78	229	173	88	74
198	156	156	198	230	188	152	194
199	157	28	70	231	189	24	66
200	200	236	236	232	232	232	232
201	201	108	108	233	233	104	104
202	216	172	228	234	248	168	224
203	217	44	100	235	249	40	96
204	204	204	204	236	236	200	200
205	205	76	76	237	237	72	72
206	220	140	196	238	252	136	192
207	221	12	68	239	253	8	64
208	138	244	174	240	170	240	170
209	139	116	46	241	171	112	42
210	154	180	166	242	186	176	162
211	155	52	38	243	187	48	34
212	142	212	142	244	174	208	138
213	143	84	14	245	175	80	10
214	158	148	134	246	190	144	130
215	159	20	6	247	191	16	2
216	202	228	172	248	234	224	168
217	203	100	44	249	235	96	40
218	218	164	164	250	250	160	160
219	219	36	36	251	251	32	32
220	206	196	140	252	238	192	136
221	207	68	12	253	239	64	8
222	222	132	132	254	254	128	128
223	223	4	4	255	255	0	0

Table 5.   List of 72 identical twin rules.

0	1	4	5	250	251	254	255
15	18	19	22	233	236	237	240
29	32	33	36	219	222	223	226
37	43	50	54	201	205	212	218
55	57	71	72	183	184	198	200
73	76	85	90	165	170	179	182
91	94	95	99	156	160	161	164
104	108	109	113	142	146	147	151
122	123	126	127	128	129	132	133

Table 6.   List of eight identical quadruplet rules.

23	51	77	105	150	178	204	232

velocity, *time*, and *sign*, respectively. Observe that local rules $\boxed{6}$, $\boxed{9}$, $\boxed{11}$, $\boxed{14}$, $\boxed{27}$, $\boxed{35}$, $\boxed{38}$, $\boxed{43}$, $\boxed{56}$, $\boxed{57}$, $\boxed{58}$, $\boxed{134}$, $\boxed{142}$, and $\boxed{184}$ have two robust Bernoulli attractors, whereas local rules $\boxed{25}$ and $\boxed{74}$ have three robust Bernoulli attractors.

Observe from Table 10 that only five rules listed in Table 10 ($\boxed{11}$, $\boxed{14}$, $\boxed{15}$, $\boxed{43}$ and $\boxed{142}$) have a *negative* sign for β. The *space-time* evolution patterns of these five rules are generated by following the same procedures as the other rules (*shift left* by σ bits if $\sigma > 0$, or *shift right* by $|\sigma|$ bits if $\sigma < 0$, every τ iterations), and then *complementing* (change color of all bits) the resulting bit string. In fact, except for rule $\boxed{15}$, only one of two Bernoulli attractors from the other four rules have a negative sign for β.

Observe that any Bernoulli (σ, τ, β) rule with $\beta < 0$ is equivalent to iterating the rule with *twice* the *velocity* and *return time without* complementation, i.e.

$$(\sigma, \tau, \beta) = (2\sigma, 2\tau, |\beta|), \quad \text{if } \beta < 0 \qquad (1)$$

For examples illustrating this equivalence, see Table 5 (pp. 2393) for $\boxed{15}$ in [Chua *et al.*, 2003], Fig. 29(a_2) for $\boxed{11}$, Fig. 29(b_2) for $\boxed{14}$, Fig. 29(d_2) for $\boxed{43}$, and Fig. 29(i_2) for $\boxed{142}$ in [Chua *et al.*, 2006].

In general, $T = \tau L$ if $T_0 \overset{\Delta}{=} \tau L/|\sigma|$ is not an integer. If T_0 is an integer, then $T = \tau L/|\sigma|$ for $|\sigma| \geq 2$. If each bit string in the period-T orbit consists of a concatenation of m identical substrings, then the period T is reduced further to T/m.

Each Bernoulli rule listed in Table 9 can possess up to three robust Bernoulli attractors, as depicted in Table 29A of [Chua *et al.*, 2006, pp. 1293–1297] for rules $\boxed{74}$, $\boxed{88}$, $\boxed{173}$ and $\boxed{229}$. Each of these attractors has a large enough basin of attraction that different random initial bit strings could converge to one of these robust Bernoulli σ_τ-shift attractors. This steady-state behavior does *not* depend on the length $L \overset{\Delta}{=} I + 1$ of the bit string. Except for local rule $\boxed{15}$ and $\boxed{170}$, whose orbits are all *isles of Eden*, all other *generic* steady states converge to a Bernoulli σ_τ-shift attractor.

1.4.5.   *There are ten complex Bernoulli and eight hyper Bernoulli shift rules*

Together, Tables 7–9, plus the period-3 rule $\boxed{62}$, made up 70, out of the 88, local rules from Table 3. The robust steady-state behaviors of these 70 local rules have been completely characterized in [Chua *et al.*, 2006]. The remaining 18 rules listed in Table 3 that have not yet been characterized are listed in Table 11, dubbed *complex Bernoulli-shift* rules, and Table 12, dubbed *hyper Bernoulli-shift rules*. It will be clear from the sequel that all of these 18 yet uncharacterized rules are also identified with *Bernoulli shifts* because they behave like Bernoulli σ_τ-shifts from Table 9 except that the number of attractors is no longer bounded by 3, but increases

Table 7.   List of 25 robust Period-1 local rules.

25 Topologically-Distinct Period-1 Rules

0	4	8	12	13
32	36	40	44	72
76	77	78	104	128
132	136	140	160	164
168	172	200	204	232

Table 8. List of 13 robust Period-2 local rules.

13 Topologically-Distinct Period-2 Rules

1	**5**	**19**	**23**
28	**29**	**33**	**37**
50	**51**	**108**	**156**
178			

Table 9. List of 30 robust Bernoulli σ_τ-shift local rules.

30 Topologically-Distinct Bernoulli σ_τ -shift Rules

2	**3**	**6**	**7**	**9**
10	**11**	**14**	**15**	**24**
25	**27**	**34**	**35**	**38**
42	**43**	**46**	**56**	**57**
58	**74**	**130**	**134**	**138**
142	**152**	**162**	**170**	**184**

with the length $L \stackrel{\Delta}{=} I + 1$ of the bit strings. The ten complex Bernoulli shift rules in Table 11 are *bilateral*, and correspond to those listed in column 1 of Table 17 of [Chua *et al.*, 2006, pp. 1176]. The eight hyper Bernoulli-shift rules in Table 12 are *nonbilateral*, and correspond to those listed in column 1 of Table 18 of [Chua *et al.*, 2006]. Table 13 gives a composition of the asymptotic behaviors of all 88 dynamically-independent local rules listed in Table 3.

In this paper (Part VII) only the ten *complex Bernoulli-shift rules* from Table 11 will be studied. The remaining eight *Hyper Bernoulli-shift rules* from Table 12 will be studied in Part VIII.

2. Basin Tree Diagrams of Ten Complex Bernoulli Shift Rules

For binary bit strings

$$\boldsymbol{x^n} = (x_0^n \quad x_1^n \quad x_2^n \quad \cdots \quad x_{L-1}^n) \tag{2}$$

at *time n* with *finite L* and periodic (or fixed) boundary conditions, the evolution

$$\boldsymbol{x}^n \mapsto \boldsymbol{\chi}^1_{\boxed{N}}(\boldsymbol{x}^n) \stackrel{\Delta}{=} \boldsymbol{x}^{n+1} \tag{3}$$

under local rule $\boxed{N}$ must converge to either a *fixed point*

$$\boldsymbol{x}^* = (x_0^{n^*} \quad x_1^{n^*} \quad x_2^{n^*} \quad \cdots \quad x_{L-1}^{n^*}) \tag{4}$$

or to a *periodic orbit* $\Gamma_T(\boxed{N})$ of period $T \leq T_{\max}$, at some *finite* time $n^* = T_{\text{transient}} + T$, where

$$\chi^1_{\boxed{N}} : \sum \to \sum \tag{5}$$

is the *time-1 characteristic function* defined in [Chua *et al.*, 2005a], and

$$T_{\max} \stackrel{\Delta}{=} 2^L \tag{6}$$

is the number of distinct binary bit strings of length L.

2.1. *Basin of attraction and basin trees*

In general, many initial bit strings can converge to one of several *period-T orbits*, including period-1 orbits (i.e. fixed points of $\chi^1_{\boxed{N}}$).

Definition 1. Basin of attraction $B\big(\Gamma_T(\boxed{N})\big)$ of $\Gamma_T(\boxed{N})$.

The *union* of all bit strings which converge to a period-T orbit $\Gamma_T(\boxed{N})$ of local rule $\boxed{N}$, *including* all bit strings belonging to $\Gamma_T(\boxed{N})$, is called the *basin of attraction* $\mathcal{B}\big(\Gamma_T(\boxed{N})\big)$ of $\Gamma_T(\boxed{N})$.

Table 10.   Bernoulli Parameters σ (Bernoulli shift *velocity*), τ (Bernoulli return *time*), and β (Bernoulli complementation *sign*) associated with the 30 Robust Bernoulli Rules from Table 9.

N	σ	τ	β	N	σ	τ	β
2	1	1	+	42	1	1	+
3	-1	2	+	43	1	1	+
6	2	2	+		-1	1	−
	-2	2	+	46	1	1	+
7	-1	2	+	56	1	1	+
9	-2	2	+		-1	1	+
	2	3	+	57	1	1	+
10	1	1	+		-1	1	+
11	1	1	+	58	1	1	+
	-1	1	−		-1	2	+
14	1	1	+	74	1	1	+
	-1	1	−		2	2	+
15	-1	1	−		-3	3	+
24	-1	1	+	130	1	1	+
25	-1	2	+	134	2	2	+
	3	3	+		-2	2	+
	2	5	+	138	1	1	+
27	-1	2	+	142	1	1	+
	2	2	+		-1	1	−
34	1	1	+	152	-1	1	+
35	-1	2	+	162	1	1	+
	1	1	+	170	1	1	+
38	2	2	+	184	1	1	+
	2	2	+		-1	1	+

Table 11.   List of ten complex Bernoulli-shift rules.

18	22	54	73	90
105	122	126	146	150

Table 12.   List of eight Hyper Bernoulli-shift rules.

26	30	41	45
60	106	110	154

Table 13.   Steady-state characterization of 88 dynamically-independent local rules.

Topological Classifications of 88 Equivalence Classes

Topologically-distinct Rules	Number
Period-1 Rules	25
Period-2 Rules	13
Period-T Rules $T > 2$	2
Bernoulli σ_τ-Shift Rules	30
Complex Bernoulli-Shift Rules	10
Hyper Bernoulli-Shift Rules	8
Total	**88**

More precisely,

$$\mathcal{B}\big(\Gamma_T(\boxed{N})\big) \triangleq \cup \left\{ x \in \sum : \rho_{\boxed{N}}^n(x) \in \Gamma_T \right\} \quad (7)$$

where $\rho_{\boxed{N}}^n(x) \triangleq \underbrace{\rho_{\boxed{N}}^1 \circ \rho_{\boxed{N}}^1 \circ \cdots \circ \rho_{\boxed{N}}^1(x)}_{n \text{ times}}$

is the *time-n map* of $\boxed{N}$ [Chua *et al.*, 2005a] $\rho_{\boxed{N}}^n$: $x^0 \mapsto x^n$, where n depends in general on $\boldsymbol{x}$.

Definition 2.   Basin Trees $\Im(\Gamma_T)$. The set of all bit strings which converges to a period-T orbit

$\Gamma_T(\boxed{N})$, *excluding* $\Gamma_T(\boxed{N})$, is called the *basin trees* of $\Gamma_T(\boxed{N})$.

More precisely,

$$\Im(\Gamma_T) \triangleq \mathcal{B}\big(\Gamma_T(\boxed{N})\big)\backslash\Gamma_T(\boxed{N}) \quad (8)$$

An example of a basin tree is shown in Fig. 3(g) of [Chua *et al.*, 2006] for rule $\boxed{62}$ with $L = 9$. In this case, $\Gamma_T = \Gamma_1(\boxed{62}) = \textcircled{0}$, and

$$\Im(\Gamma_T) = \Im\big(\Gamma_1(\boxed{62})\big)$$
$$= \big\{ \textcircled{73}, \textcircled{85}, \textcircled{146}, \textcircled{149}, \textcircled{165}, \textcircled{169},$$
$$\textcircled{170}, \textcircled{292}, \textcircled{298}, \textcircled{330}, \textcircled{338}, \textcircled{340}, \textcircled{511} \big\}$$

$$\mathcal{B}\big(\Gamma_T(\boxed{62})\big) \triangleq \Im(\Gamma_1) \cup \textcircled{0} \quad (9)$$

Observe from Fig. 3(g) that the *digraph* of $\Im\big(\Gamma_1(\boxed{62})\big)$ is a *directed tree* from *graph theory*.

Another example of a basin tree is shown in Fig. 6 of [Chua *et al.*, 2006]. Consider the period-3 orbit

$$\Gamma_3(\boxed{62}) \triangleq \big\{ \textcircled{3}, \textcircled{38}, \textcircled{61} \big\} \quad (10)$$

in Fig. 6(a)-i. The basin tree of $\Gamma_3(\boxed{62})$ is the set of bit strings

$$\Im(\Gamma_3) \triangleq \big\{ \textcircled{40}, \textcircled{23}, \textcircled{60}, \textcircled{1}, \textcircled{35}, \textcircled{22} \big\} \quad (11)$$

In this case, one can associate the basin tree $\Im(\Gamma_3)$ as two subtrees $\{\textcircled{40}\}$ and $\{\textcircled{23}, \textcircled{60}, \textcircled{1}, \textcircled{35}, \textcircled{22}\}$ emerging from the period-3 orbit $\Gamma_3(\boxed{62})$, which is analogous to a *cluster of roots*. For large L, a basin tree in general is made of many topologically similar subtrees, such as Fig. 11 of [Chua *et al.*, 2006]. In this case, we have a period-14 orbit

$$\Gamma(\boxed{62}) = \big\{ \textcircled{59}, \textcircled{102}, \textcircled{93}, \textcircled{51}, \textcircled{110}, \textcircled{89}, \textcircled{55},$$
$$\textcircled{108}, \textcircled{91}, \textcircled{54}, \textcircled{109}, \textcircled{27}, \textcircled{118}, \textcircled{77} \big\} \quad (12)$$

and the basin tree $\Im\big(\Gamma_{14}(\boxed{62})\big)$ of $\Gamma_{14}(\boxed{62})$ is made of seven subtrees having identical topologies.

2.2. *Garden of Eden*

Definition 3.   Garden of Eden. A bit string

$$\boldsymbol{x} = (x_0 \quad x_1 \quad x_2 \quad \ldots \quad x_{L-1})$$

is said to be a *garden of Eden* of a local rule $\boxed{N}$ iff its *preimage* is an empty set.

More precisely,[4] a bit string $\boldsymbol{x}$ is a *garden of Eden* of $\boxed{N}$ iff it has no *predecessors* in the sense

---

[4]Under Definition 3, a *fixed point* x^* of $\chi_{\boxed{N}}^1$, i.e. a *period-1 orbit*, is not a garden of Eden of $\boxed{N}$ because $\chi^{-1}(x^*) = x^*$.

that there *does not exist* a bit string y such that $\boldsymbol{x} = \chi^{1}_{\boxed{N}}(y)$.

Many examples of gardens of Eden can be found in [Chua *et al.*, 2006]. In particular, all *gardens of Eden* of $\boxed{62}$ are identified by a *pink* color in Figs. 3, 6, 8, 9, 11–14, in [Chua *et al.*, 2006]. Observe that they are just the *termini* of subtrees.

2.3. *Isle of Eden*

A cursory inspection of the basin of attractions of the period-3 orbits $\Gamma_3\left(\boxed{62}\right)$ of rule $\boxed{62}$ in Figs. 5(a)–5(f) in [Chua *et al.*, 2006] reveals that there are no basin trees converging to any node (i.e. bit string) belonging to these period-3 orbits! Such orbits are indeed special, and except for rules $\boxed{15}$, $\boxed{85}$, $\boxed{45}$, $\boxed{105}$, $\boxed{150}$, $\boxed{154}$, $\boxed{170}$, and $\boxed{240}$, they are isolated period-T orbits which are *buried* amidst neighboring bit strings belonging to basin trees of other periodic orbits. We will see in Part VIII that for large L, these isolated period-T orbits could have extremely long periods and hence are very, very hard to find,[5] like well-hidden *Easter eggs*! Moreover, such rare objects *cannot* exist in $\mathbb{R}^n$ in view of the *Zubov–Ura–Kimura Theorem* [Garay & Hofbauer, 2003], which implies that "no compact isolated invariant sets in $\mathbb{R}^n$ can be an isle of Eden". These objects can be either *isolated* or *dense*, and are called *Isles of Eden* in [Chua *et al.*, 2005b] and [Chua *et al.*, 2006]. It's time to give a formal definition.

Definition 4. Isle of Eden.
A bit string

$$\boldsymbol{x} = (x_0 \quad x_1 \quad x_2 \quad \cdots \quad x_{L-1})$$

is said to be a *period-n isle of Eden* of a local rule $\boxed{N}$ iff its *preimage* under $\chi^{n}_{\boxed{N}}$ is *itself*, where $\chi^{n}_{\boxed{N}}$ is the *time-n characteristic function* of $\boxed{N}$.

More precisely, $\boldsymbol{x}$ is a *period-n isle of Eden* of a local rule $\boxed{N}$ iff

$$\chi^{-n}_{\boxed{N}}(\boldsymbol{x}) = \boldsymbol{x} \tag{13}$$

Proposition 1. *A bit string $\boldsymbol{x}$ is a period-n isle of Eden of $\boxed{N} \Leftrightarrow \boldsymbol{x}$ belongs to a period-n orbit $\Gamma_n\left(\boxed{N}\right)$ with an empty basin tree; i.e.*

$$\Im\left(\Gamma_n\left(\boxed{N}\right)\right) = \varnothing \tag{14}$$

when $\varnothing$ denotes the empty set.

Proof. Follows directly from Definitions 2 and 4. ∎

Corollary 1. *A bit string $\boldsymbol{x}$ is a period-n isle of Eden of $\boxed{N} \Leftrightarrow$ the orbit through $\boldsymbol{x}$ is a period-n orbit $\Gamma_n\left(\boxed{N}\right)$ where each bit string $\boldsymbol{x}$, $\chi^{1}_{\boxed{N}}(\boldsymbol{x}), \chi^{2}_{\boxed{N}}(\boldsymbol{x}), \ldots, \chi^{n-1}_{\boxed{N}}(\boldsymbol{x})$ has a unique preimage.*

Proof. Follows from Eq. (14) and Proposition 1. ∎

Remarks

1. To avoid clutter, we will usually refer to *all* bit strings belonging to the orbit of a *period-n isle of Eden* also as an *isle of Eden*.
2. Every bit string belonging to a period-n isle of Eden has exactly one *incoming* and one *outgoing* bit string, for all $n \geq 2$.

2.4. *Gallery of basin tree diagrams*

The collection of all *period-n orbits* $\Gamma_n\left(\boxed{N}\right)$ of all possible periods $n = 1, 2, \ldots$ and their associated *basin trees* $\Im\left(\Gamma_n\left(\boxed{N}\right)\right)$ of an L-bit cellular automata under local rule $\boxed{N}$ is called a *basin tree diagram* of local rule $\boxed{N}$. An examination of such diagrams, even for a relatively small L, can reveal certain characteristic qualitative behaviors of the space-time patterns of many local rules. These empirical characteristics can sometimes be proved to be true in general, as will be illustrated for the *complex Bernoulli shift* rules $\boxed{105}$ and $\boxed{150}$ in this paper, and for the *hyper Bernoulli shift* rules $\boxed{45}$ and $\boxed{154}$ in Part VIII.

A *gallery* of such *basin tree diagrams* for the ten *complex Bernoulli shift* rules listed in Table 11 is exhibited in Tables 14–23 for $L = 3, 4, 5, 6, 7$ and 8, respectively. Each table displays the periodic orbits and their basin trees, where each bit string is displayed in color along with its *decimal* identification number, calculated from the decimal equivalent of the binary bit string as in Fig. 6 of [Chua *et al.*, 2006]. For example, for $L = 3$, the two binary bit strings ■■■ and ■■■ in Gallery 18-1 from Table 14 would be identified by the decimal numbers[6]

$$1 \bullet 2^2 + 0 \bullet 2^1 + 0 \bullet 2^0 = 4$$

Table 14.   Basin tree diagrams for rule $\boxed{18}$.

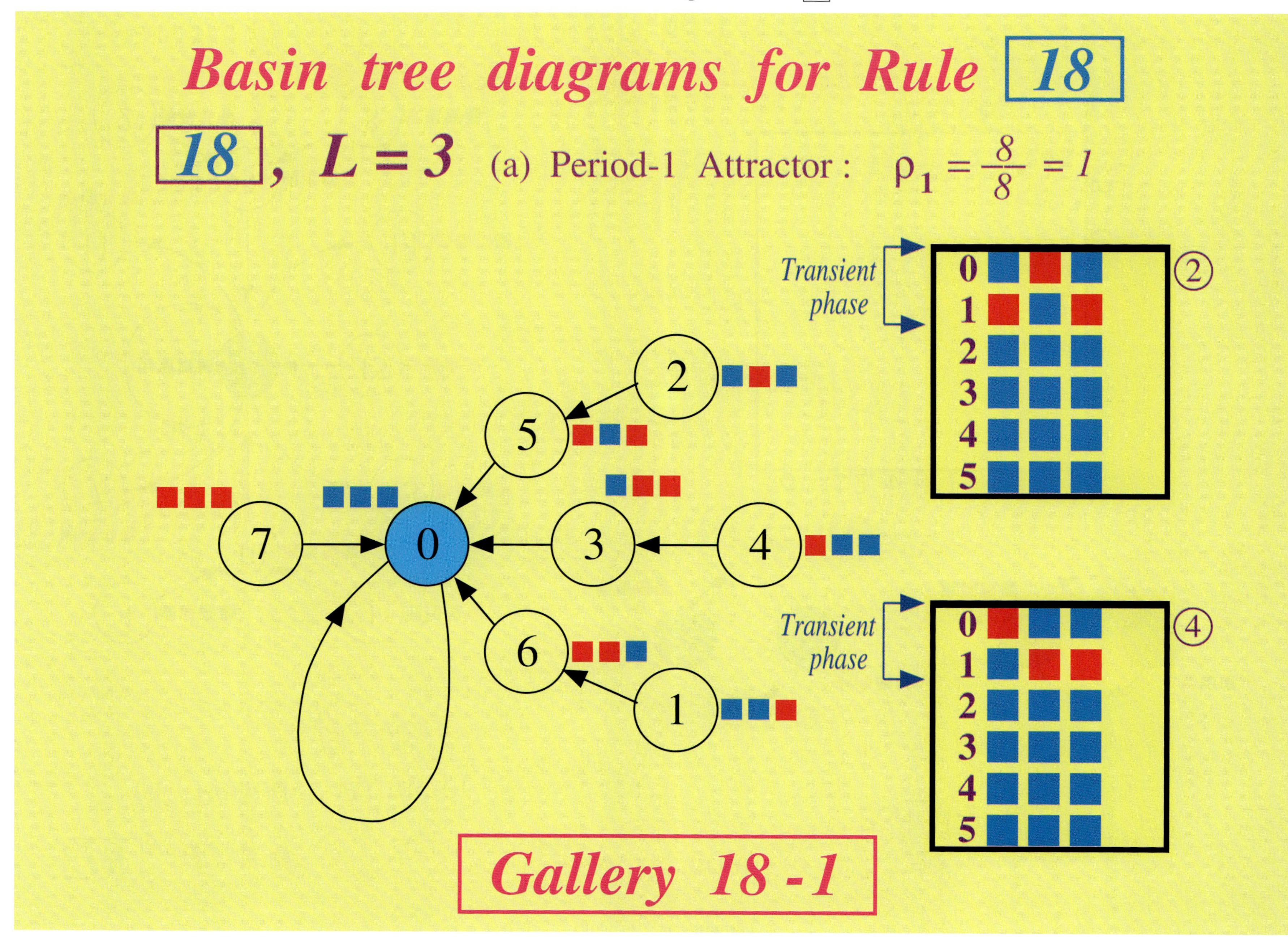

Table 14. (*Continued*)

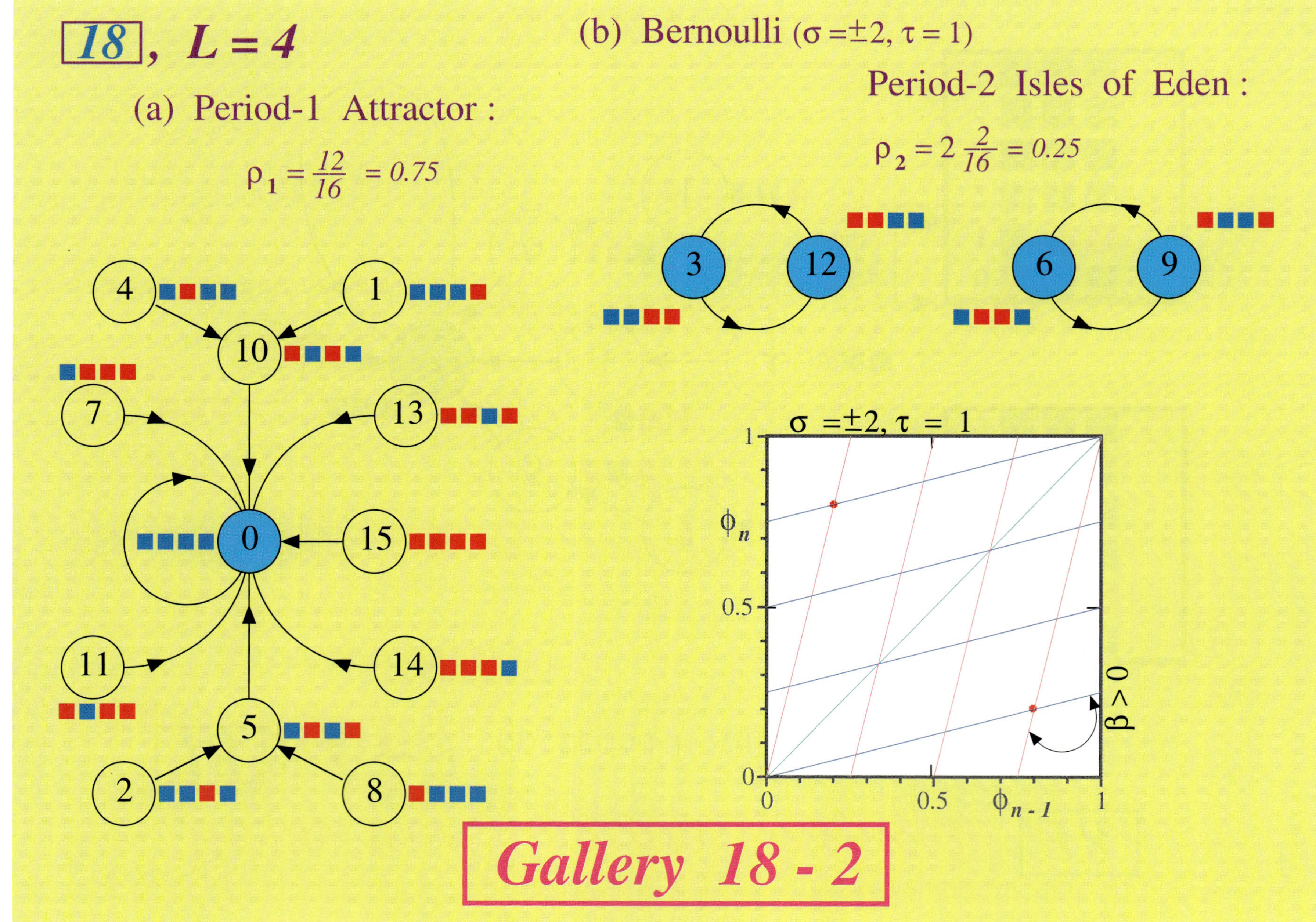

$\boxed{18}$, $L = 5$

(a) Period-1 Attractor : $\rho_1 = \frac{12}{32} = 0.375$

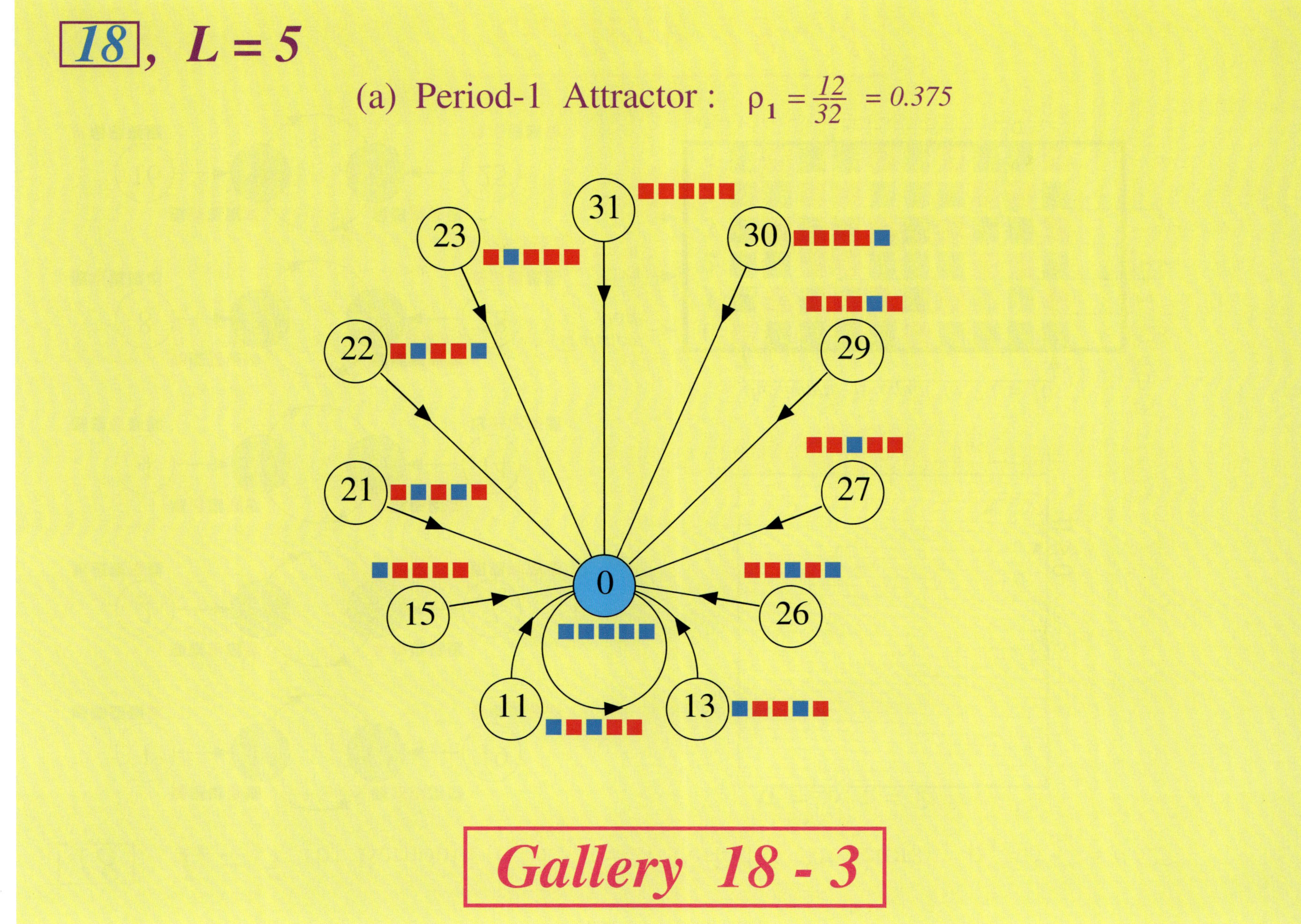

Gallery 18 - 3

18, **L = 5** (b) Bernoulli ($\sigma = \pm 5$, $\tau = 2$) Period-2 Attractors : $\rho_2 = 5\frac{4}{32} = 0.625$

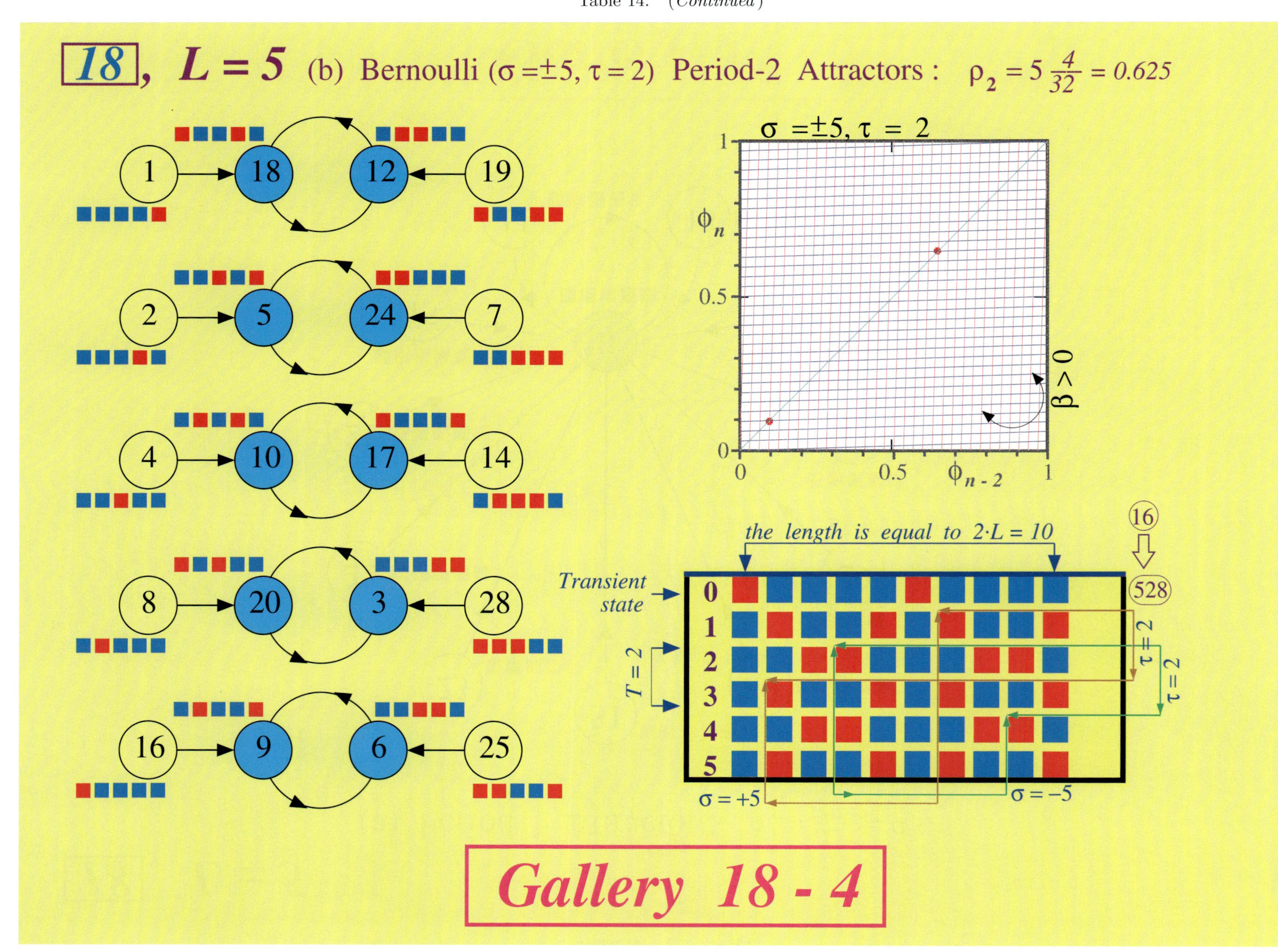

Gallery 18 - 4

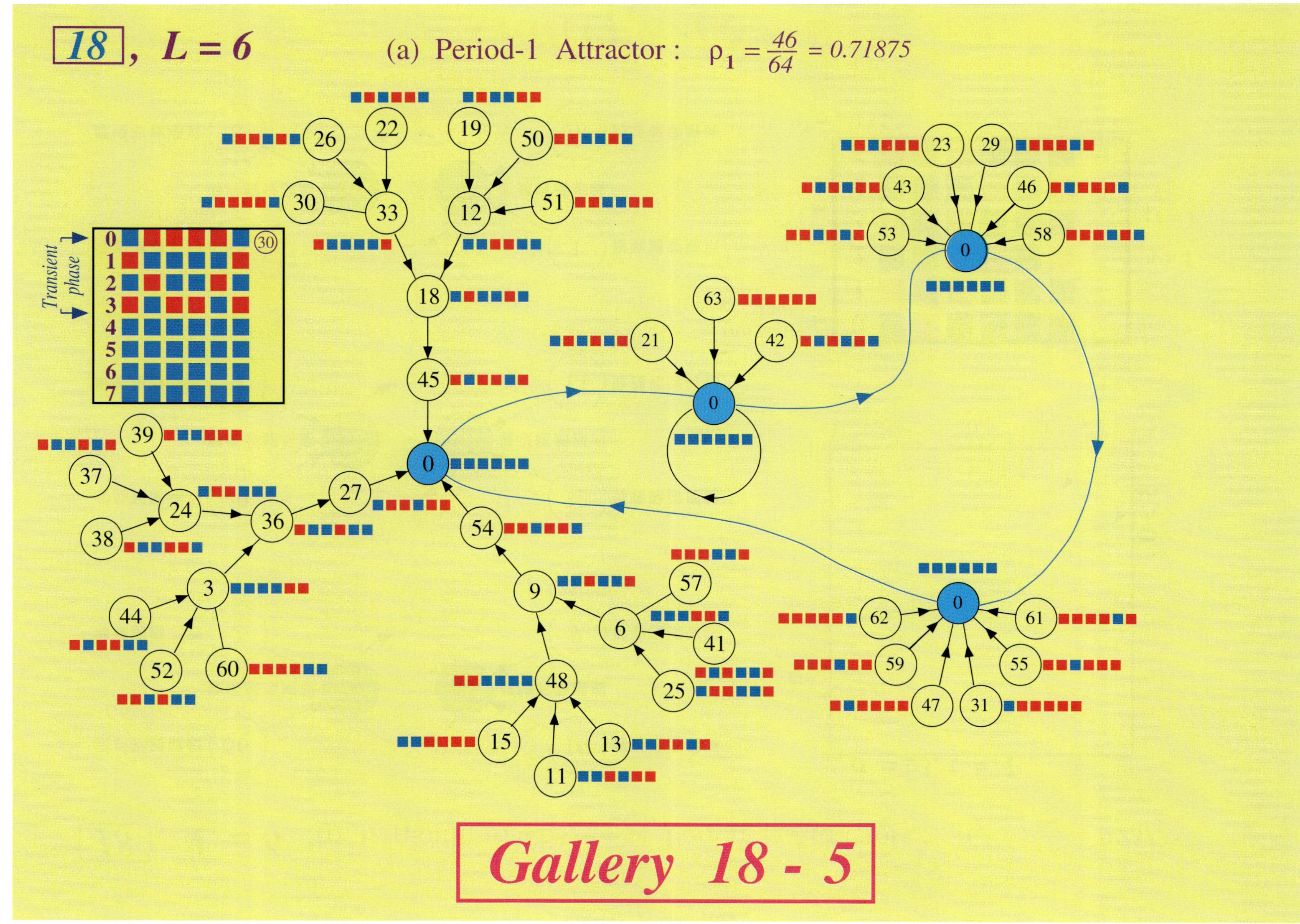
18 , L = 6   (a) Period-1 Attractor : $\rho_1 = \frac{46}{64} = 0.71875$
Transient phase
0 1 2 3 4 5 6 7
Gallery 18 - 5

$\boxed{18}$, **$L = 6$** (b) Bernoulli ($\sigma = \pm 3, \tau = 1$) Period-2 Attractors : $\rho_2 = 3\frac{6}{64} = 0.28125$

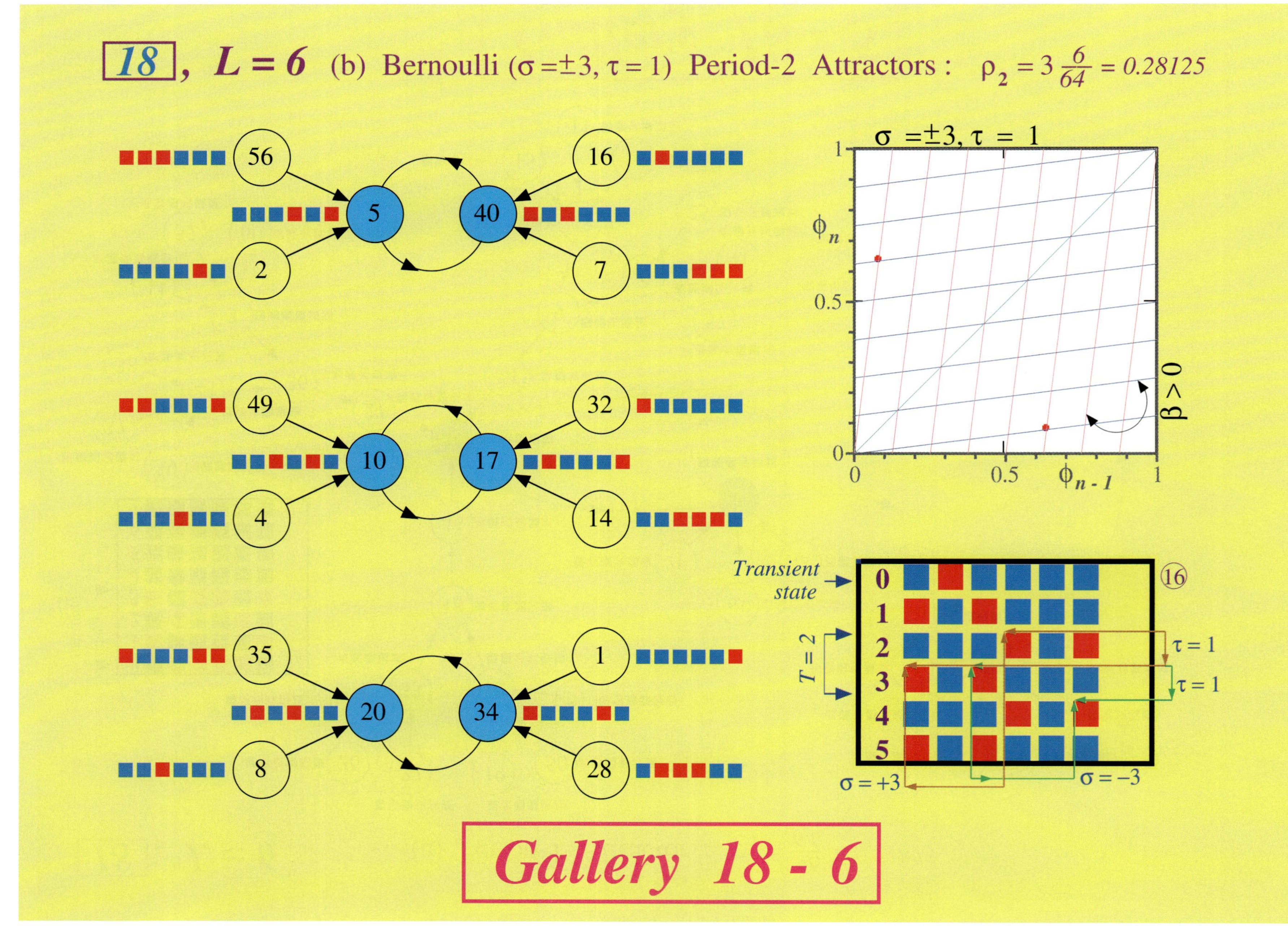

Gallery 18 - 6

Table 14. (*Continued*)

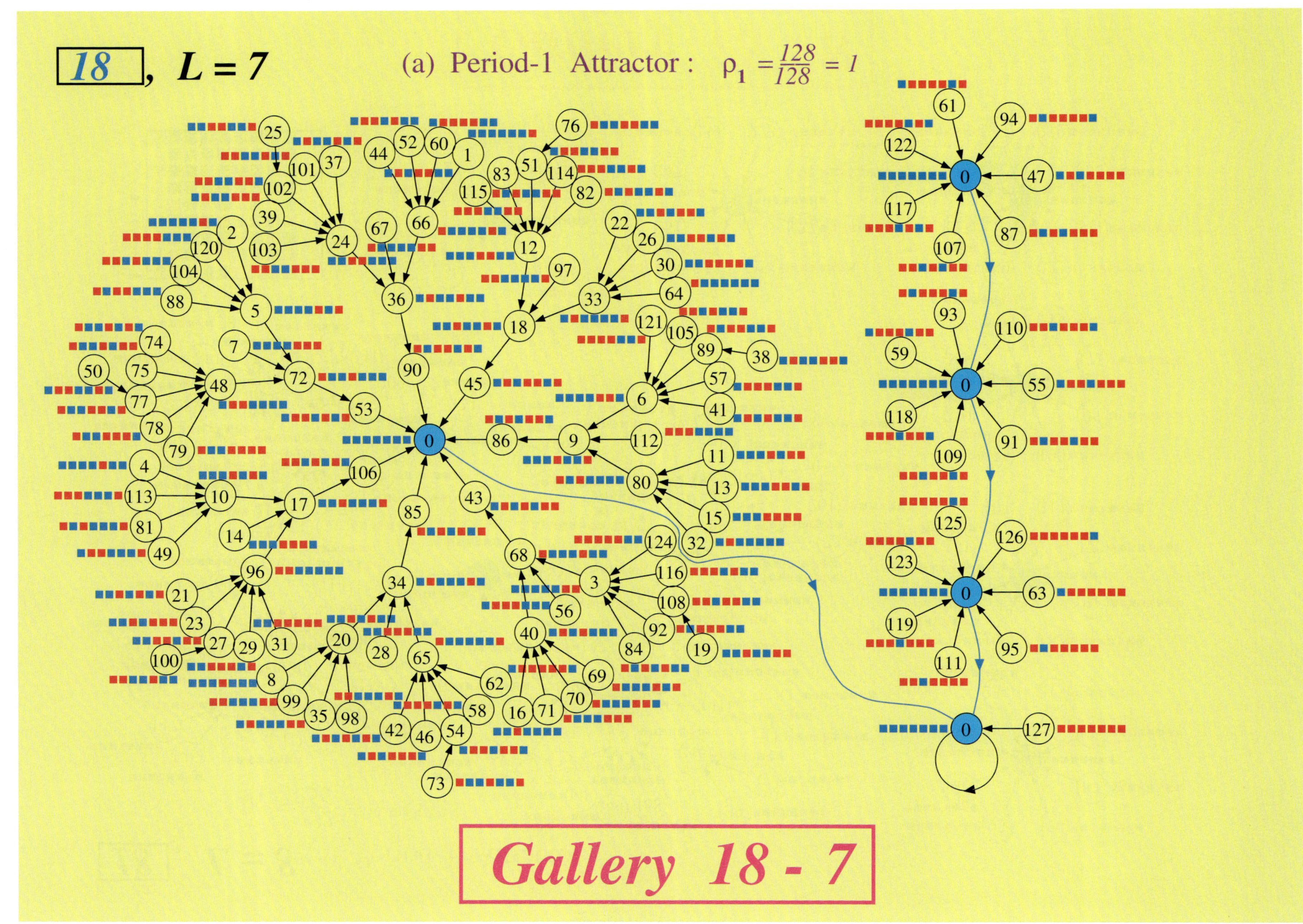

Table 14.   (*Continued*)

18 , $L = 8$ (a) Period-1 Attractor :

$$\rho_1 = \frac{132}{256} = 0.515625$$

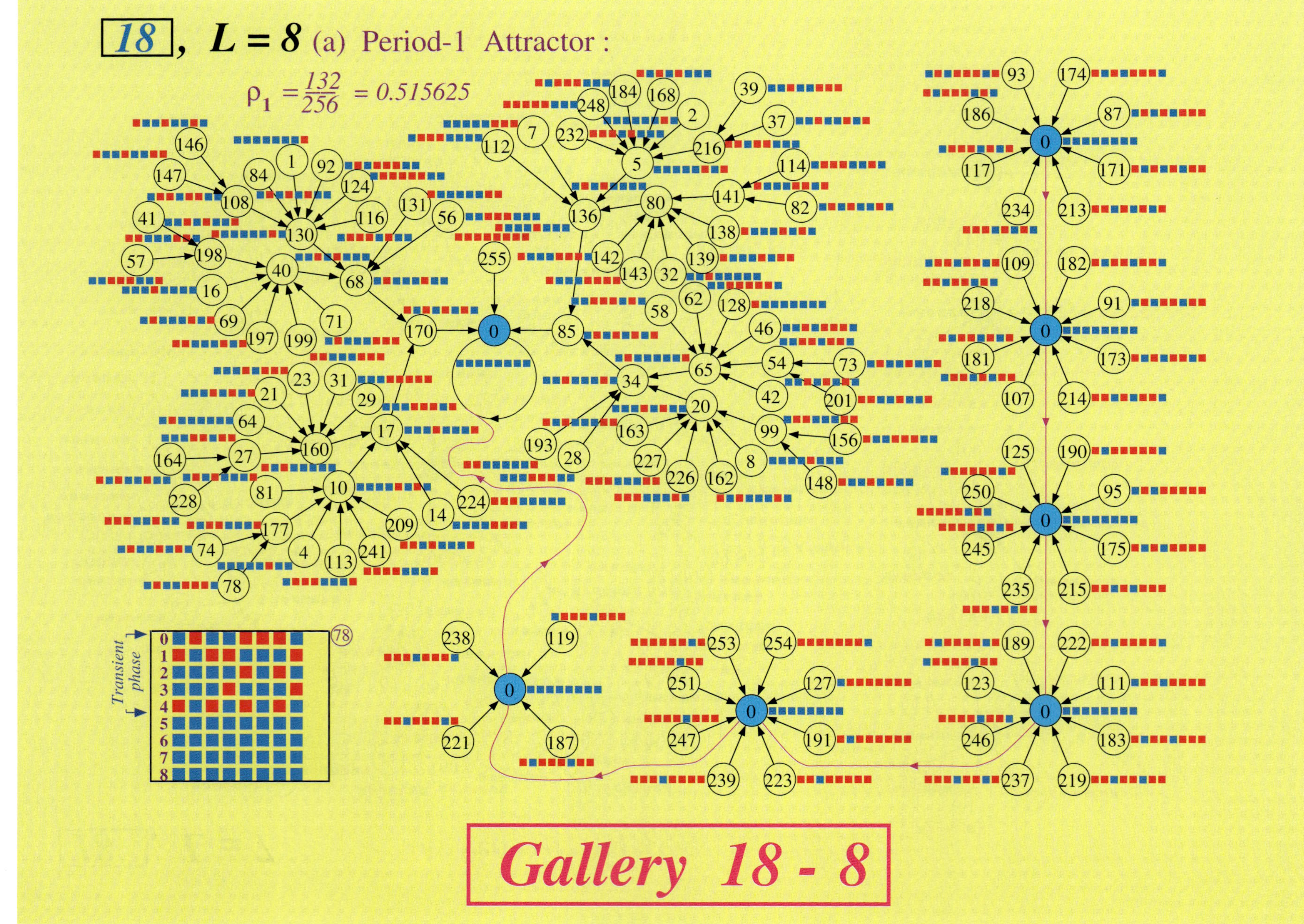

Gallery 18 - 8

26

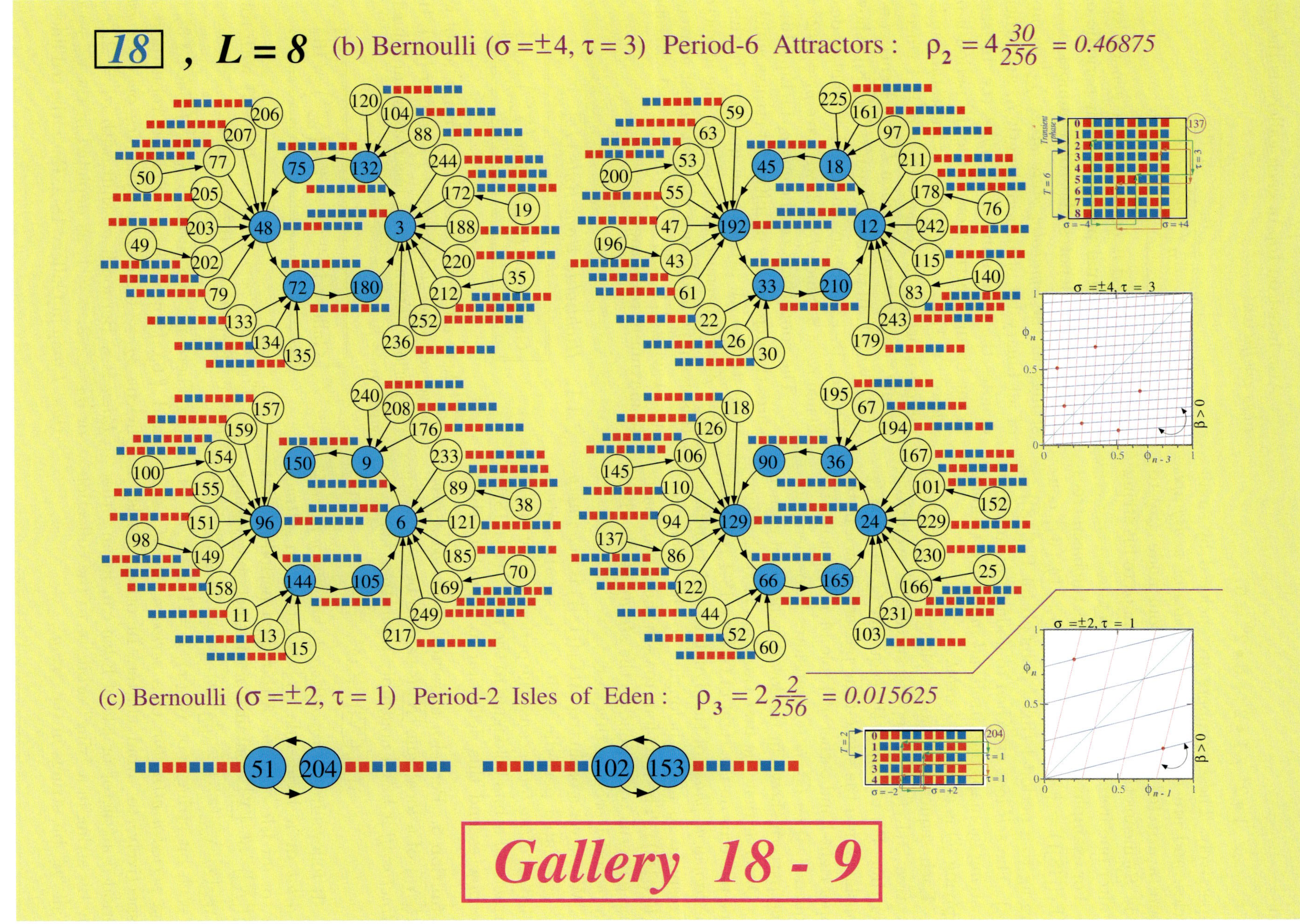

Table 14. (Continued)
18 , L = 8   (b) Bernoulli (σ =±4, τ = 3) Period-6 Attractors :   ρ₂ = 4 30/256 = 0.46875
(c) Bernoulli (σ =±2, τ = 1) Period-2 Isles of Eden :   ρ₃ = 2 2/256 = 0.015625
Gallery 18 - 9

and

$$1 \bullet 2^2 + 1 \bullet 2^1 + 0 \bullet 2^0 = 6$$

respectively. These numbers are enclosed by small circles, and are represented as *nodes* of a *digraph* where a *directed* edge pointing from node $\widehat{S_1}$ to node $\widehat{S_2}$ means that bit string S_1 maps to bit string S_2 after one iteration under rule $\boxed{N}$.

For example, Gallery 18-1 shows the basin trees $\Im\left(\Gamma_1\left(\boxed{18}\right)\right) = \{\widehat{7}; \widehat{2}\ \widehat{5}; \widehat{4}, \widehat{3}; \widehat{1}, \widehat{6}\}$ converging to a period-1 (fixed point) orbit $\Gamma_1\left(\boxed{18}\right) = \{\widehat{0}\}$. The *self-loop* attached to node $\widehat{0}$ means that bit string $\widehat{0}$ maps into itself, *ad infinitum*, thereby implying $\widehat{0}$ is a *period-1 orbit*.

Each sequence of nodes along each branch of the tree $\Im\left(\Gamma_1\left(\boxed{18}\right)\right)$ depicts successive evolutions over time. For example, the sequence $\widehat{2} \to \widehat{5} \to \widehat{0}$ translates into the space-time pattern shown in the upper right-hand corner of Table 14-1. Similarly, the sequence $\widehat{4} \to \widehat{3} \to \widehat{0}$ translates into the space-time pattern shown in the lower right-hand corner.

Observe that the first two rows in both space-time patterns on the right of Gallery 18-1 represent the *transient phase* of the dynamic evolution; they correspond to nodes belonging to the basin tree[7] $\Im\left(\Gamma_1\left(\boxed{18}\right)\right)$. The next four rows in these two space-time patterns correspond to the *steady state*, which is a period-1 orbit in this case.

Whenever a basin tree $\Im\left(\Gamma_1\left(\boxed{18}\right)\right)$ is *not empty*, the associated periodic orbit $\widehat{0}$ in steady state is called an *attractor* because the period-1 bit string $\widehat{0}$ attracts all orbits belonging to the tree $\Im\left(\Gamma_1\left(\boxed{18}\right)\right)$. We now extend this definition to period-n orbits.

Definition 5. Period-n attractor. A *period-n orbit* $\Gamma_n\left(\boxed{N}\right)$ of a local rule $\boxed{N}$ is said to be *a period-n attractor iff* it has a nonempty basin tree, i.e.

$$\Im\left(\Gamma_n\left(\boxed{N}\right)\right) \neq \varnothing \qquad (15)$$

It follows from *Proposition 1* that *every period-n orbit* of a local rule $\boxed{N}$ is either an *attractor*, or an *isle of Eden*. Although a period-n orbit $\Gamma_n\left(\boxed{N}\right)$ of $\boxed{N}$ contains n distinct bit strings $\boldsymbol{x}, \chi^1_{\boxed{N}}(\boldsymbol{x}), \chi^2_{\boxed{N}}(\boldsymbol{x}), \ldots \chi^{n-1}_{\boxed{N}}(\boldsymbol{x})$, we will usually refer to each bit string $\boldsymbol{x}$ or $\chi^k_{\boxed{N}}(\boldsymbol{x})$, $k = 1, 2, \ldots, n-1$, as a *period-n attractor*, or a *period-n isle of Eden*, respectively, to avoid clutter. In other words, a period-n attractor or isle of Eden can mean either any bit string in a "ring" orbit, or to the collection of all "n" bit strings in the "ring".

Also listed on top of each gallery is the *robustness coefficient*

$$\rho_i = \frac{n_i}{n\left(\sum^L\right)} \overset{\triangle}{=} \frac{n_i}{2^L} \qquad (16)$$

of the ith period-n orbit (n is a generic symbol denoting the actual period of each periodic orbit) where $n(\sum^L)$ denotes the total number of all bit strings in the *symbolic state space* $\sum^L$ composed of all binary bit strings of length L, and where n_i denotes the total number of nodes (i.e. bit strings) in the basin of attraction of the ith period-n orbit, where $i = 1, 2, \ldots, m$, and m is the total number of period-n orbits. In Gallery 18-1, $m = 1$ since there is only "one" attractor when $L = 3$. Hence, $i = 1$ in Gallery 18-1. In the basin tree $\Im\left(\Gamma_1\left(\boxed{18}\right)\right)$ shown in Gallery 18-1, there are all together eight nodes and hence $n_i = 8$. Since $L = 3$, we have $\rho_1 = 8/2^3 = 1$.

The robustness coefficient ρ_i in Eq. (16) measures the percentage of initial bit strings which converge to the ith attractor in question. In this case $\rho_i = \rho_1 = 1$ because there is only one attractor in this example and hence *all* orbits must converge to $\widehat{0}$. In general, $0 < \rho_i \leq 1$, where $\rho_i = 1$ corresponds to maximum robustness.

2.4.1.   *Highlights from Rule* $\boxed{18}$

$$\boxed{\text{Gallery 18-1} : L = 3, n\left(\sum^3\right) = 8}$$

There are seven basin-tree strings, all of which converge to the *global period-1 attractor* $\{\widehat{0}\}$. Hence the period-1 attractor $\widehat{0}$ has maximum robustness with $\rho_1 = 1$.

$$\boxed{\text{Gallery 18-2} : L = 4, n\left(\sum^4\right) = 16}$$

(a) There is a *period-1 attractor* $\{\widehat{0}\}$ with robustness coefficient $\rho_1 = 0.75$.

(b) There are two *period-2 isles of Eden* $\{\widehat{3}, \widehat{12}\}$, and $\{\widehat{6}, \widehat{9}\}$ with a combined robustness coefficient $\rho_2 = 0.25$. The dynamics on each *isle of Eden* is a Bernoulli σ_τ-shift with $\sigma_1 = 2$, $\tau = 1$, or $\sigma_2 = -2$, $\tau = 1$, as depicted in the $\phi_n \mapsto \phi_{n-1}$ *time-1 map* in Gallery 18-2. Here, the red lines have

---

[7] Note that our definition of a basin tree $\Im\left(\Gamma_n\left(\boxed{N}\right)\right)$ does *not* include bit strings belonging to the associated period-n orbit $\Gamma_n\left(\boxed{N}\right)$.

slope equal to $2^{\sigma_1} = 4$, and the blue lines have slope equal to $2^{\sigma_2} = 1/4$. Both sets of parallel lines have a positive slope, implying that $\beta > 0$.

Observe that the two period-2 "red" dots correspond to the *decimal representation*

$$\phi = \sum_{i=0}^{L-1} 2^{-(i+1)} x_i \qquad (17)$$

(defined in Eq. (2) of [Chua *et al.*, 2006]) of bit string ③ and ⑫ of the *isle of Eden* {③, ⑫} on the left; namely,

③ $\mapsto 1 \bullet 2^{-3} + 1 \bullet 2^{-4} = 0.1875$ (left red dot)

⑫ $\mapsto 1 \bullet 2^{-1} + 1 \bullet 2^{-2} = 0.75$ (right red dot)

Observe that the two red dots lie at the intersection of corresponding pairs of red and blue "Bernoulli" lines, thereby confirming that the dynamics on this isle of Eden can be described by a *left shift* of two bits ($\sigma = 2$) or, equivalently, by a *right shift* of two bits ($\sigma = -2$), per iteration ($\tau = 1$), as extensively illustrated in [Chua *et al.*, 2005a] and [Chua *et al.*, 2006].

Gallery 18-3, 18-4 : $L = 5, n\left(\sum^5\right) = 32$

(a) There is a *period-1 attractor* {⓪} with robustness coefficient $\rho_1 = 0.375$.

(b) There are five *period-2 attractors* with a combined robustness coefficient $\rho_2 = 0.625$. The dynamics on each *attractor* is a Bernoulli σ_τ-shift with $\sigma_1 = 5$, $\tau = 2$, or $\sigma_2 = -5$, $\tau = 2$, as depicted in the $\phi_{n-2} \mapsto \phi_n$ *time-2 map*.

The time-2 map $\phi_{n-2} \mapsto \phi_n$ consists of $\beta = 2^{\sigma_1} = 32$ parallel red Bernoulli lines with slope $2^{\sigma_1} = 32$, or equivalently, to $\beta = 2^{|-\sigma_2|} = 32$ parallel blue Bernoulli lines with slope $2^{\sigma_2} = 1/32$. Observe that the two red dots now fall on the diagonal of the *time-2* map, as expected of *period-2* orbits. Again, $\beta > 0$ because the slope of each red (or blue) Bernoulli line is positive.

For ease of visualization, we have displayed the space-time pattern using bit strings with double the length, namely, $2L = 10$, which corresponds to shifting around the period-2 ring twice. Note that the decimal code of the 5-bit basin tree ⑯ translates into the corresponding 10-bit string ㊄㉘ shown in Gallery 18-4.

Observe that all basin subtrees contain only one bit string, implying that all basin trees of rule ⑱ are *gardens of Eden*, when $L = 5$.

Gallery 18-5, 18-6 : $L = 6, n\left(\sum^6\right) = 64$

(a) There is a *period-1 attractor* {⓪} with robustness coefficient $\rho_1 = 0.71875$. Note that there are three blue lines joining bit string ⓪ at three locations in the basin tree diagram. This is done to avoid clutter. The reader should interpret all three nodes labeled ⓪ as representing the same node. Observe also from the basin tree diagram that the longest transient regime is four iterations, such as the one depicted in the space-time pattern originating from string ㉚ in Gallery 18-5. The shortest transient regime is one iteration; they correspond to the 15 gardens of Eden in the three "translated" subtrees joined by blue lines.

(b) There are three *period-2 attractors* with a combined robustness coefficient $\rho_2 = 0.28125$. The dynamics on each attractor is a Bernoulli σ_τ-shift with $\sigma_1 = 3$, $\tau = 1$, or $\sigma_2 = -3$, $\tau = 1$. In this case, all basin trees are gardens of Eden.

Gallery 18-7 : $L = 7, n\left(\sum^7\right) = 128$

There are 127 basin tree strings, all of which converge to the *global period-1 attractor* {⓪}. It follows that we have maximum robustness with $\rho_1 = 1$, as in Gallery 18-1.

Gallery 18-8, 18-9 : $L = 8, n\left(\sum^8\right) = 256$

(a) There is a *period-1 attractor* {⓪} with robustness coefficient $\rho_1 = 0.515625$. The transient regime ranges from one iteration (corresponding to subtrees composed of garden of Edens) to five iterations, as illustrated in a typical space-time diagram starting from bit string ㊲ in Gallery 18-8.

(b) There are four *period-6 attractors* with a combined robustness coefficient $\rho_2 = 0.46875$. The dynamics on each attractor is a Bernoulli σ_τ-shift with $\sigma_1 = 4$, $\tau = 3$, or $\sigma_2 = -4$, $\tau = 3$.

The time-3 map $\phi_{n-3} \mapsto \phi_n$ shows $\beta = 2^4 = 16$ parallel Bernoulli "red" lines with slope $2^{\sigma_1} = 16$, or equivalently, 16 parallel Bernoulli "blue" lines with slope $2^{\sigma_2} = 1/16$. Observe that there are six red dots in the time-3 map, implying a *period-6* attractor. Again, $\beta > 0$ because both red and blue lines have a positive slope.

(c) There are two *period-2 isles of Eden* with a combined robustness coefficient $\rho_3 = 0.015625$. The dynamics on each isle of Eden is a Bernoulli σ_τ-shift $\sigma_1 = 2$, $\tau = 1$, or $\sigma_2 = -2$, $\tau = 1$.

The qualitative properties of local rule $\boxed{18}$ extracted from the above basin-tree Galleries 18-1 to 18-9 are summarized below:

Summary of Qualitative properties of local rule $\boxed{18}$ *extracted from Gallery 18 for Rule* $\boxed{18}$

L	ID Number i	Number of Period-n attractors	Number of Period-n Isles of Eden	Period n	Bernoulli Parameters						Robustness coefficient ρ
					σ_1	τ_1	β_1	σ_2	τ_2	β_2	
3	1	1		1	0	1	+				$\rho_1 = 1$
4	1	1		1	0	1	+				$\rho_1 = 0.75$
	2		2	2	2	1	+	-2	1	+	$\rho_2 = 0.25$
5	1	1		1	0	1	+				$\rho_1 = 0.375$
	2	5		2	5	2	+	-5	2	+	$\rho_2 = 0.625$
6	1	1		1	0	1	+				$\rho_1 = 0.71875$
	2	3		2	3	1	+	-3	1	+	$\rho_2 = 0.28125$
7	1	1		1	0	1	+				$\rho_1 = 1$
8	1	1		1	0	1	+				$\rho_1 = 0.515625$
	2	4		6	4	3	+	-4	3	+	$\rho_2 = 0.46875$
	3		2	2	2	1	+	-2	1	+	$\rho_3 = 0.015625$

Basin tree diagrams for Rule 22

22 , $L = 3$ (a) Period-1 Attractor : $\rho_1 = \dfrac{8}{8} = 1$

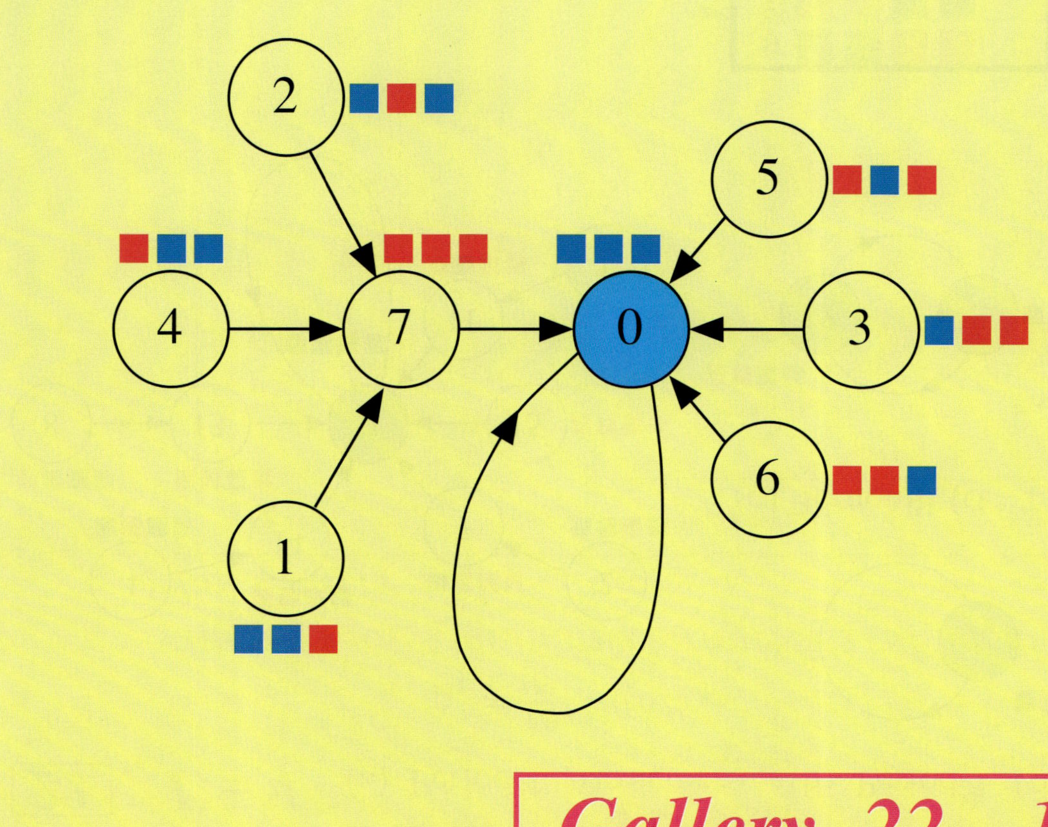

Gallery 22 - 1

Table 15. (*Continued*)

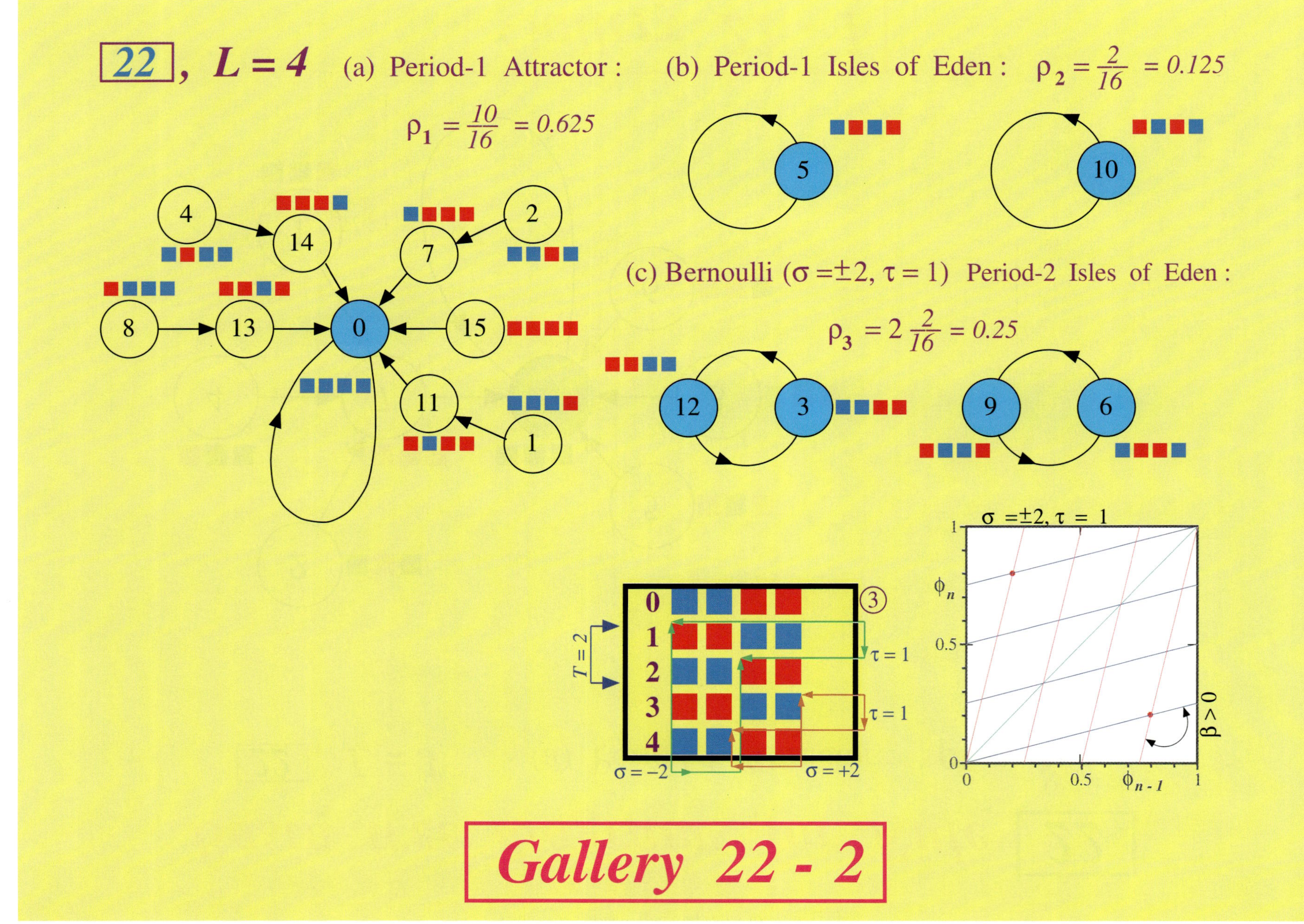

Gallery 22 - 2

32

22, $L = 5$ (a) Period-1 Attractor : $\rho_1 = \dfrac{32}{32} = 1$

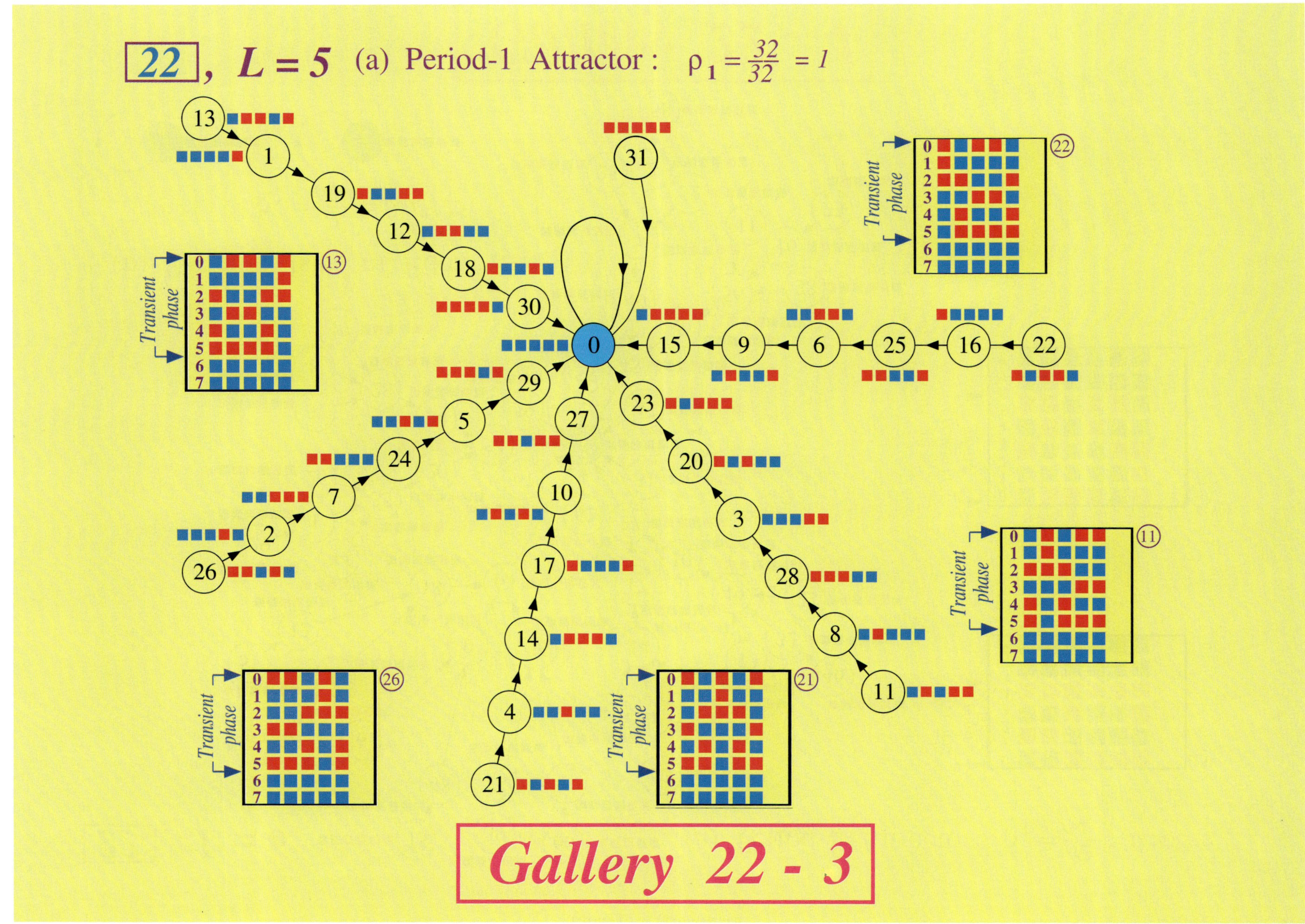

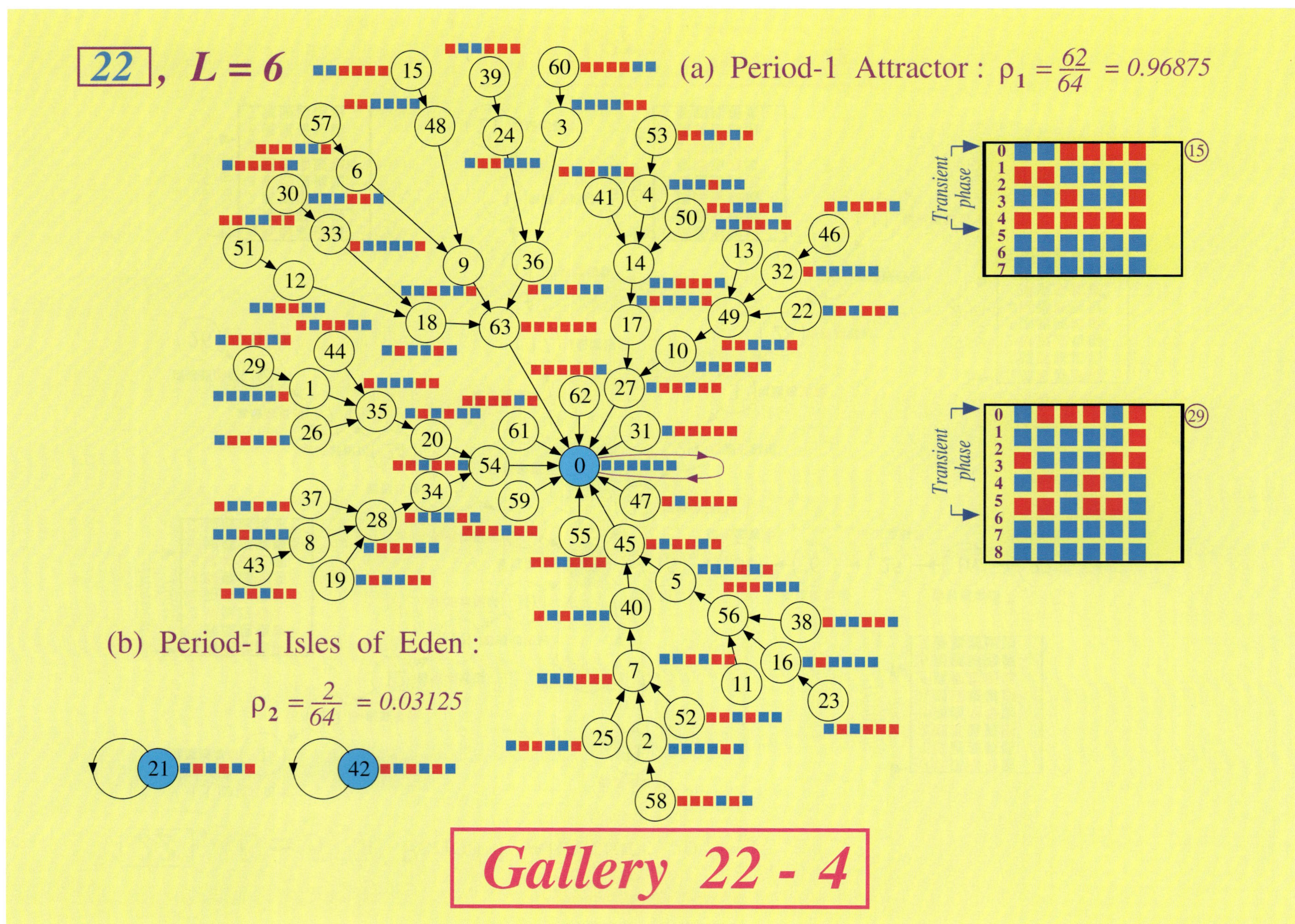

22 , L = 6
(a) Period-1 Attractor : ρ₁ = 62/64 = 0.96875
Transient phase
15
29
(b) Period-1 Isles of Eden :
ρ₂ = 2/64 = 0.03125
Gallery 22 - 4

Table 15.   (Continued)

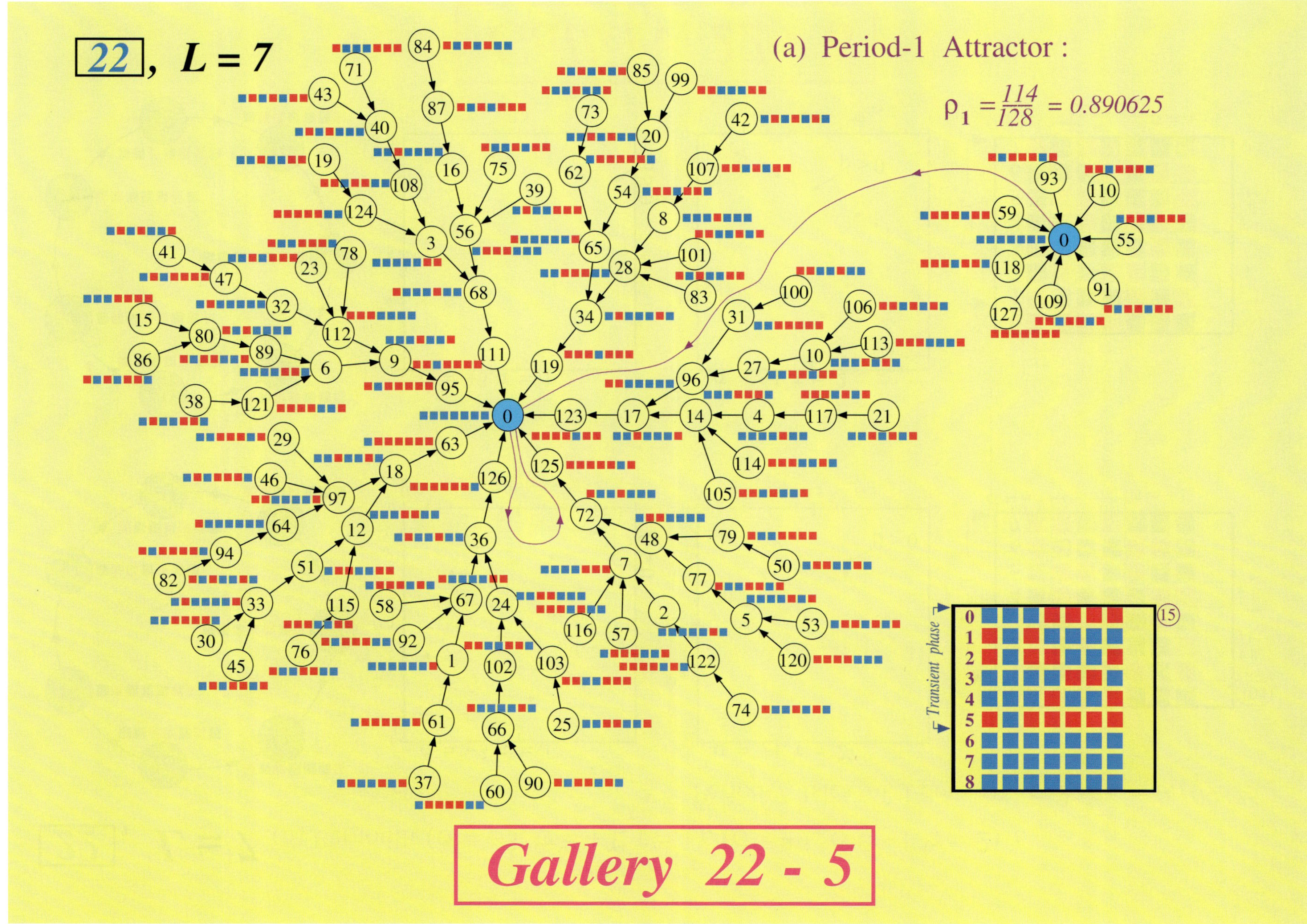

35

22, **L = 7** (b) Bernoulli ($\sigma = -3$, $\sigma = +3$, $\tau = 1$), ($\sigma = +1$, $\sigma = -1$, $\tau = 2$)

Period-7 Isles of Eden : $\rho_3 = 2\frac{7}{128} = 0.109375$

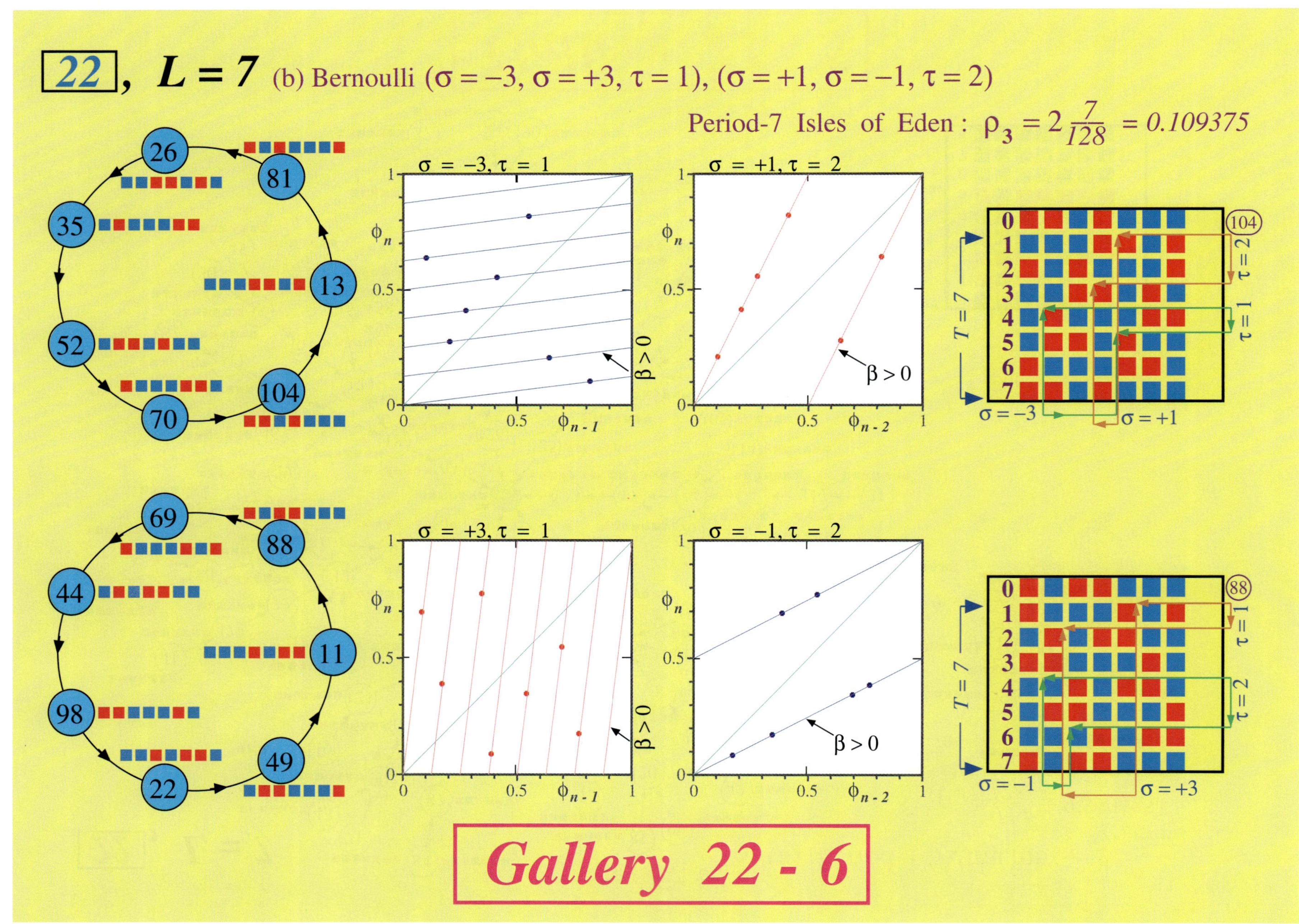

Table 15. (*Continued*)

22, *L = 8*   (b) Bernoulli : ($\sigma = \pm 4$, $\tau = 3$) Period-6 Attractors : $\rho_2 = 4\frac{14}{256} = 0.21875$

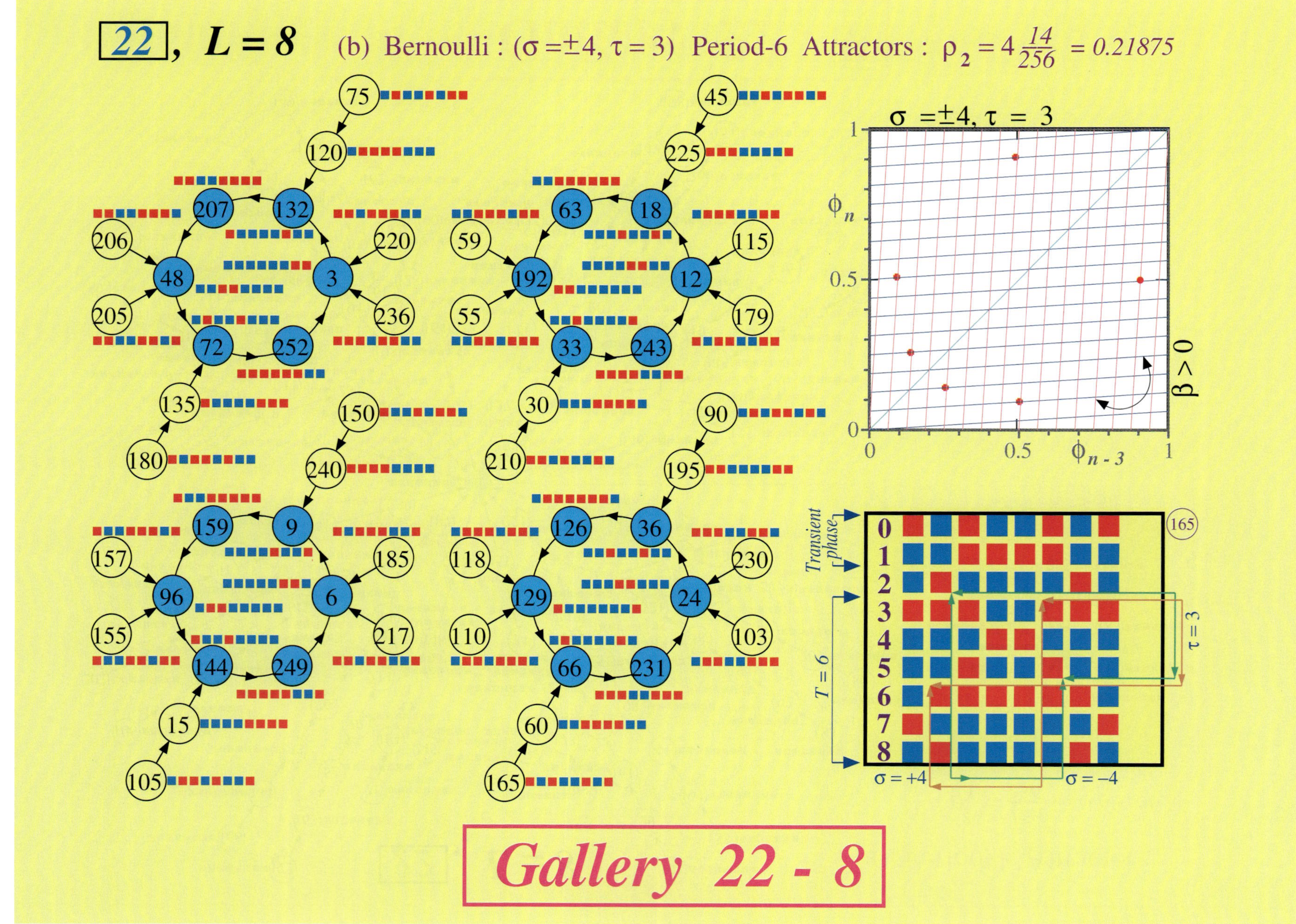

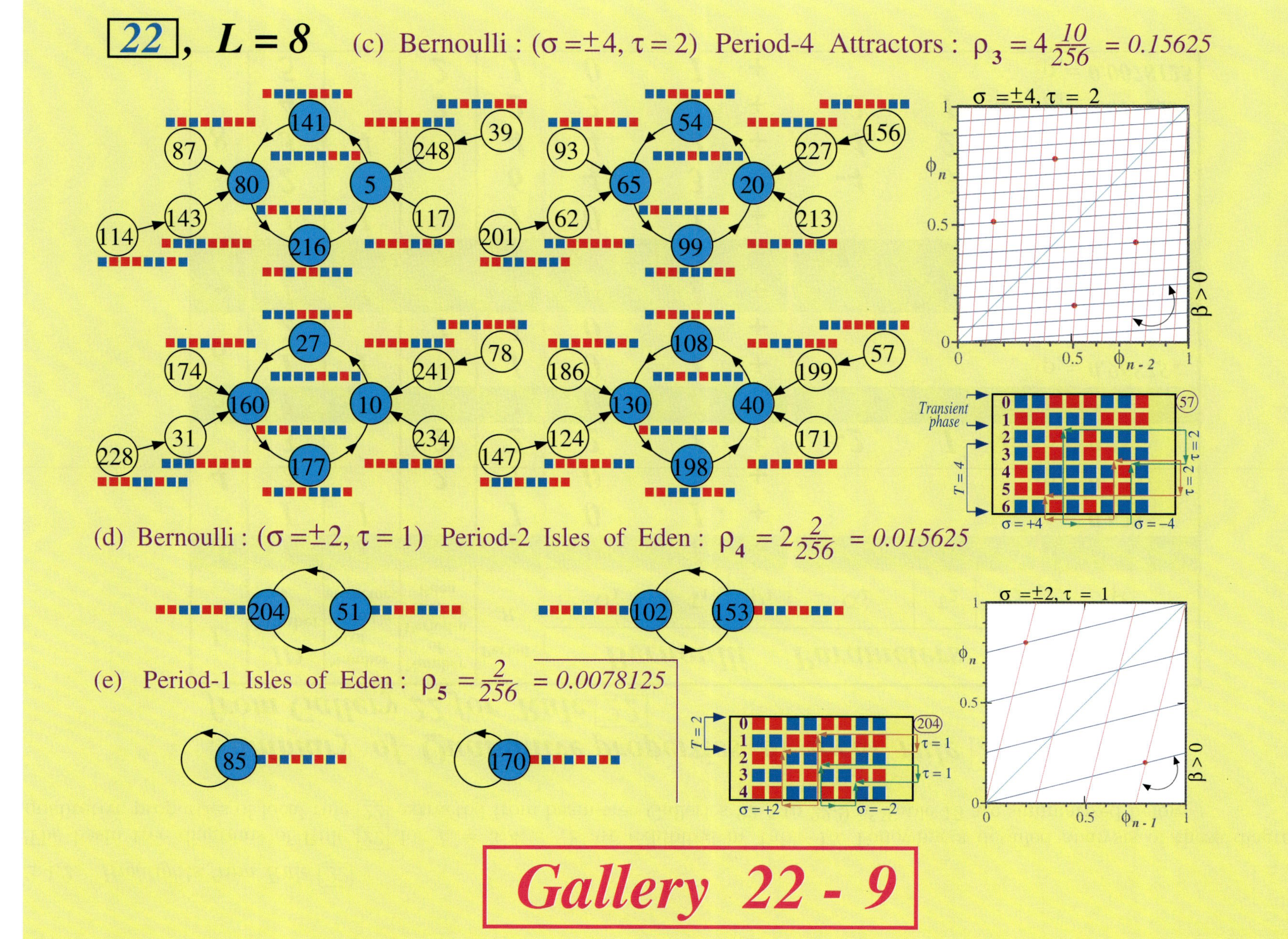

22 , L = 8   (c) Bernoulli : (σ = ±4, τ = 2) Period-4 Attractors : $\rho_3 = 4\frac{10}{256} = 0.15625$
σ = ±4, τ = 2
φ_n
φ_{n-2}
β > 0
Transient phase
T = 4
τ = 2 τ = 2
σ = +4   σ = −4
57
(d) Bernoulli : (σ = ±2, τ = 1) Period-2 Isles of Eden : $\rho_4 = 2\frac{2}{256} = 0.015625$
σ = ±2, τ = 1
φ_n
φ_{n-1}
β > 0
(e) Period-1 Isles of Eden : $\rho_5 = \frac{2}{256} = 0.0078125$
T = 2
τ = 1
τ = 1
σ = +2   σ = −2
204
Gallery 22 - 9

2.4.2. *Highlights from Rule 22*

The basin tree diagrams of Rule 22 for $L = 3, 4, \ldots, 8$ are exhibited in Table 15. Following a detailed analysis of these diagrams, the qualitative properties of local rule 22 extracted from basin-tree Galleries 22-1 to 22-9 of Table 15 are summarized below:

Summary of Qualitative properties of local rule 22 extracted from Gallery 22 for Rule 22

L	ID Number i	Number of Period-n attractors	Number of Period-n Isles of Eden	Period n	Bernoulli			Parameters			Robustness coefficient ρ
					σ_1	τ_1	β_1	σ_2	τ_2	β_2	
3	1	1		1	0	1	+				$\rho_1 = 1$
4	1	1		1	0	1	+				$\rho_1 = 0.625$
	2		2	1	0	1	+				$\rho_2 = 0.125$
	3		2	2	2	1	+	-2	1	+	$\rho_3 = 0.25$
5	1	1		1	0	1	+				$\rho_1 = 1$
6	1	1		1	0	1	+				$\rho_1 = 0.96875$
	2		2	1	0	1	+				$\rho_2 = 0.03125$
7	1	1		1	0	1	+				$\rho_1 = 0.890625$
	2		2	7	∓ 3	1	+	± 1	2	+	$\rho_2 = 0.109375$
8	1	1		1	0	1	+				$\rho_1 = 0.6015625$
	2	4		6	4	3	+	-4	3	+	$\rho_2 = 0.21875$
	3	4		4	4	2	+	-4	2	+	$\rho_3 = 0.15625$
	4		2	2	2	1	+	-2	1	+	$\rho_4 = 0.015625$
	5		2	1	0	1	+				$\rho_5 = 0.0078125$

Basin tree diagrams for Rule 54

54 , $L = 3$ (a) Period-1 Attractor : $\rho_1 = \frac{8}{8} = 1$

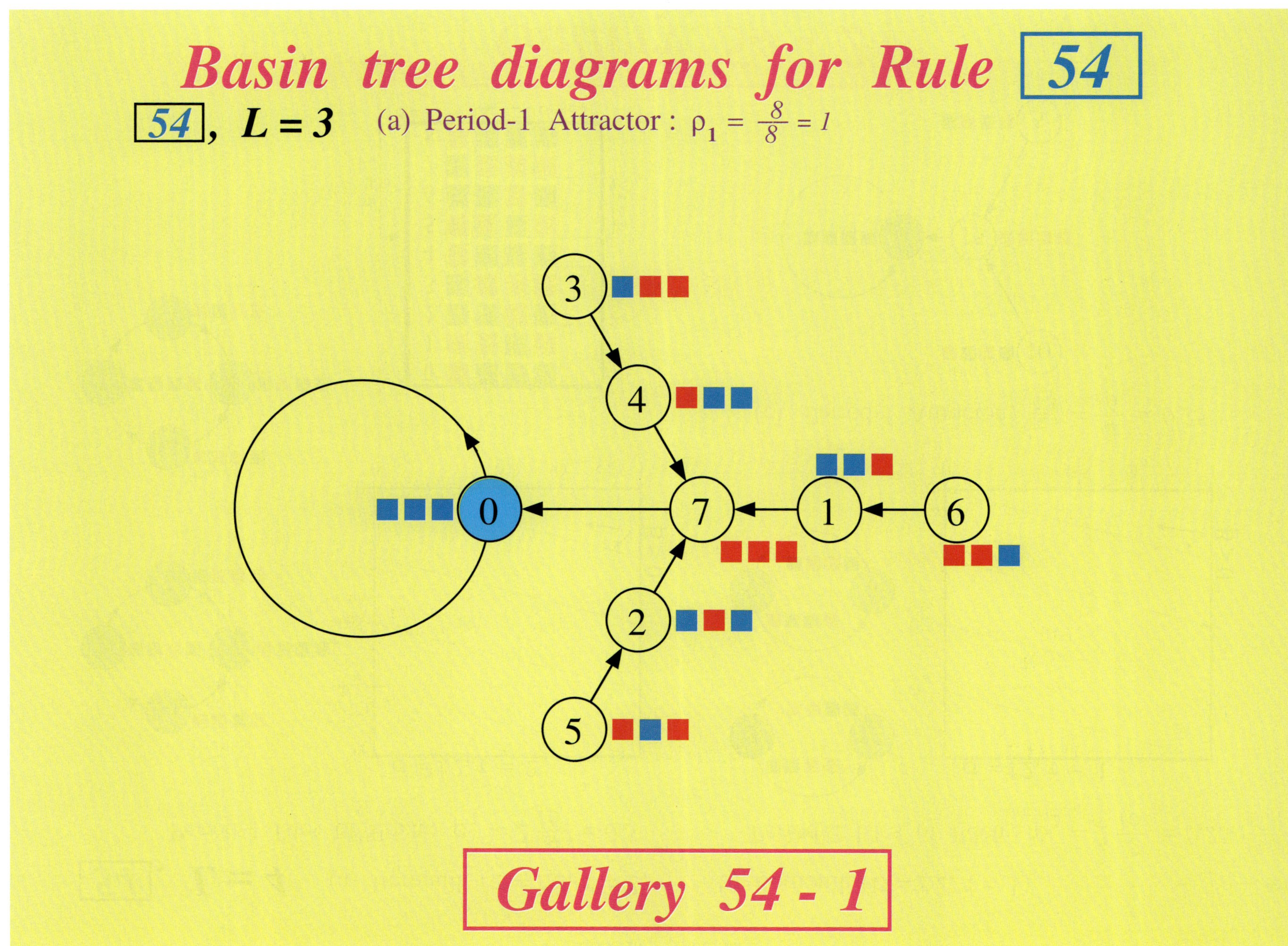

Gallery 54 - 1

54, **L = 4** (a) Bernoulli ($\sigma = \pm 2, \tau = 2$)   (b) Bernoulli ($\sigma = \pm 2, \tau = 1$)

Period-4 Isles of Eden : $\rho_1 = 2\frac{4}{16} = 0.5$ Period-2 Isles of Eden : $\rho_2 = 2\frac{2}{16} = 0.25$

(c) Period-1 Attractor : $\rho_3 = \frac{4}{16} = 0.25$

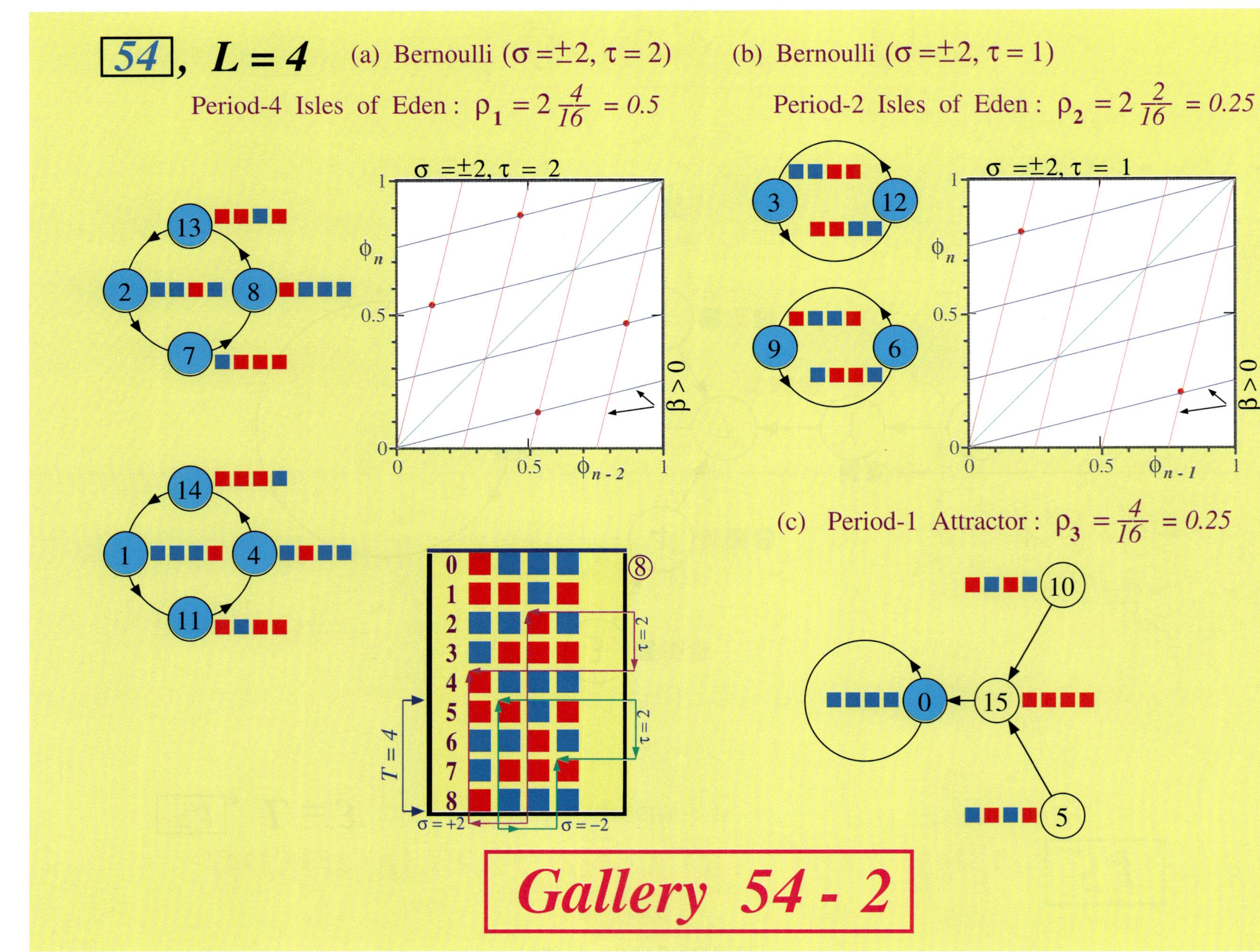

$$\boxed{\textit{Gallery \ 54 - 2}}$$

Table 16.   (*Continued*)

$\boxed{54}$, $L = 5$ (a) Period-1 Attractor : $\rho_1 = \frac{32}{32} = 1$

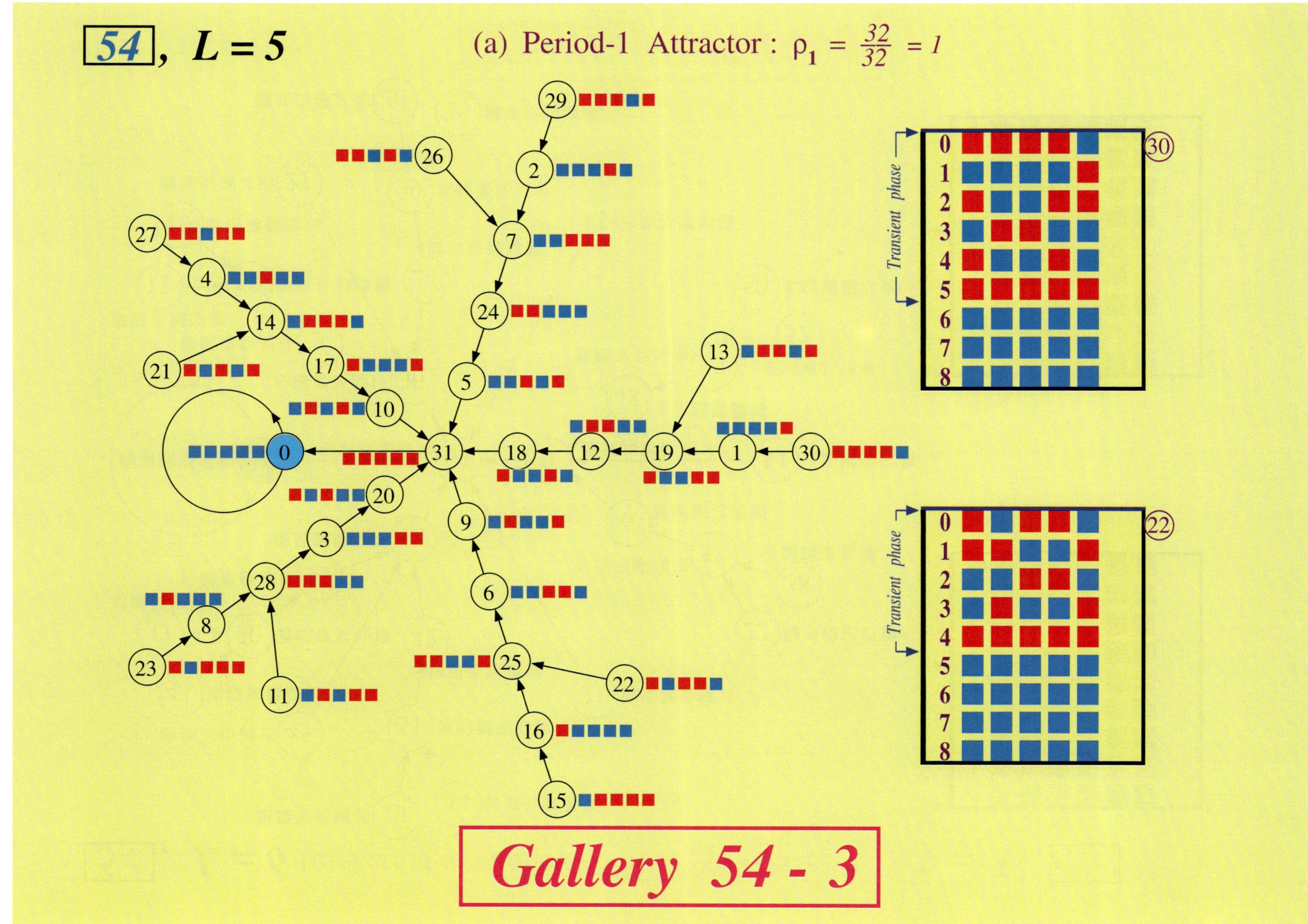

Table 16. (*Continued*)

54, $L = 6$ (a) Period-1 Attractor : $\rho_1 = \frac{34}{64} = 0.53125$

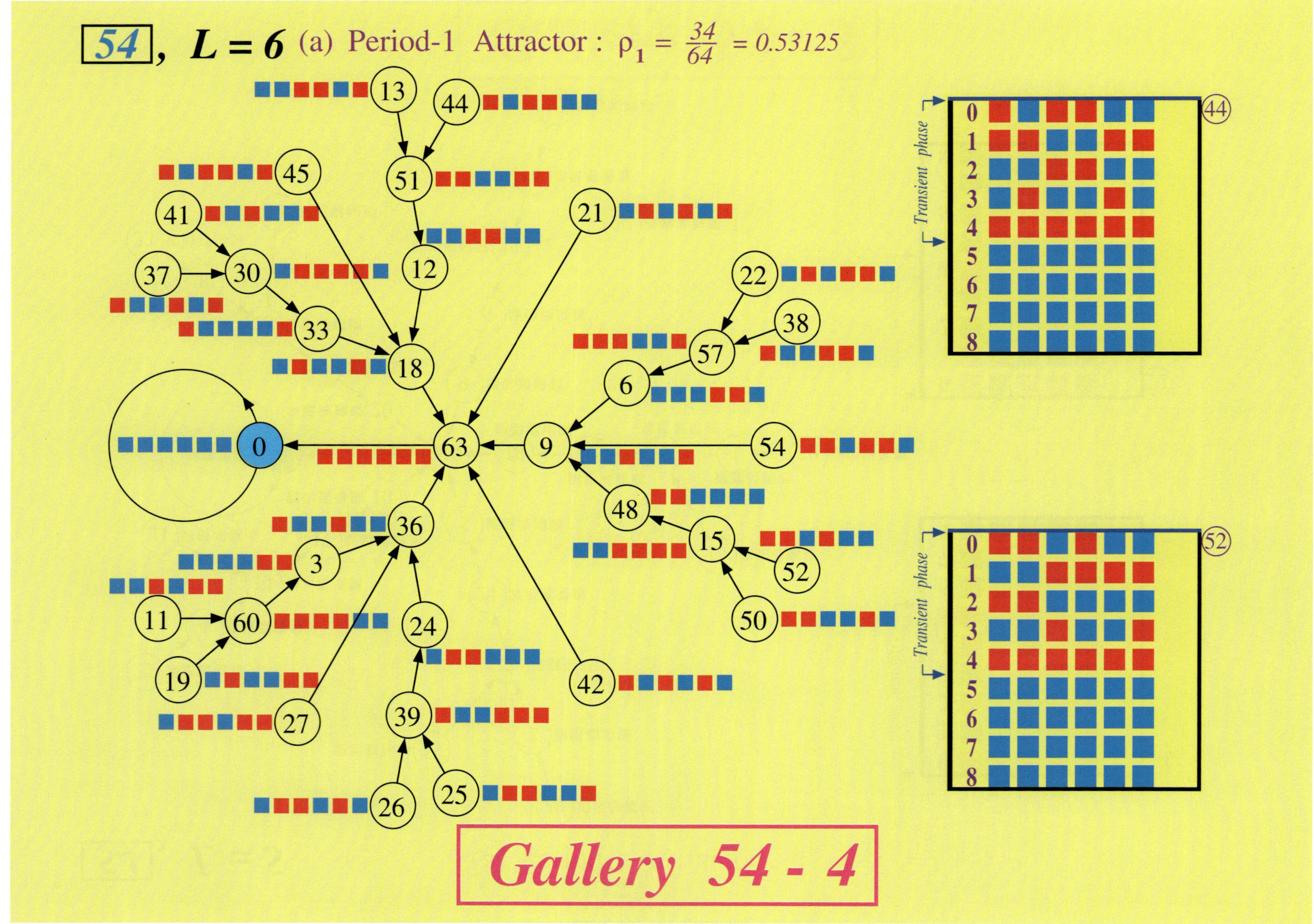

44

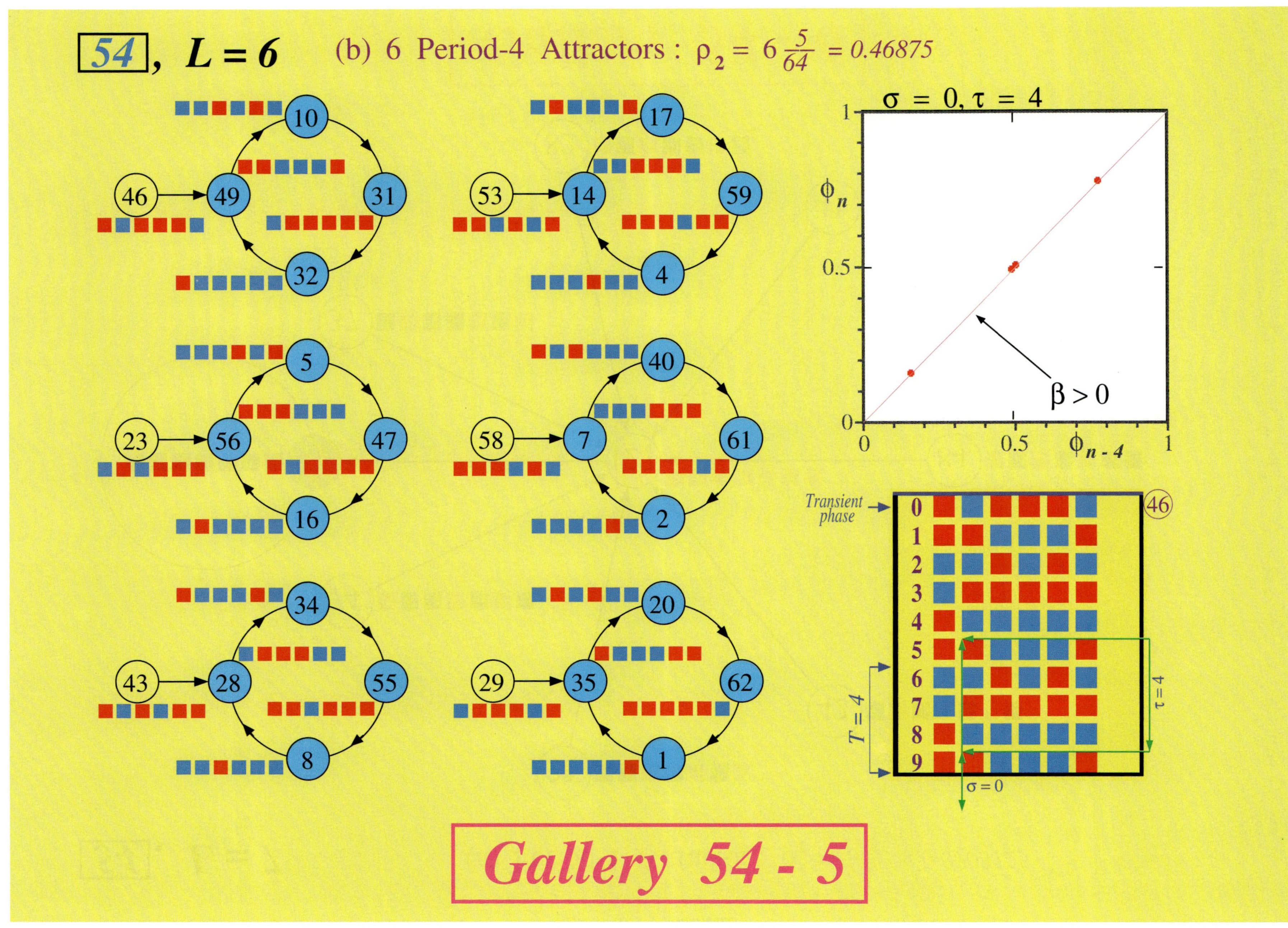

54, $L = 6$
(b) 6 Period-4 Attractors : $\rho_2 = 6\frac{5}{64} = 0.46875$
$\sigma = 0, \tau = 4$
ϕ_n
ϕ_{n-4}
$\beta > 0$
Transient phase
$T = 4$
$\tau = 4$
$\sigma = 0$
Gallery 54 - 5

Table 16. (*Continued*)

$\boxed{54}$, $L = 7$ (a) Period-1 Attractor : $\rho_1 = \frac{9}{128} = 0.0703125$

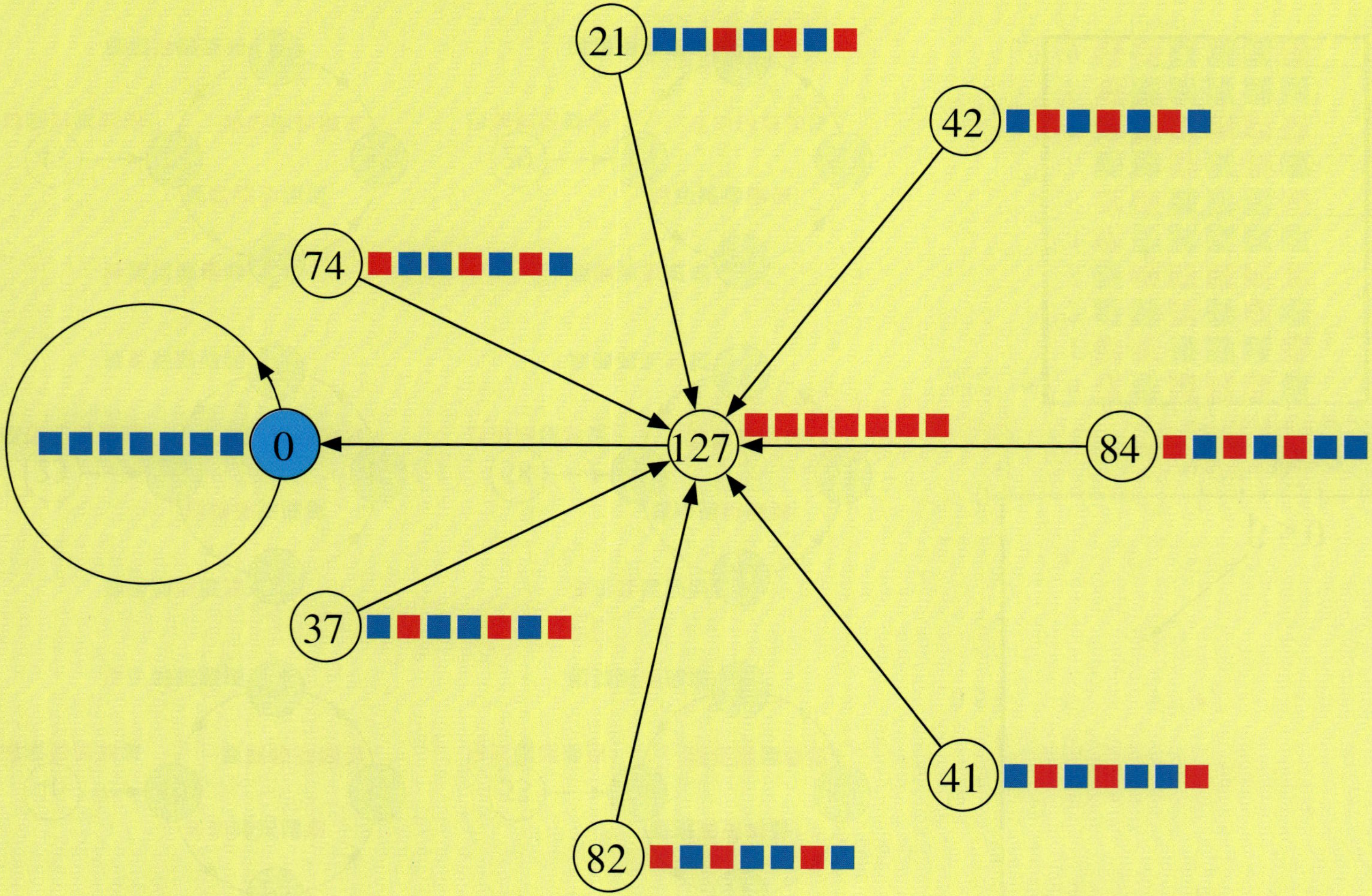

46

47

$\boxed{54}$, $L = 7$ (b) 7 Period-4 Attractors : $\rho_2 = 7\frac{17}{128} = 0.9296875$

Transient phase
$T = 4$
$\tau = 4$
$\sigma = 0$

Gallery 54 - 7

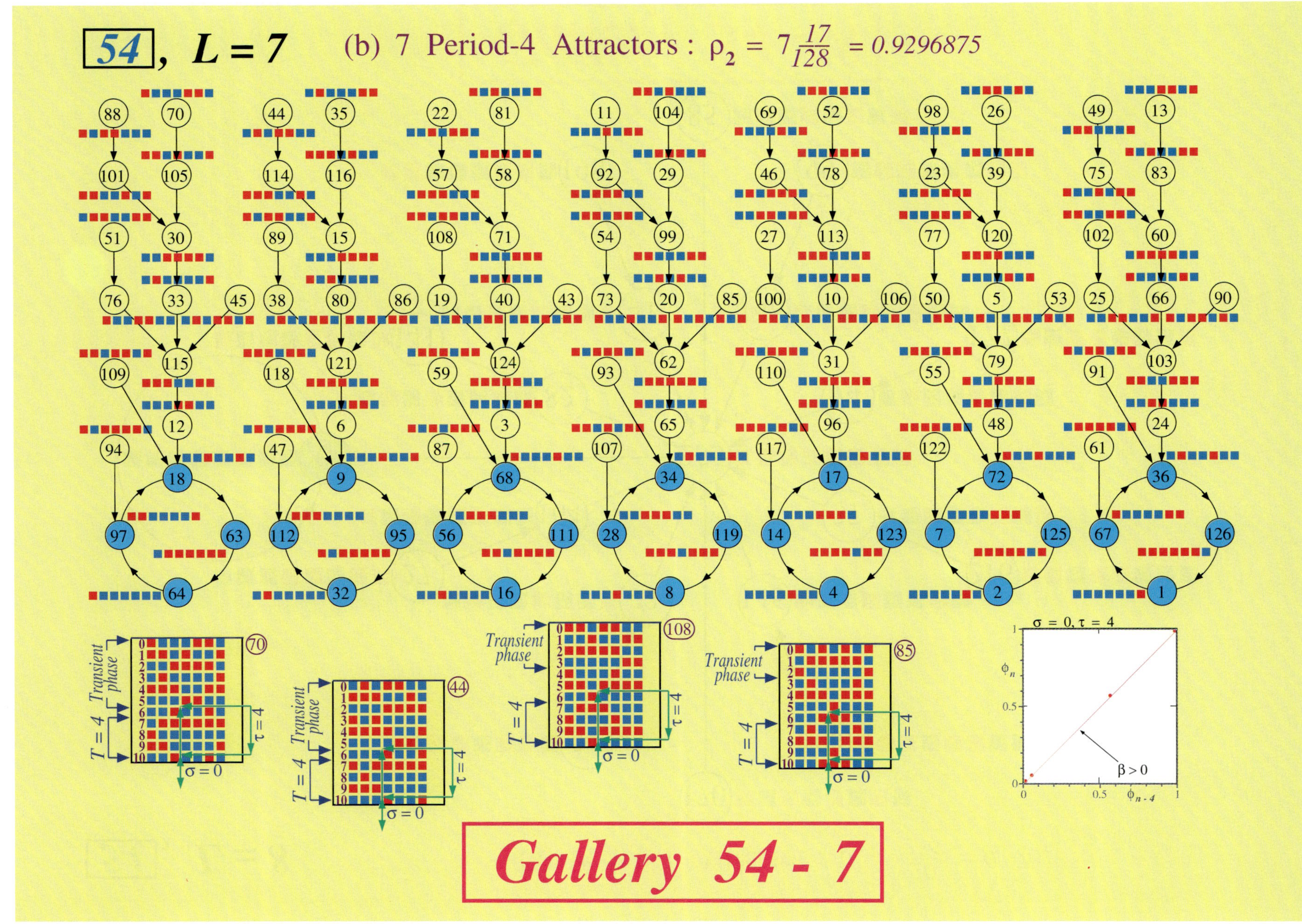

Table 16.   (*Continued*)

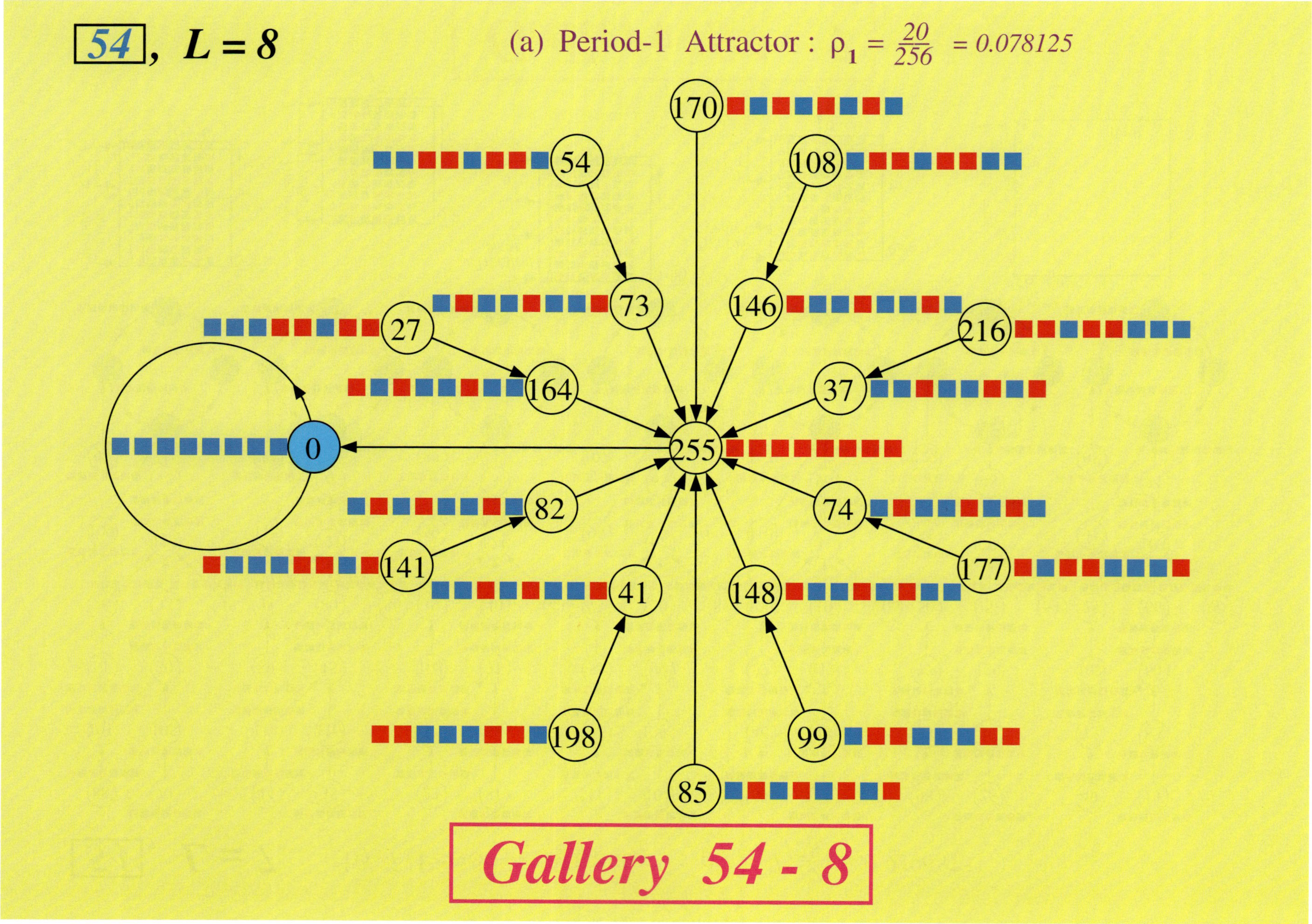

$\boxed{54}$, $L = 8$

(a) Period-1 Attractor : $\rho_1 = \frac{20}{256} = 0.078125$

Gallery 54 - 8

Table 16. (*Continued*)

54, **L = 8**   (b) Bernoulli ($\sigma = \pm 2$, $\tau = 2$) 2 Period-4 Attractors : $\rho_2 = 2\frac{24}{256} = 0.1875$

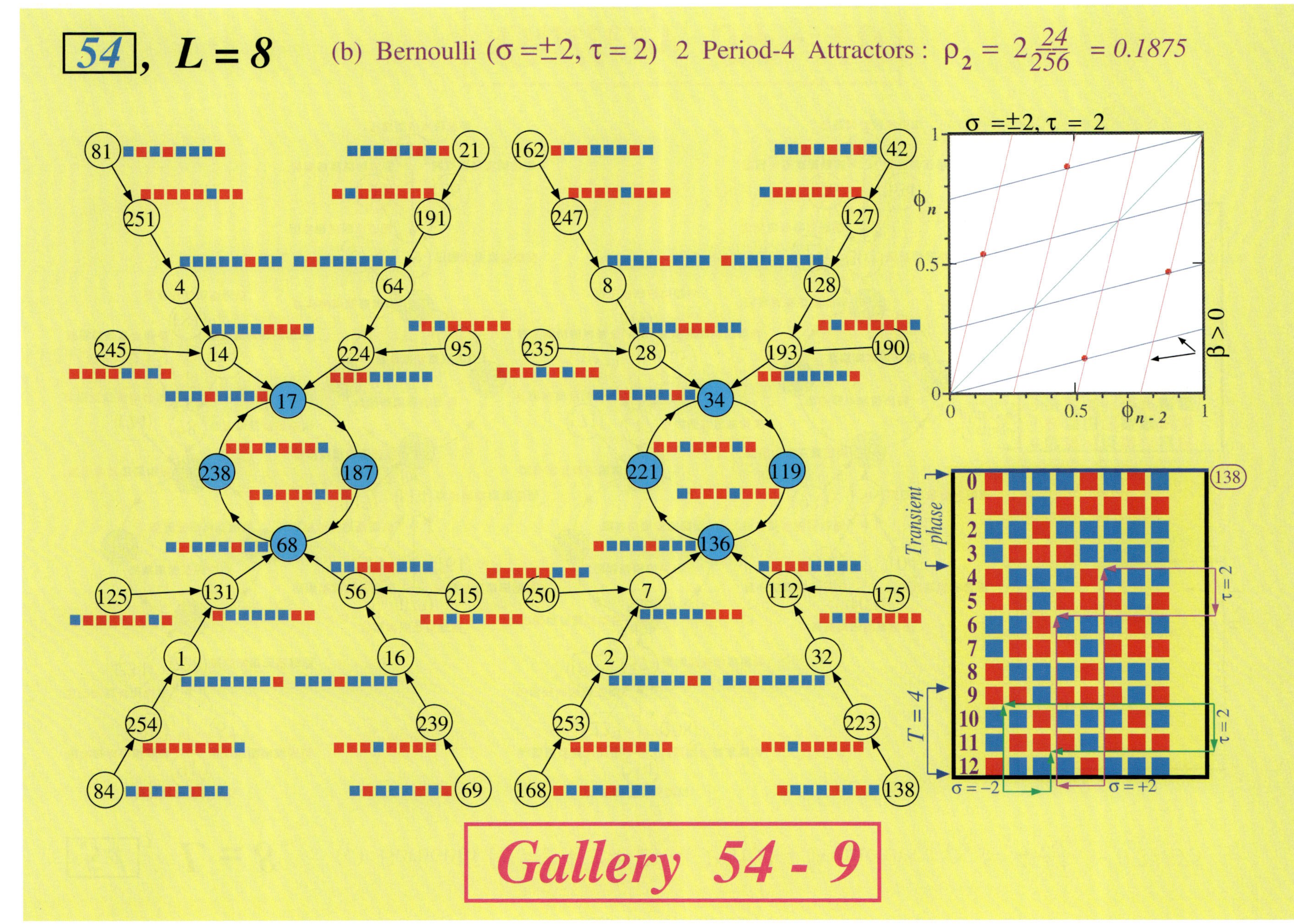

Gallery 54 - 9

49

54, **L = 8**   (c) Bernoulli ($\sigma = \pm 4$, $\tau = 2$) 4 Period-4 Attractors : $\rho_3 = 4\frac{14}{256} = 0.21875$

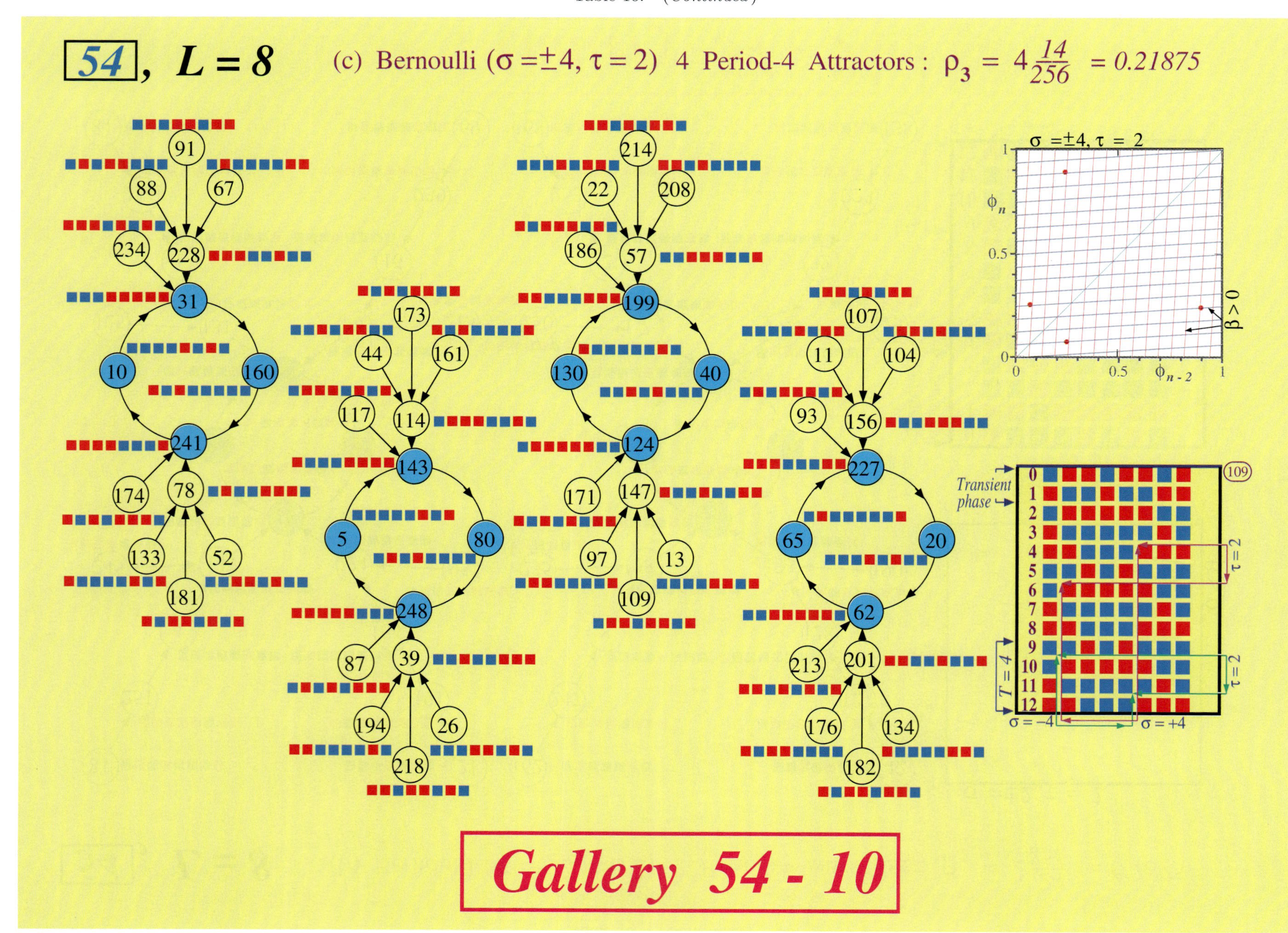

Gallery 54 - 10

Table 16.   (*Continued*)

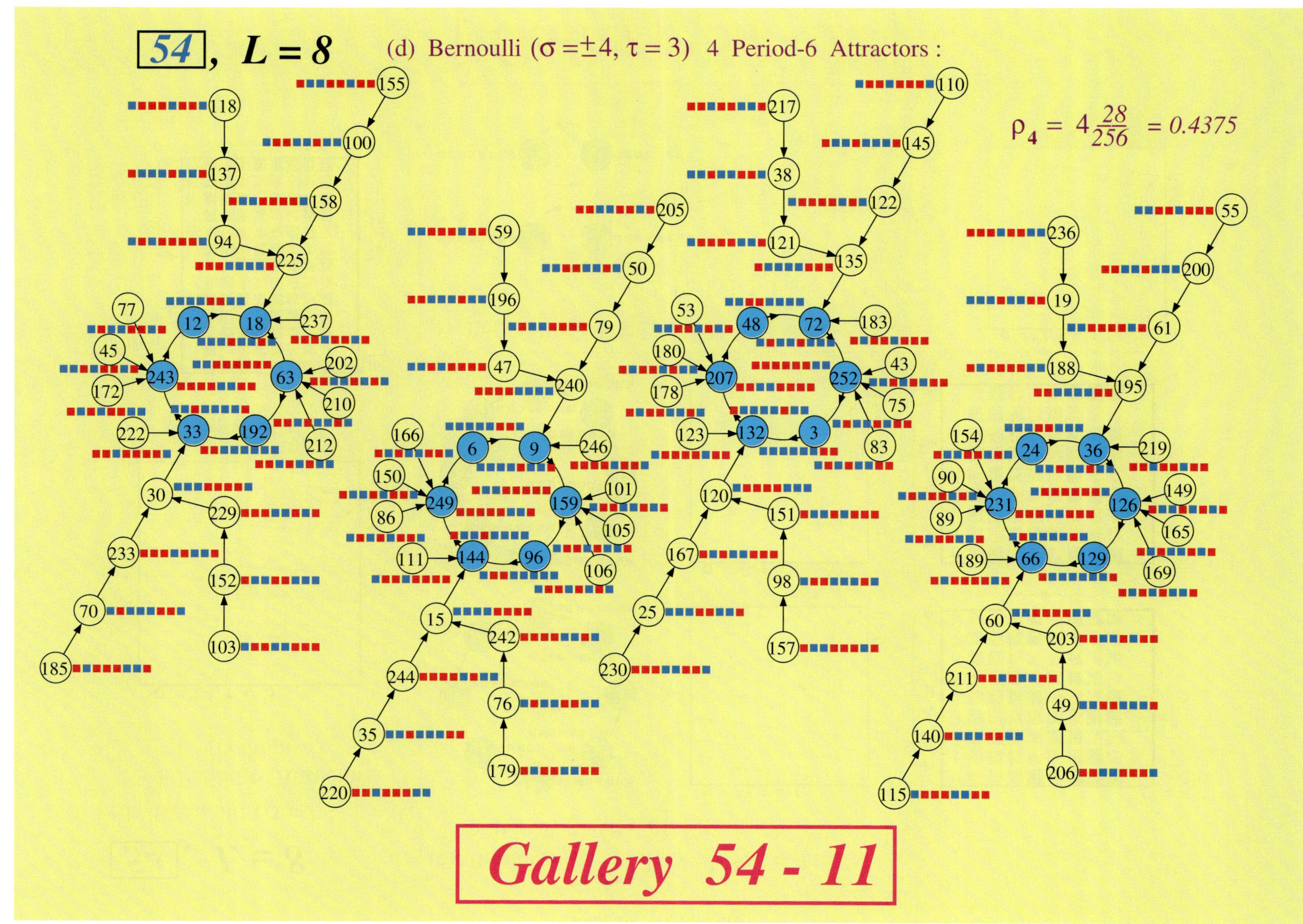

51

Table 16. (*Continued*)

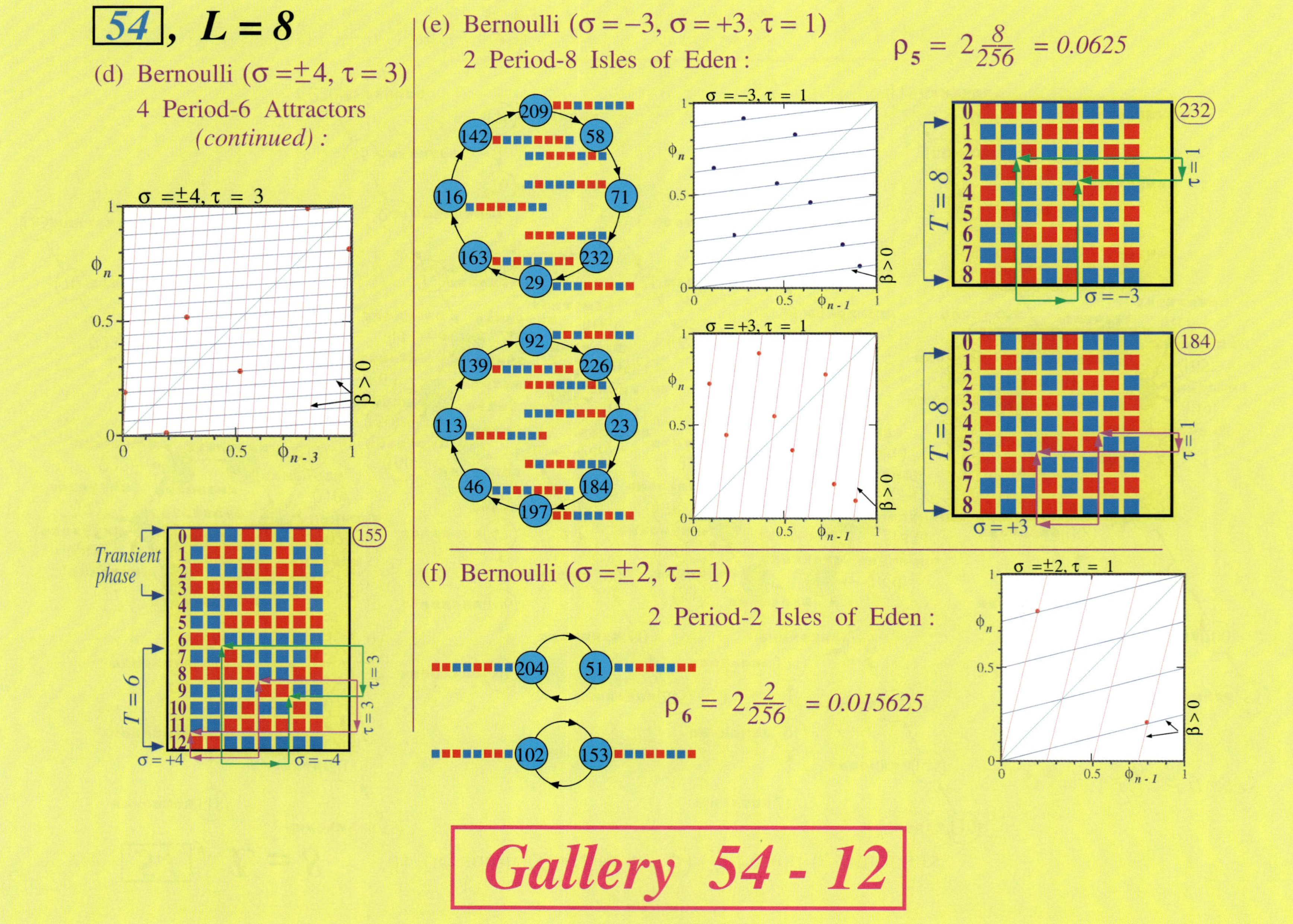

2.4.3. *Highlights from Rule* 54

The basin tree diagrams of Rule 54 for $L = 3, 4, \ldots, 8$ are exhibited in Table 16. Following a detailed analysis of these diagrams, the qualitative properties of local rule 54 extracted from basin-tree Galleries 54-1 to 54-12 of Table 16 are summarized below:

Summary of Qualitative properties of local rule 54 *extracted from Gallery 54 for Rule* 54

L	ID Number i	Number of Period-n attractors	Number of Period-n Isles of Eden	Period n	Bernoulli	Parameters					Robustness coefficient ρ
					σ_1	τ_1	β_1	σ_2	τ_2	β_2	
3	1	1		1	0	1	+				$\rho_1 = 1$
4	1		2	4	2	2	+	-2	2	+	$\rho_1 = 0.5$
	2		2	2	2	1	+	-2	1	+	$\rho_2 = 0.25$
	3	1		1	0	1	+				$\rho_3 = 0.25$
5	1	1		1	0	1	+				$\rho_1 = 1$
6	1	1		1	0	1	+				$\rho_1 = 0.53125$
	2	6		4	0	4	+				$\rho_2 = 0.46875$
7	1	1		1	0	1	+				$\rho_1 = 0.0703125$
	2	7		4	0	4	+				$\rho_2 = 0.9296875$
8	1	1		1	0	1	+				$\rho_1 = 0.078125$
	2	2		4	2	2	+	-2	2	+	$\rho_2 = 0.1875$
	3	4		4	4	2	+	-4	2	+	$\rho_3 = 0.21875$
	4	4		6	4	3	+	-4	3	+	$\rho_4 = 0.4375$
	5		2	8	3	1	+	-3	1	+	$\rho_5 = 0.0625$
	6		2	2	2	1	+	-2	1	+	$\rho_6 = 0.015625$

Basin tree diagrams for Rule 73

73 , **L = 3**   (a)   Period-2 Attractor :   (b)   Period-1 Isles of Eden :

$$\rho_1 = \frac{5}{8} = 0.625 \qquad\qquad \rho_2 = \frac{3}{8} = 0.375$$

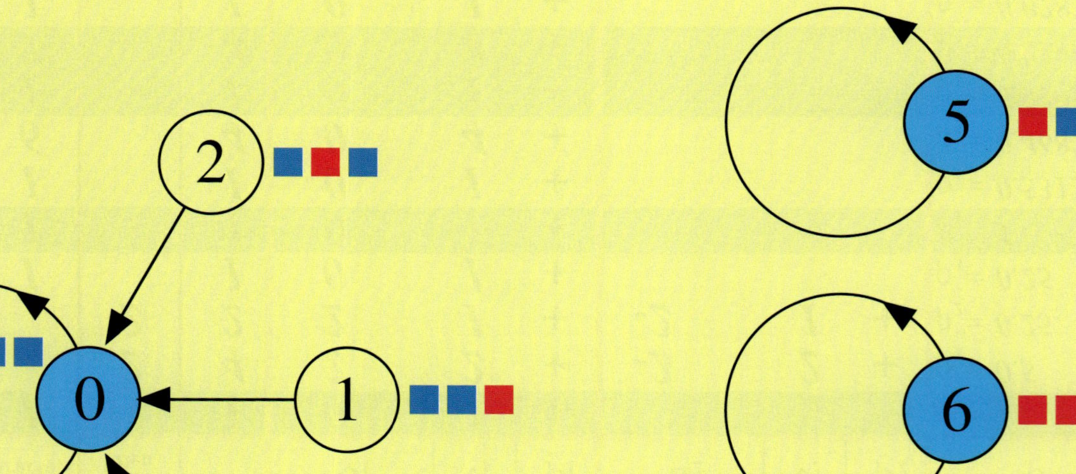
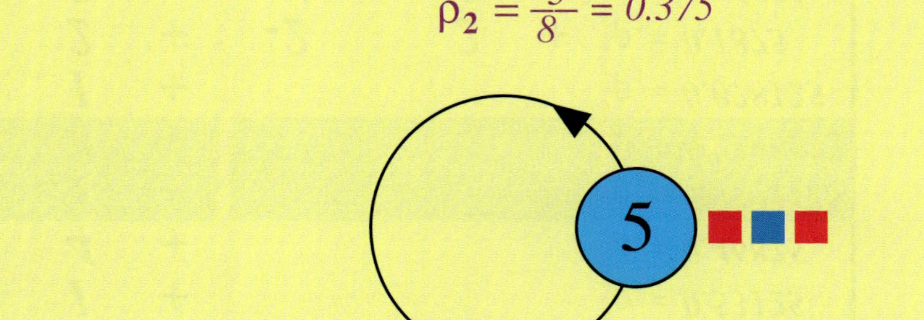
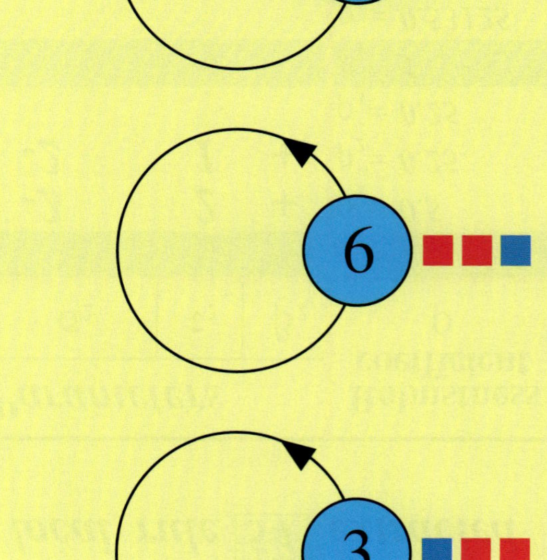
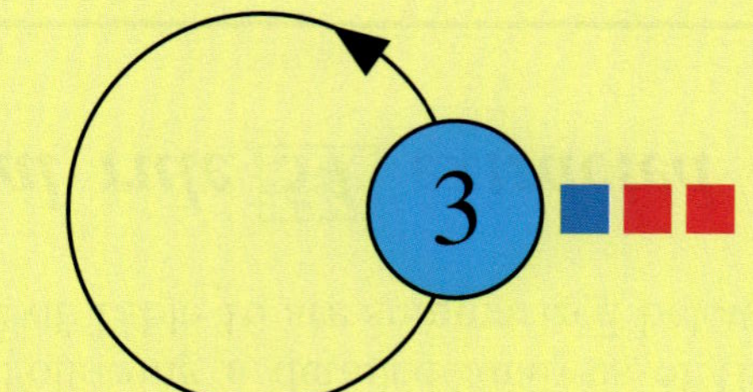

Gallery 73 - 1

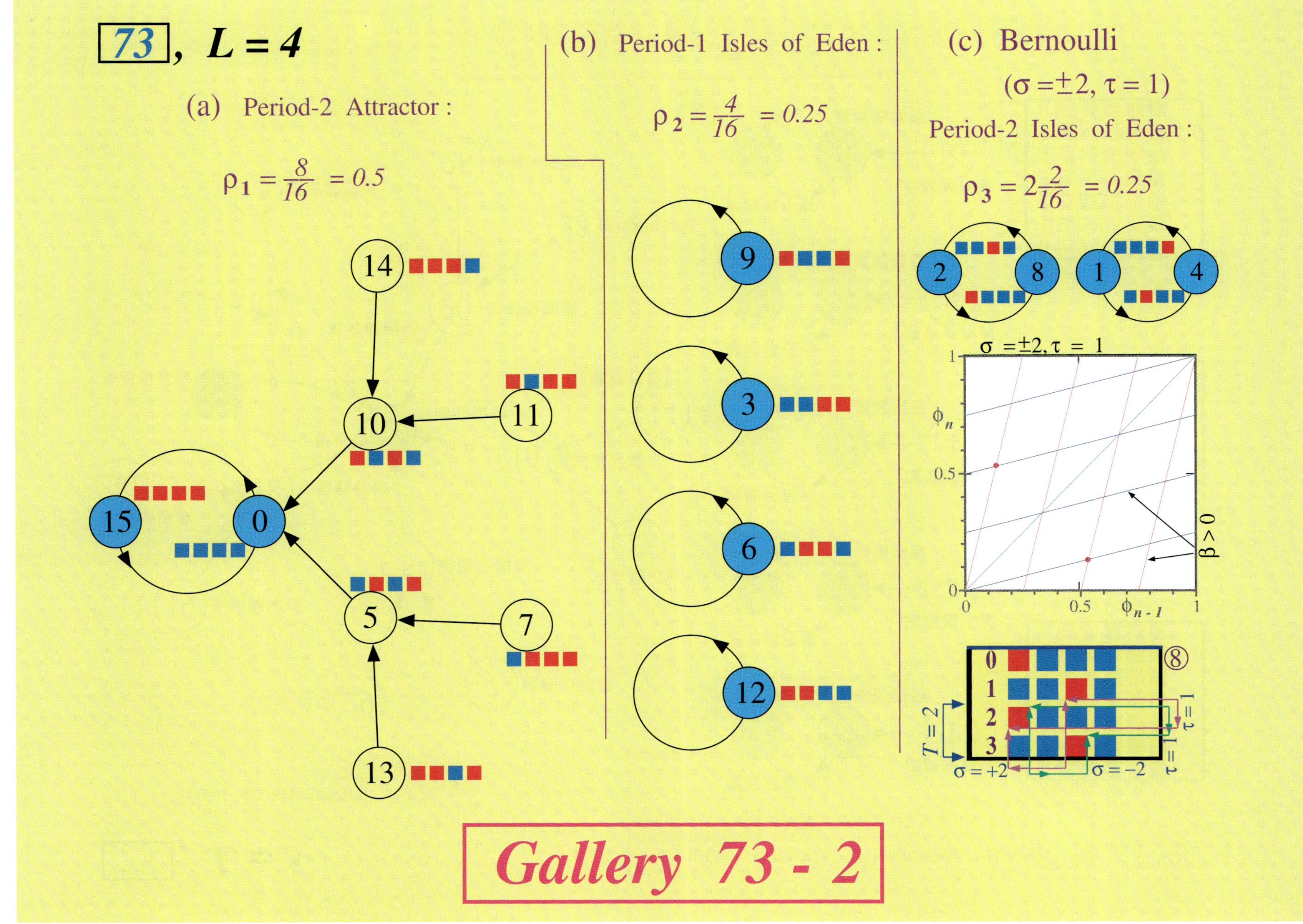

73, L = 4
(a) Period-2 Attractor :
$\rho_1 = \frac{8}{16} = 0.5$
14
10
11
15
0
5
7
13
(b) Period-1 Isles of Eden :
$\rho_2 = \frac{4}{16} = 0.25$
9
3
6
12
(c) Bernoulli
$(\sigma = \pm 2, \tau = 1)$
Period-2 Isles of Eden :
$\rho_3 = 2\frac{2}{16} = 0.25$
2
8
1
4
$\sigma = \pm 2, \tau = 1$
ϕ_n
1
0.5
0
$\beta > 0$
0
0.5
ϕ_{n-1}
1
⑧
0
1
2
3
$T = 2$
$\tau = 1$
$\tau = 1$
$\sigma = +2$
$\sigma = -2$
Gallery 73 - 2

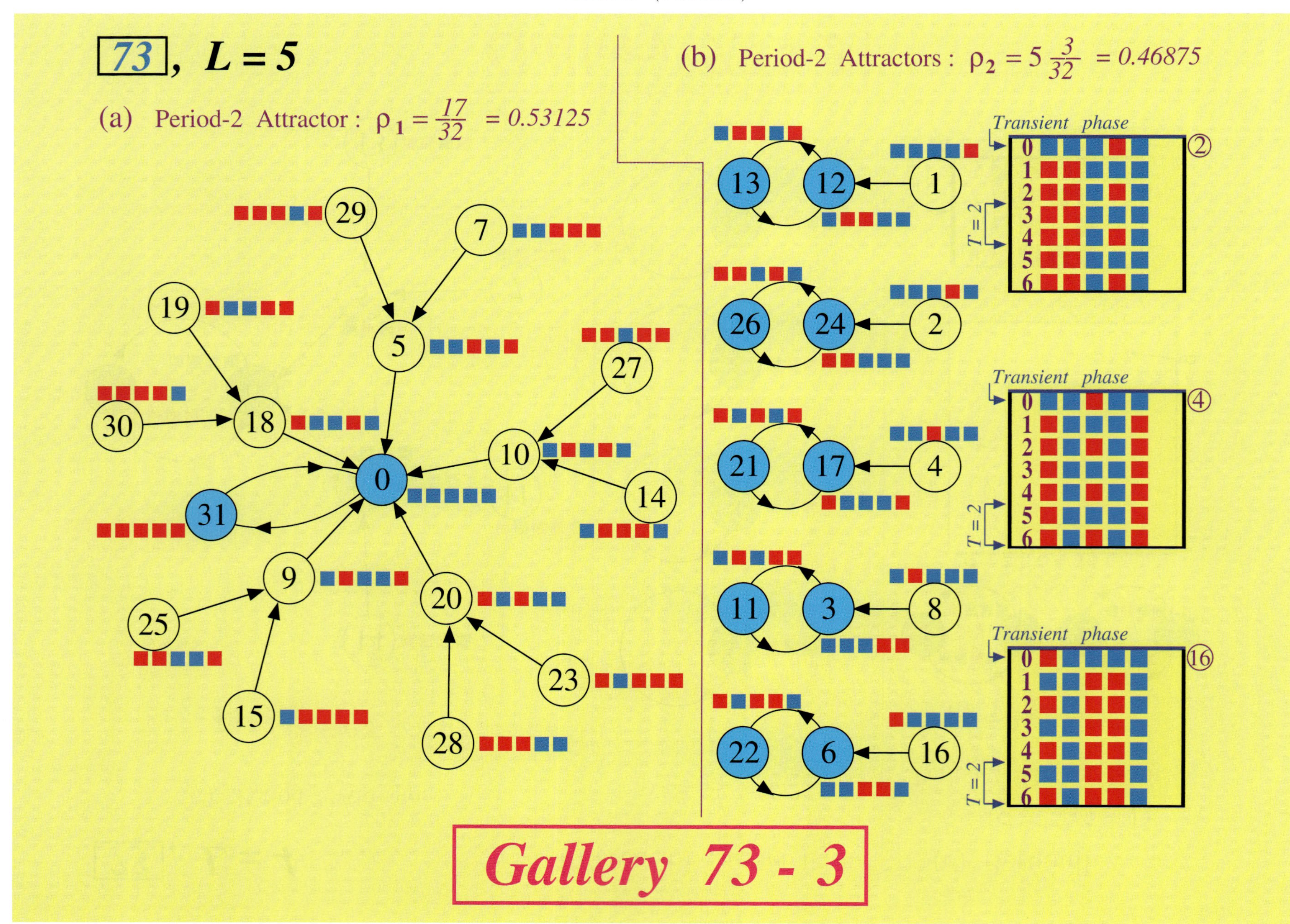
73 , L = 5
(a) Period-2 Attractor : $\rho_1 = \frac{17}{32} = 0.53125$
(b) Period-2 Attractors : $\rho_2 = 5\frac{3}{32} = 0.46875$
Transient phase
T = 2
Gallery 73 - 3

73, **L = 6** (a) Period-2 Attractor : $\rho_1 = \dfrac{43}{64} = 0.671875$ (b) Period-1 Attractors :

$$\rho_2 = 3\,\frac{7}{64} = 0.328125$$

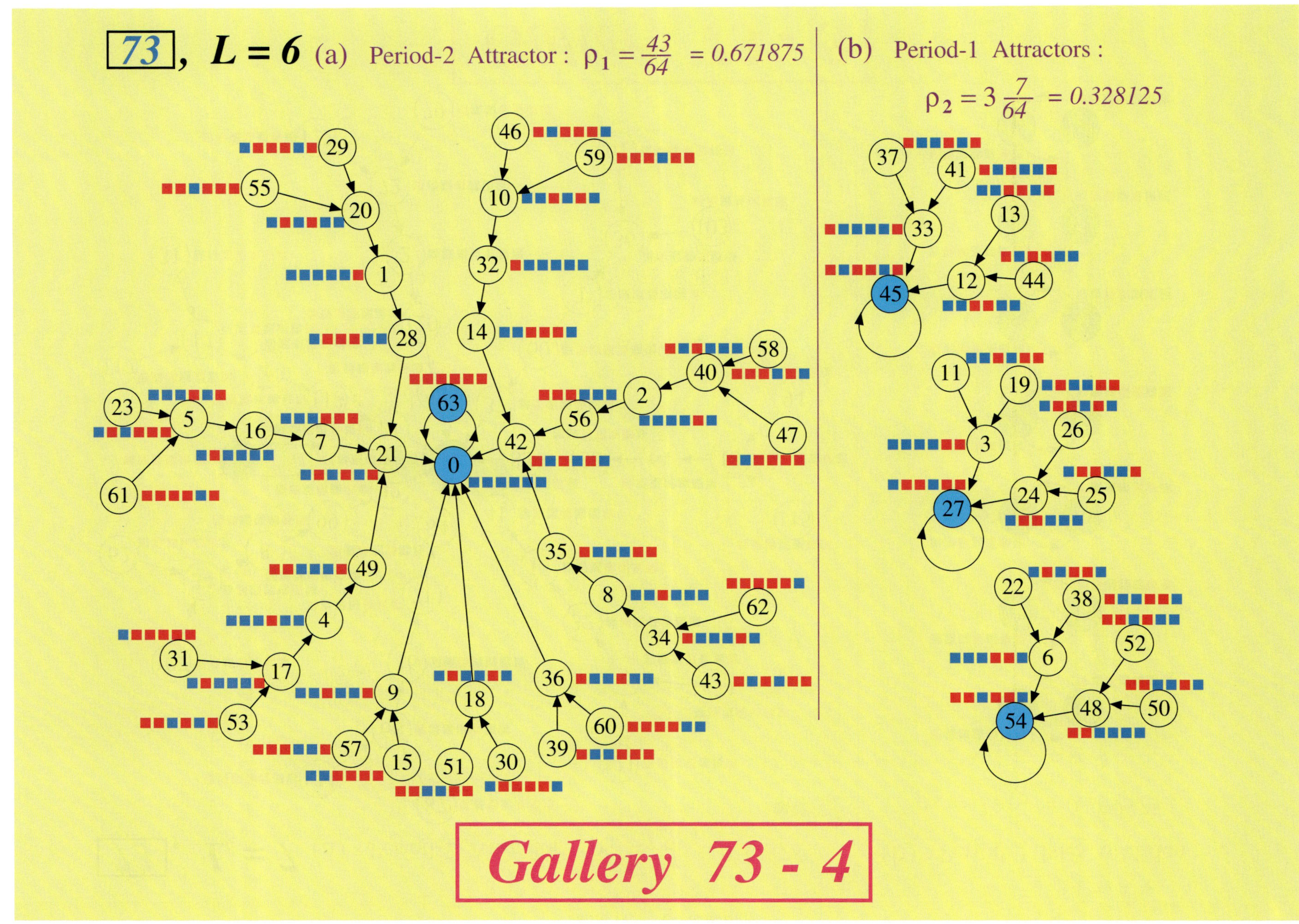

Gallery 73 - 4

Table 17.   (*Continued*)

73, **L = 7** (a) Period-2 Attractor: $\rho_1 = \frac{44}{128} = 0.34375$ (b) Period-2 Isles of Eden: $\rho_2 = 7\frac{2}{128} = 0.109375$

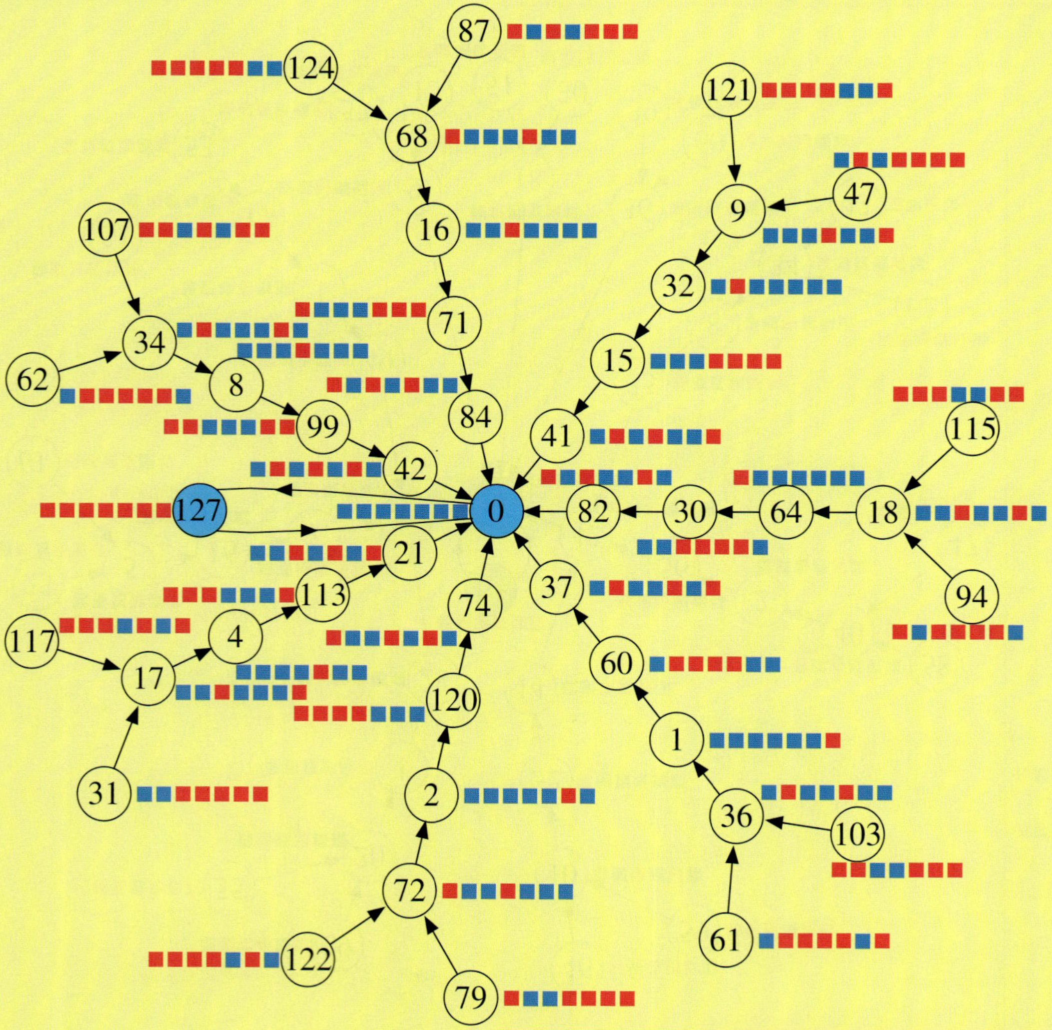

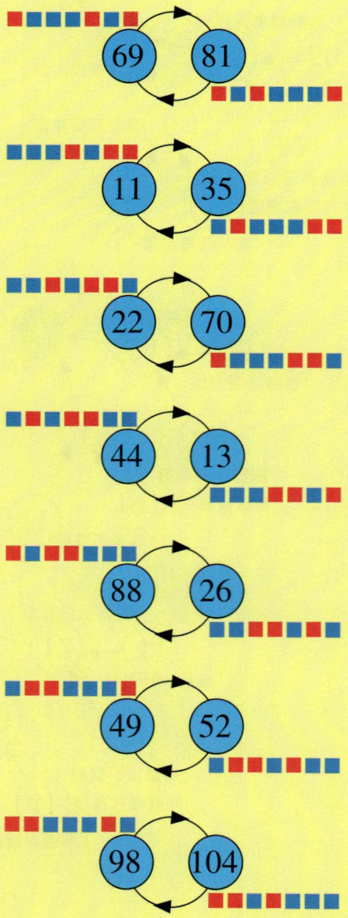

Gallery 73 - 5

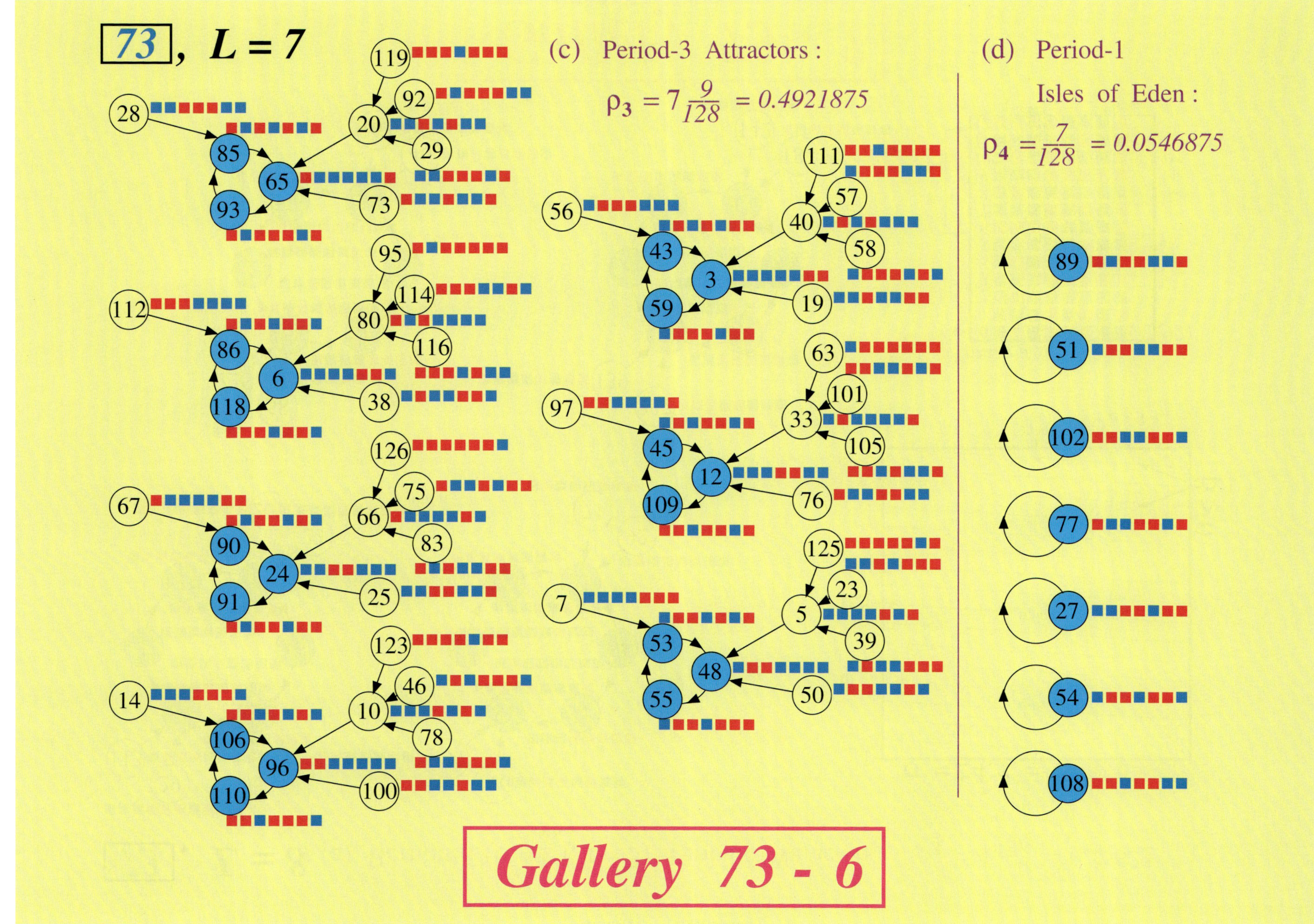
73, L = 7
(c) Period-3 Attractors :
$\rho_3 = 7\frac{9}{128} = 0.4921875$
(d) Period-1
Isles of Eden :
$\rho_4 = \frac{7}{128} = 0.0546875$
Gallery 73 - 6

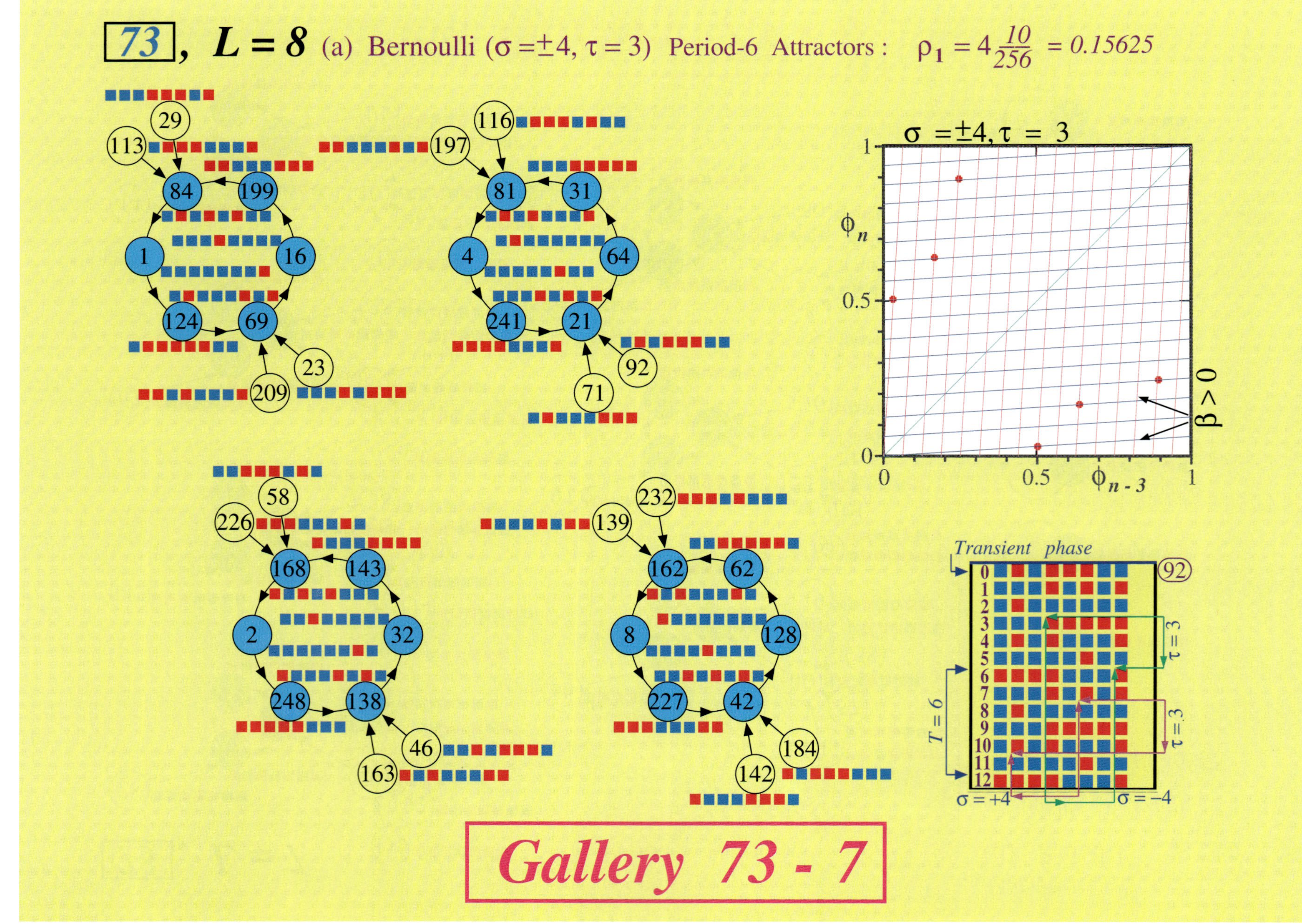

73 , L = 8 (a) Bernoulli (σ = ±4, τ = 3) Period-6 Attractors :   ρ₁ = 4 10/256 = 0.15625
σ = ±4, τ = 3
φ_n
φ_n - 3
β > 0
Transient phase
T = 6
τ = 3
τ = 3
σ = +4
σ = −4
92
Gallery 73 - 7

$\boxed{73}$, $L = 8$ (b) Bernoulli ($\sigma = 0$, $\tau = 3$) Period-3 Attractors : $\rho_2 = 8\frac{5}{256} = 0.15625$

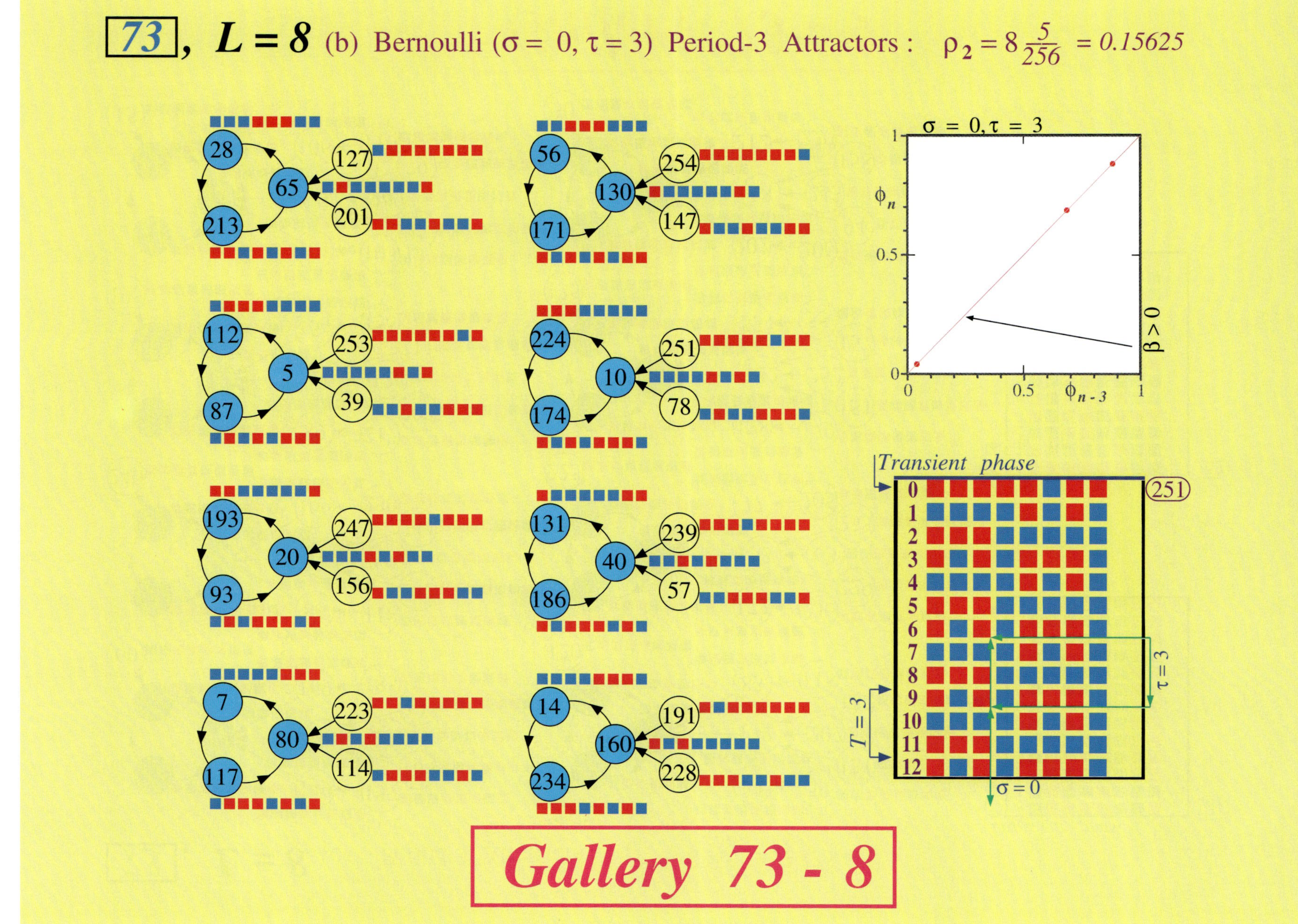

Gallery 73 - 8

Table 17. (*Continued*)

$\boxed{73}$, $L = 8$ (c) Period-3 Attractors : $\rho_3 = 8\frac{12}{256} = 0.375$

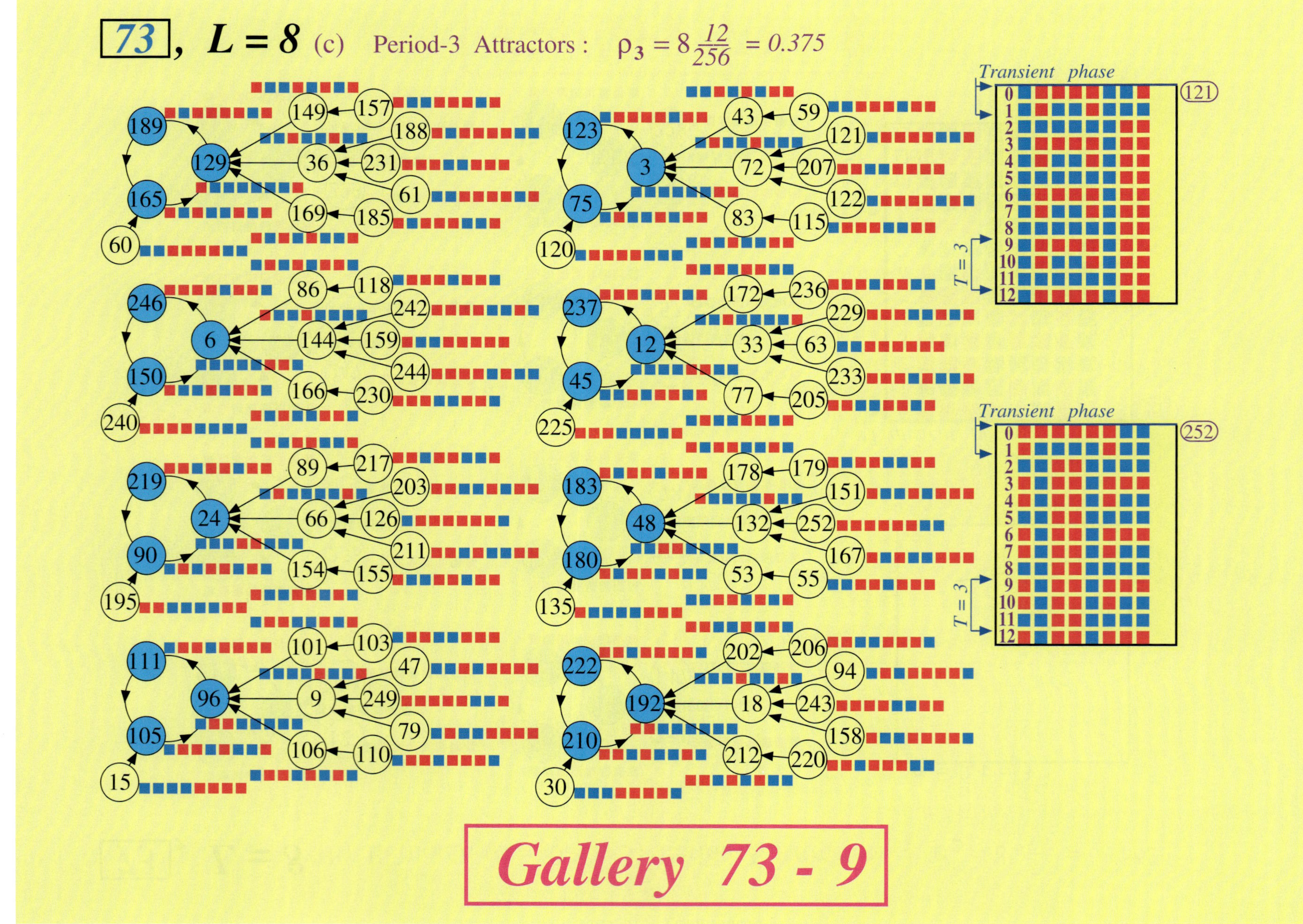

Gallery 73 - 9

73, $L = 8$　(d)　Period-2 Attractors :　$\rho_4 = 8\frac{6}{256} = 0.1875$　　(e)　Period-1 Isles of Eden :

$$\rho_5 = \frac{4}{256} = 0.015625$$

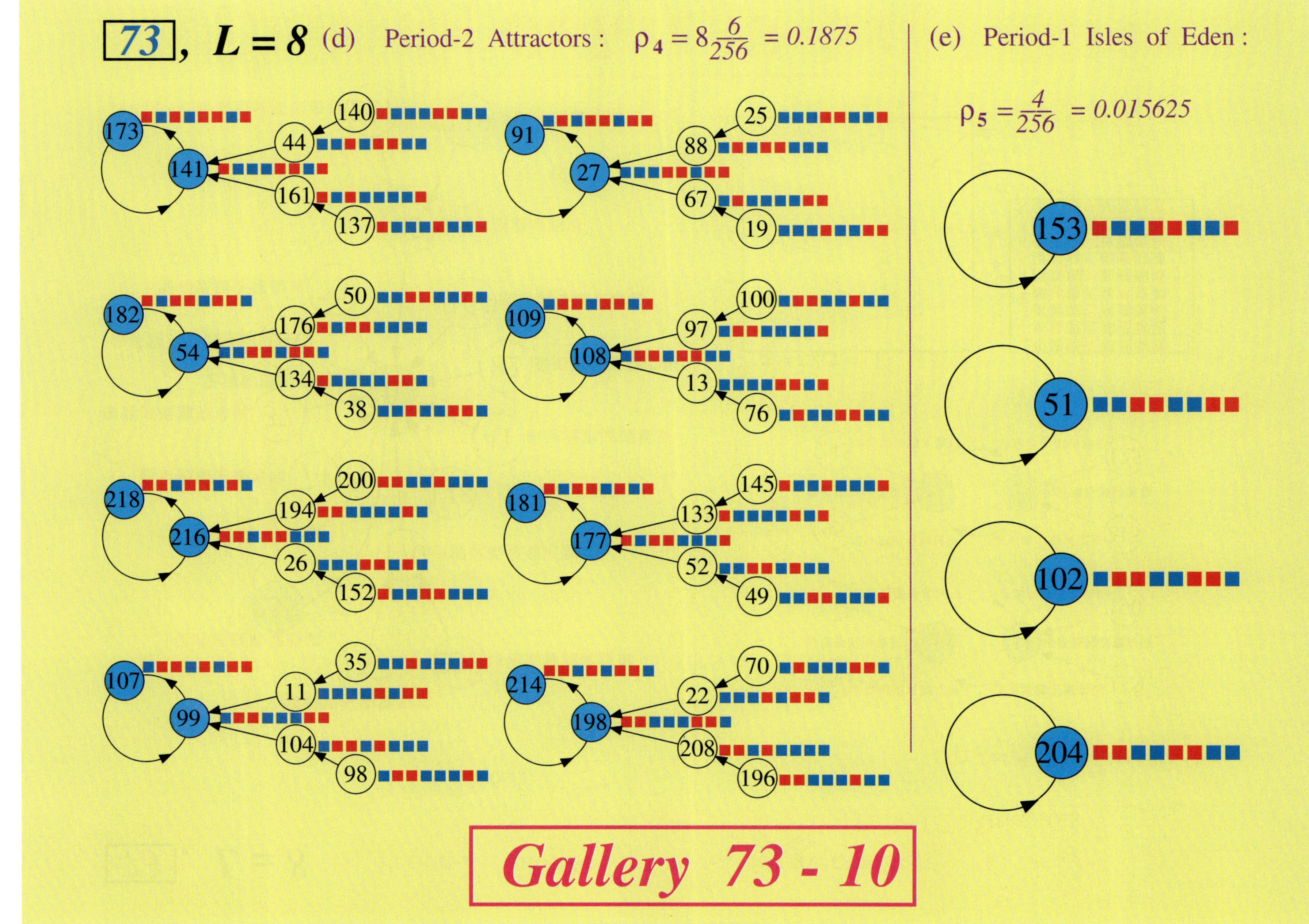

Gallery 73 - 10

$\boxed{73}$, $L = 8$ (f) Period-2 Attractor :

$$\rho_6 = \frac{16}{256} = 0.0625$$

(g) Bernoulli ($\sigma = \pm 2$, $\tau = 1$)

Period-2 Attractors :

$$\rho_7 = 2\,\frac{6}{256} = 0.046875$$

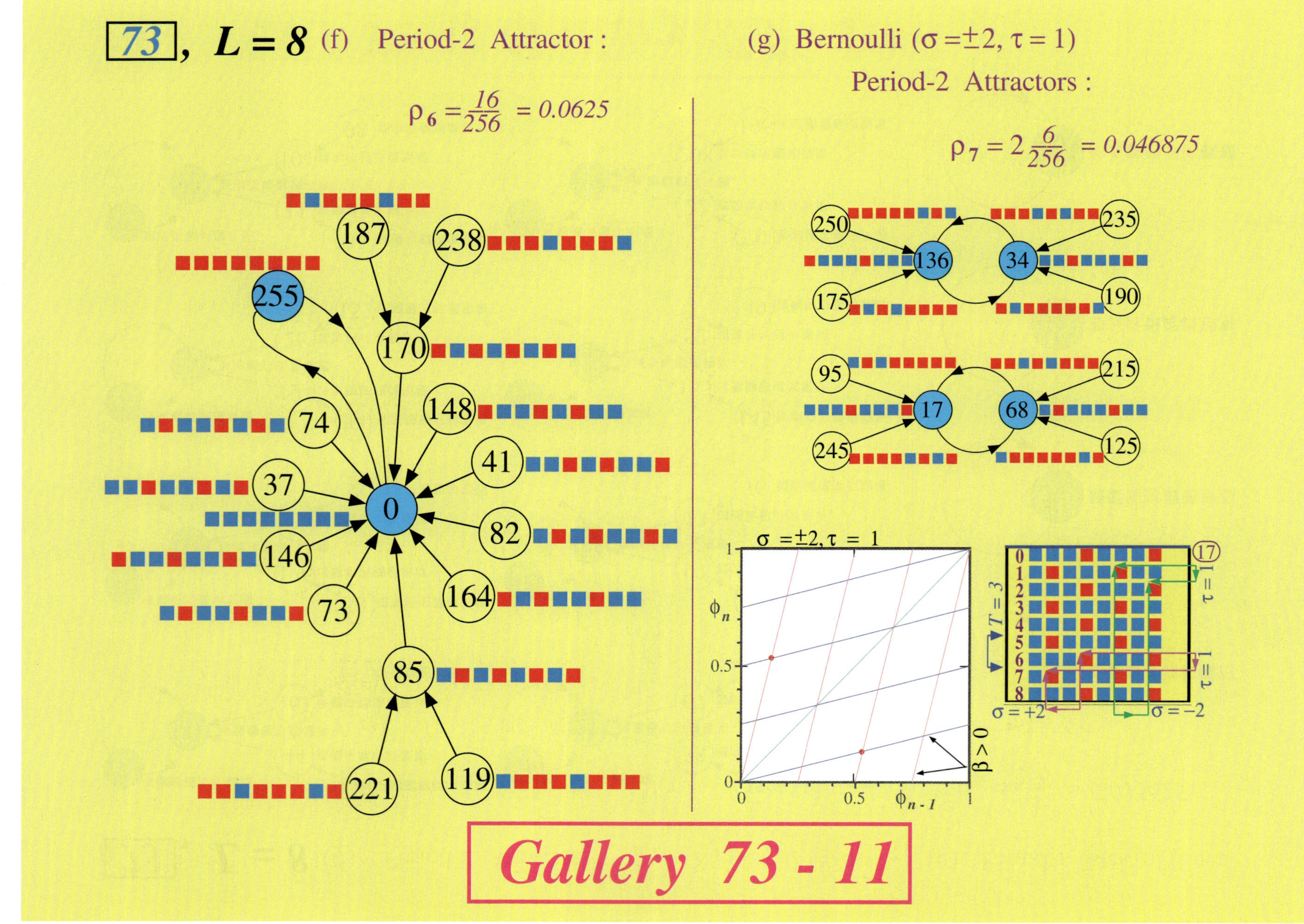

Gallery 73 - 11

2.4.4. *Highlights from Rule* 73

The basin tree diagrams of Rule 73 for $L = 3, 4, \ldots, 8$ are exhibited in Table 17. Following a detailed analysis of these diagrams, the qualitative properties of local rule 73 extracted from basin-tree Galleries 73-1 to 73-11 of Table 17 are summarized below:

Summary of Qualitative properties of local rule 73 extracted from Gallery 73 for Rule 73

L	ID Number i	Number of Period-n attractors	Number of Period-n Isles of Eden	Period n	σ_1	τ_1	β_1	σ_2	τ_2	β_2	Robustness coefficient ρ
3	1	1		2	0	1	−				$\rho_1 = 0.625$
	2		3	1	0	1	+				$\rho_2 = 0.375$
4	1	1		2	0	1	−				$\rho_1 = 0.5$
	2		4	1	0	1	+				$\rho_2 = 0.25$
	3		2	2	2	1	+	−2	1	+	$\rho_3 = 0.25$
5	1	1		2	0	1	−				$\rho_1 = 0.53125$
	2	5		2	0	2	+				$\rho_2 = 0.46875$
6	1	1		2	0	1	−				$\rho_1 = 0.671875$
	2	3		1	0	1	+				$\rho_2 = 0.328115$
7	1	1		1	0	1	−				$\rho_1 = 0.34375$
	2		7	2	0	2	+				$\rho_2 = 0.109375$
	3	7		3	0	3	+				$\rho_3 = 0.4921875$
	4		7	1	0	1	+				$\rho_4 = 0.0546875$
8	1	4		6	4	3	+	−4	3	+	$\rho_1 = 0.15625$
	2	8		3	0	3	+				$\rho_2 = 0.15625$
	3	8		3	0	3	+				$\rho_3 = 0.375$
	4	8		2	0	2	+				$\rho_4 = 0.1875$
	5		4	1	0	1	+				$\rho_5 = 0.015625$
	6	1		2	0	1	−				$\rho_6 = 0.0625$
	7	2		2	2	1	+	−2	1	+	$\rho_7 = 0.046875$

Basin tree diagrams for Rule 90
90 , L = 3   (a)   Period-1 Attractors : $\rho_1 = 4\frac{2}{8} = 1$
0
7
3
4
5
2
6
1
Gallery 90 - 1

90, **L = 4**   (a) Period-1 Attractor : $\rho_1 = \frac{16}{16} = 1$

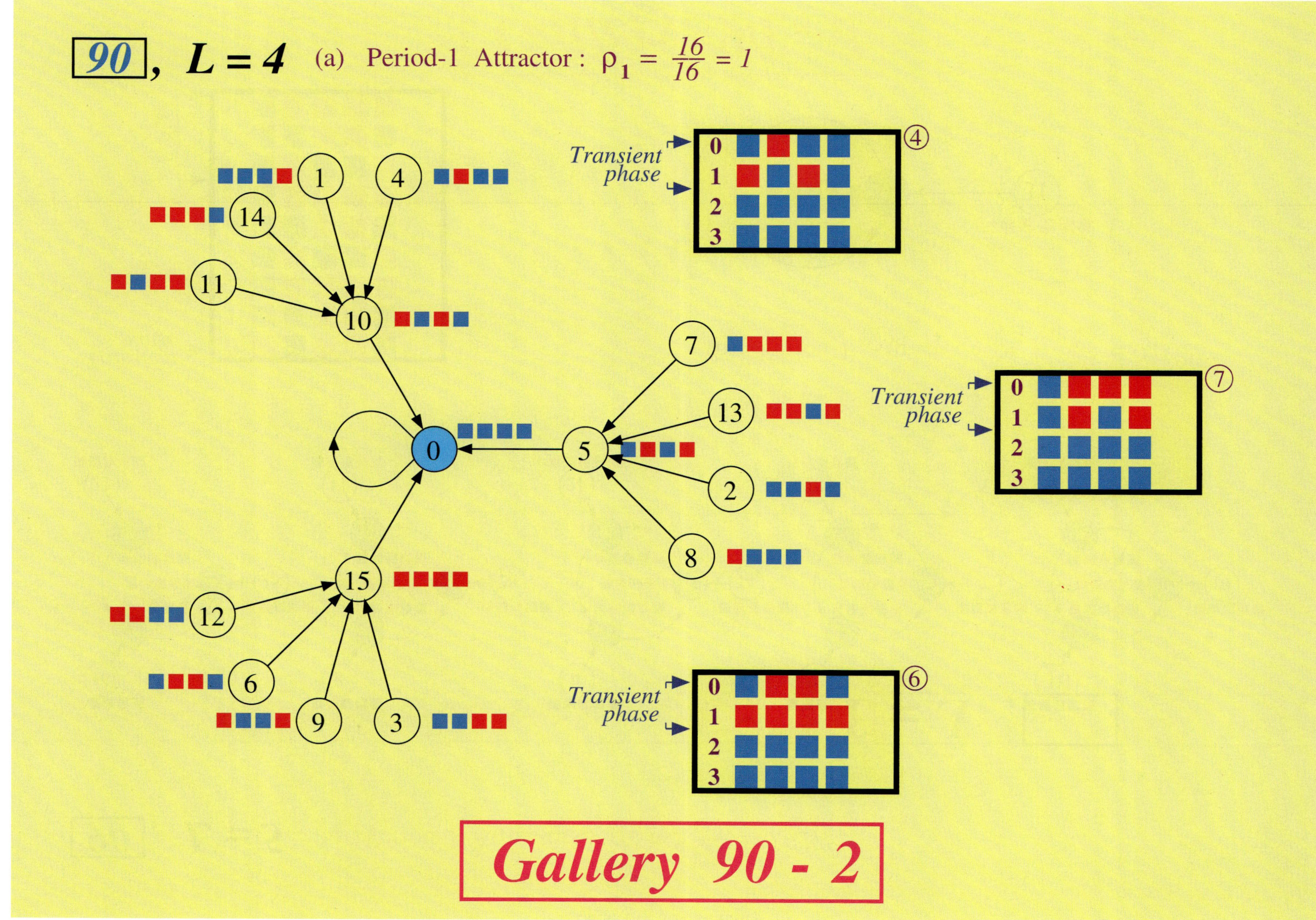

90, **L = 5**   (a) Period-3 Attractors : $\rho_1 = 5\frac{6}{32} = 0.9375$

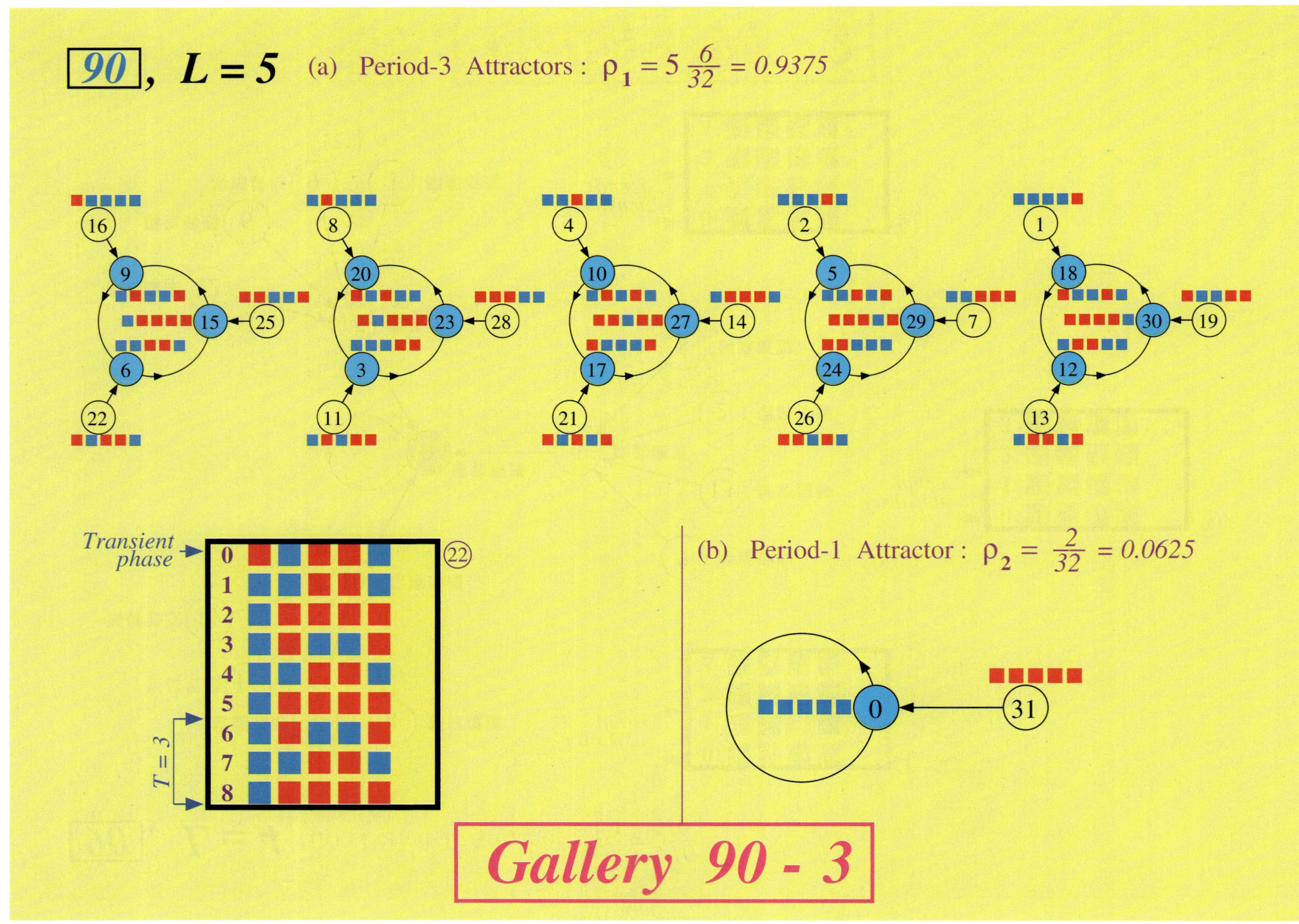

(b) Period-1 Attractor : $\rho_2 = \frac{2}{32} = 0.0625$

Gallery 90 - 3

$\boxed{90}$, $L = 6$ (a) Period-1 Attractors: $\rho_1 = 4\frac{4}{64} = 0.25$

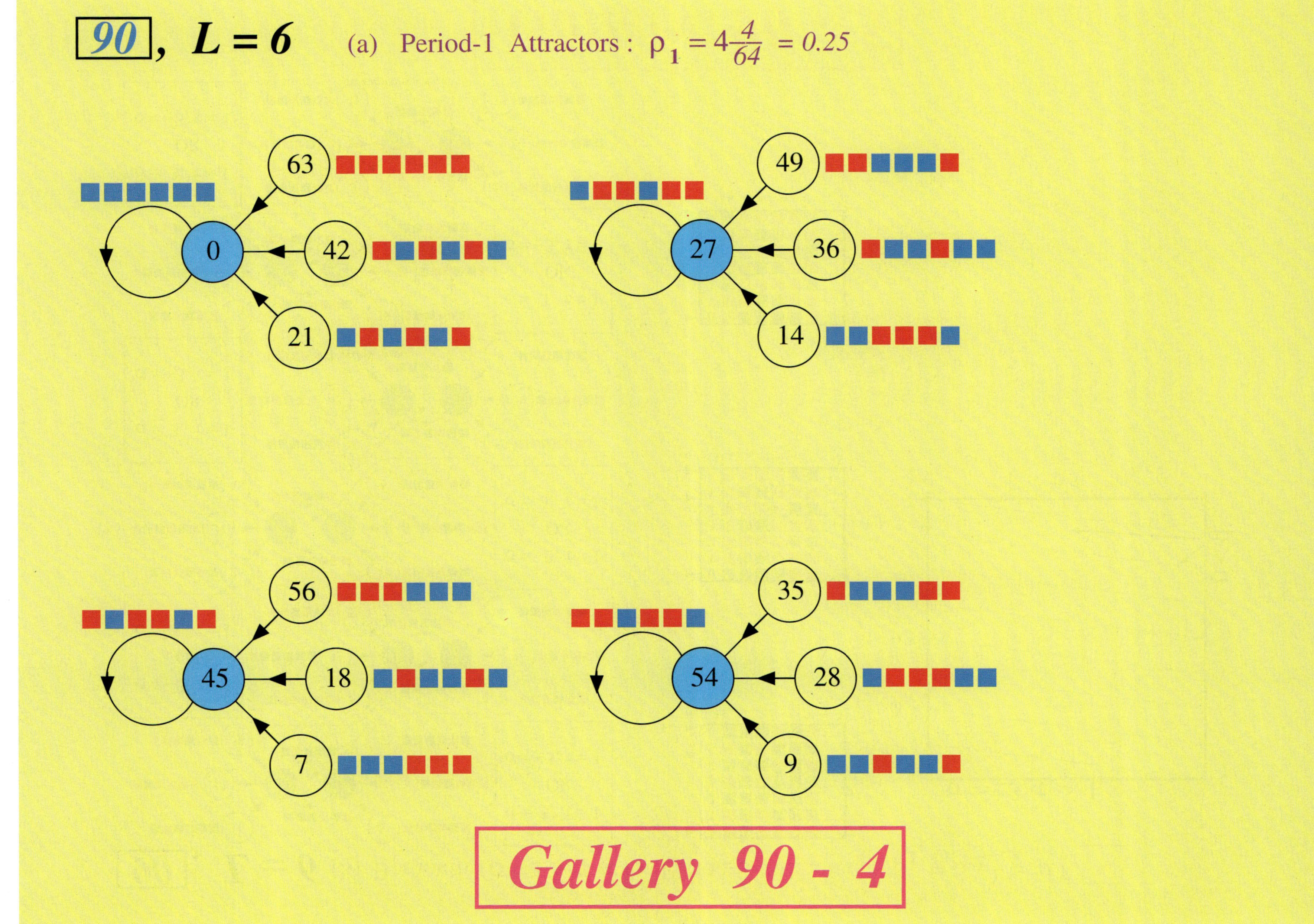

Table 18. (*Continued*)

$\boxed{90}$, $L = 6$ (b) Bernoulli ($\sigma = \pm 3$, $\tau = 1$) Period-2 Attractors : $\rho_2 = 6\frac{8}{64} = 0.75$

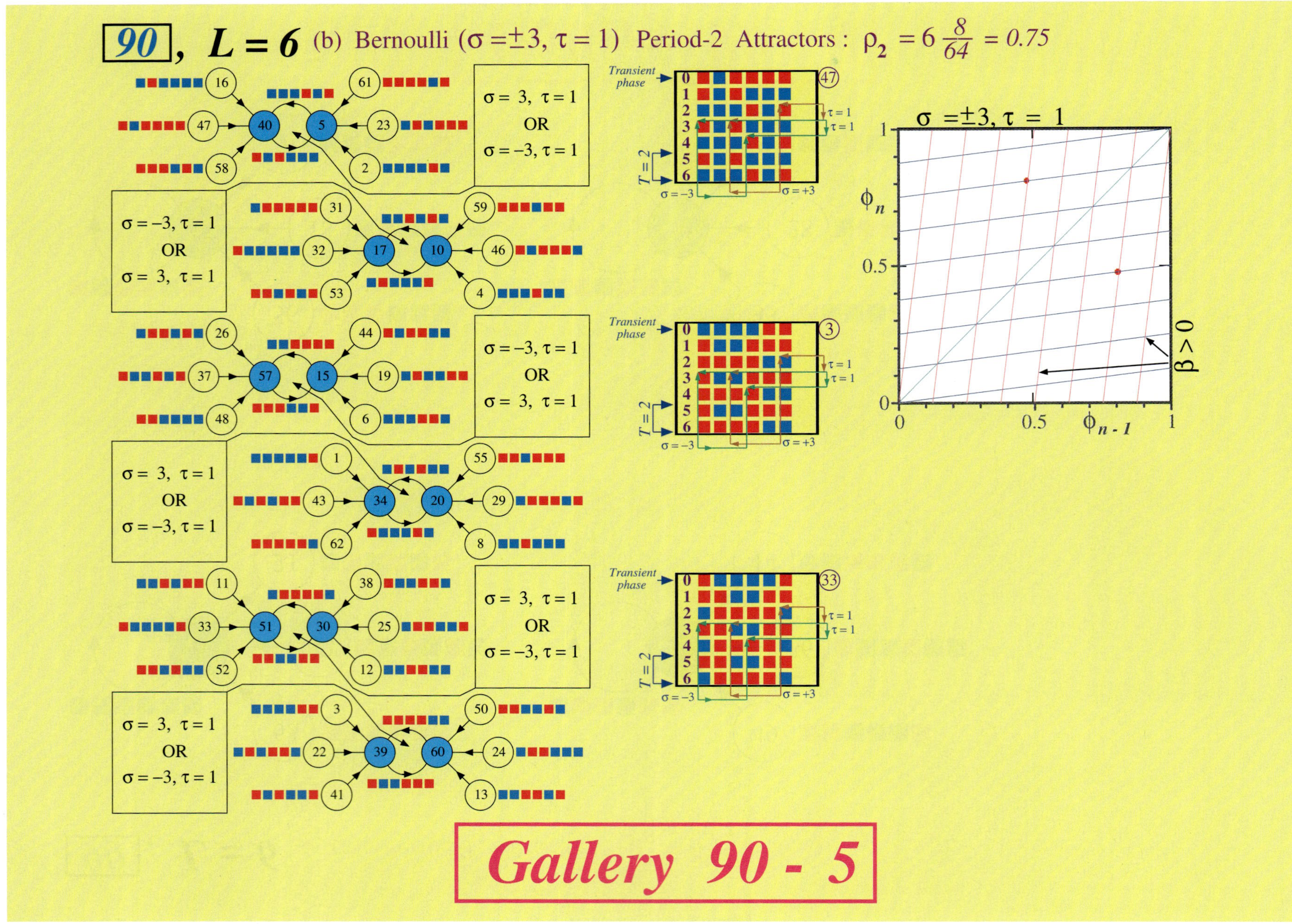

Gallery 90 - 5

70

$\boxed{90}$, $L = 7$ (a) Bernoulli ($\sigma = 0$, $\tau = 7$) Period-7 Attractors : $\rho_1 = 7\frac{14}{128} = 0.765625$

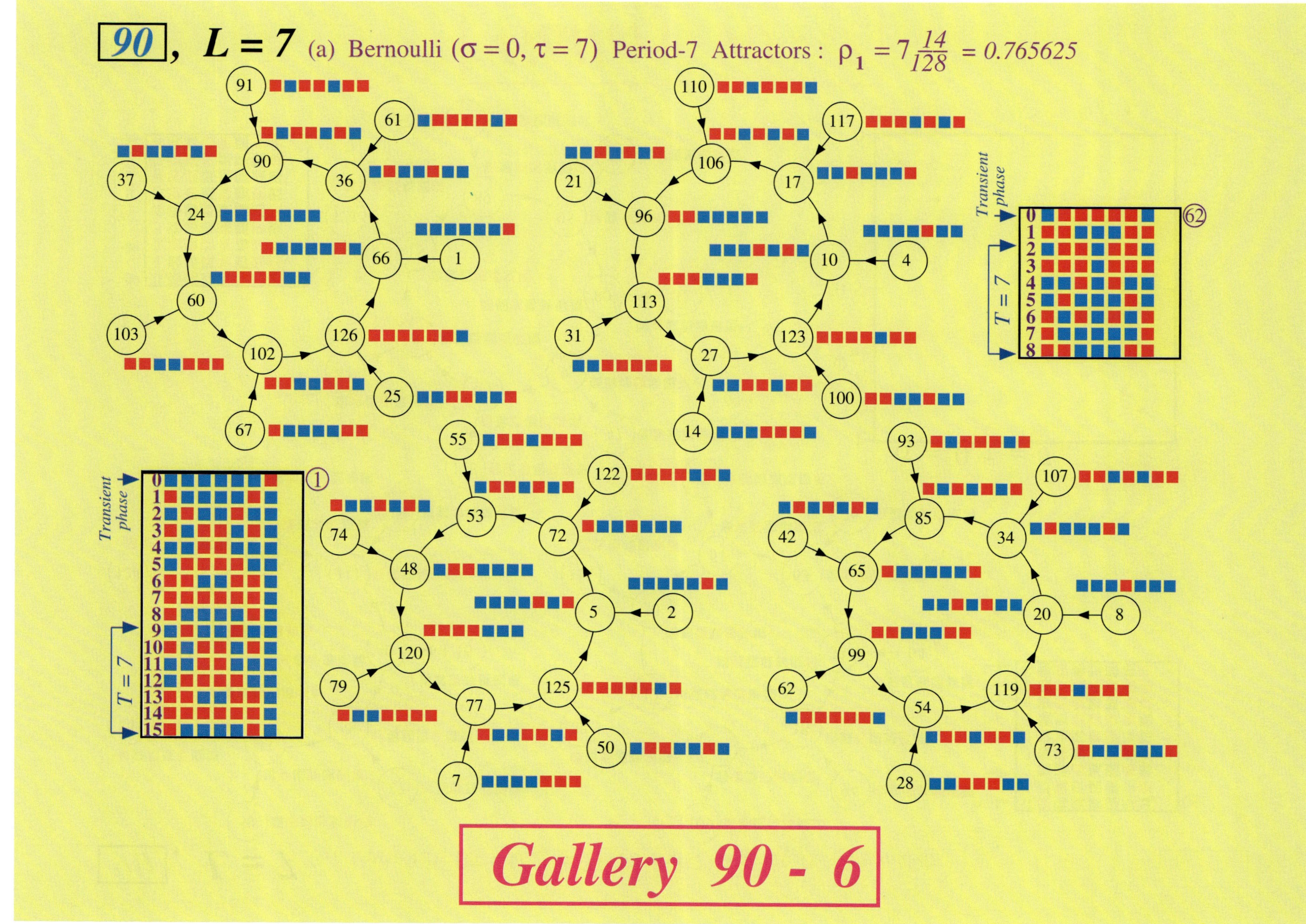

Table 18. (Continued)

90, **L = 7** (a) Bernoulli ($\sigma = 0$, $\tau = 7$) Period-7 Attractors *(continued)*

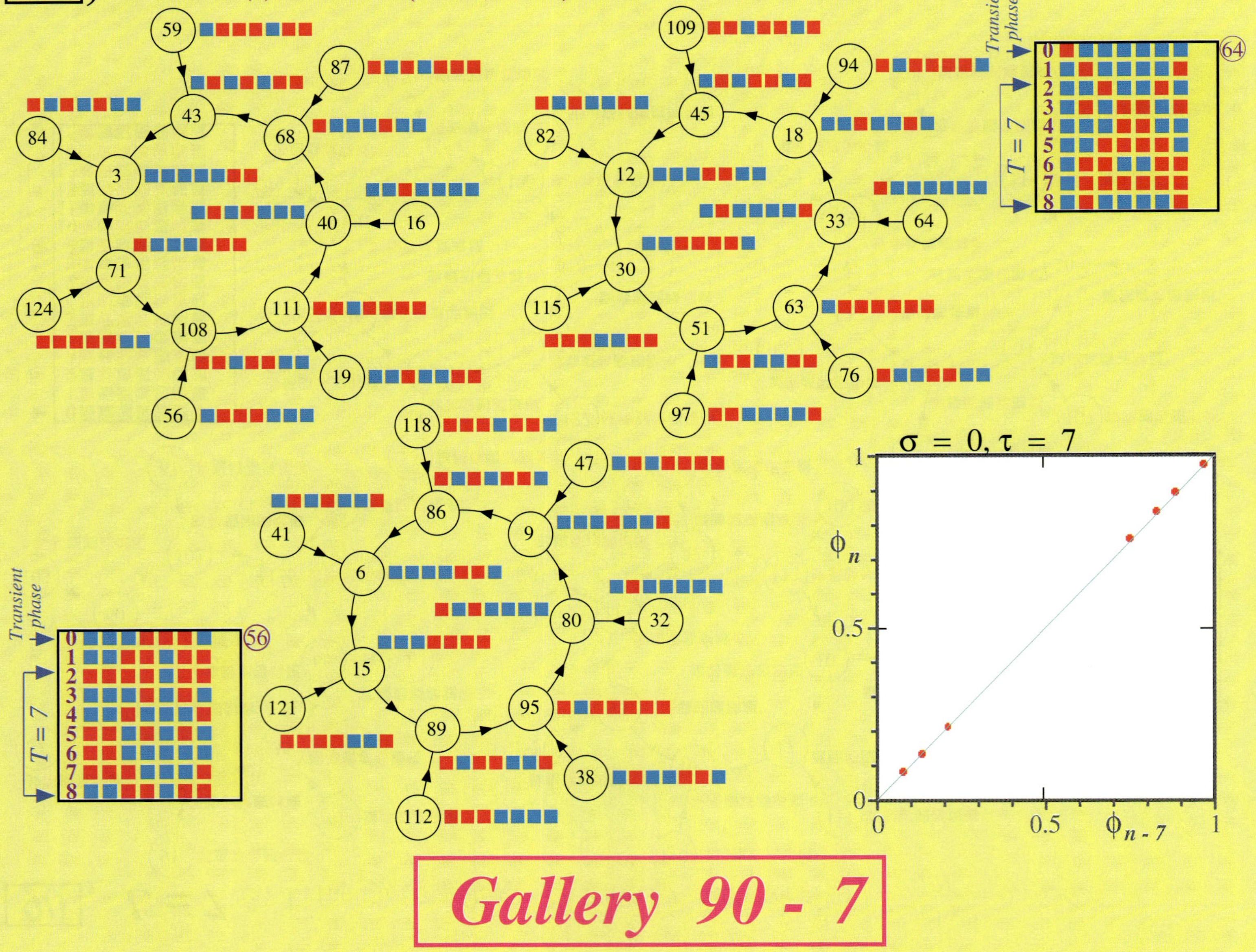

$\sigma = 0, \tau = 7$

Gallery 90 - 7

72

Table 18. (*Continued*)

90, $L = 7$ (b) Bernoulli ($\sigma = \pm 2$, $\sigma = \mp 5$, $\tau = 1$) Period-7 Attractors :

$$\rho_2 = 2\frac{14}{128} = 0.21875$$

(c) Period-1 Attractor :

$$\rho_3 = \frac{2}{128}$$

$$= 0.015625$$

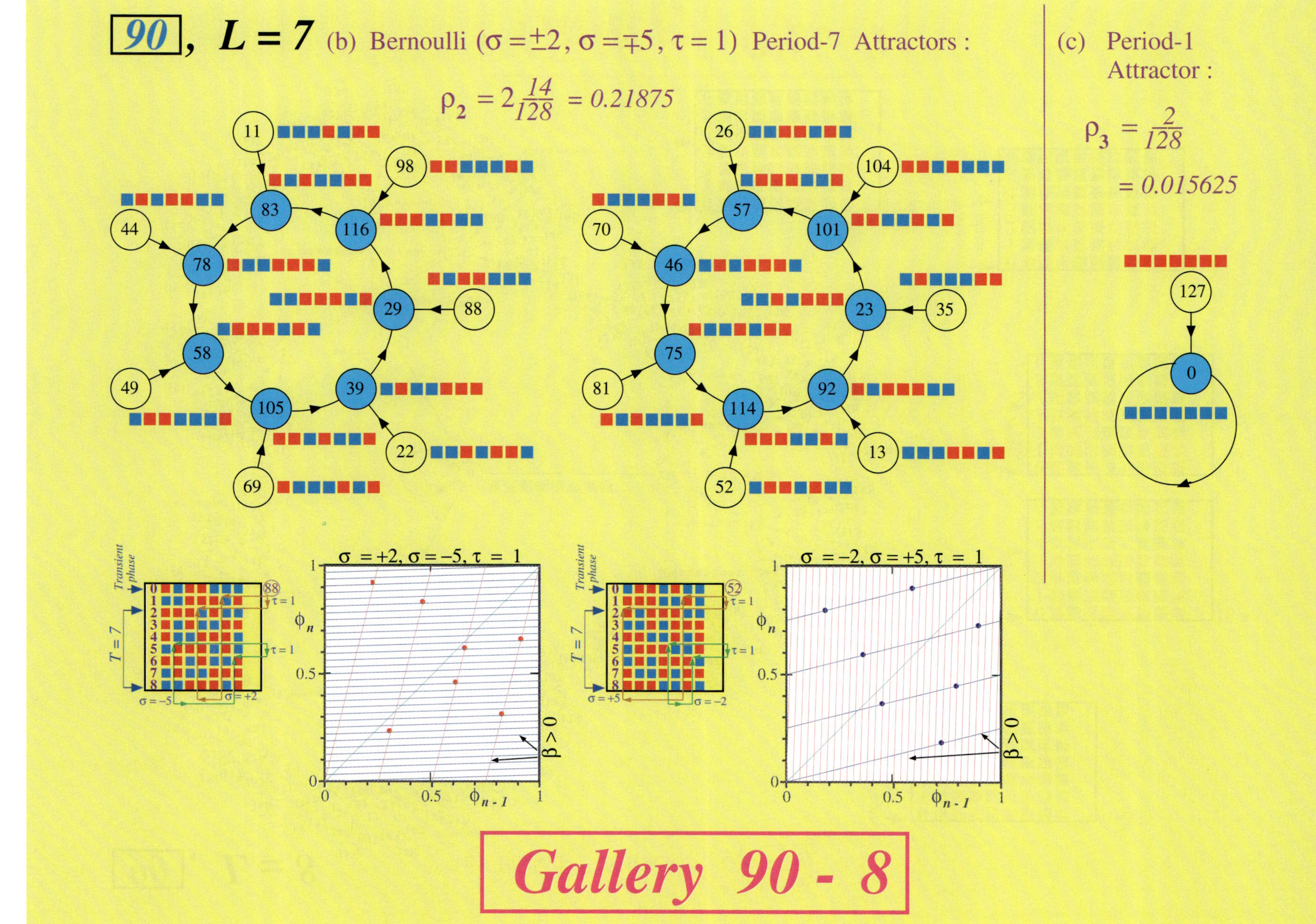

Gallery 90 - 8

73

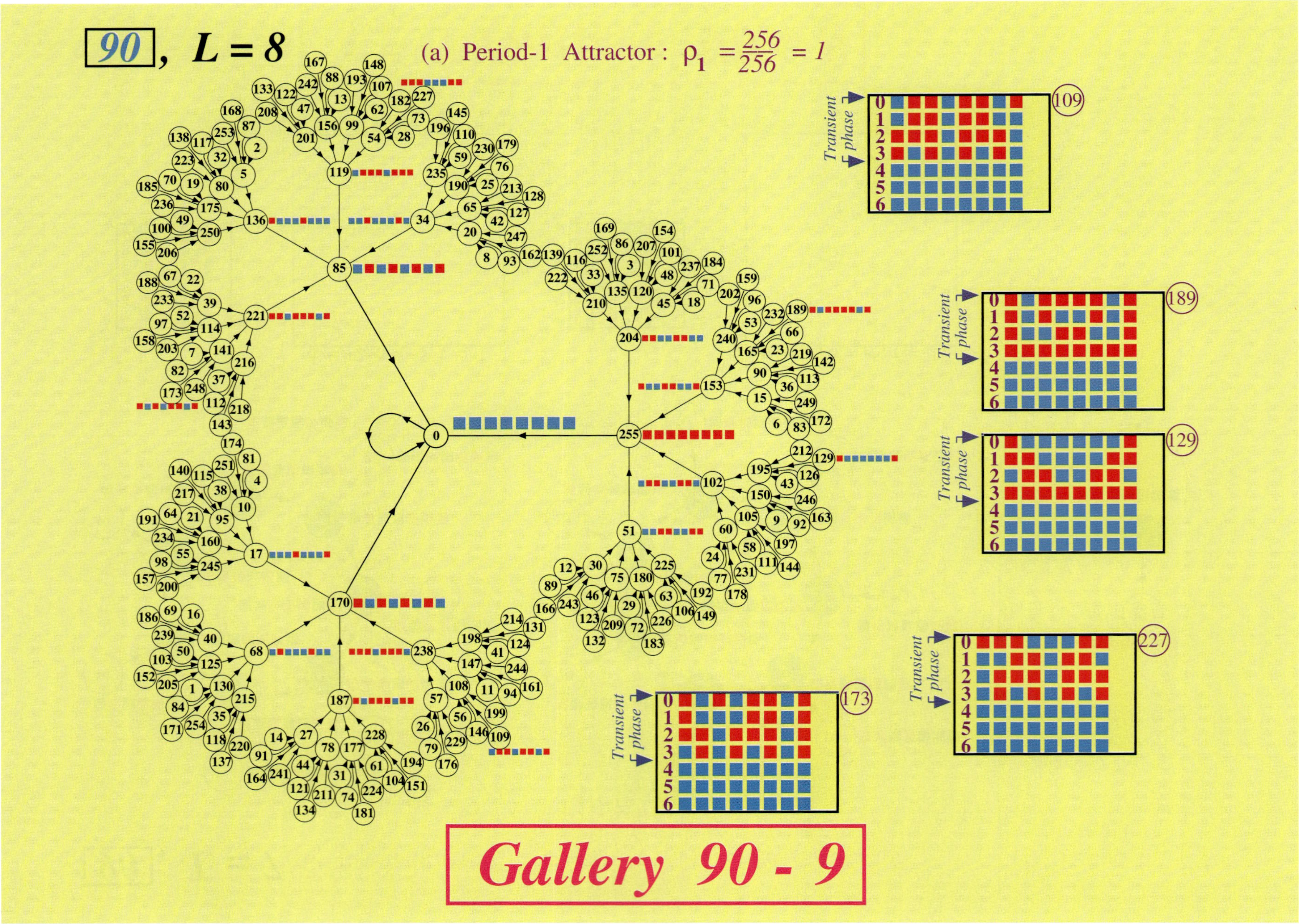

90, $L = 8$ (a) Period-1 Attractor: $\rho_1 = \frac{256}{256} = 1$

Gallery 90 - 9

2.4.5. *Highlights from Rule* 90

The basin tree diagrams of Rule 90 for $L = 3, 4, \ldots, 8$ are exhibited in Table 18. Following a detailed analysis of these diagrams, the qualitative properties of local rule 90 extracted from basin-tree Galleries 90-1 to 90-9 of Table 18 are summarized below:

Summary of Qualitative properties of local rule 90 *extracted from Gallery 90 for Rule* 90

L	ID Number i	Number of Period-n attractors	Number of Period-n Isles of Eden	Period n	Bernoulli Parameters						Robustness coefficient ρ
					σ_1	τ_1	β_1	σ_2	τ_2	β_2	
3	1	4		1	0	1	+				$\rho_1 = 1$
4	1	1		1	0	1	+				$\rho_1 = 1$
5	1	5		3	0	3	+				$\rho_1 = 0.9375$
	2	1		1	0	1	+				$\rho_2 = 0.0625$
6	1	4		1	0	1	+				$\rho_1 = 0.25$
	2	6		2	3	1	+	-3	1	+	$\rho_2 = 0.75$
7	1	7		7	0	1	+				$\rho_1 = 0.765625$
	2	2		7	± 2	1	+	∓ 5	1	+	$\rho_2 = 0.21875$
	3	1		1	0	1	+				$\rho_3 = 0.015625$
8	1	1		1	0	1	+				$\rho_1 = 1$

Basin tree diagrams for Rule 105

105, **L = 3**   (a) Period-2 Attractor : $\rho_1 = \frac{8}{8} = 1$

Gallery 105 - 1

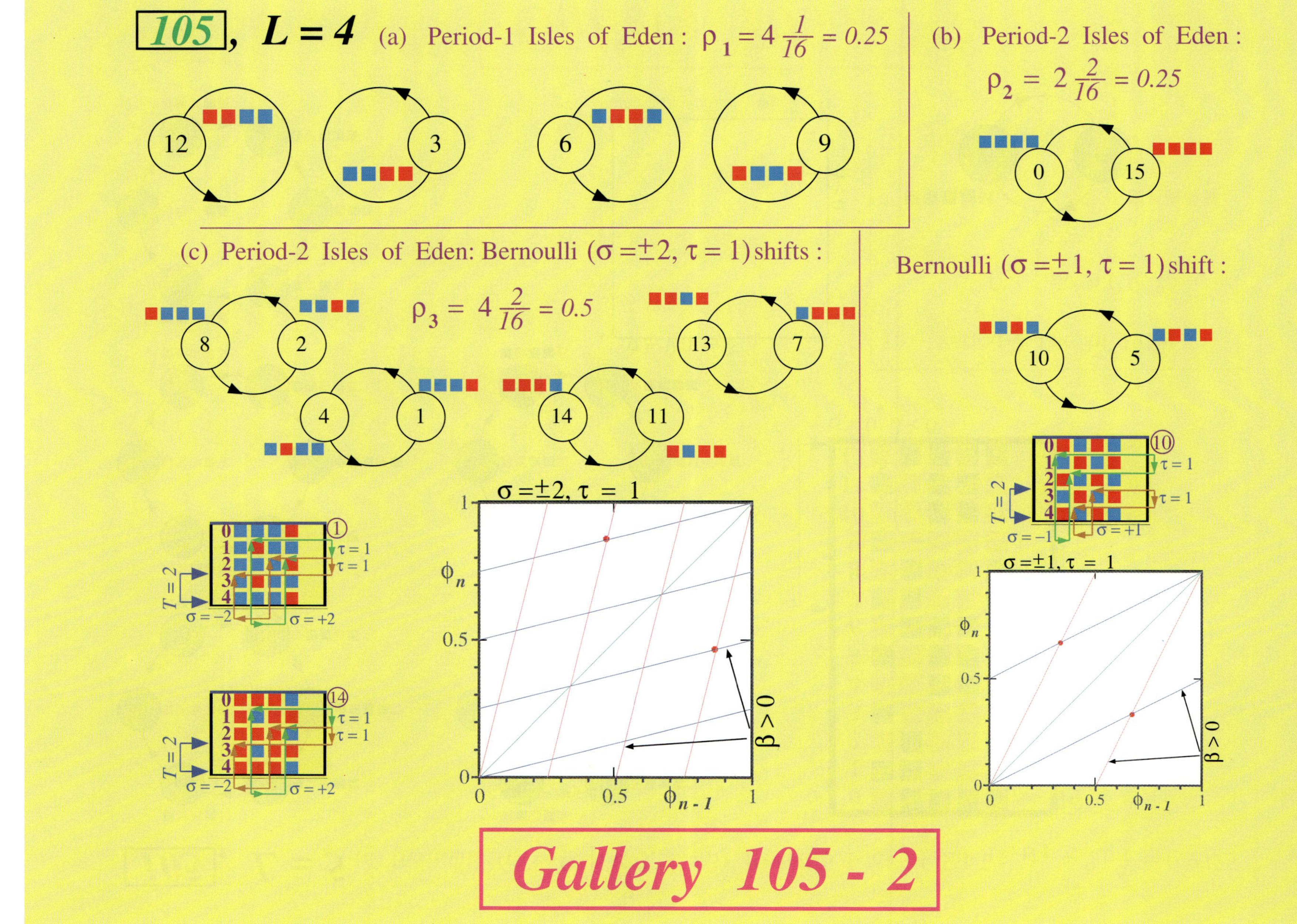

105, L = 4
(a) Period-1 Isles of Eden : $\rho_1 = 4\frac{1}{16} = 0.25$
(b) Period-2 Isles of Eden : $\rho_2 = 2\frac{2}{16} = 0.25$
12   3   6   9
0   15
(c) Period-2 Isles of Eden: Bernoulli ($\sigma = \pm 2,\ \tau = 1$) shifts :
$\rho_3 = 4\frac{2}{16} = 0.5$
Bernoulli ($\sigma = \pm 1,\ \tau = 1$) shift :
8   2   4   1   13   7   14   11
10   5
$\sigma = \pm 2,\ \tau = 1$
ϕ_n
ϕ_{n-1}
$\beta > 0$
$\sigma = \pm 1,\ \tau = 1$
ϕ_n
ϕ_{n-1}
$\beta > 0$
$T = 2$ $\sigma = -2$ $\sigma = +2$ $\tau = 1$ $\tau = 1$
$T = 2$ $\sigma = -1$ $\sigma = +1$ $\tau = 1$ $\tau = 1$
Gallery 105 - 2

Table 19. (*Continued*)

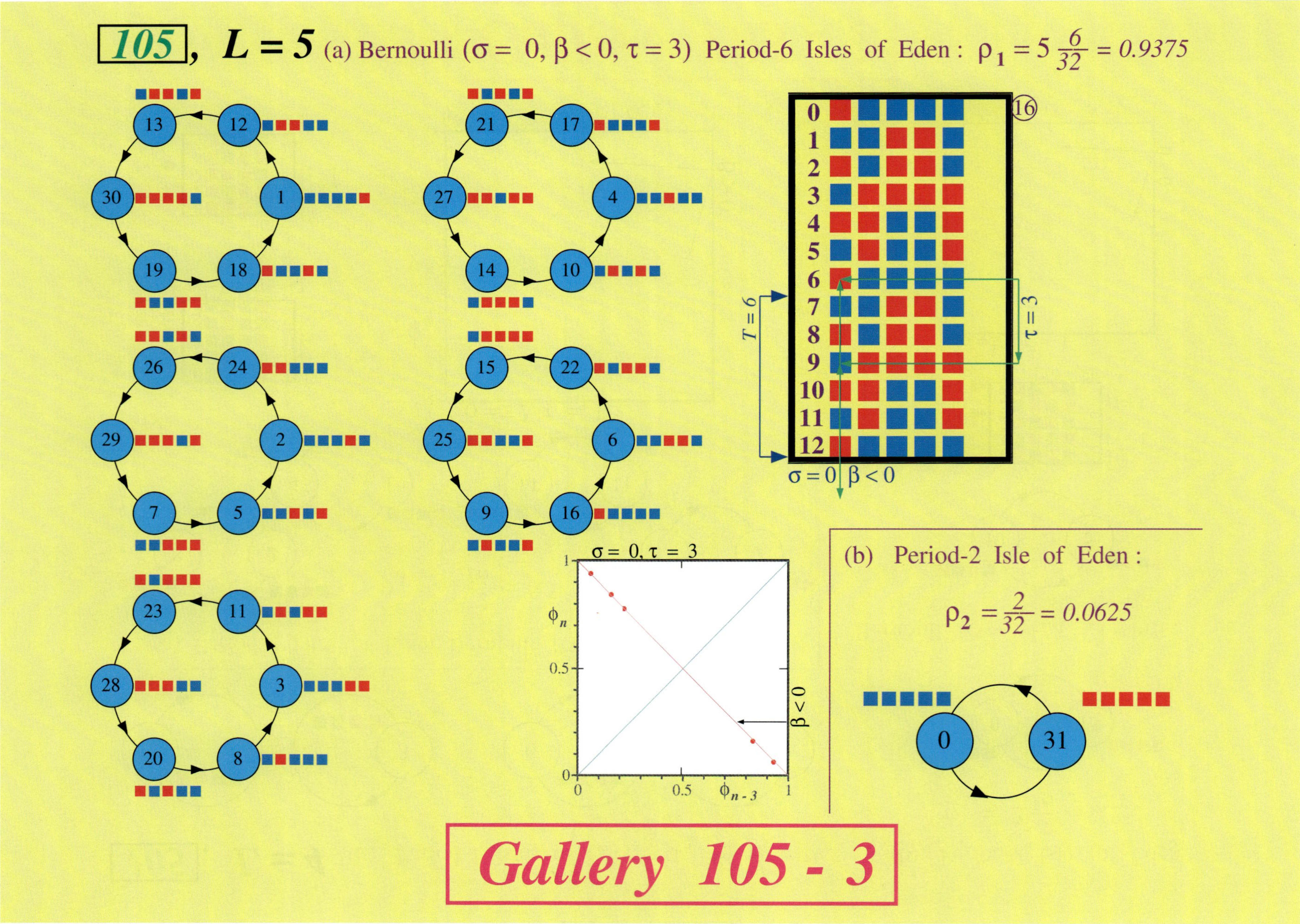
105, L = 5 (a) Bernoulli (σ = 0, β < 0, τ = 3) Period-6 Isles of Eden : ρ1 = 5 6/32 = 0.9375
13
12
30
1
19
18
21
17
27
4
14
10
26
24
29
2
7
5
15
22
25
6
9
16
23
11
28
3
20
8
0
1
2
3
4
5
6
7
8
9
10
11
12
16
T = 6
τ = 3
σ = 0
β < 0
σ = 0, τ = 3
φn
φn - 3
0
0.5
1
β < 0
(b) Period-2 Isle of Eden :
ρ2 = 2/32 = 0.0625
0
31
Gallery 105 - 3

105, $L = 6$ — (a) Period-2 Attractor : $\rho_1 = \frac{32}{64} = 0.5$

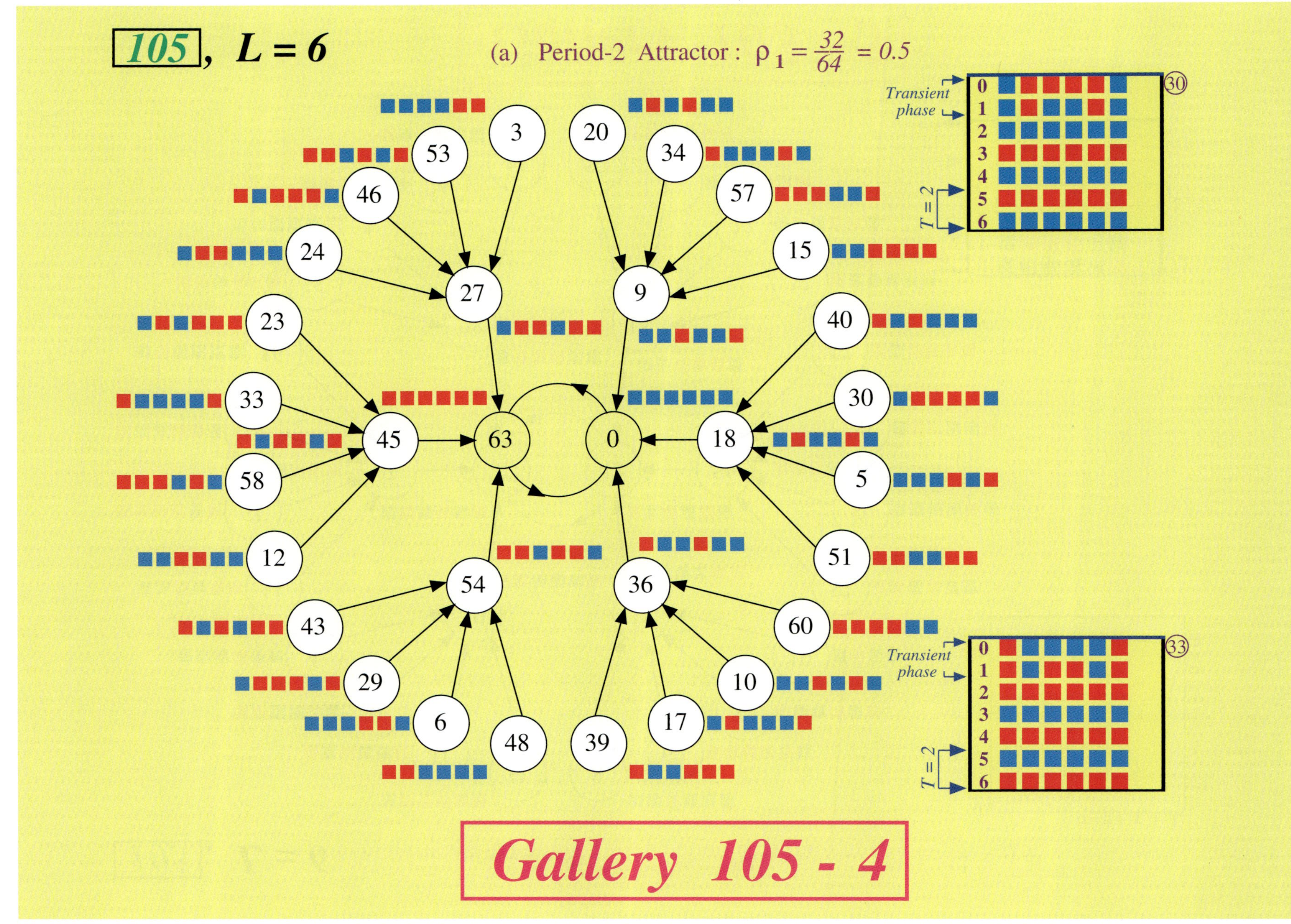

Table 19. (*Continued*)

105, $L = 6$ (b) Bernoulli ($\sigma = \pm 1$, $\tau = 1$) Period-2 Attractor : $\rho_2 = \frac{32}{64} = 0.5$

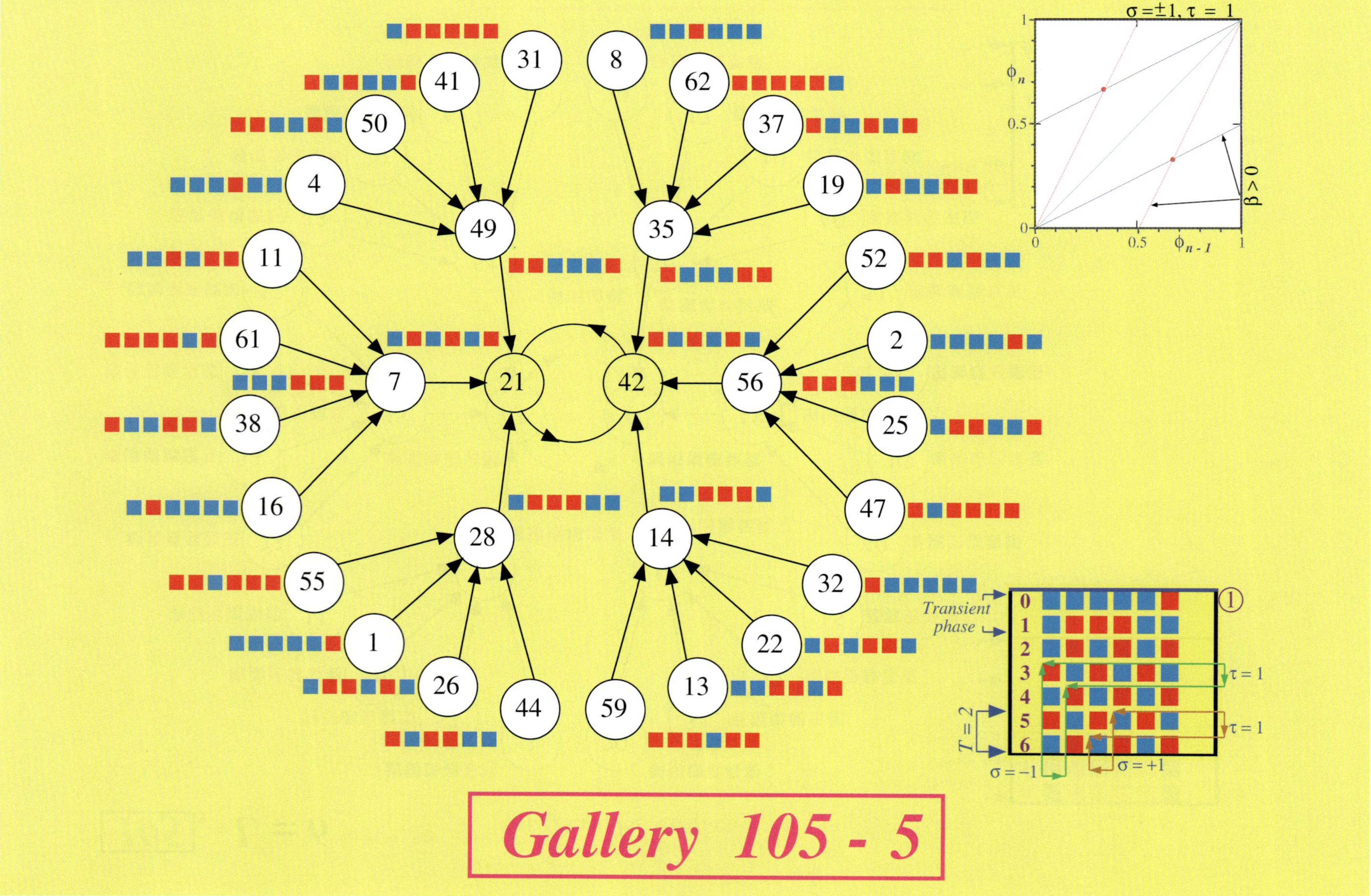

Gallery 105 - 5

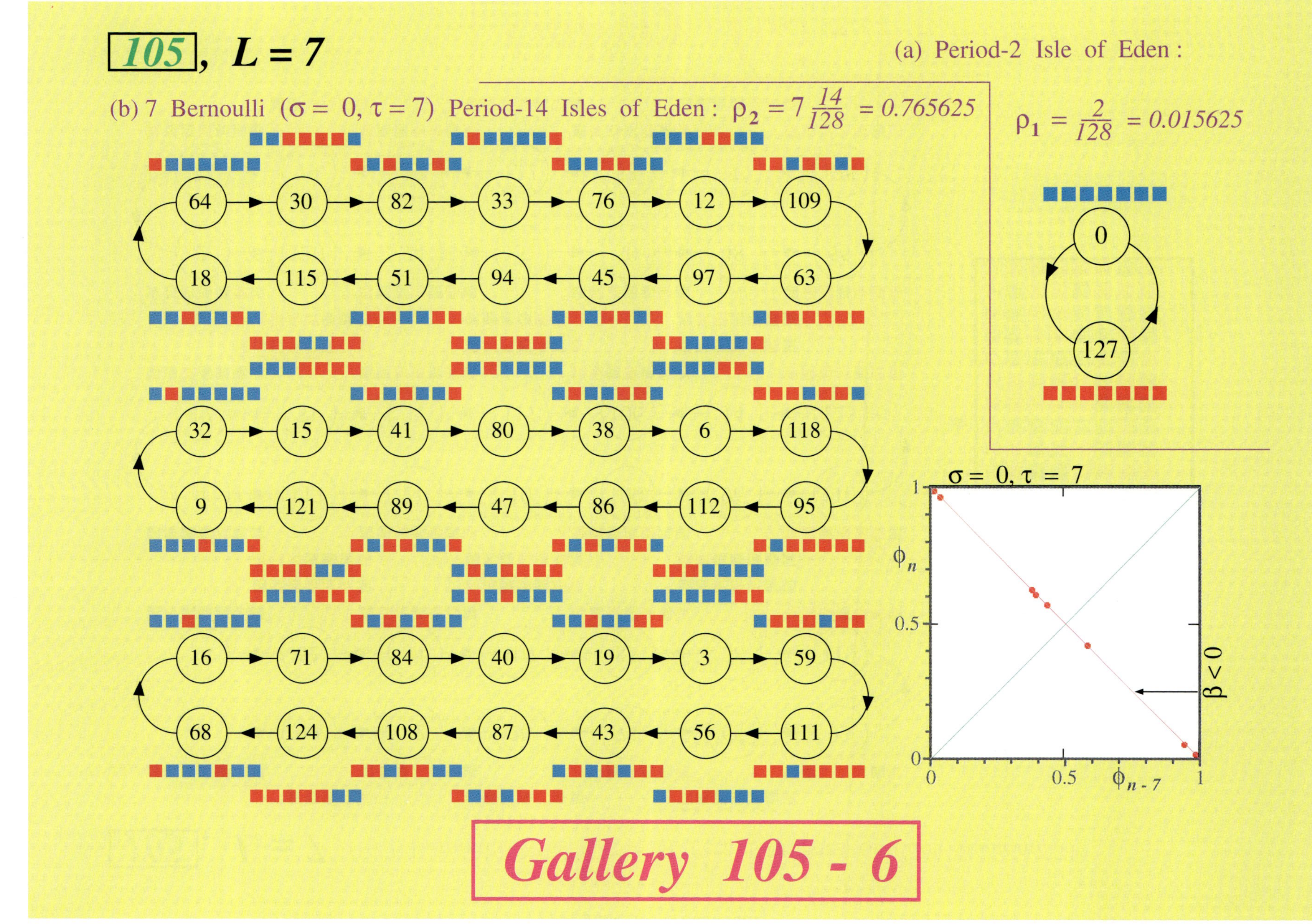
105, L = 7
(a) Period-2 Isle of Eden :
(b) 7 Bernoulli (σ = 0, τ = 7) Period-14 Isles of Eden : ρ₂ = 7 14/128 = 0.765625
ρ₁ = 2/128 = 0.015625
64 30 82 33 76 12 109
18 115 51 94 45 97 63
32 15 41 80 38 6 118
9 121 89 47 86 112 95
16 71 84 40 19 3 59
68 124 108 87 43 56 111
0
127
σ = 0, τ = 7
φ_n
φ_{n-7}
β < 0
Gallery 105 - 6

Table 19. (*Continued*)

105, **L = 7** (b) 7 Bernoulli $(\sigma = 0, \tau = 7)$ Period-14 Isles of Eden (*continued*) :

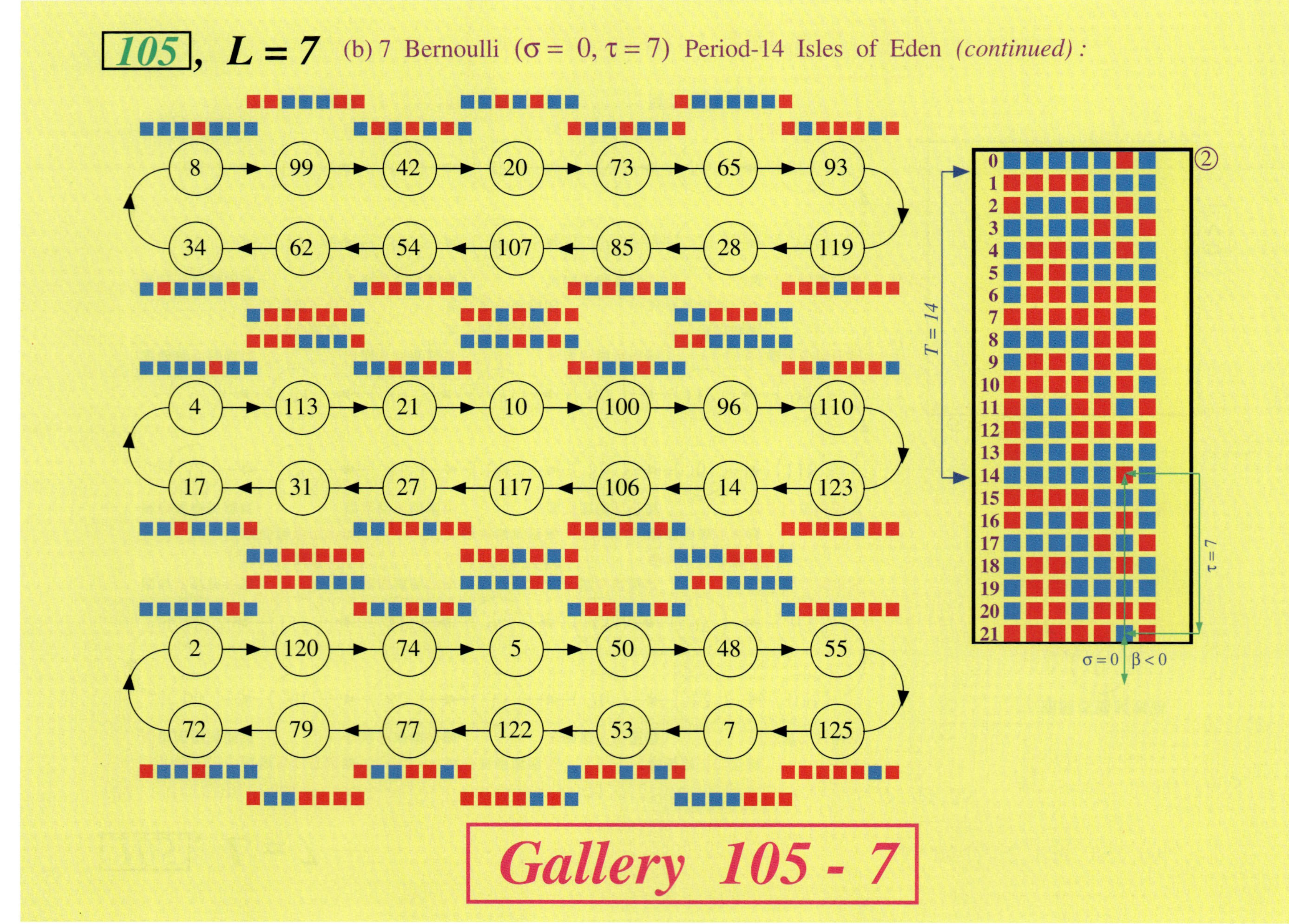

82

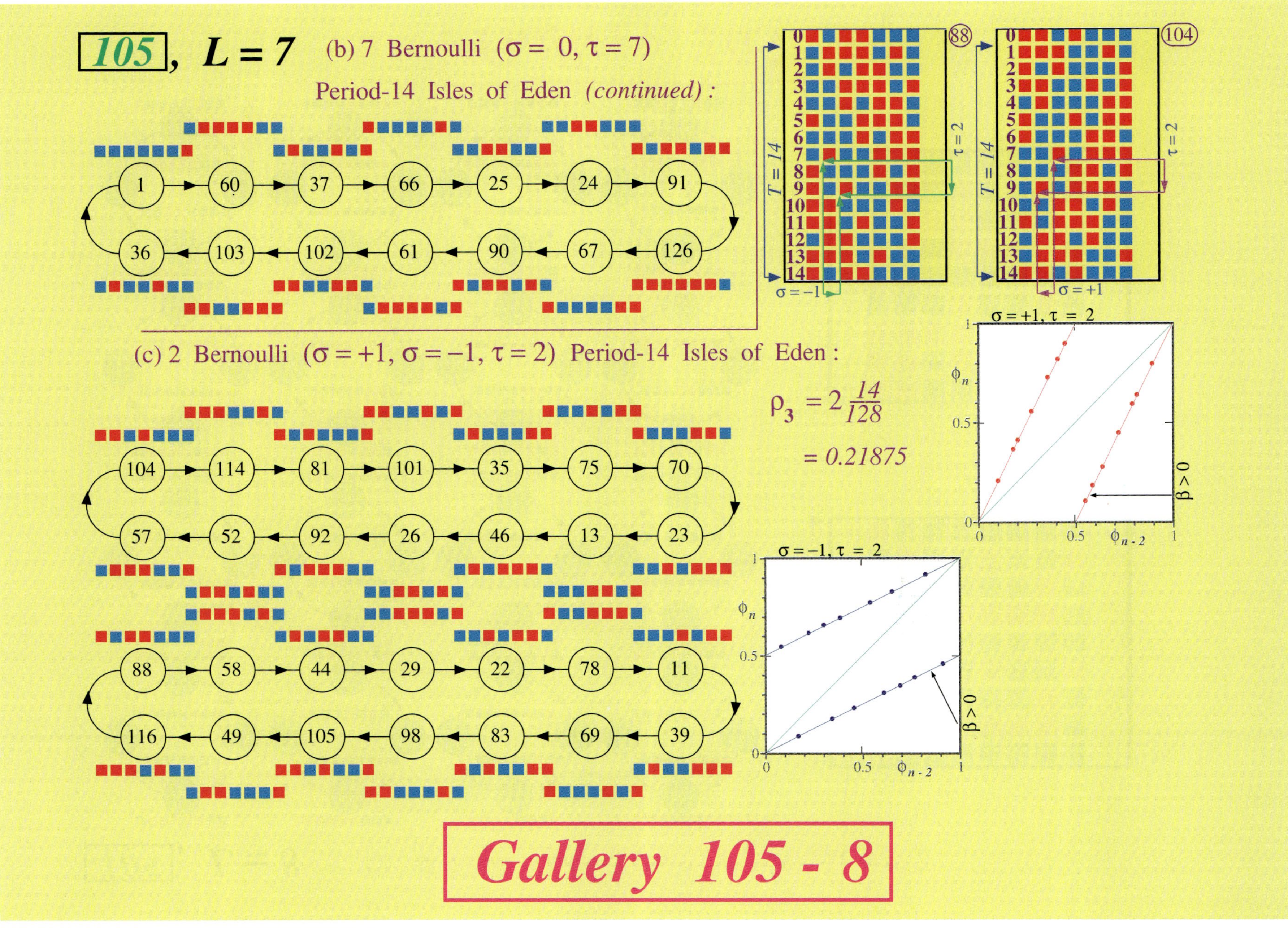

105 , L = 7   (b) 7 Bernoulli (σ = 0, τ = 7)
Period-14 Isles of Eden (continued) :
1 → 60 → 37 → 66 → 25 → 24 → 91
36 ← 103 ← 102 ← 61 ← 90 ← 67 ← 126
T = 14   τ = 2   σ = −1   88
T = 14   τ = 2   σ = +1   104
(c) 2 Bernoulli (σ = +1, σ = −1, τ = 2) Period-14 Isles of Eden :
104 → 114 → 81 → 101 → 35 → 75 → 70
57 ← 52 ← 92 ← 26 ← 46 ← 13 ← 23
88 → 58 → 44 → 29 → 22 → 78 → 11
116 ← 49 ← 105 ← 98 ← 83 ← 69 ← 39
ρ₃ = 2 14/128
= 0.21875
σ = +1, τ = 2
φₙ   φₙ₋₂   β > 0
σ = −1, τ = 2
φₙ   φₙ₋₂   β > 0
Gallery 105 - 8

Table 19.   (Continued)

$\boxed{105}$, **$L = 8$**   (a)   48 Period-4 Isles of Eden : $\rho_1 = 48\frac{4}{256} = 0.75$

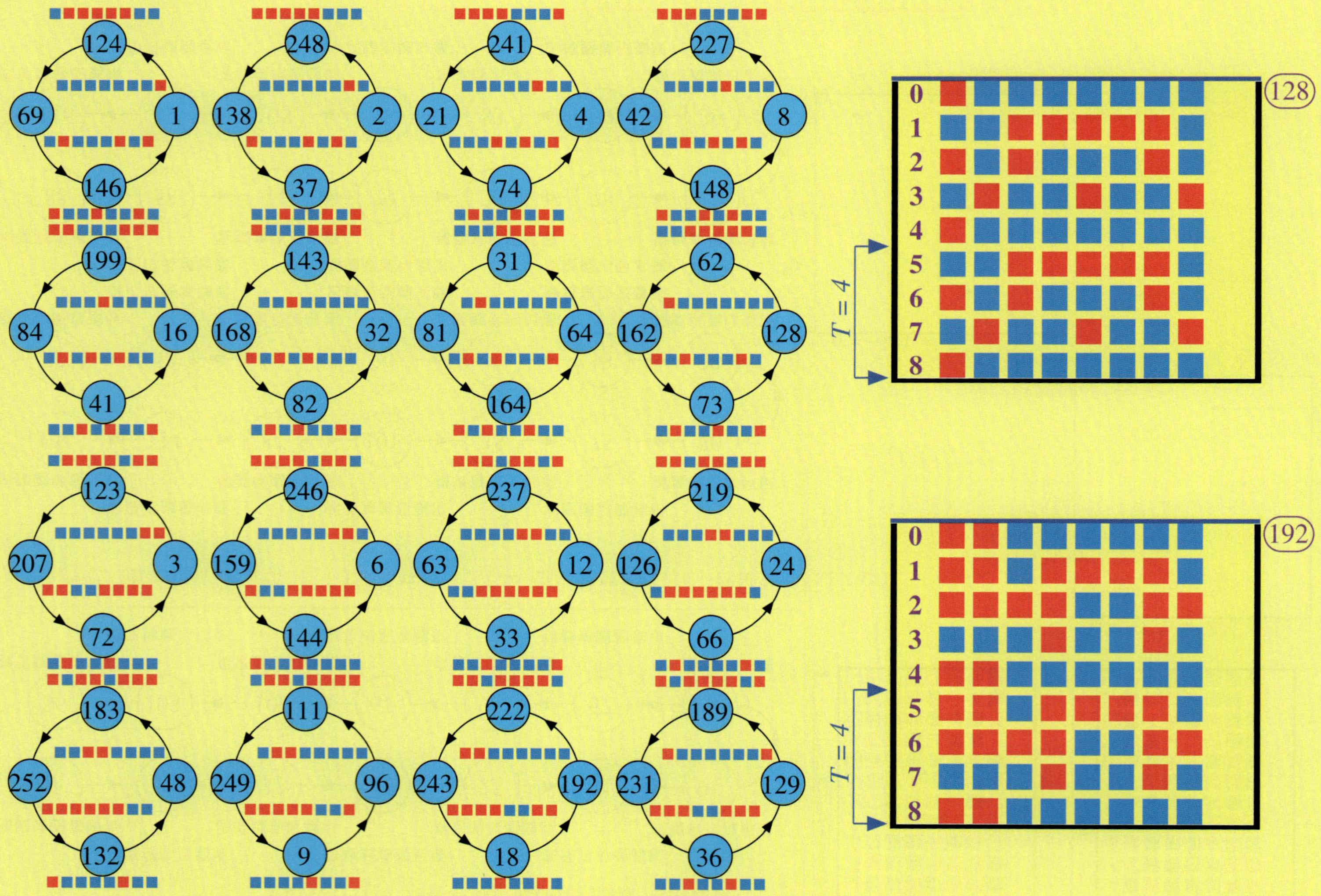

Gallery 105 - 9

105, $L = 8$ (a) 48 Period-4 Isles of Eden (*continued*) :

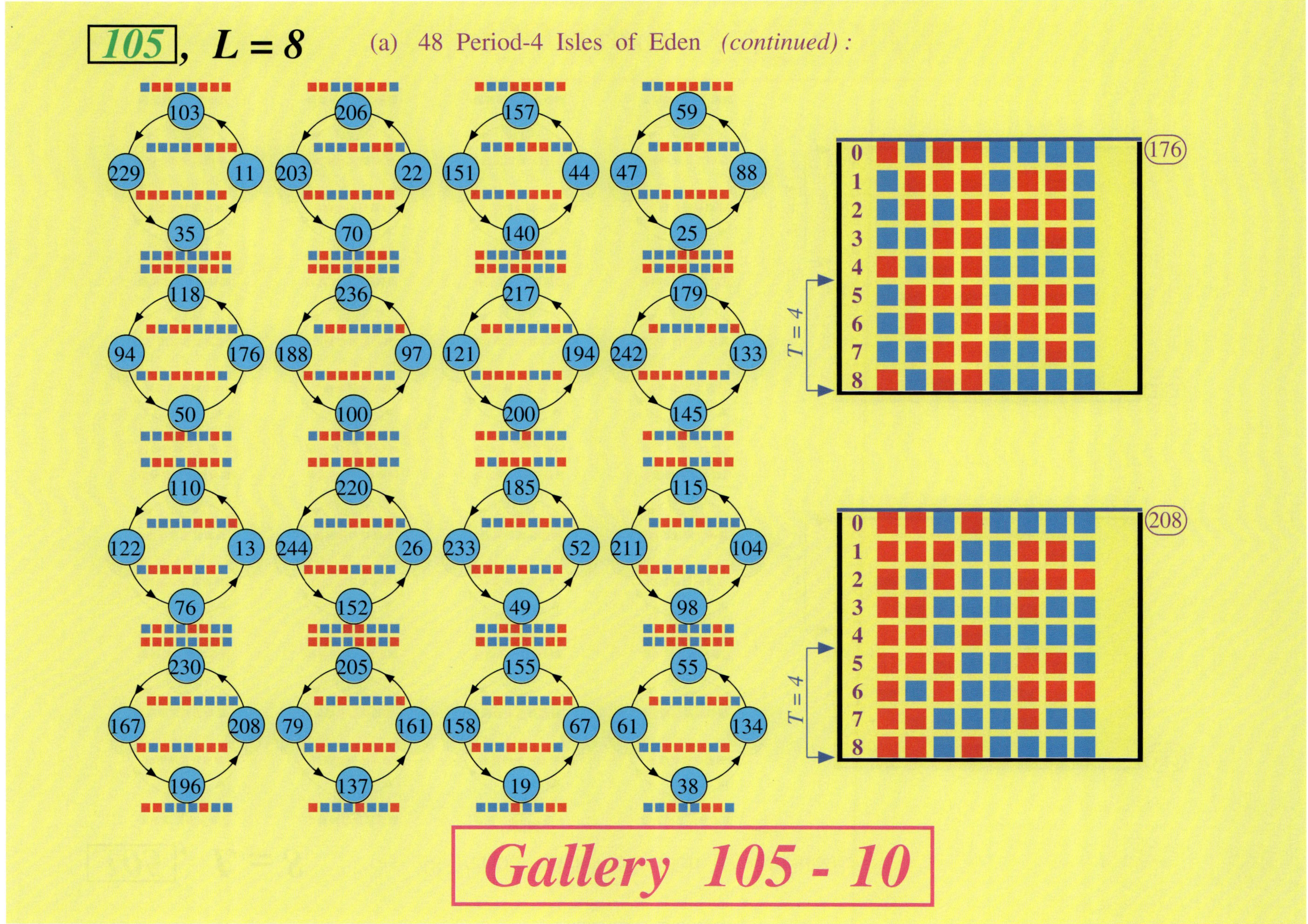

Gallery 105 - 10

Table 19. (Continued)

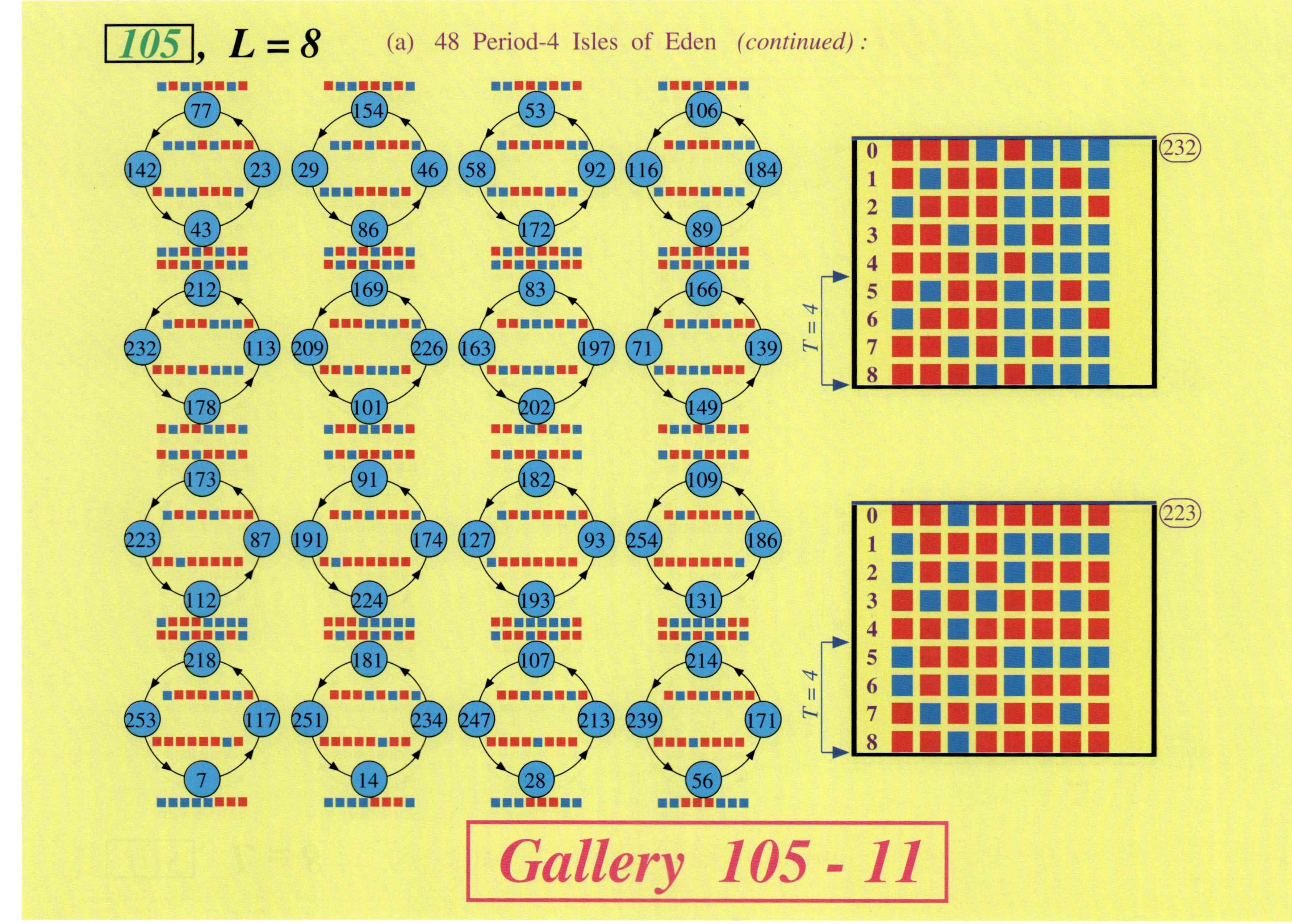

Table 19. (*Continued*)

105, $L = 8$ (b) 12 Bernoulli ($\sigma = \pm 4,\ \tau = 2$) Period-4

Isles of Eden $\rho_2 = 12\frac{4}{256} = 0.1875$

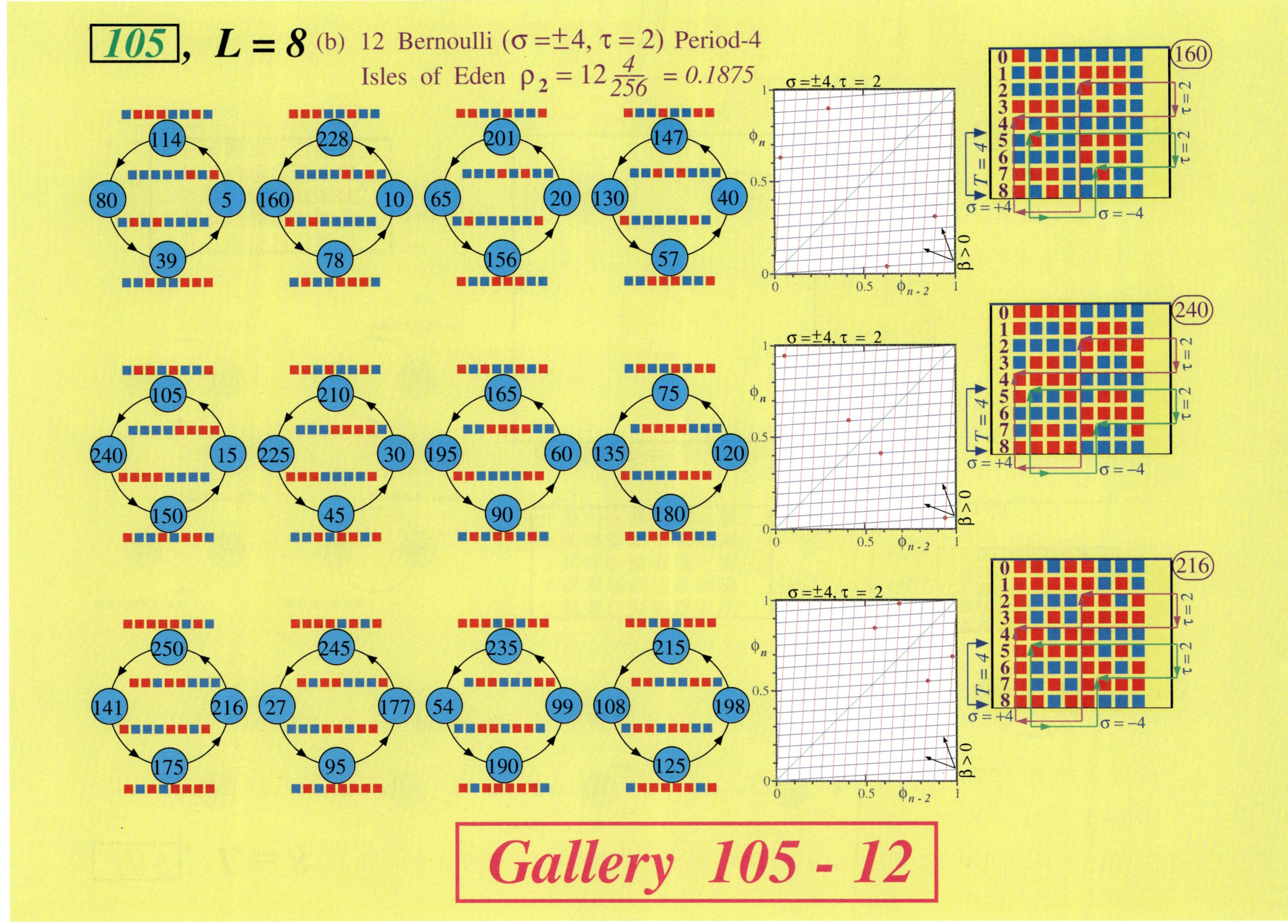

105, $L = 8$ (c) Period-1 Isles of Eden $\rho_3 = \frac{4}{256} = 0.015625$

(d) 2 Bernoulli ($\sigma = \pm 2, \tau = 1$) Period-2 Isles of Eden

$$\rho_4 = 4\frac{2}{256} = 0.03125$$

(e) Bernoulli ($\sigma = \pm 1, \tau = 1$)

Period-2 Isle of Eden

$$\rho_5 = \frac{2}{256} = 0.0078125$$

(f) Period-2 Isle of Eden

$$\rho_6 = \frac{2}{256} = 0.0078125$$

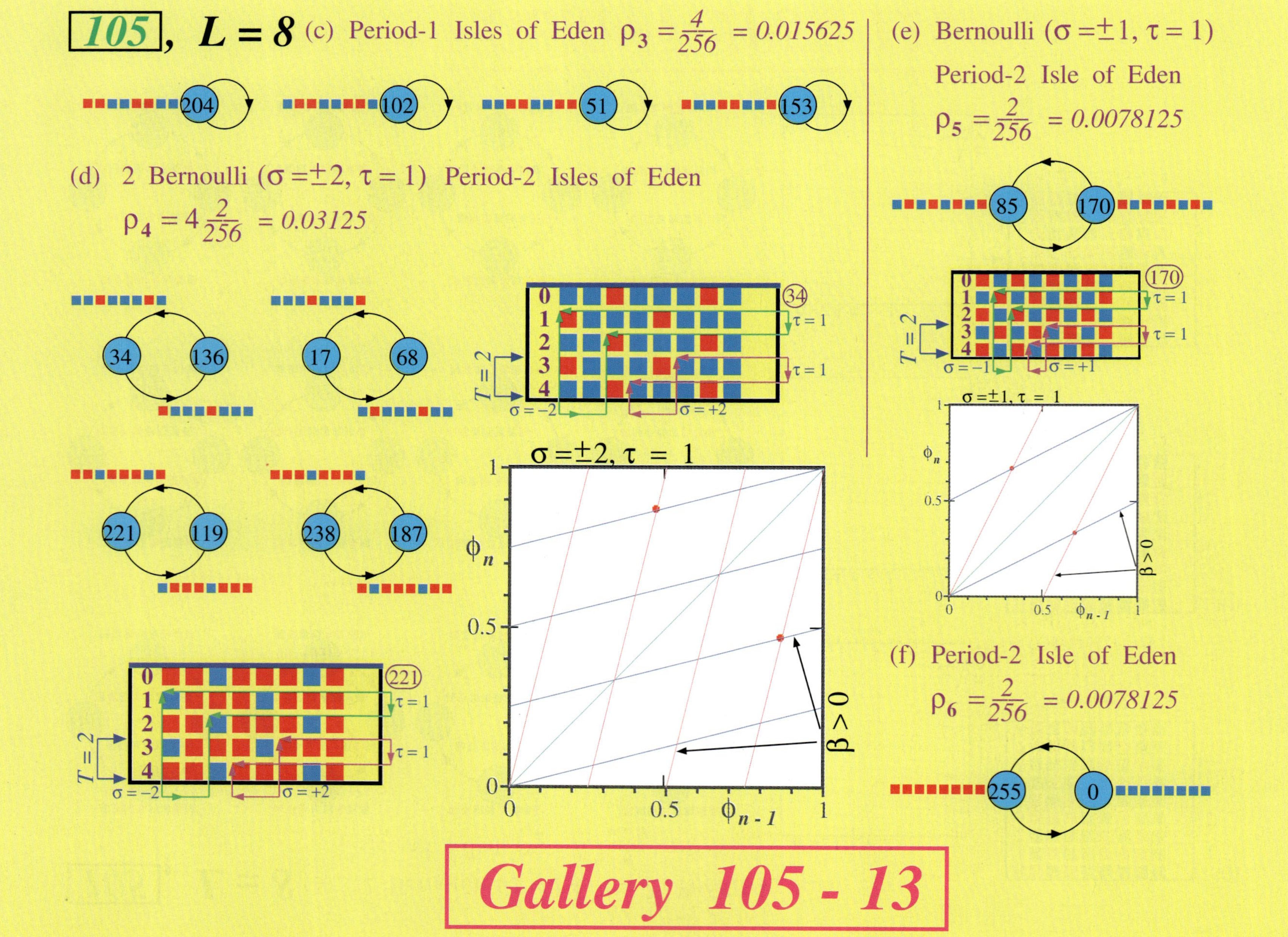

Gallery 105 - 13

2.4.6. *Highlights from Rule* $\boxed{105}$

The basin tree diagrams of Rule $\boxed{105}$ for $L = 3, 4, \ldots, 8$ are exhibited in Table 19. Following a detailed analysis of these diagrams, the qualitative properties of local rule $\boxed{105}$ extracted from basin-tree Galleries 105-1 to 105-13 of Table 19 are summarized below:

Summary of Qualitative properties of local rule $\boxed{105}$ extracted from Gallery 105 for Rule $\boxed{105}$

L	ID Number i	Number of Period-n attractors	Number of Period-n Isles of Eden	Period n	Bernoulli Parameters						Robustness coefficient ρ
					σ_1	τ_1	β_1	σ_2	τ_2	β_2	
3	1	1		2	0	1	−				$\rho_1 = 1$
4	1		4	1	0	1	+				$\rho_1 = 0.25$
	2		2	2	1	1	+	-1	1	+	$\rho_2 = 0.25$
	3		4	2	2	1	+	-2	1	+	$\rho_3 = 0.5$
5	1		5	6	0	3	−				$\rho_1 = 0.9375$
	2		1	2	0	1	−				$\rho_2 = 0.0625$
6	1	1	2	2	0	1	−				$\rho_1 = 0.5$
	2	1	2	2	1	1	+	-1	1	+	$\rho_2 = 0.5$
7	1		1	2	0	1	−				$\rho_1 = 0.015625$
	2		7	14	0	7	+				$\rho_2 = 0.765625$
	3		2	14	1	2	+	-1	2	+	$\rho_3 = 0.21875$
8	1		48	4	0	1	+				$\rho_1 = 0.75$
	2		12	4	4	2	+	-4	2	+	$\rho_2 = 0.1875$
	3		4	1	0	1	+				$\rho_3 = 0.015625$
	4		4	2	2	1	+	-2	1	+	$\rho_4 = 0.03125$
	5		1	2	1	1	+	-1	1	+	$\rho_5 = 0.0078125$
	6		1	2	0	1	−				$\rho_6 = 0.0078125$

Table 20. Basin tree diagrams for rule 122 .

Basin tree diagrams for Rule 122

122 , *L = 3* (a) Period-1 Attractor : $\rho_1 = \frac{8}{8} = 1$

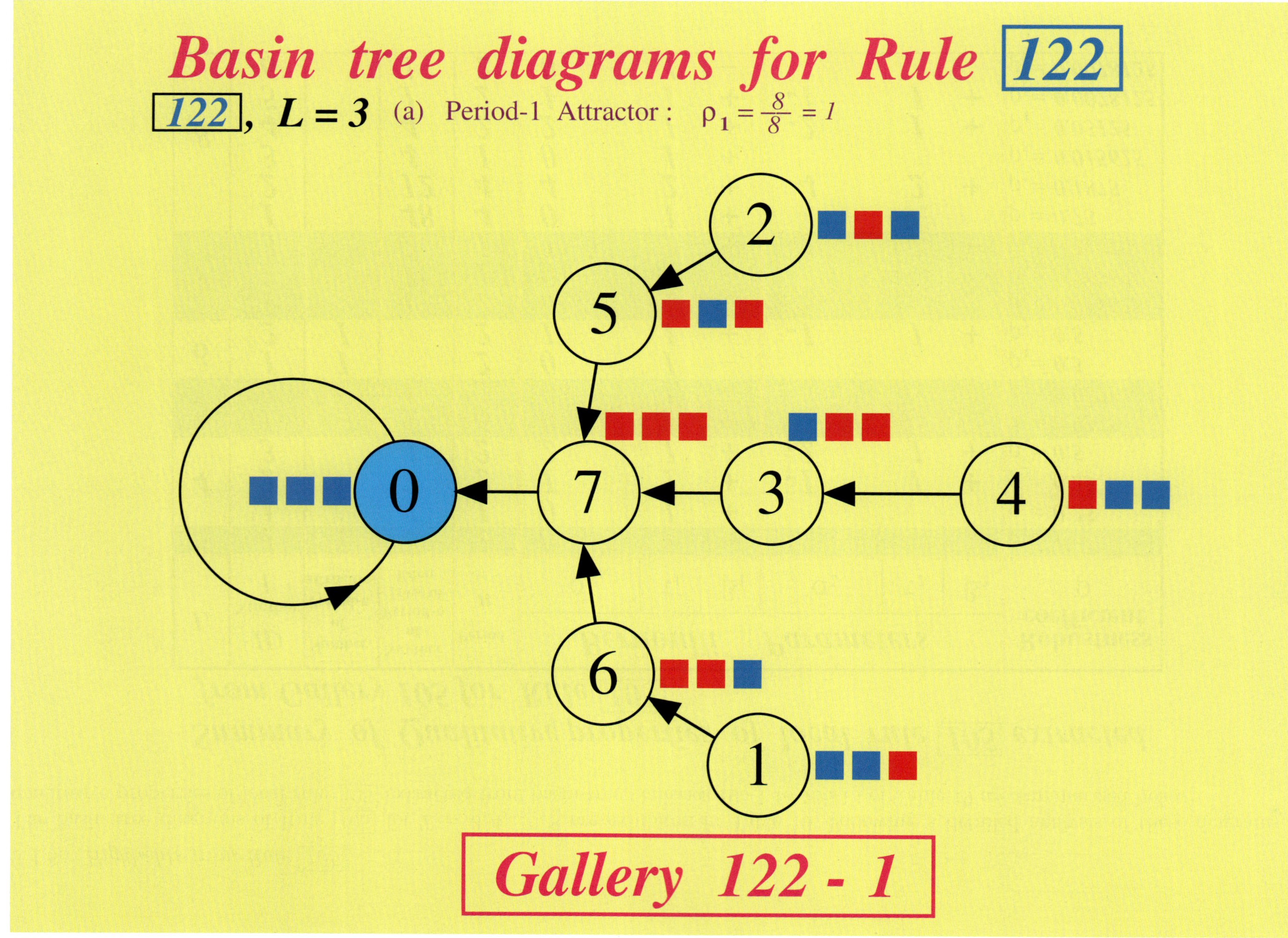

Gallery 122 - 1

122, $L = 4$

(a) Bernoulli ($\sigma = \pm 2$, $\tau = 1$)

Period-2 Isles of Eden :

$$\rho_1 = 2\,\frac{2}{16} = 0.25$$

(b) Bernoulli ($\sigma = \pm 1$, $\tau = 1$)

Period-2 Attractor :

$$\rho_2 = \frac{6}{16} = 0.375$$

(c) Period-1 Attractor :

$$\rho_3 = \frac{6}{16} = 0.375$$

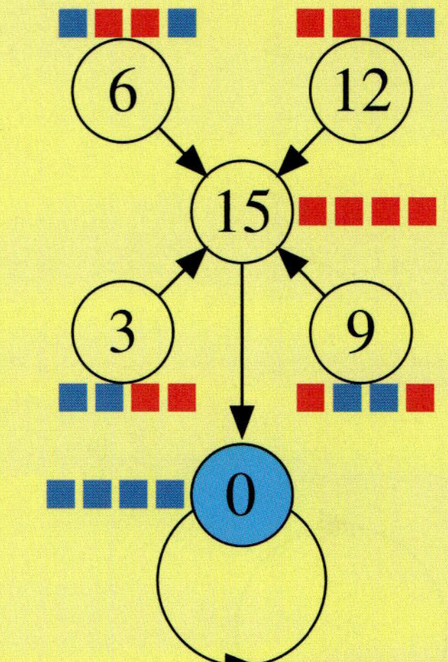

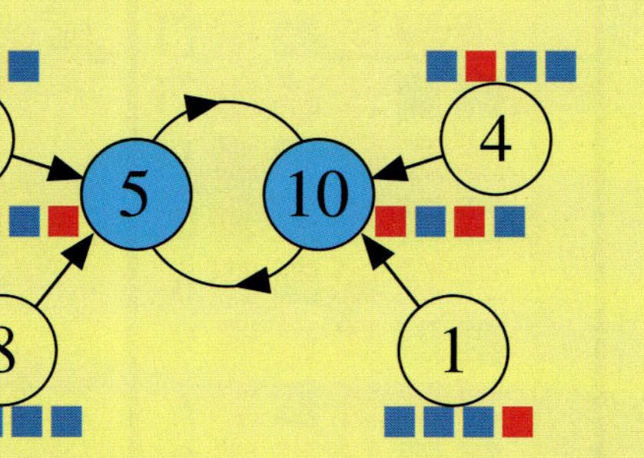

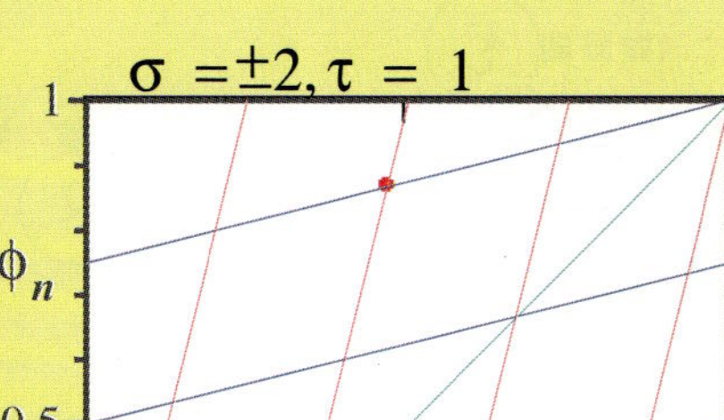

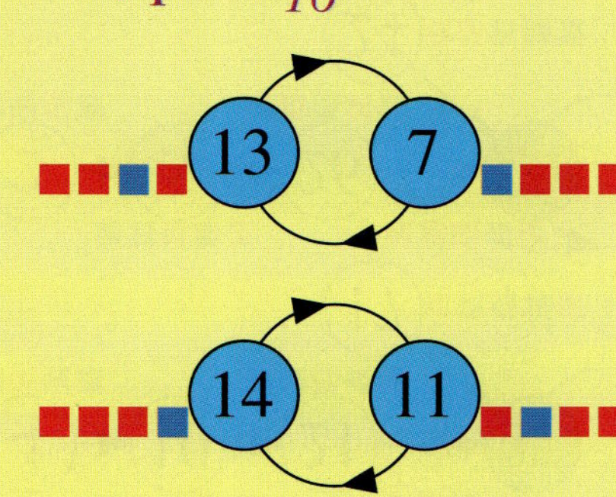

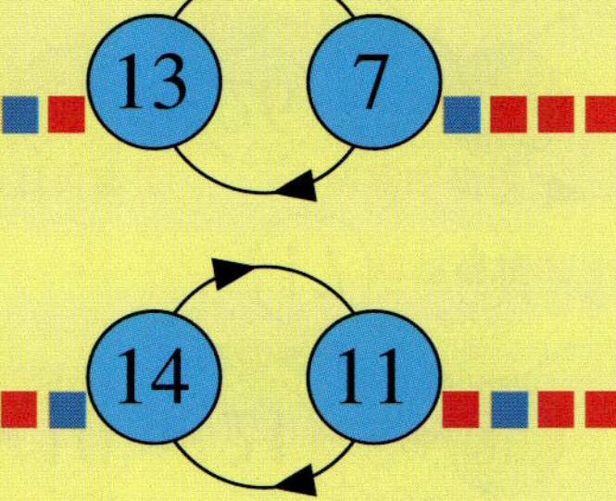

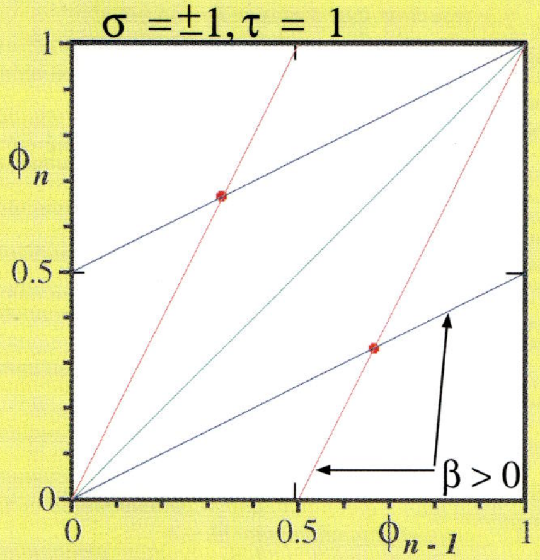

Gallery 122 - 2

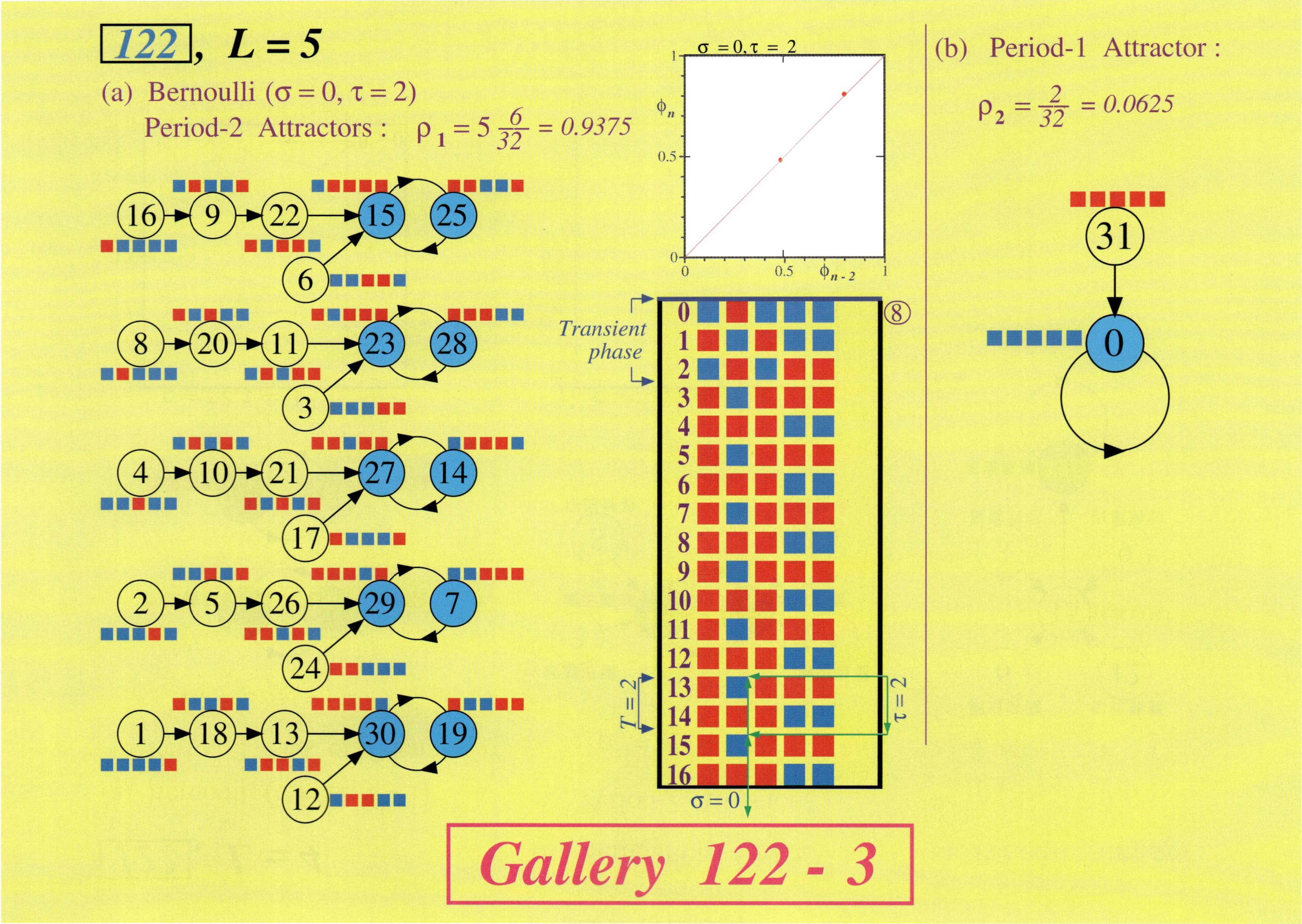

Table 20. (*Continued*)

122, $L = 5$

(a) Bernoulli ($\sigma = 0$, $\tau = 2$)

Period-2 Attractors : $\rho_1 = 5 \frac{6}{32} = 0.9375$

(b) Period-1 Attractor :

$\rho_2 = \frac{2}{32} = 0.0625$

Gallery 122 - 3

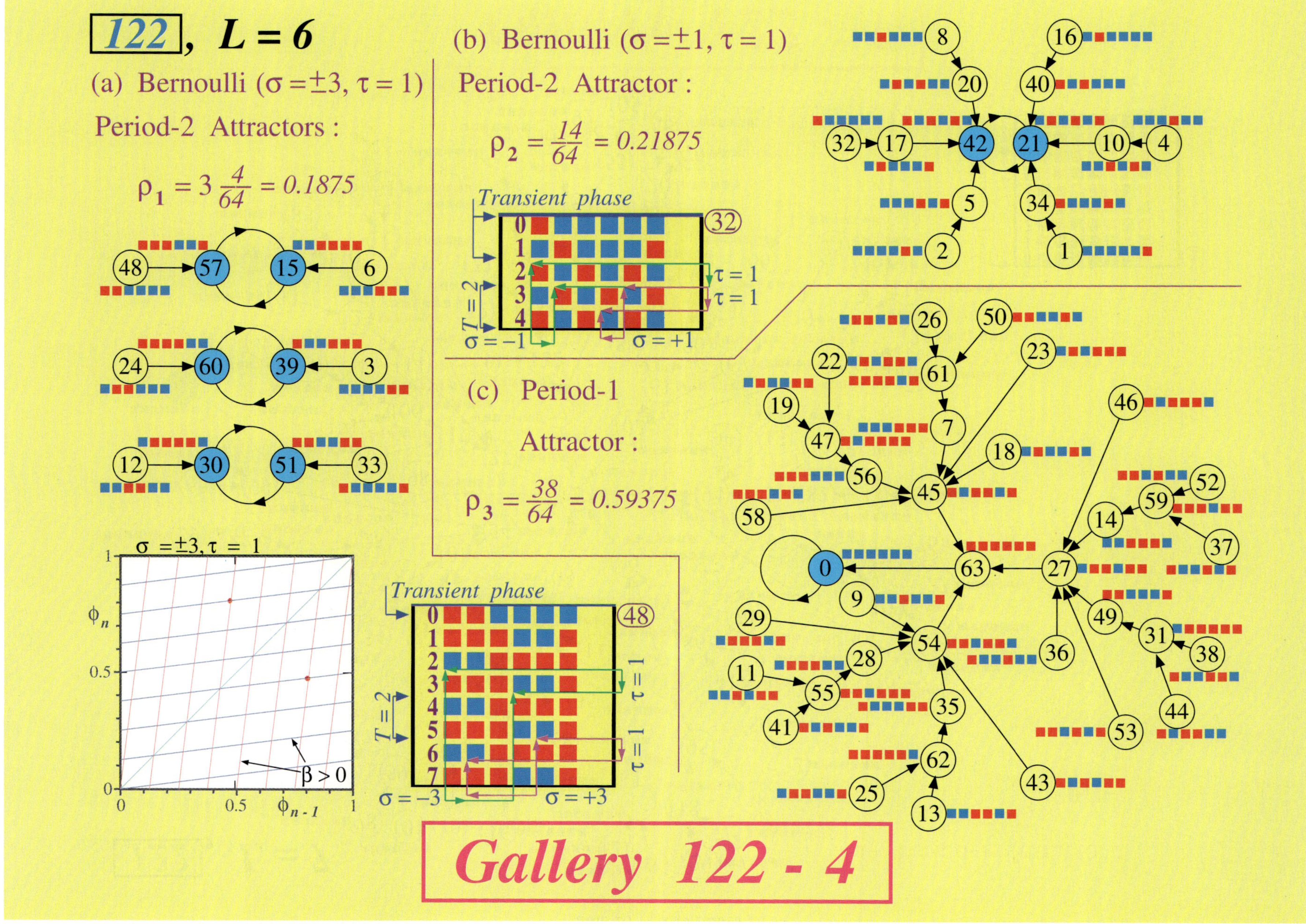

122, L = 6
(a) Bernoulli ($\sigma = \pm 3$, $\tau = 1$)
Period-2 Attractors :
$\rho_1 = 3\frac{4}{64} = 0.1875$
(b) Bernoulli ($\sigma = \pm 1$, $\tau = 1$)
Period-2 Attractor :
$\rho_2 = \frac{14}{64} = 0.21875$
Transient phase
$T = 2$
$\tau = 1$
$\tau = 1$
$\sigma = -1$
$\sigma = +1$
(c) Period-1
Attractor :
$\rho_3 = \frac{38}{64} = 0.59375$
Transient phase
$T = 2$
$\tau = 1$
$\tau = 1$
$\sigma = -3$
$\sigma = +3$
$\sigma = \pm 3, \tau = 1$
ϕ_n
$\beta > 0$
ϕ_{n-1}
Gallery 122 - 4

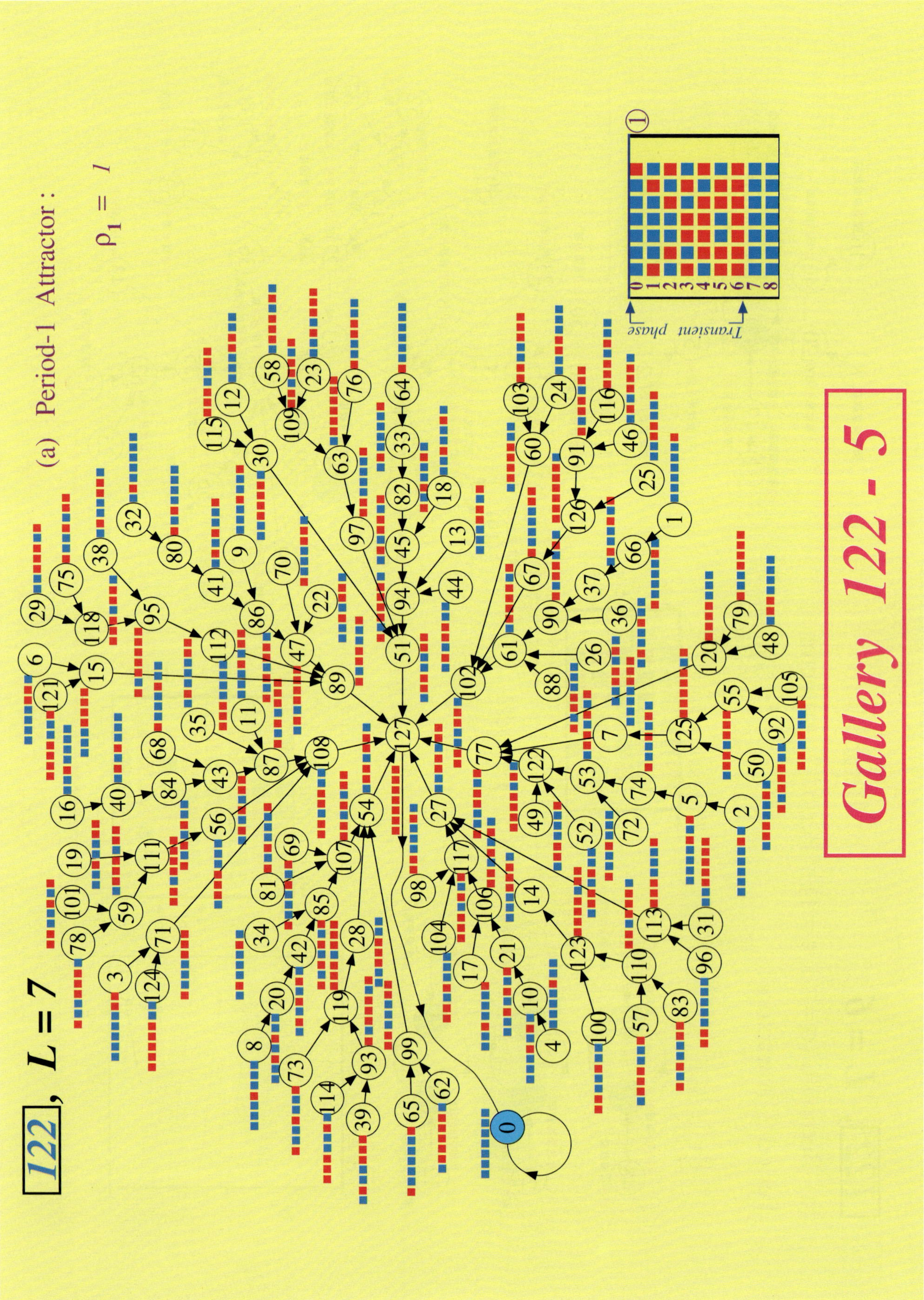

Table 20.   (*Continued*)

122, **L = 7**

(a)   Period-1 Attractor :

$\rho_1 = 1$

Gallery 122 - 5

$\boxed{122}$, $L = 8$ (a) Bernoulli ($\sigma = \pm 4$, $\tau = 3$) Period-6 Attractors : $\rho_1 = 4\frac{40}{256} = 0.625$

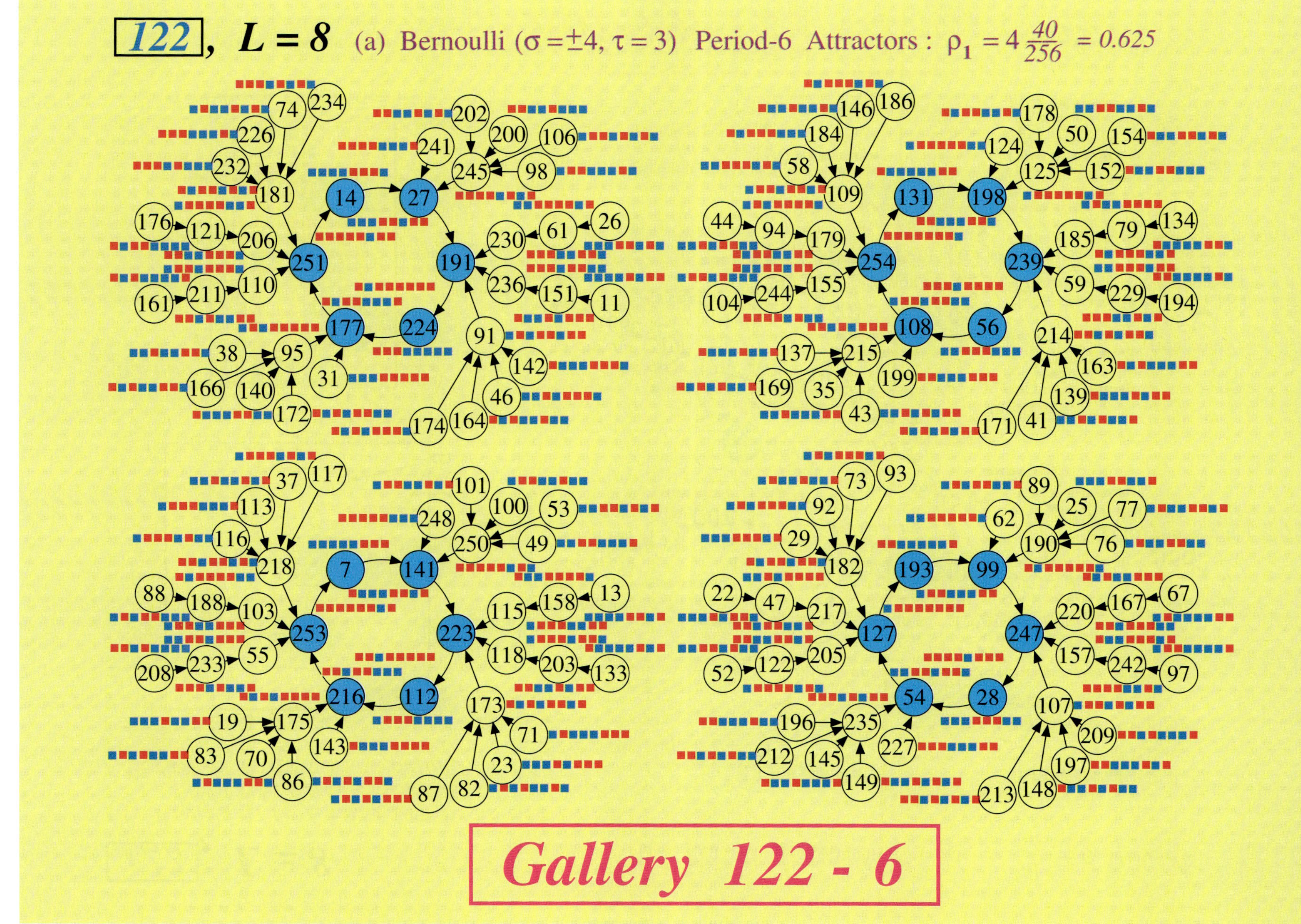

122, **$L = 8$**

(a) Bernoulli ($\sigma = \pm 4, \tau = 3$)

Period-6 Attractors *(continued)* :

(b) Period-1 Attractor : $\rho_2 = \dfrac{54}{256} = 0.2109375$

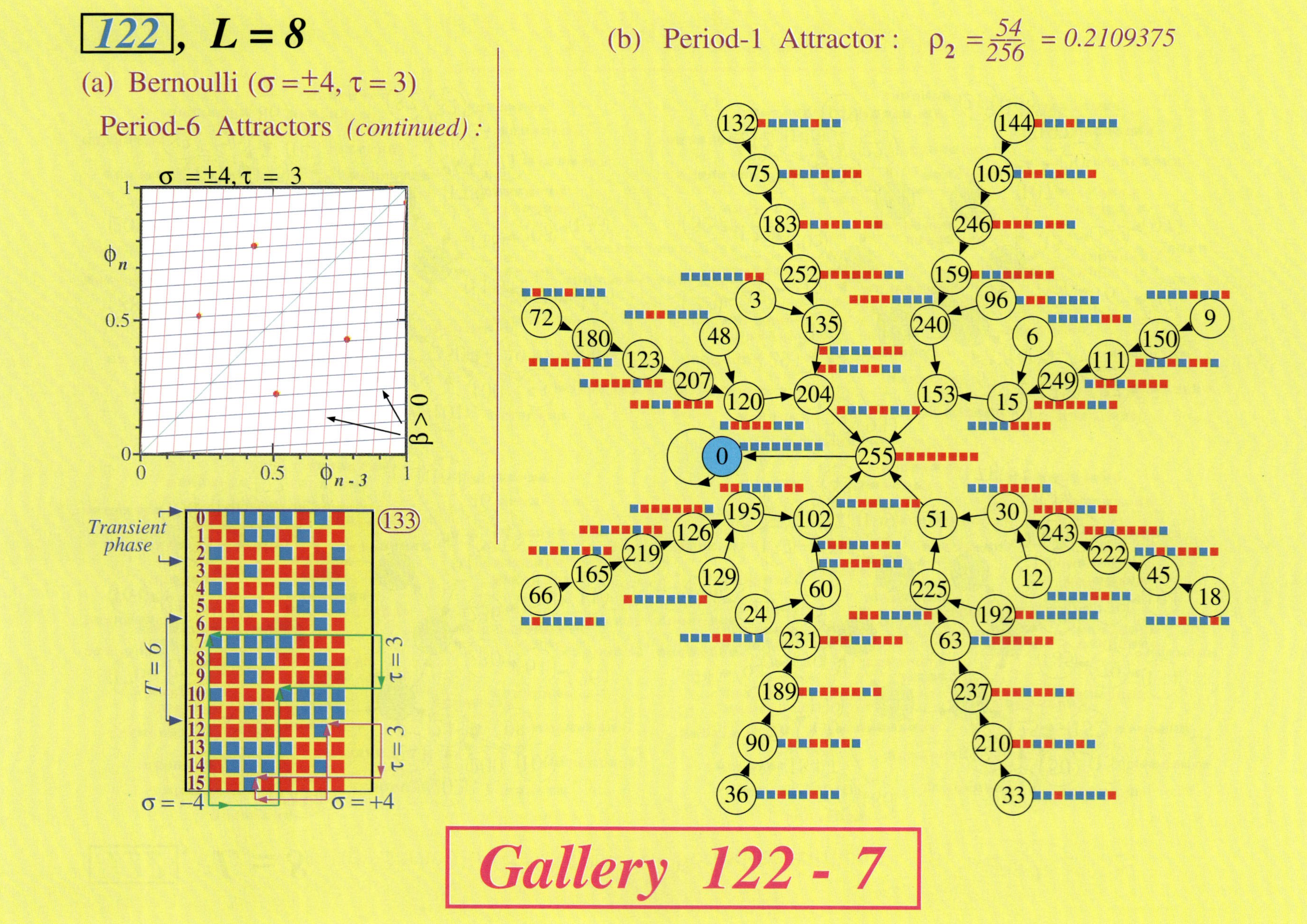

Gallery 122 - 7

122, $L = 8$

(c) Bernoulli ($\sigma = \pm 1$, $\tau = 1$)

Period-2 Attractor $\rho_3 = \frac{30}{256} = 0.1171875$

(d) Bernoulli ($\sigma = \pm 2$, $\tau = 1$)

Period-2 Attractors : $\rho_4 = 2\frac{6}{256} = 0.046875$

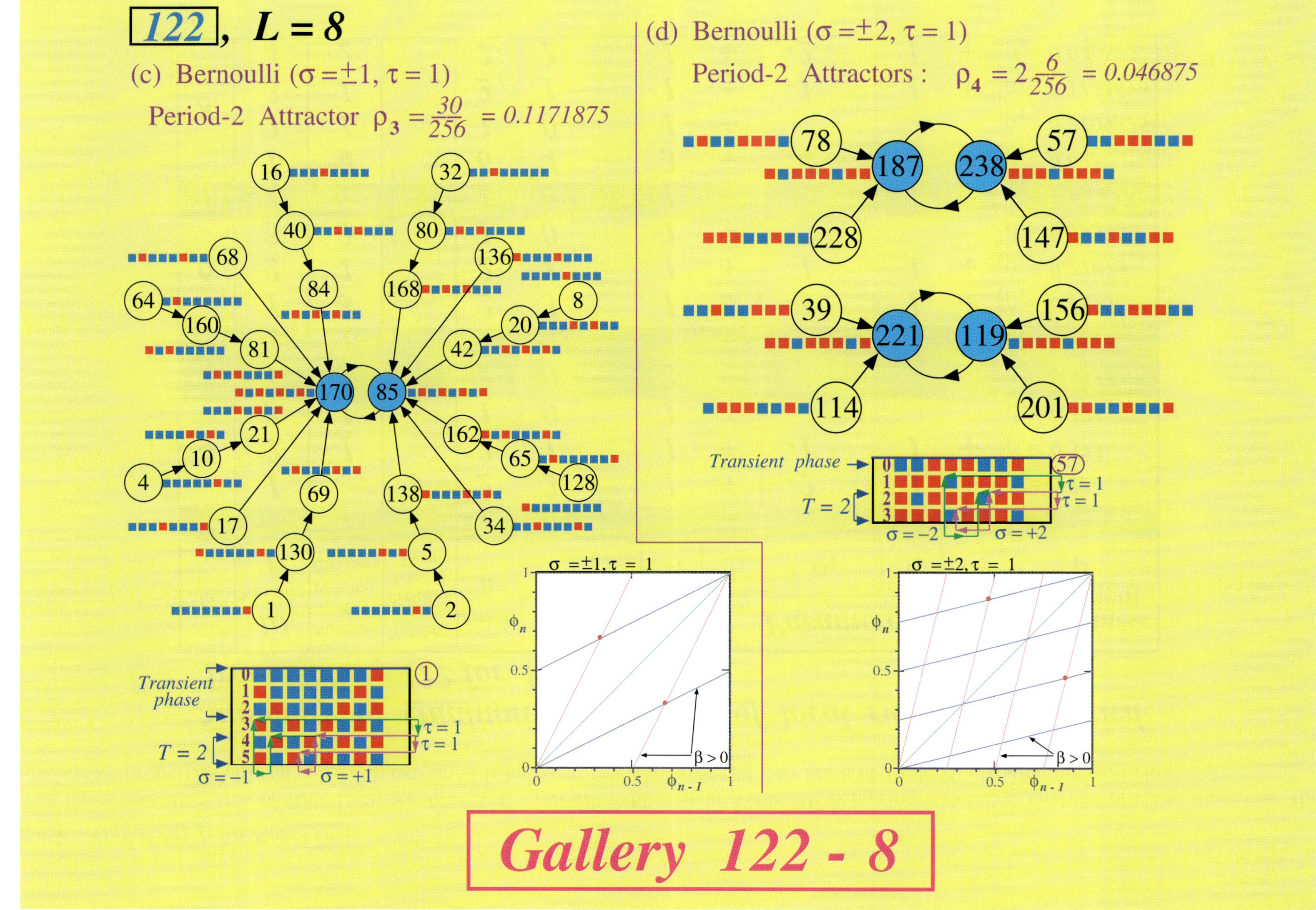

Gallery 122 - 8

2.4.7. *Highlights from Rule* 122

The basin tree diagrams of Rule 122 for $L = 3, 4, \ldots, 8$ are exhibited in Table 20. Following a detailed analysis of these diagrams, the qualitative properties of local rule 122 extracted from basin-tree Galleries 122-1 to 122-8 of Table 20 are summarized below:

Summary of Qualitative properties of local rule 122 extracted from Gallery 122 for Rule 122

L	ID Number i	Number of Period-n attractors	Number of Period-n Isles of Eden	Period n	Bernoulli Parameters						Robustness coefficient ρ
					σ_1	τ_1	β_1	σ_2	τ_2	β_2	
3	1	1		1	0	1	+				$\rho_1 = 1$
4	1		2	2	2	1	+	-2	1	+	$\rho_1 = 0.25$
	2	1		2	1	1	+	-1	1	+	$\rho_2 = 0.375$
	3	1		1	0	1	+				$\rho_3 = 0.375$
5	1	5		2	0	2	+				$\rho_1 = 0.9375$
	2	1		1	0	1	+				$\rho_2 = 0.0625$
6	1	3		2	3	1	+	-3	1	+	$\rho_1 = 0.1875$
	2	1		2	1	1	+	-1	1	+	$\rho_2 = 0.21875$
	3	1		1	0	1	+				$\rho_3 = 0.59375$
7	1	1		1	0	1	+				$\rho_1 = 1$
8	1	4		6	4	3	+	-4	3	+	$\rho_1 = 0.625$
	2	1		1	0	1	+				$\rho_2 = 0.2109375$
	3	1		2	1	1	+	-1	1	+	$\rho_3 = 0.1171875$
	4	2		2	2	1	+	-2	1	+	$\rho_4 = 0.046875$

Table 21. Basin tree diagrams for rule 126.

Basin tree diagrams for Rule 126

126, $L = 3$

(a) Period-1 Attractor : $\rho_1 = \frac{8}{8} = 1$

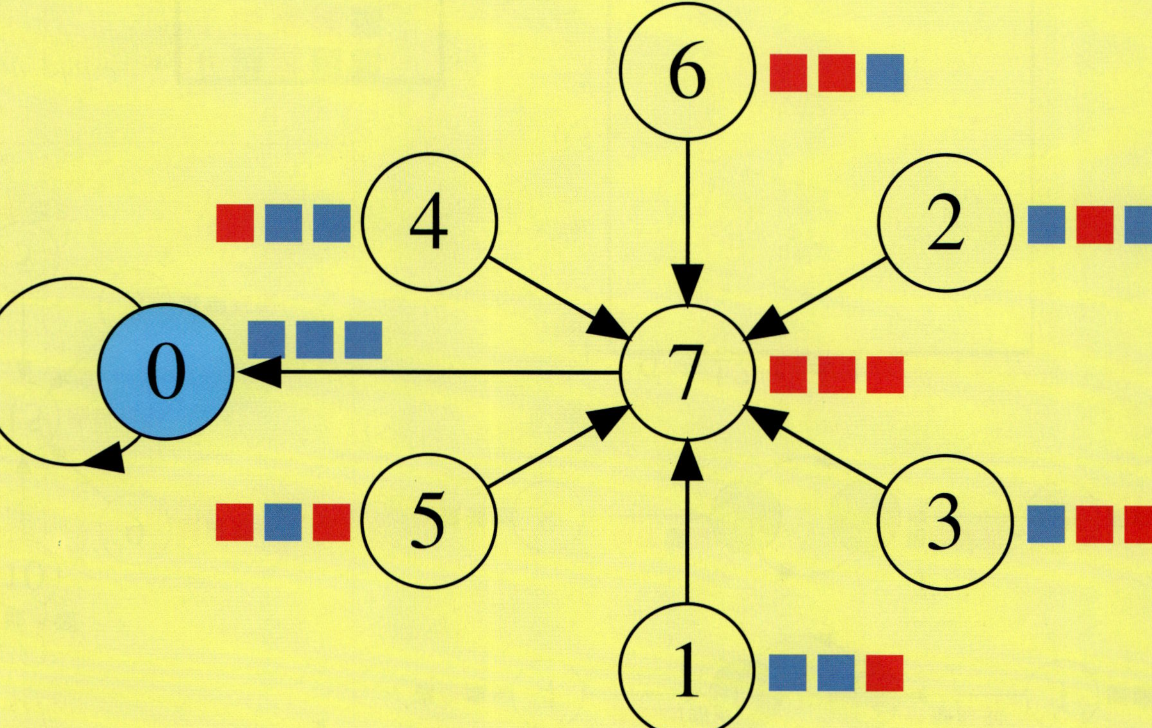

Gallery 126 - 1

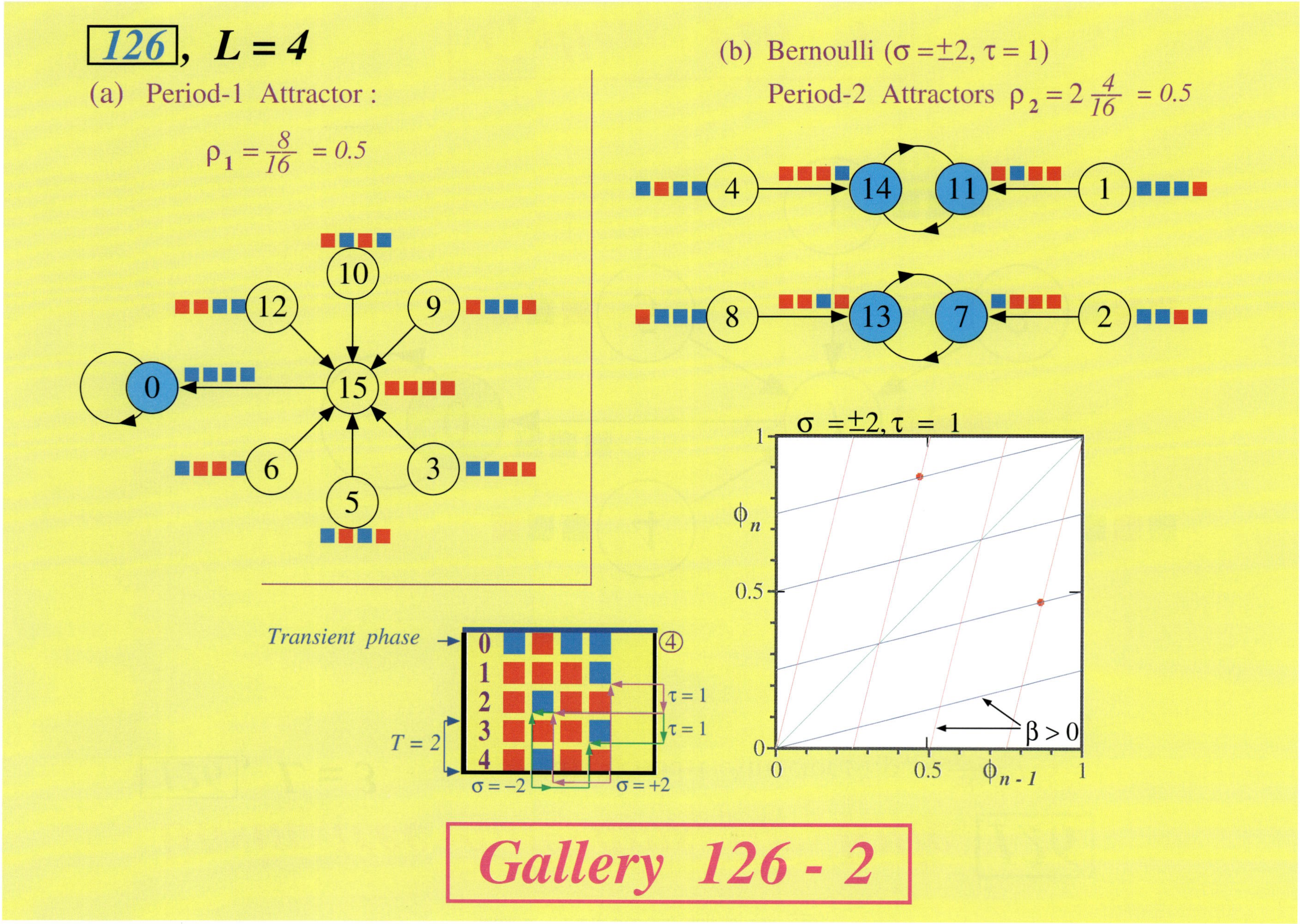

Table 21. (*Continued*)

126, $L = 4$

(a) Period-1 Attractor :

$$\rho_1 = \frac{8}{16} = 0.5$$

(b) Bernoulli ($\sigma = \pm 2,\ \tau = 1$)

Period-2 Attractors $\rho_2 = 2\,\frac{4}{16} = 0.5$

Gallery 126 - 2

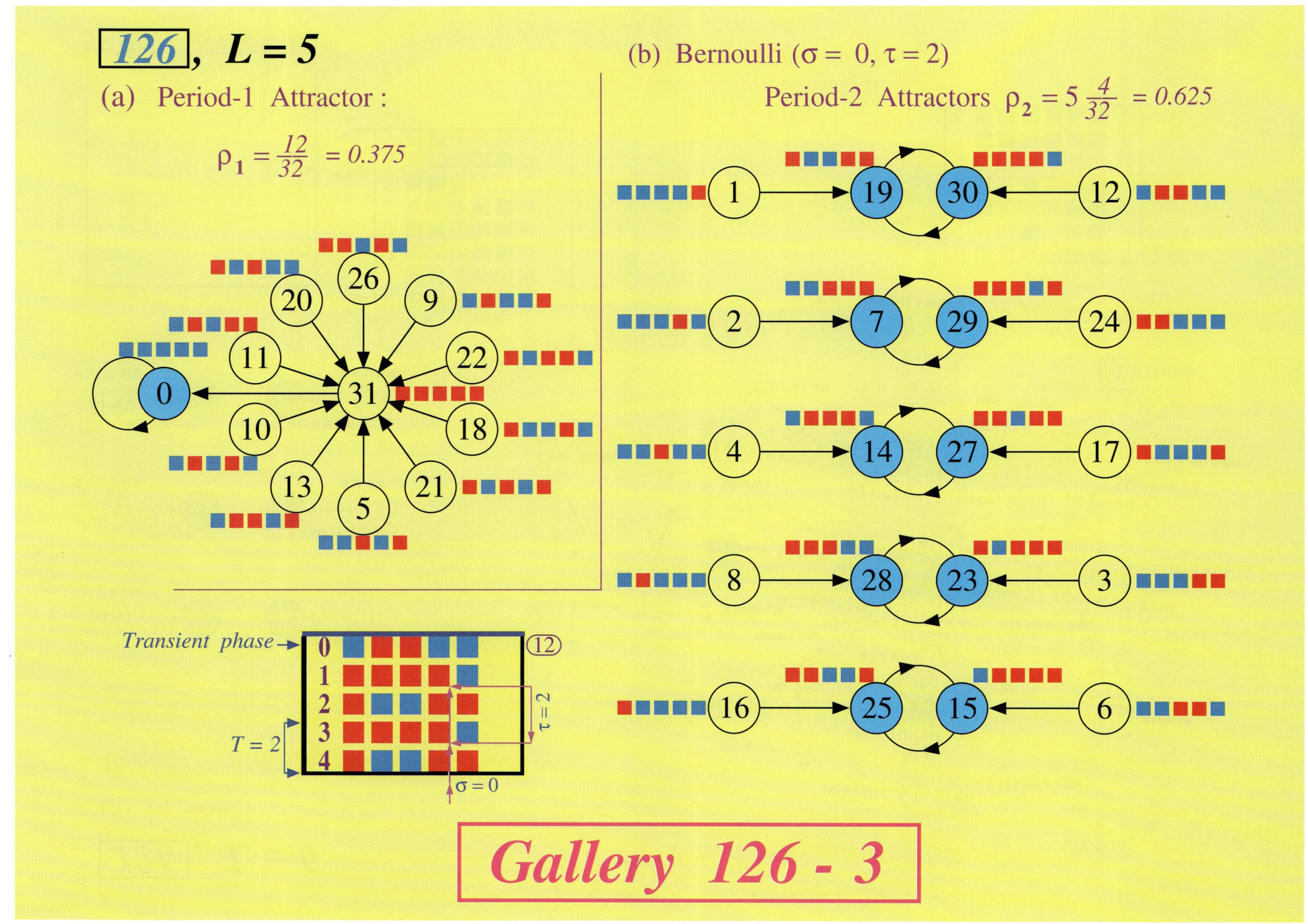

126, L = 5
(a) Period-1 Attractor :
$\rho_1 = \frac{12}{32} = 0.375$
(b) Bernoulli ($\sigma = 0$, $\tau = 2$)
Period-2 Attractors $\rho_2 = 5\frac{4}{32} = 0.625$
Transient phase
T = 2
$\tau = 2$
$\sigma = 0$
0 1 2 3 4
Gallery 126 - 3

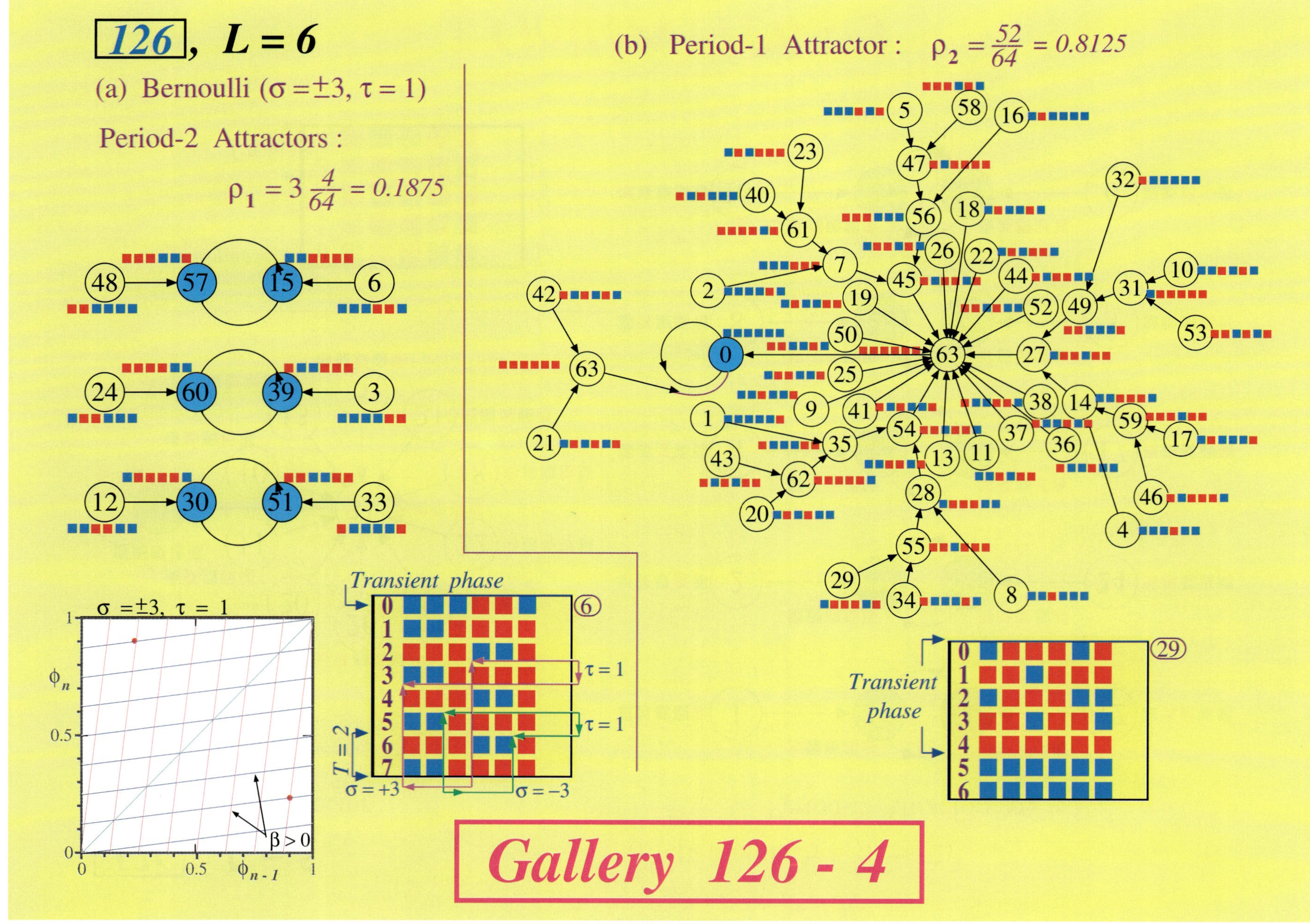
126, L = 6
(a) Bernoulli (σ = ±3, τ = 1)
Period-2 Attractors :
ρ₁ = 3 4/64 = 0.1875
(b) Period-1 Attractor : ρ₂ = 52/64 = 0.8125
σ = ±3, τ = 1
Transient phase
τ = 1
τ = 1
T = 2
σ = +3
σ = −3
Transient phase
Gallery 126 - 4

126, *L = 7*

(a) Period-1 Attractor :

$$\rho_1 = \frac{128}{128} = 1$$

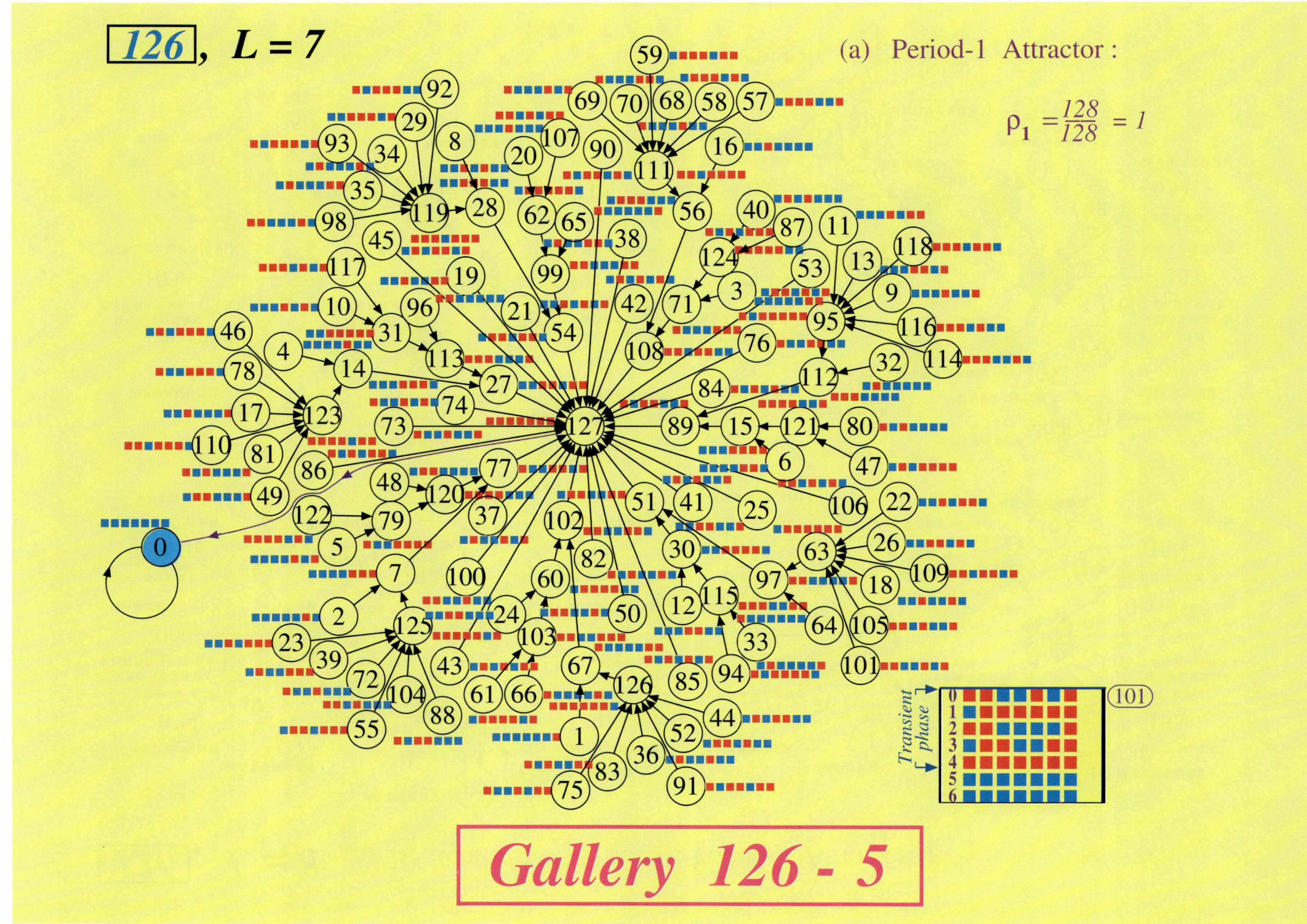

Gallery 126 - 5

126, **$L = 8$** (a) Bernoulli ($\sigma = \pm 4$, $\tau = 3$) Period-6 Attractors : $\rho_1 = 4\frac{28}{256} = 0.4375$

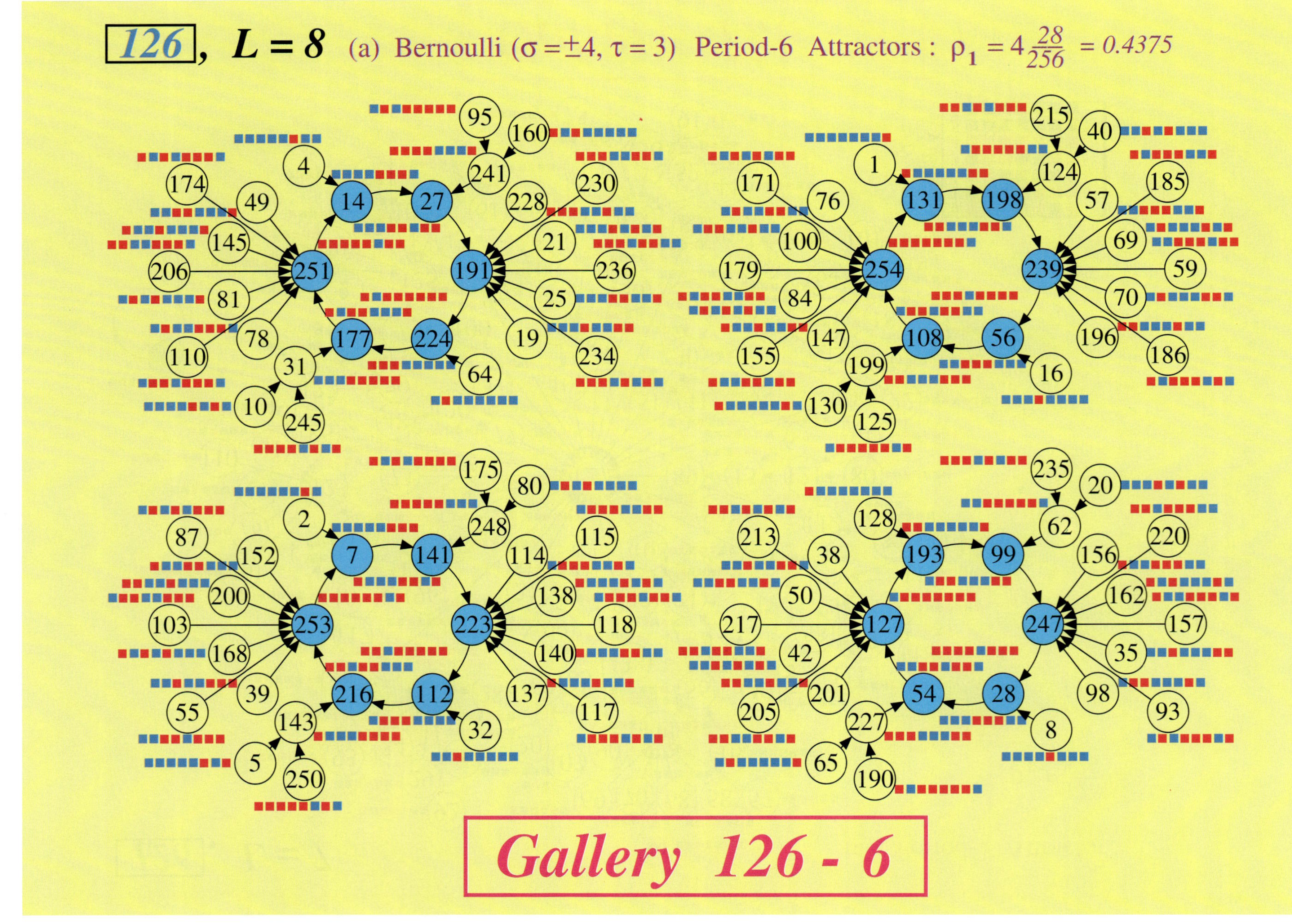

Gallery 126 - 6

126, $L = 8$

(a) Bernoulli ($\sigma = \pm 4, \tau = 3$)

Period-6 Attractors *(continued)* :

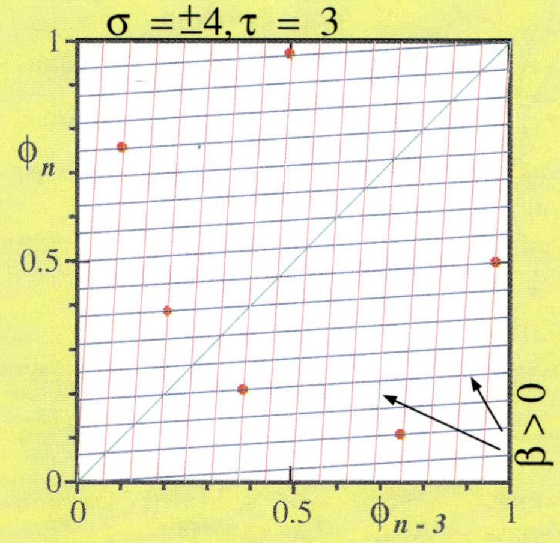

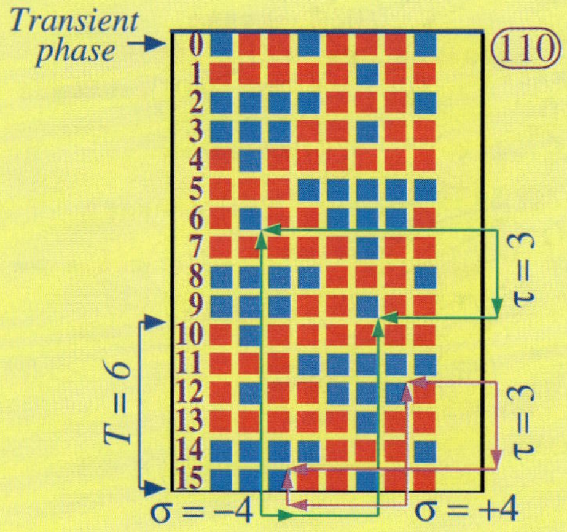

(b) Bernoulli ($\sigma = \pm 2, \tau = 1$) Period-2 Attractors :

$$\rho_2 = 2\frac{4}{256} = 0.03125$$

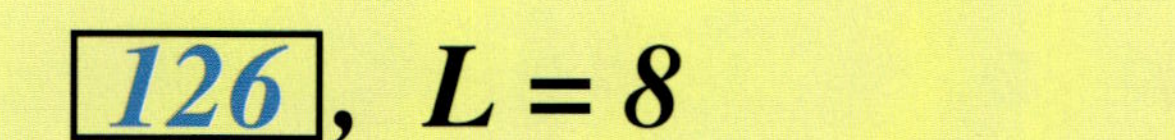

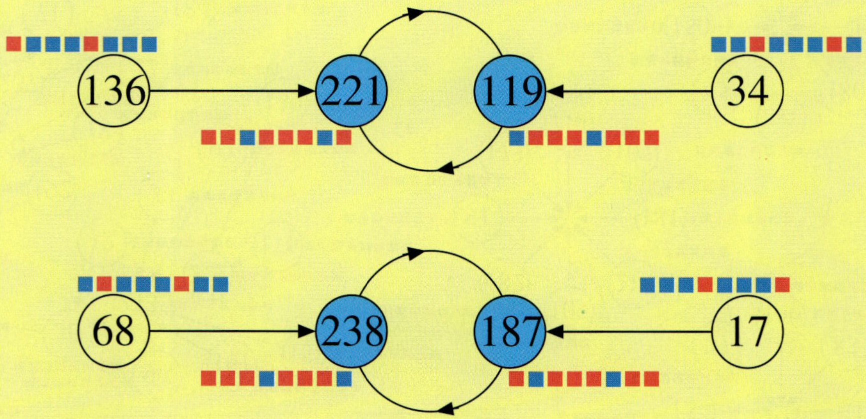

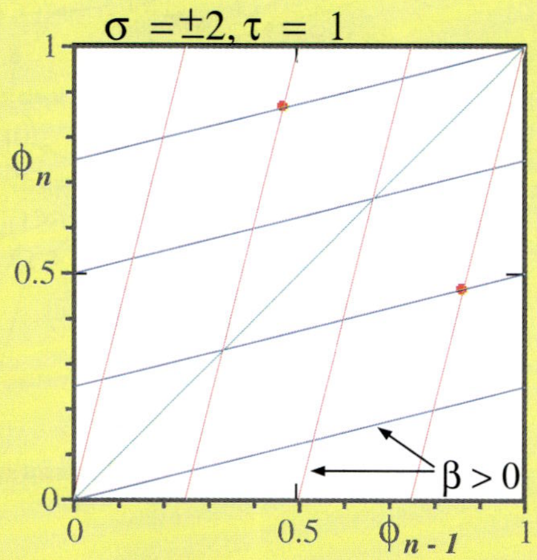

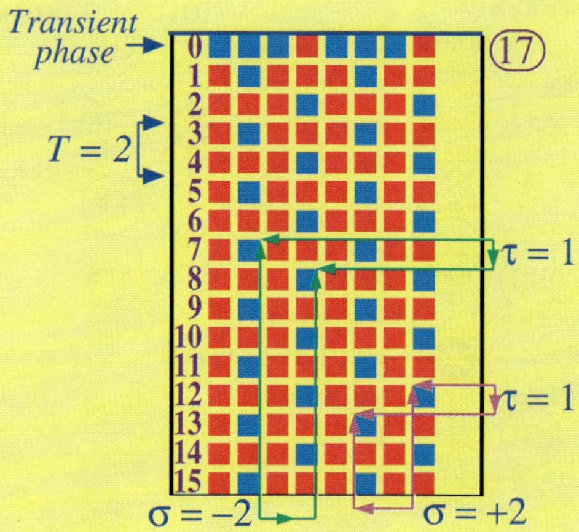

Gallery 126 - 7

126, $L = 8$

(c) Period-1 Attractor : $\rho_3 = \frac{136}{256} = 0.53125$

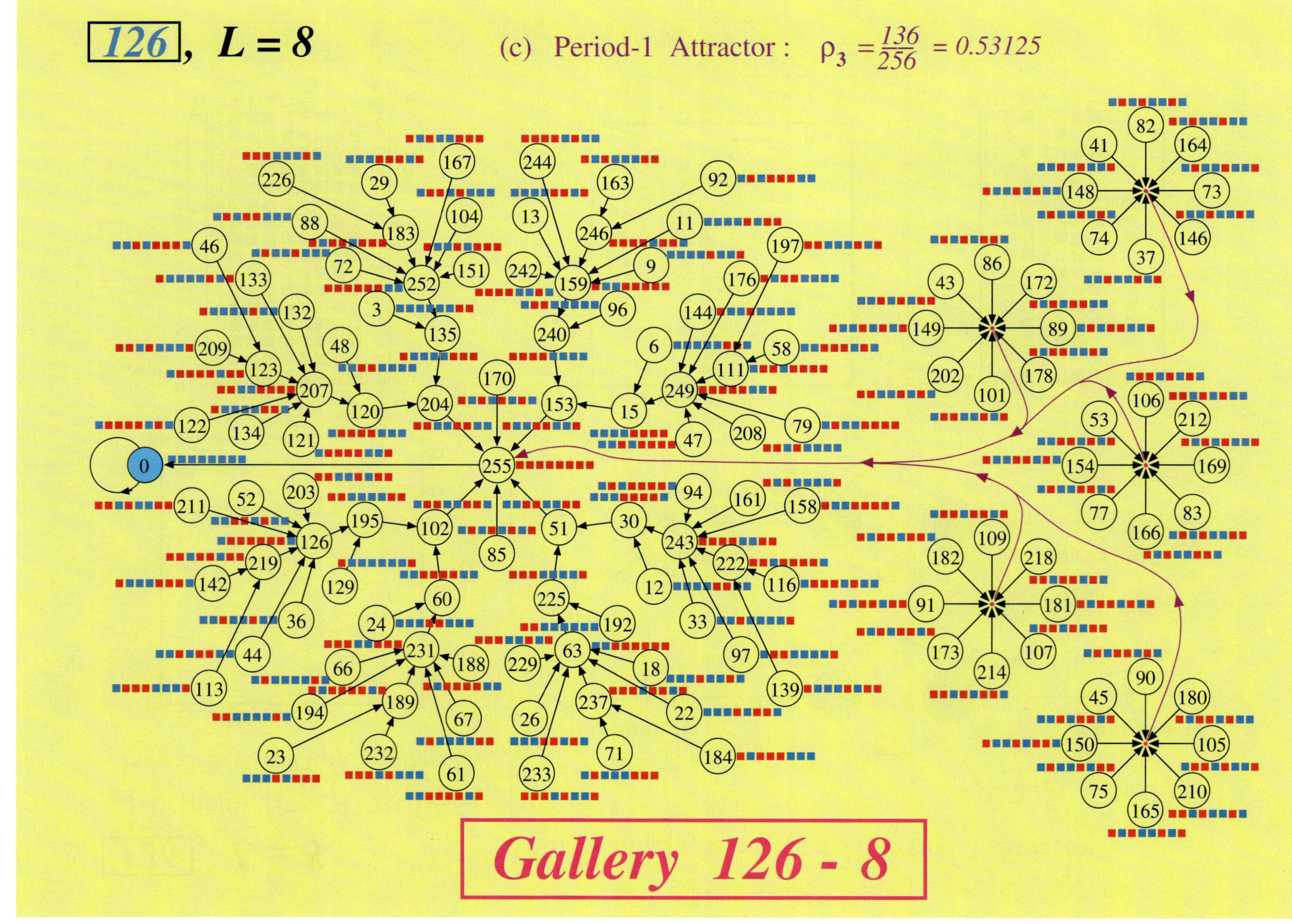

Gallery 126 - 8

2.4.8. *Highlights from Rule* 126

The basin tree diagrams of Rule 126 for $L = 3, 4, \ldots, 8$ are exhibited in Table 21. Following a detailed analysis of these diagrams, the qualitative properties of local rule 126 extracted from basin-tree Galleries 126-1 to 126-8 of Table 21 are summarized below:

Summary of Qualitative properties of local rule 126 extracted from Gallery 126 for Rule 126

L	ID Number i	Number of Period-n attractors	Number of Period-n Isles of Eden	Period n	Bernoulli Parameters						Robustness coefficient ρ
					σ_1	τ_1	β_1	σ_2	τ_2	β_2	
3	1	1		1	0	1	+				$\rho_1 = 1$
4	1	1		1	0	1	+				$\rho_1 = 0.5$
	2	2		2	2	1	+	-2	1	+	$\rho_2 = 0.5$
5	1	1		1	0	1	+				$\rho_1 = 0.375$
	2	5		2	0	2	+				$\rho_2 = 0.625$
6	1	3		2	3	1	+	-3	1	+	$\rho_1 = 0.1875$
	2	1		1	0	1	+				$\rho_2 = 0.8125$
7	1	1		1	0	1	+				$\rho_1 = 1$
8	1	4		6	4	3	+	-4	3	+	$\rho_1 = 0.4375$
	2	2		2	2	1	+	-2	1	+	$\rho_2 = 0.03125$
	3	1		1	0	1	+				$\rho_3 = 0.53125$

Table 22. Basin tree diagrams for rule 146.

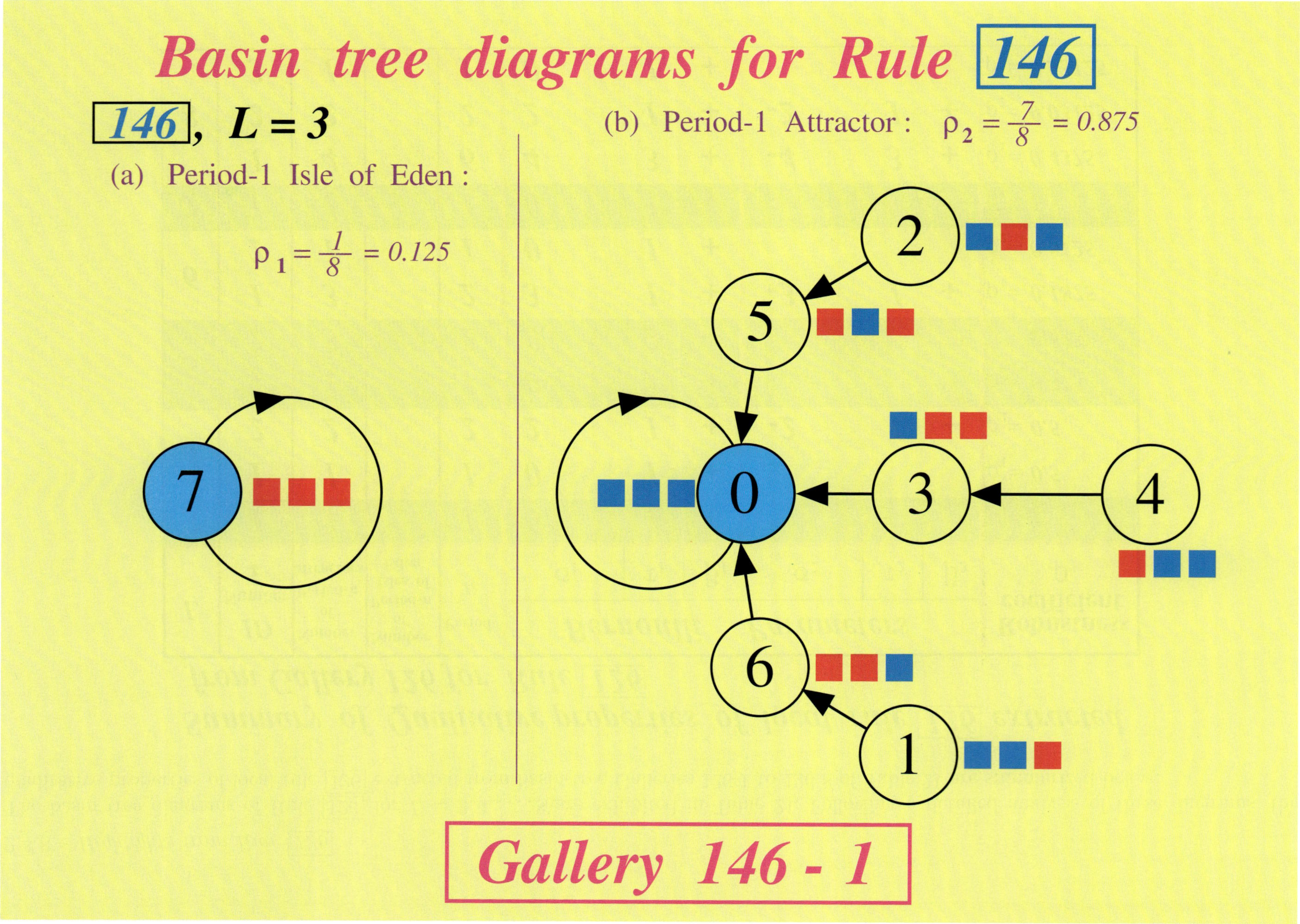

Basin tree diagrams for Rule 146
146, L = 3
(a) Period-1 Isle of Eden :
ρ1 = 1/8 = 0.125
7
(b) Period-1 Attractor : ρ2 = 7/8 = 0.875
0
1
2
3
4
5
6
Gallery 146 - 1

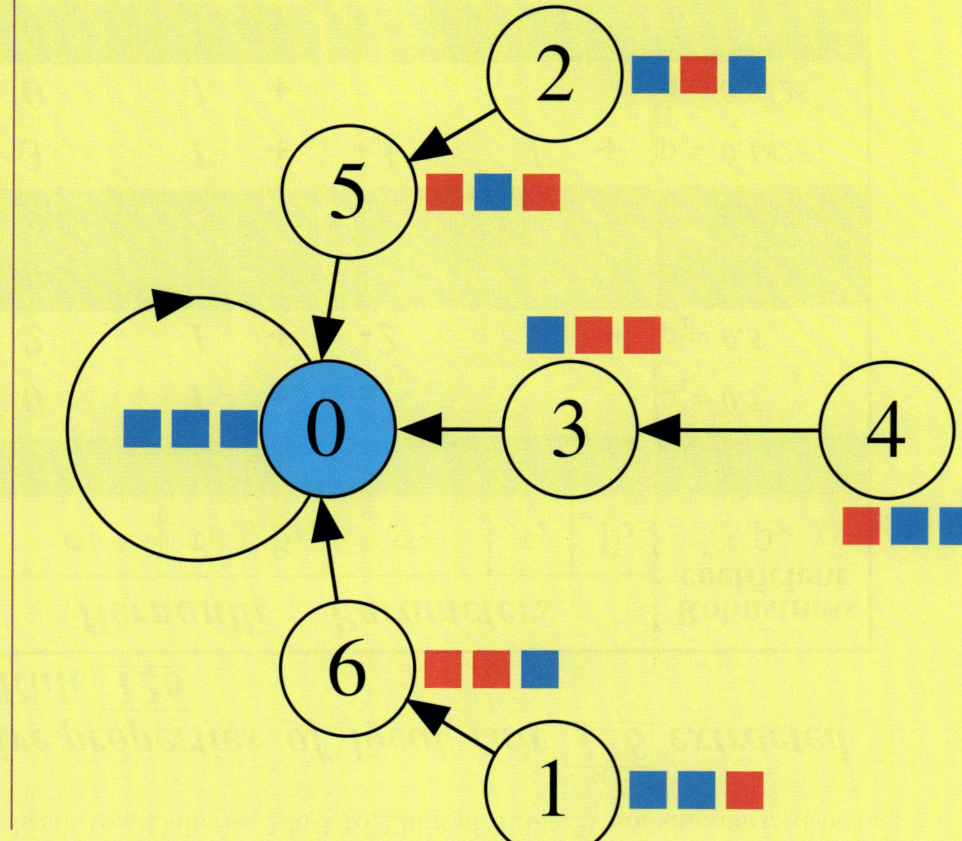

Table 22. (*Continued*)

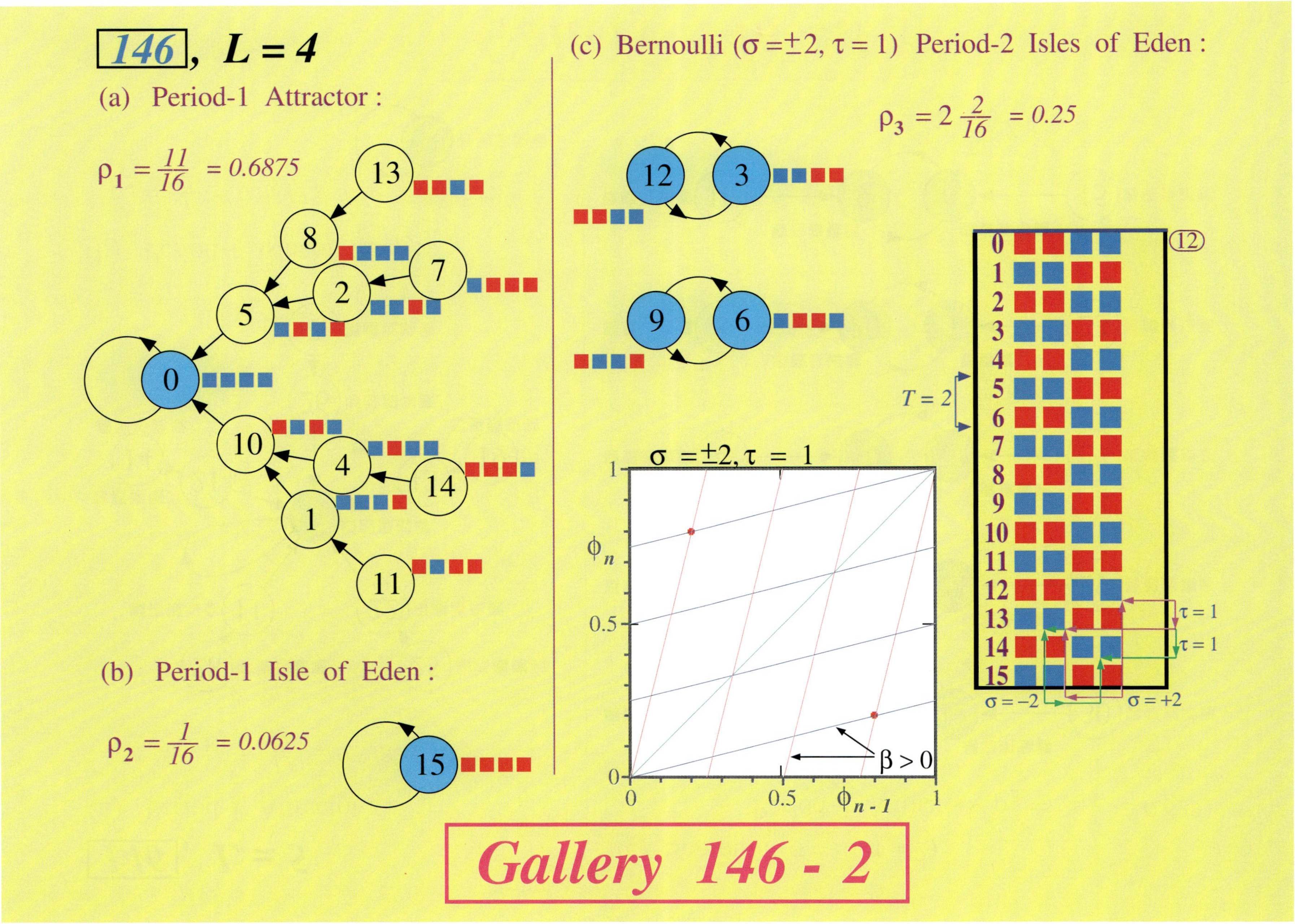
146, L = 4
(a) Period-1 Attractor :
ρ1 = 11/16 = 0.6875
(b) Period-1 Isle of Eden :
ρ2 = 1/16 = 0.0625
(c) Bernoulli (σ = ±2, τ = 1) Period-2 Isles of Eden :
ρ3 = 2 2/16 = 0.25
σ = ±2, τ = 1
φn
φn-1
β > 0
T = 2
τ = 1
τ = 1
σ = −2
σ = +2
Gallery 146 - 2

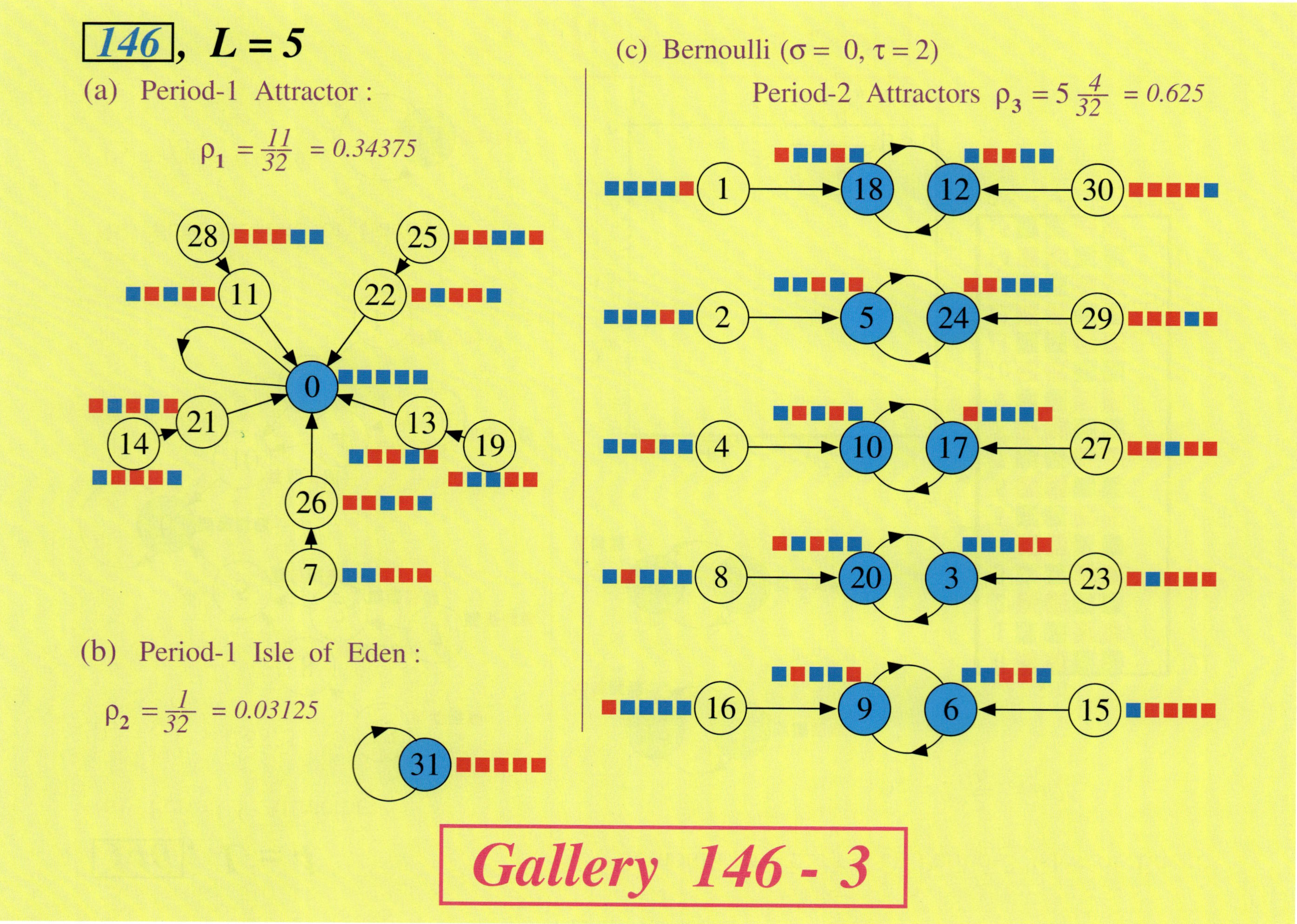

146 , L = 5
(a) Period-1 Attractor :
$\rho_1 = \frac{11}{32} = 0.34375$
(b) Period-1 Isle of Eden :
$\rho_2 = \frac{1}{32} = 0.03125$
(c) Bernoulli ($\sigma = 0$, $\tau = 2$)
Period-2 Attractors $\rho_3 = 5\frac{4}{32} = 0.625$
Gallery 146 - 3

146, **L = 6**

(a) Bernoulli ($\sigma = \pm 3$, $\tau = 1$)

Period-2 Attractors : $\rho_1 = 3\frac{6}{64} = 0.28125$

(b) Period-1 Attractor :

$\rho_2 = \frac{45}{64} = 0.703125$

(c) Period-1 Isle of Eden :

$\rho_3 = \frac{1}{64} = 0.015625$

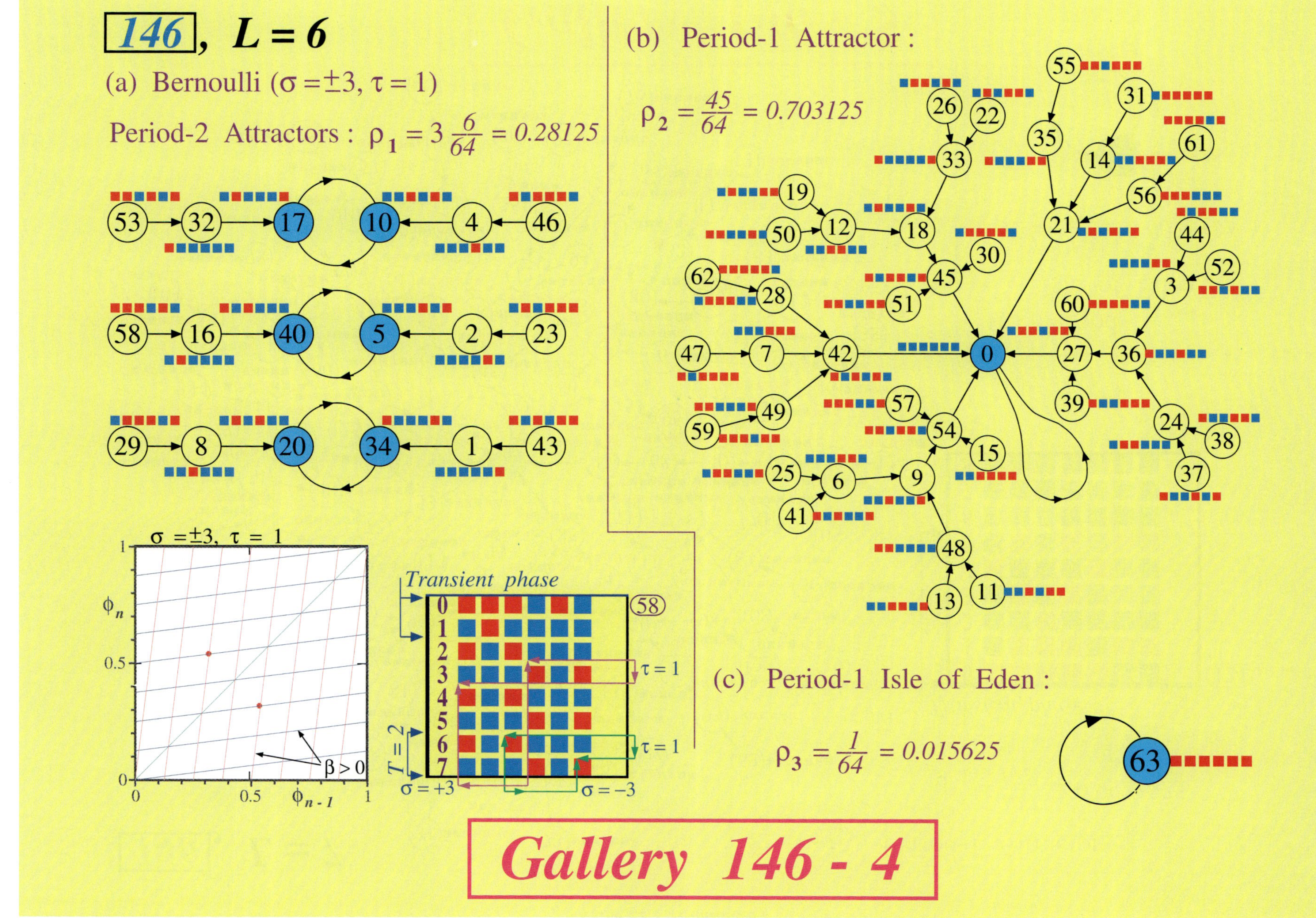

Gallery 146 - 4

Table 22.   (*Continued*)

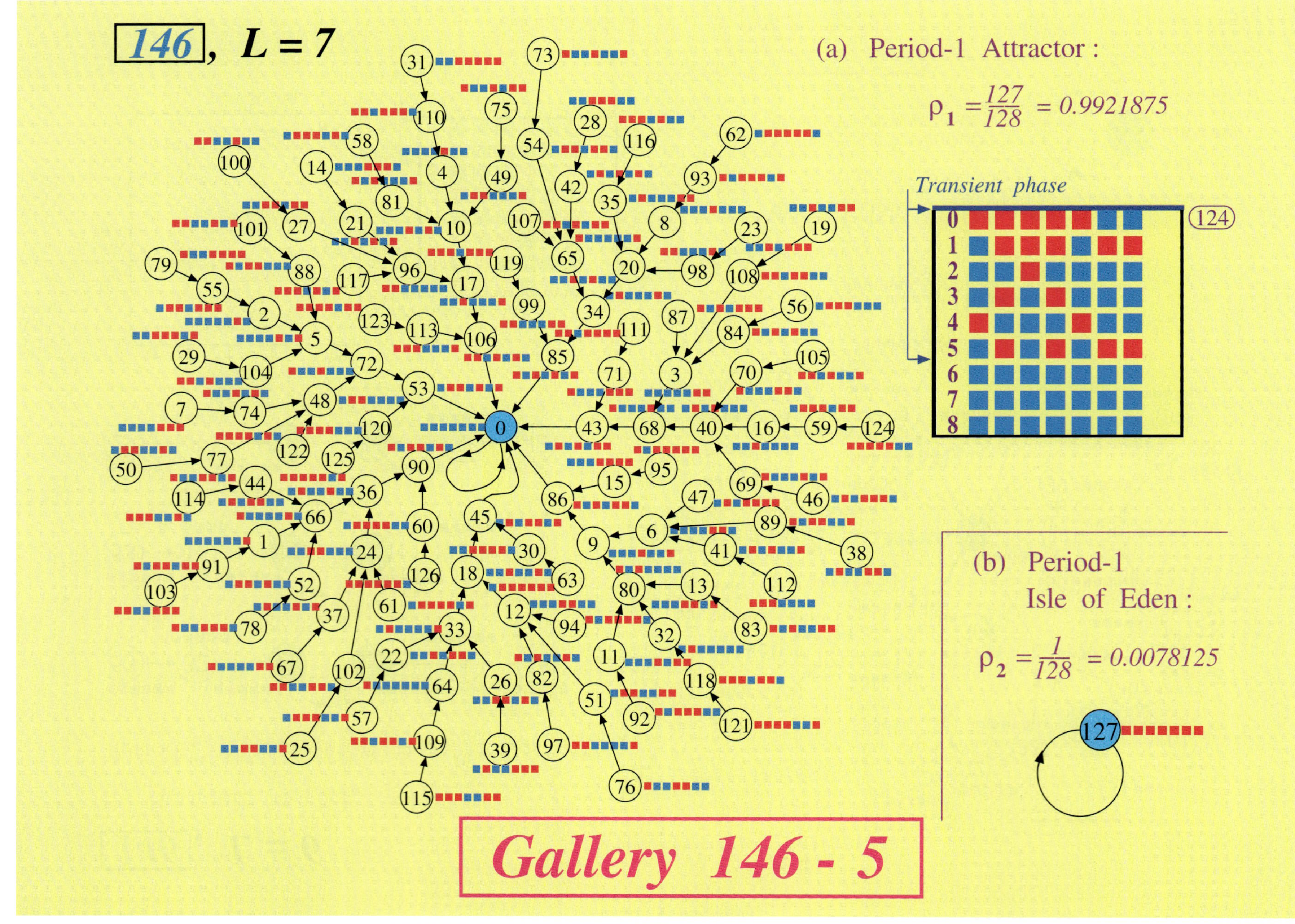

146, **$L = 8$** (a) Bernoulli ($\sigma = \pm 4$, $\tau = 3$) Period-6 Attractors :

$$\rho_1 = 4\frac{28}{256} = 0.4375$$

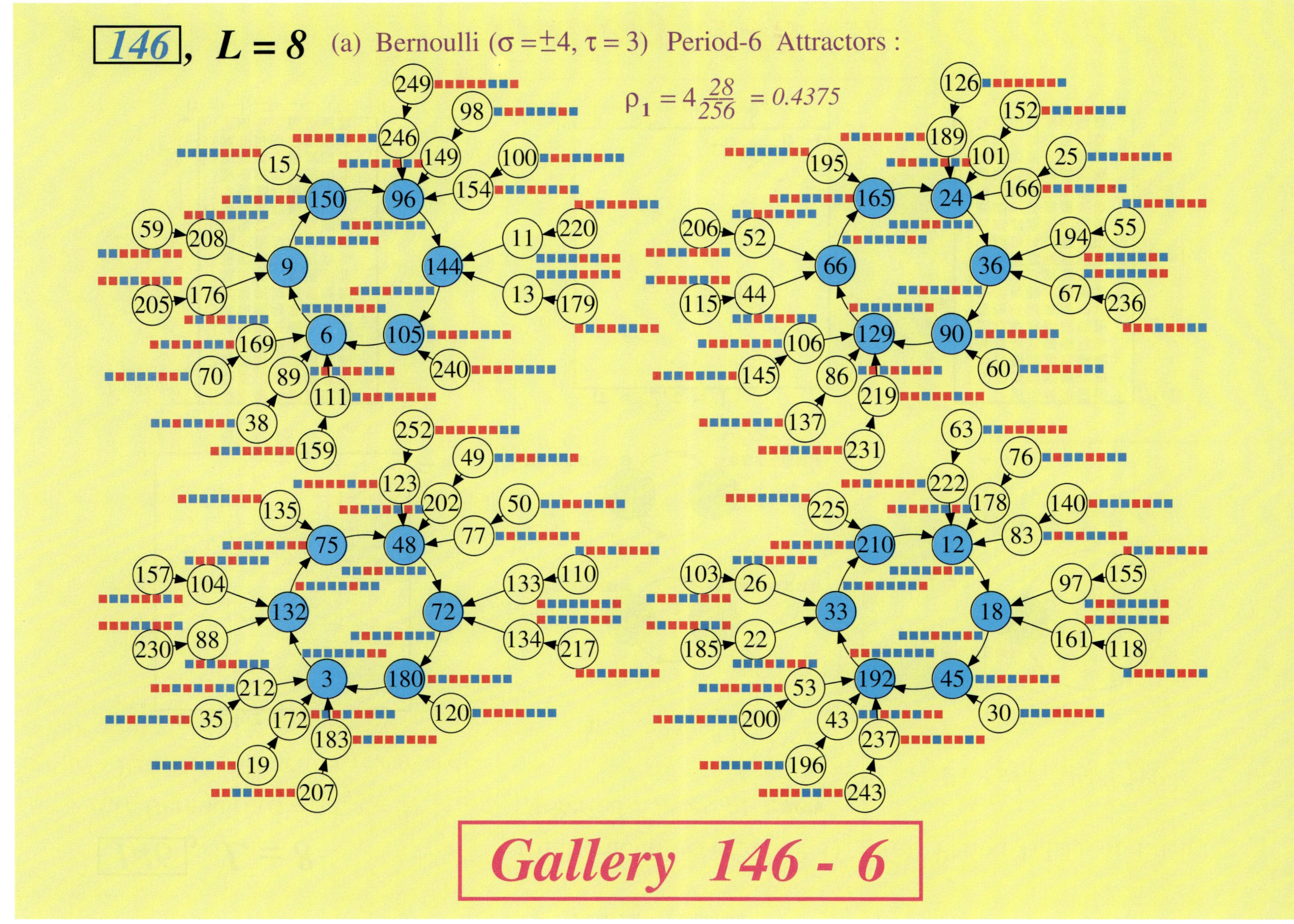

Gallery 146 - 6

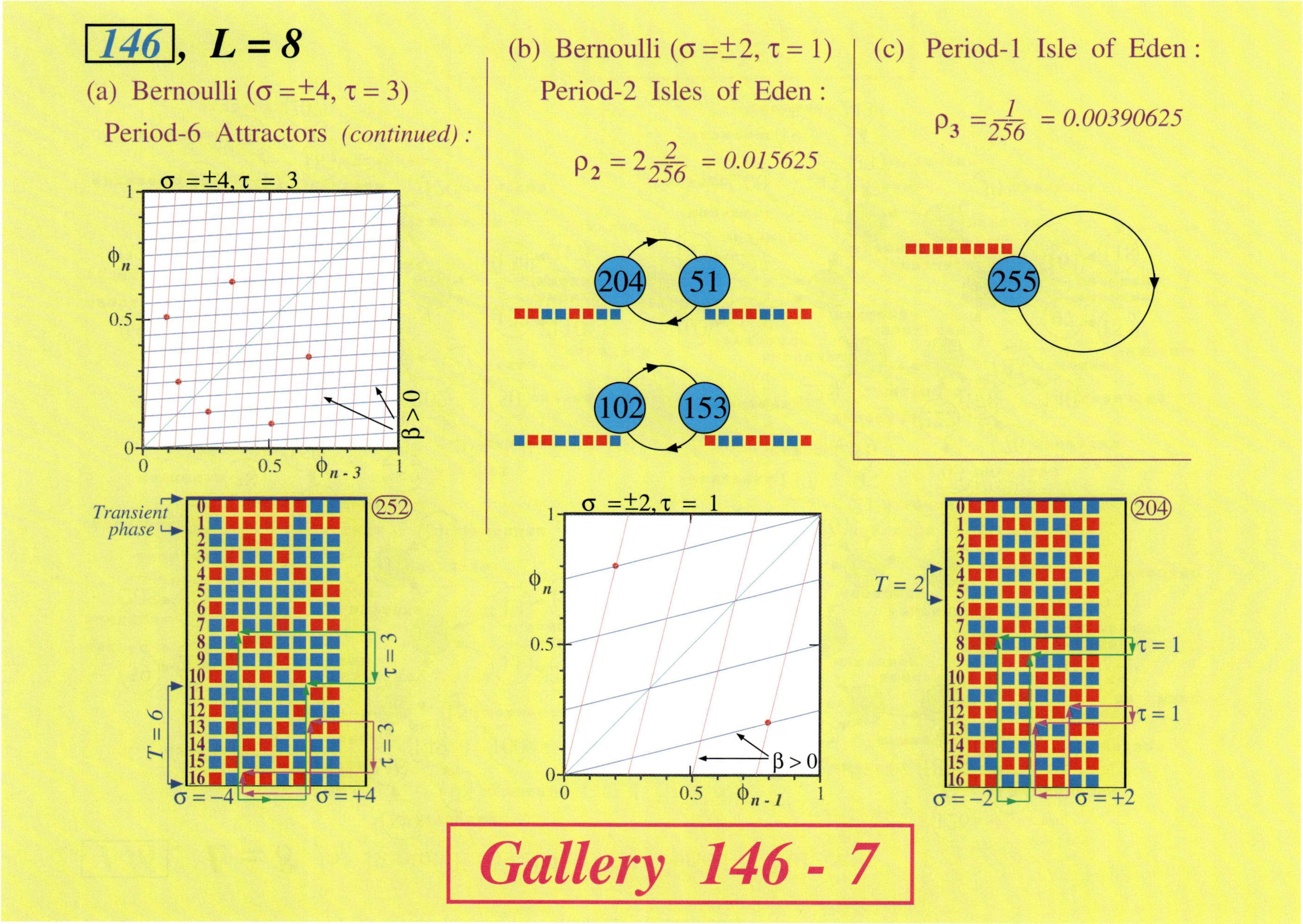

146, *L = 8*

(a) Bernoulli ($\sigma = \pm 4$, $\tau = 3$)

Period-6 Attractors *(continued)* :

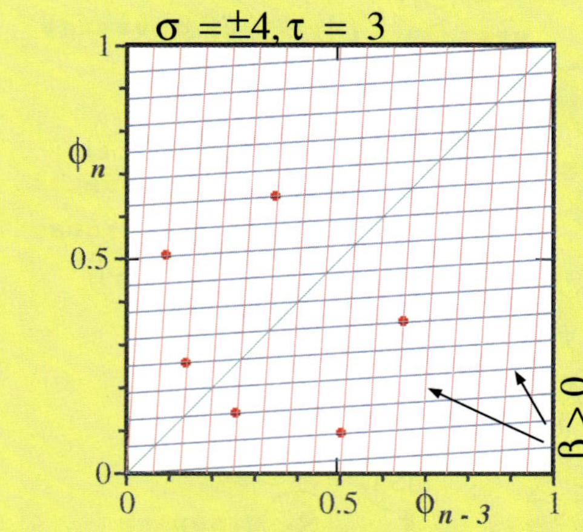

(b) Bernoulli ($\sigma = \pm 2$, $\tau = 1$)

Period-2 Isles of Eden :

$$\rho_2 = 2\frac{2}{256} = 0.015625$$

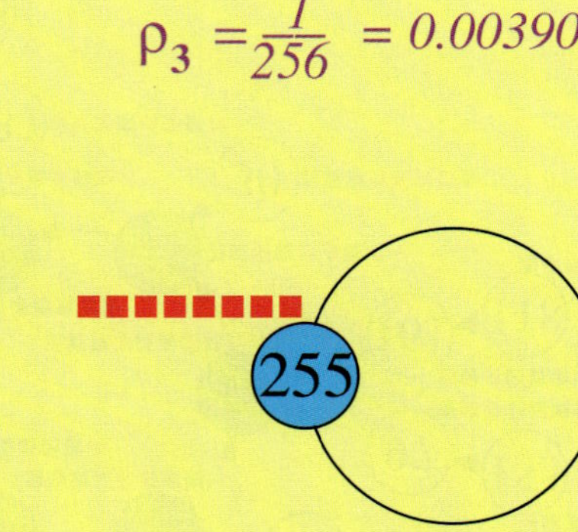

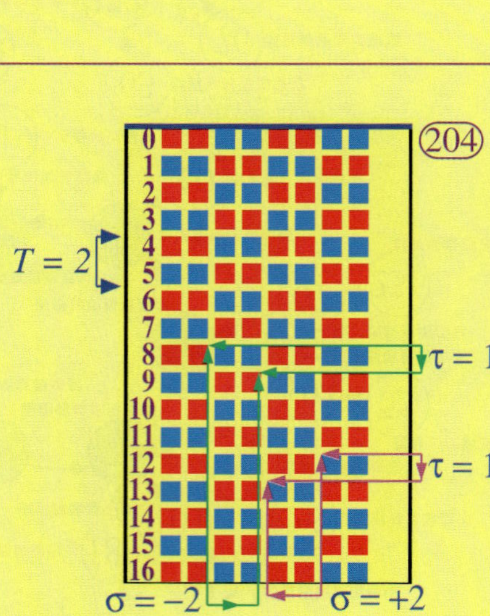

(c) Period-1 Isle of Eden :

$$\rho_3 = \frac{1}{256} = 0.00390625$$

Gallery 146 - 7

146, $L = 8$

(d) Period-1 Attractor :

$$\rho_4 = \frac{139}{256} = 0.54296875$$

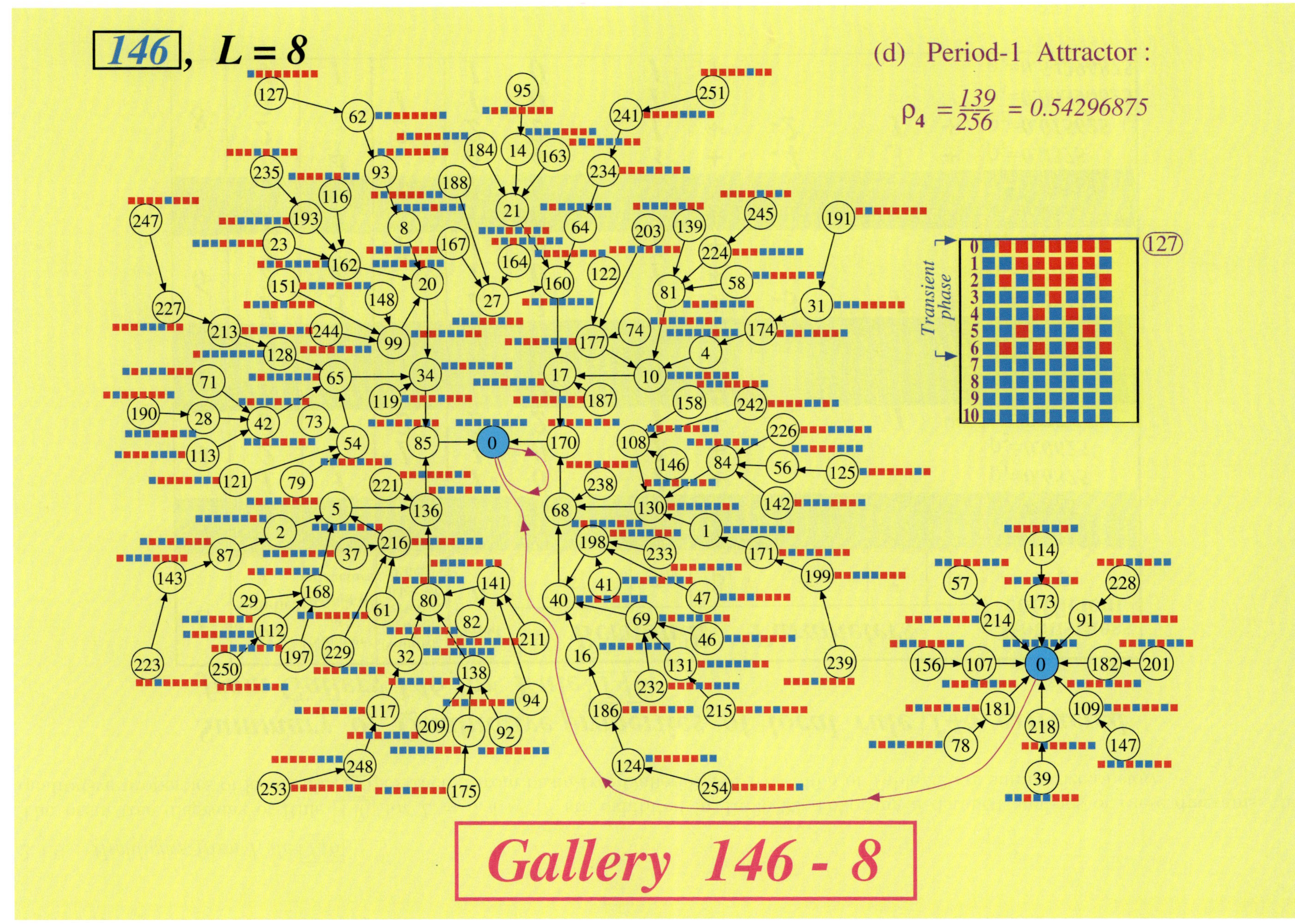

2.4.9. *Highlights from Rule* 146

The basin tree diagrams of Rule 146 for $L = 3, 4, \ldots, 8$ are exhibited in Table 22. Following a detailed analysis of these diagrams, the qualitative properties of local rule 146 extracted from basin-tree Galleries 146-1 to 146-8 of Table 22 are summarized below:

Summary of Qualitative properties of local rule 146 *extracted from Gallery 146 for Rule* 146

L	ID Number i	Number of Period-n attractors	Number of Period-n Isles of Eden	Period n	Bernoulli σ_1	τ_1	β_1	Parameters σ_2	τ_2	β_2	Robustness coefficient ρ
3	1		1	1	0	1	+				$\rho_1 = 0.125$
	2	1		1	0	1	+				$\rho_2 = 0.875$
4	1	1	1	1	0	1	+				$\rho_1 = 0.6875$
	2		1	1	0	1	+				$\rho_2 = 0.0625$
	3		2	2	2	1	+	-2	1	+	$\rho_3 = 0.25$
5	1	1	1	1	0	1	+				$\rho_1 = 0.34375$
	2		1	1	0	1	+				$\rho_2 = 0.03125$
	3	5		2	0	2	+				$\rho_3 = 0.625$
6	1	3		2	3	1	+	-3	1	+	$\rho_1 = 0.28125$
	2	1		1	0	1	+				$\rho_2 = 0.703125$
	3		1	1	0	1	+				$\rho_3 = 0.015625$
7	1	1		1	0	1	+				$\rho_1 = 0.9921875$
	2		1	1	0	1	+				$\rho_2 = 0.0078125$
8	1	4		6	4	3	+	-4	3	+	$\rho_1 = 0.4375$
	2		2	2	2	1	+	-2	1	+	$\rho_2 = 0.015625$
	3		1	1	0	1	+				$\rho_3 = 0.00390625$
	4	1		1	0	1	+				$\rho_4 = 0.54296875$

Basin tree diagrams for Rule 150

150 , **L = 3** (a) Period-1 Attractors : $\rho_1 = 2\,\frac{4}{8} = 1$

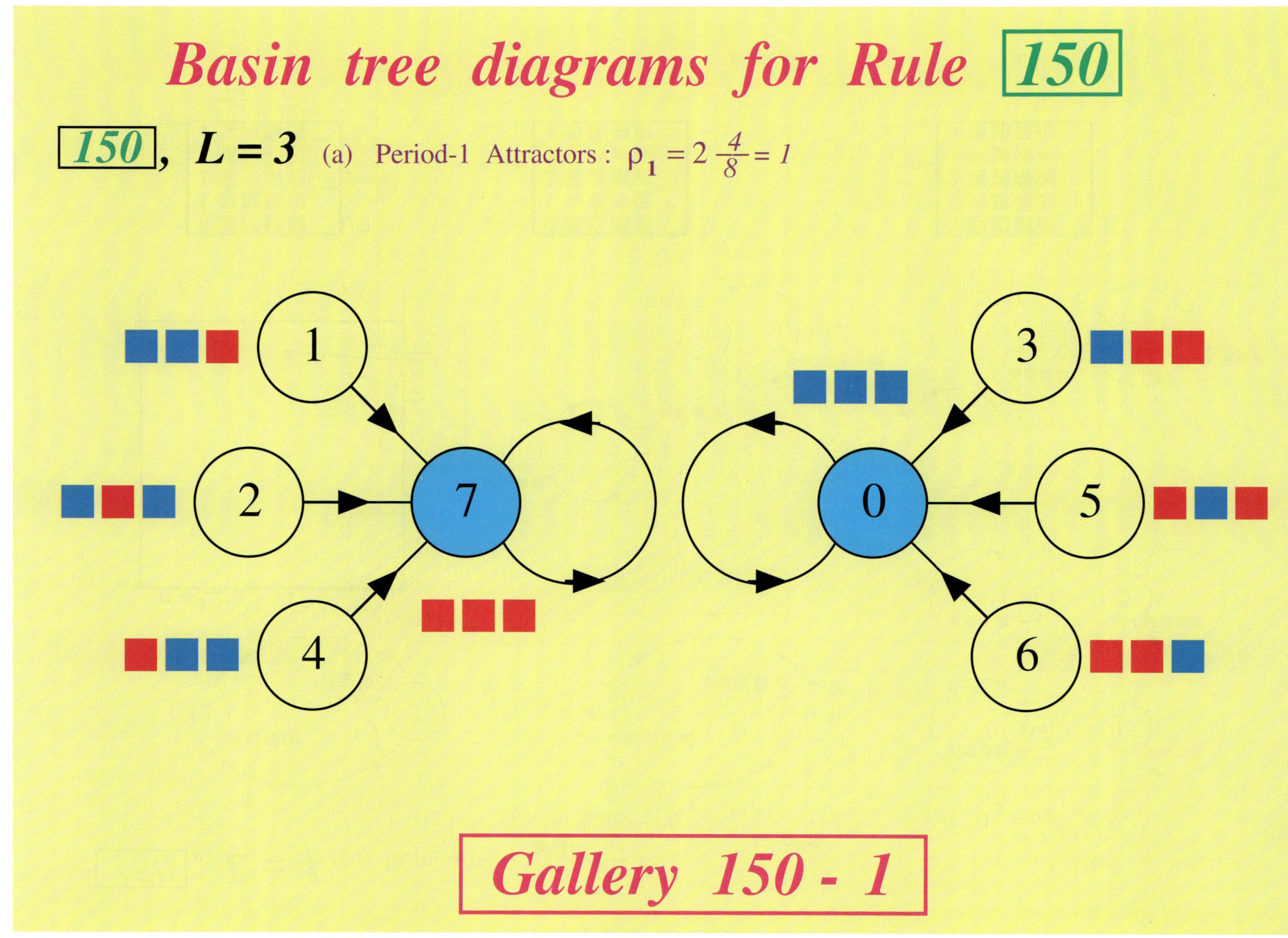

Gallery 150 - 1

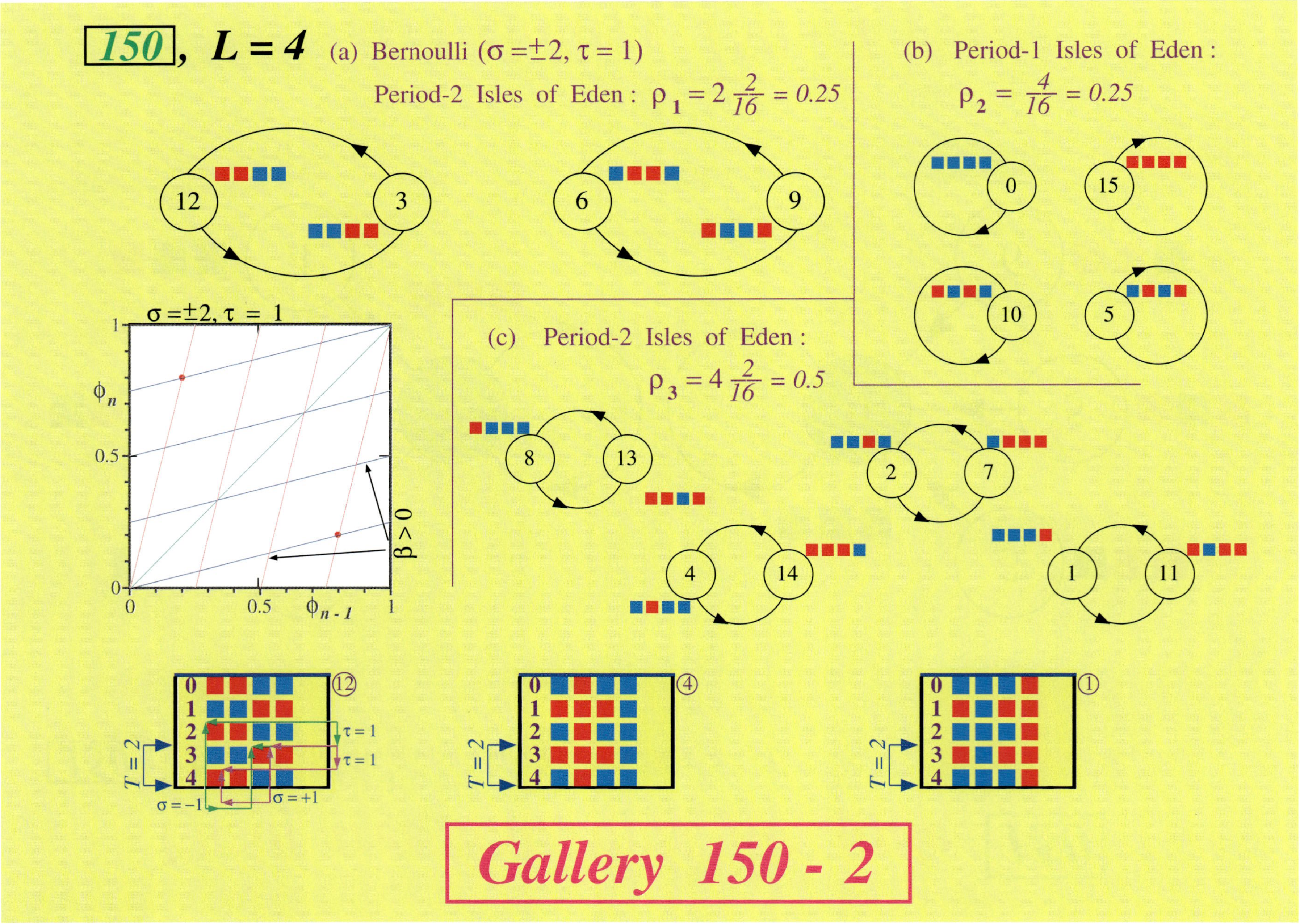
150, L = 4
(a) Bernoulli (σ = ±2, τ = 1)
Period-2 Isles of Eden : ρ1 = 2 2/16 = 0.25
12
3
6
9
σ = ±2, τ = 1
φn
φn-1
β > 0
0
0.5
1
(b) Period-1 Isles of Eden :
ρ2 = 4/16 = 0.25
0
15
10
5
(c) Period-2 Isles of Eden :
ρ3 = 4 2/16 = 0.5
8
13
2
7
4
14
1
11
T = 2
τ = 1
σ = −1
σ = +1
Gallery 150 - 2

Table 23.   (*Continued*)

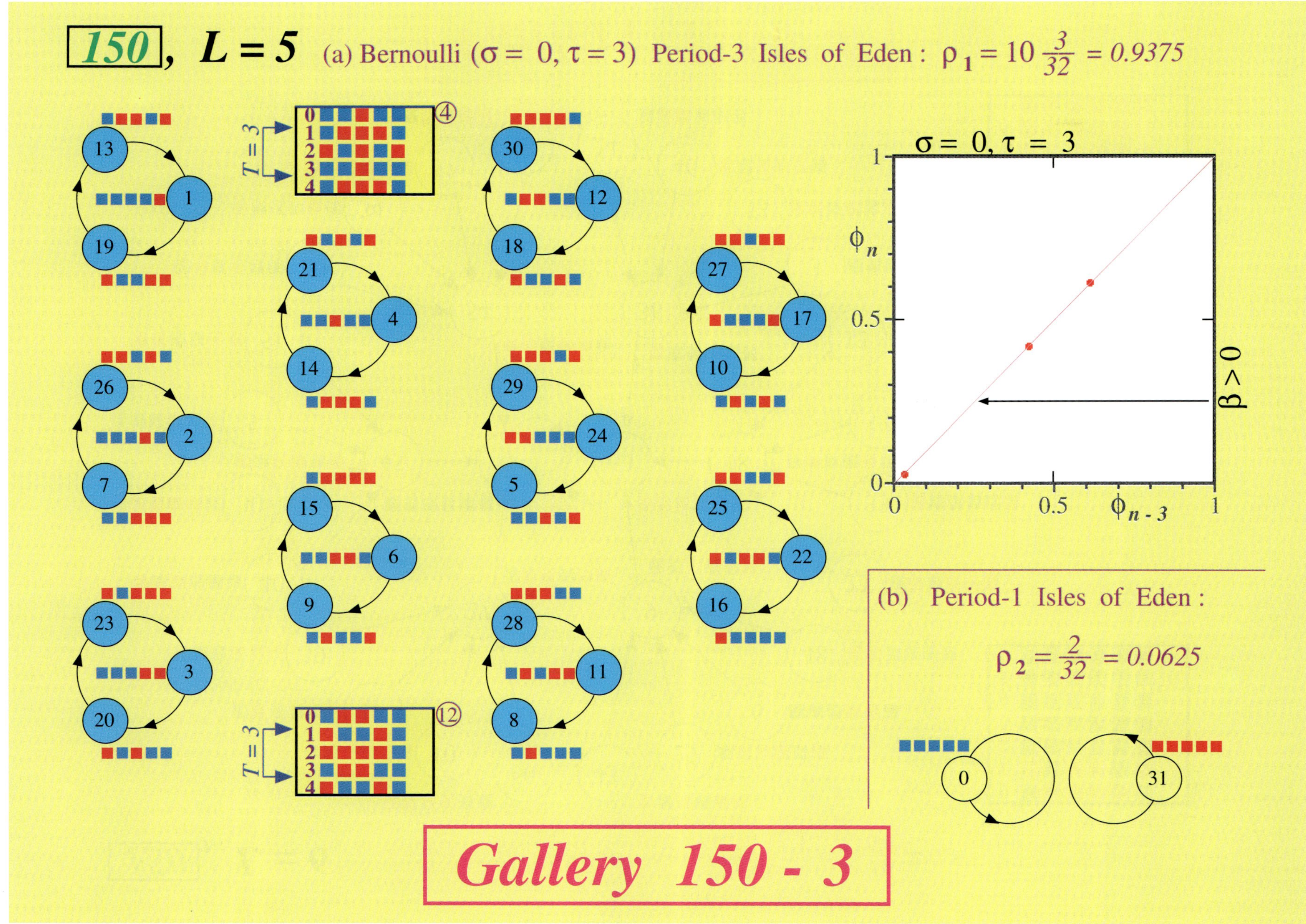

Table 23. (*Continued*)

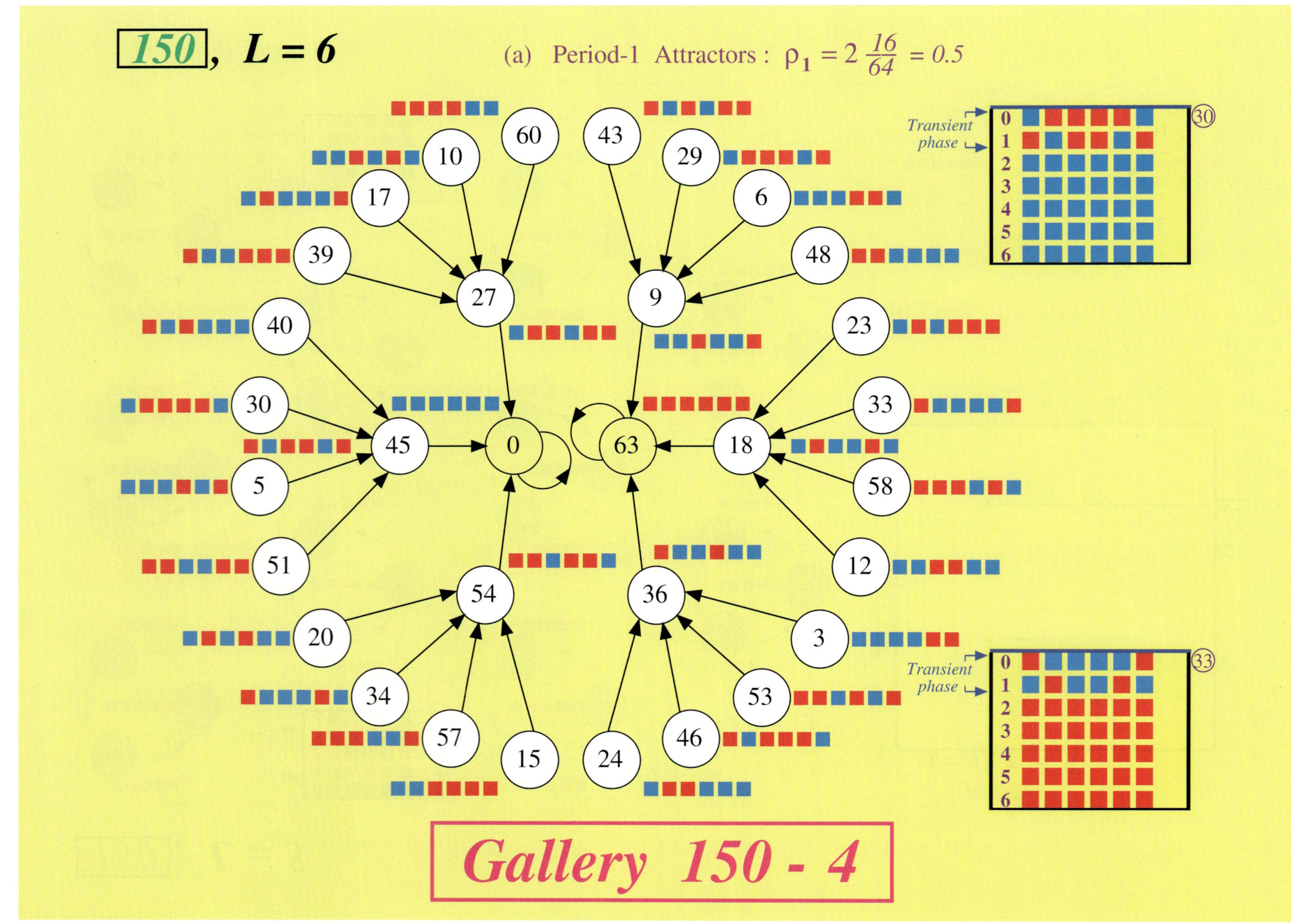

Table 23.   (*Continued*)

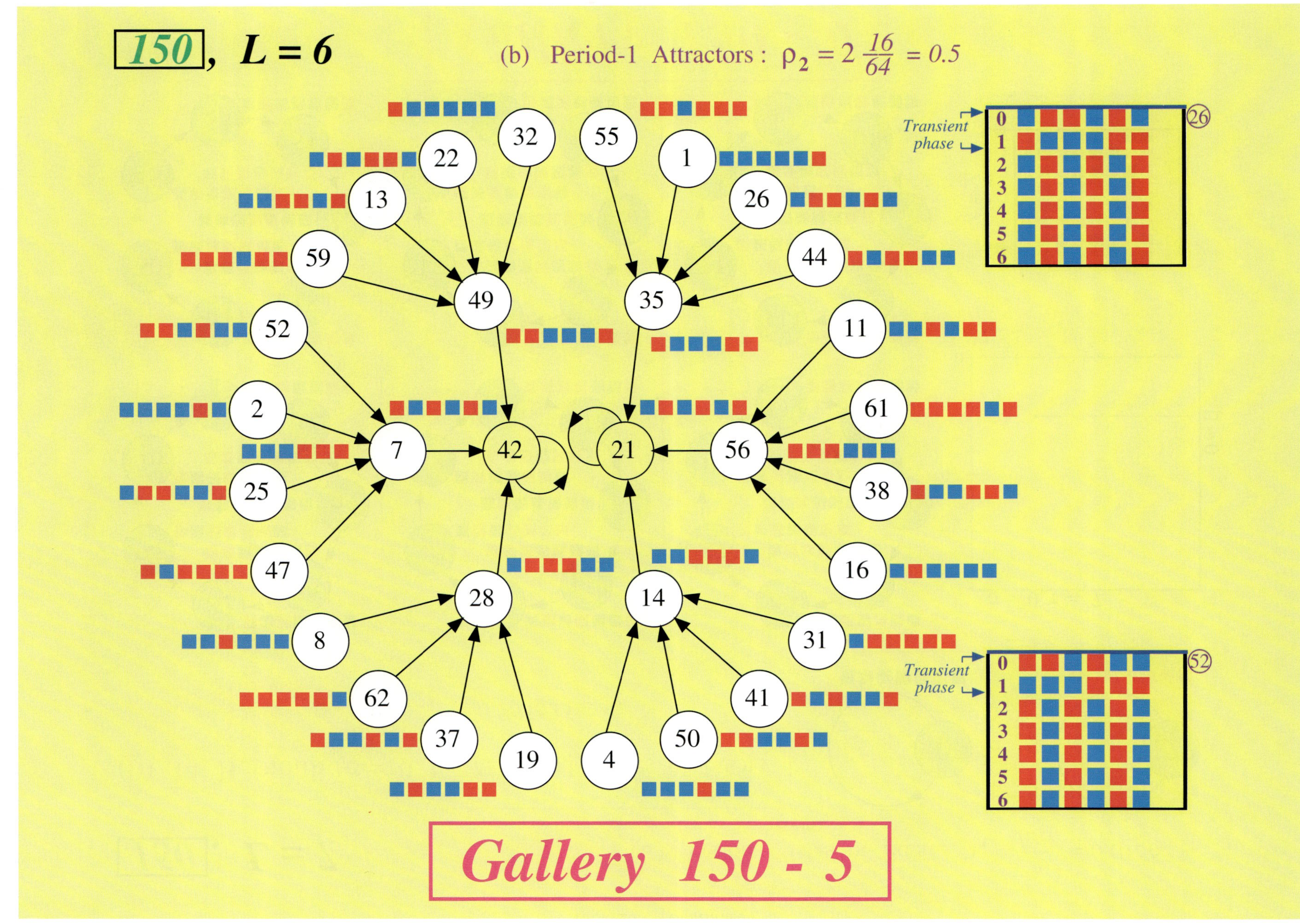

$\boxed{150}$, **L = 6**

(b) Period-1 Attractors : $\rho_2 = 2\frac{16}{64} = 0.5$

150, *L = 7*

(a) Period-1 Isles of Eden : $\rho_1 = \frac{2}{128} = 0.015625$

(b) 14 Bernouli ($\sigma = 0, \tau = 7$) Period-7 Isles of Eden :

$$\rho_2 = 14\frac{7}{128} = 0.765625$$

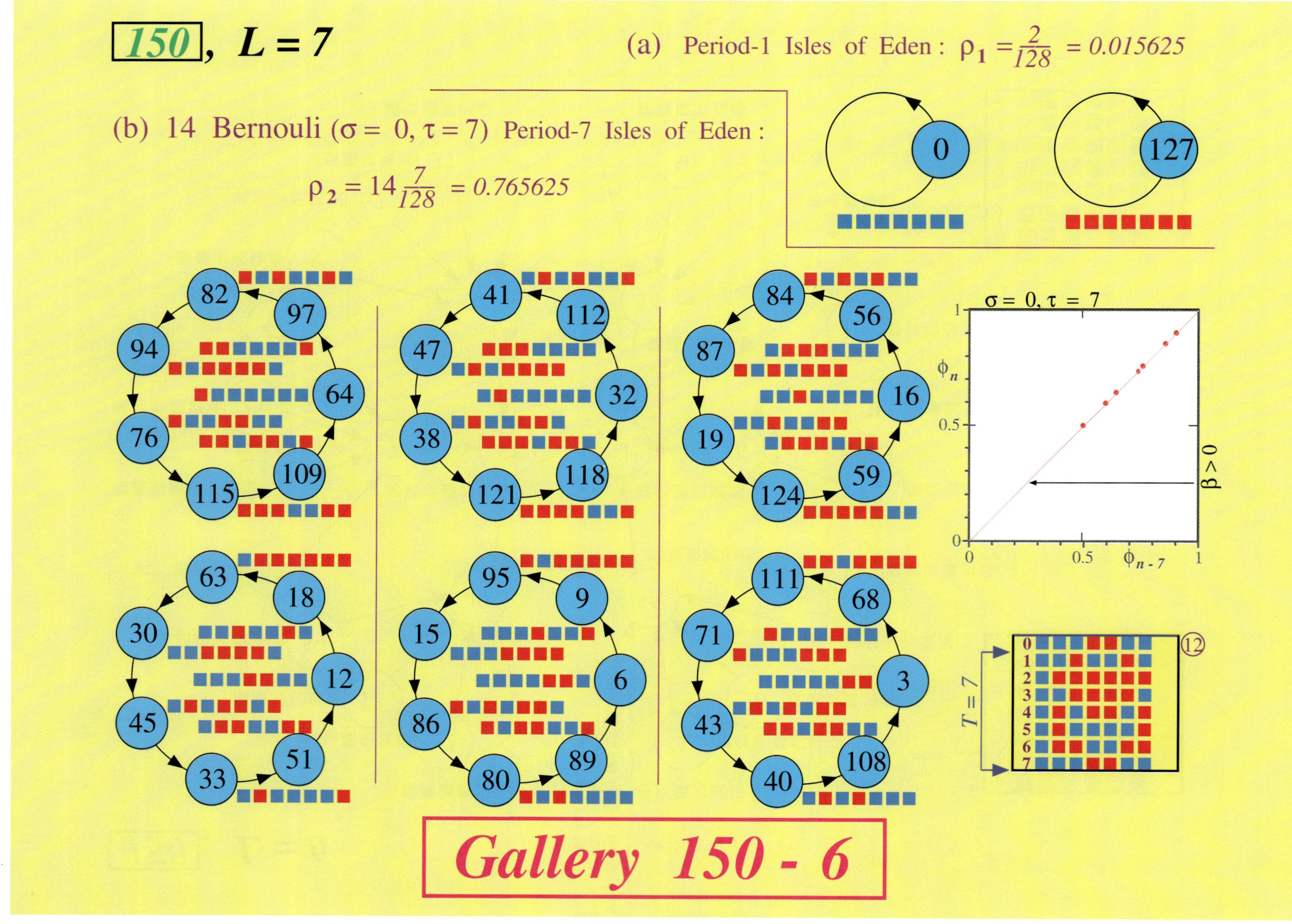

150, *L = 7*   (b) 14 Bernoulli ($\sigma = 0$, $\tau = 7$) Period-7 Isles of Eden *(continued)* :

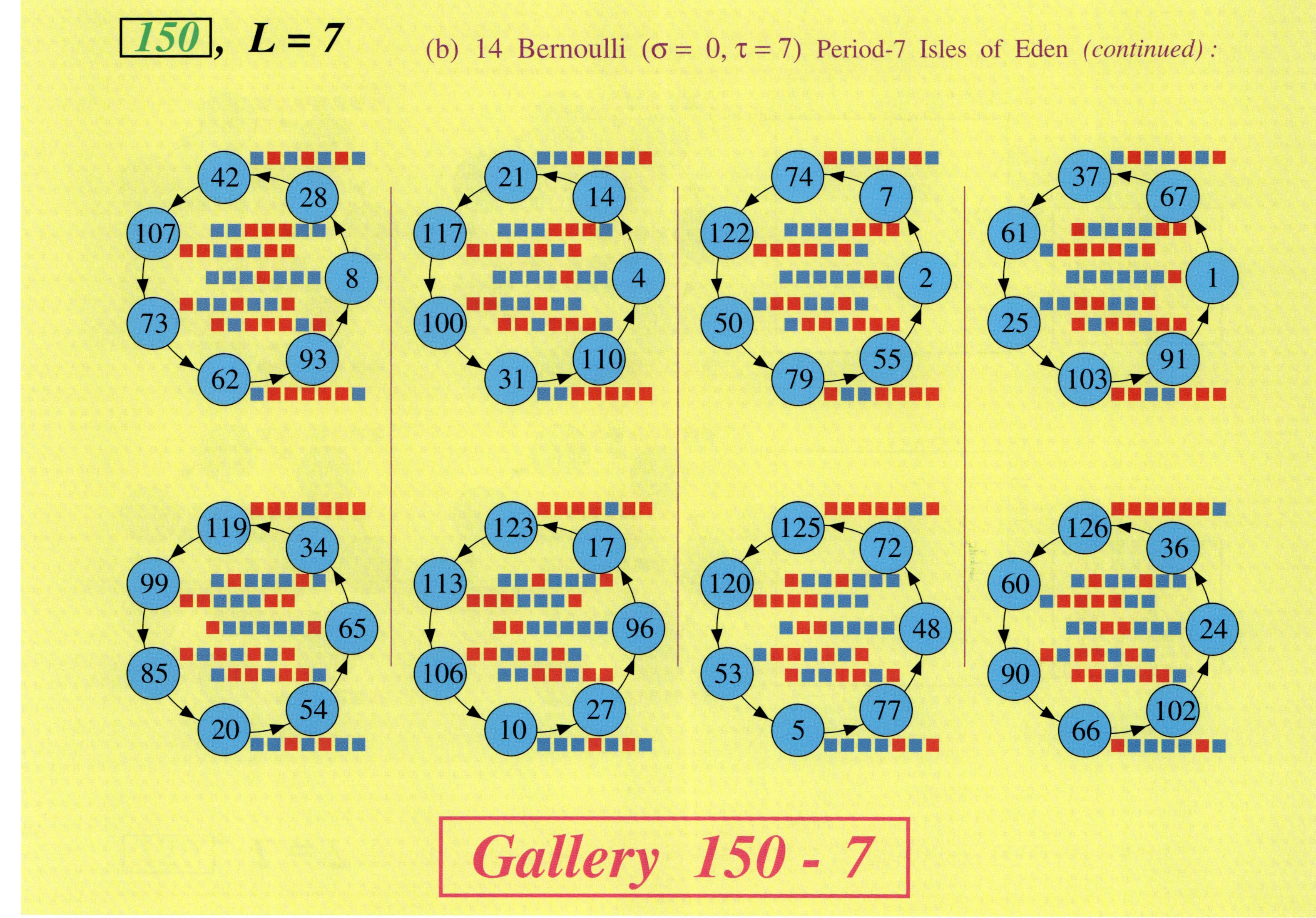

Gallery 150 - 7

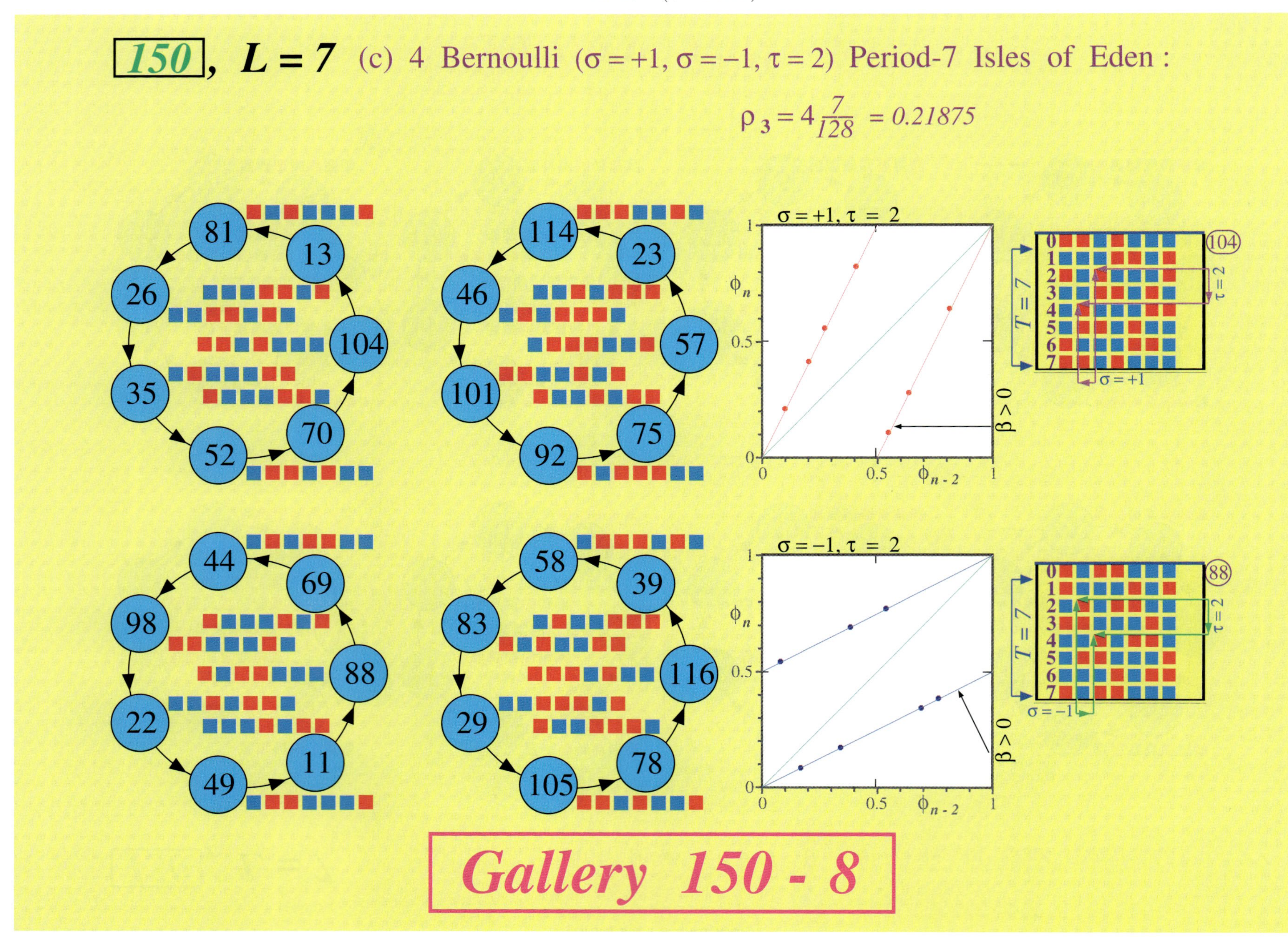

150, L = 7 (c) 4 Bernoulli (σ = +1, σ = −1, τ = 2) Period-7 Isles of Eden :
ρ_3 = 4 7/128 = 0.21875
σ = +1, τ = 2
σ = −1, τ = 2
β > 0
φ_n
φ_{n−2}
T = 7
τ = 2
σ = +1
σ = −1
104
88
Gallery 150 - 8
81 13 104 70 52 35 26
114 23 57 75 92 101 46
44 69 88 11 49 22 98
58 39 116 78 105 29 83

$\boxed{150}$, $L = 8$ (a)   48 Period-4 Isles of Eden : $\rho_1 = 48\frac{4}{256} = 0.75$

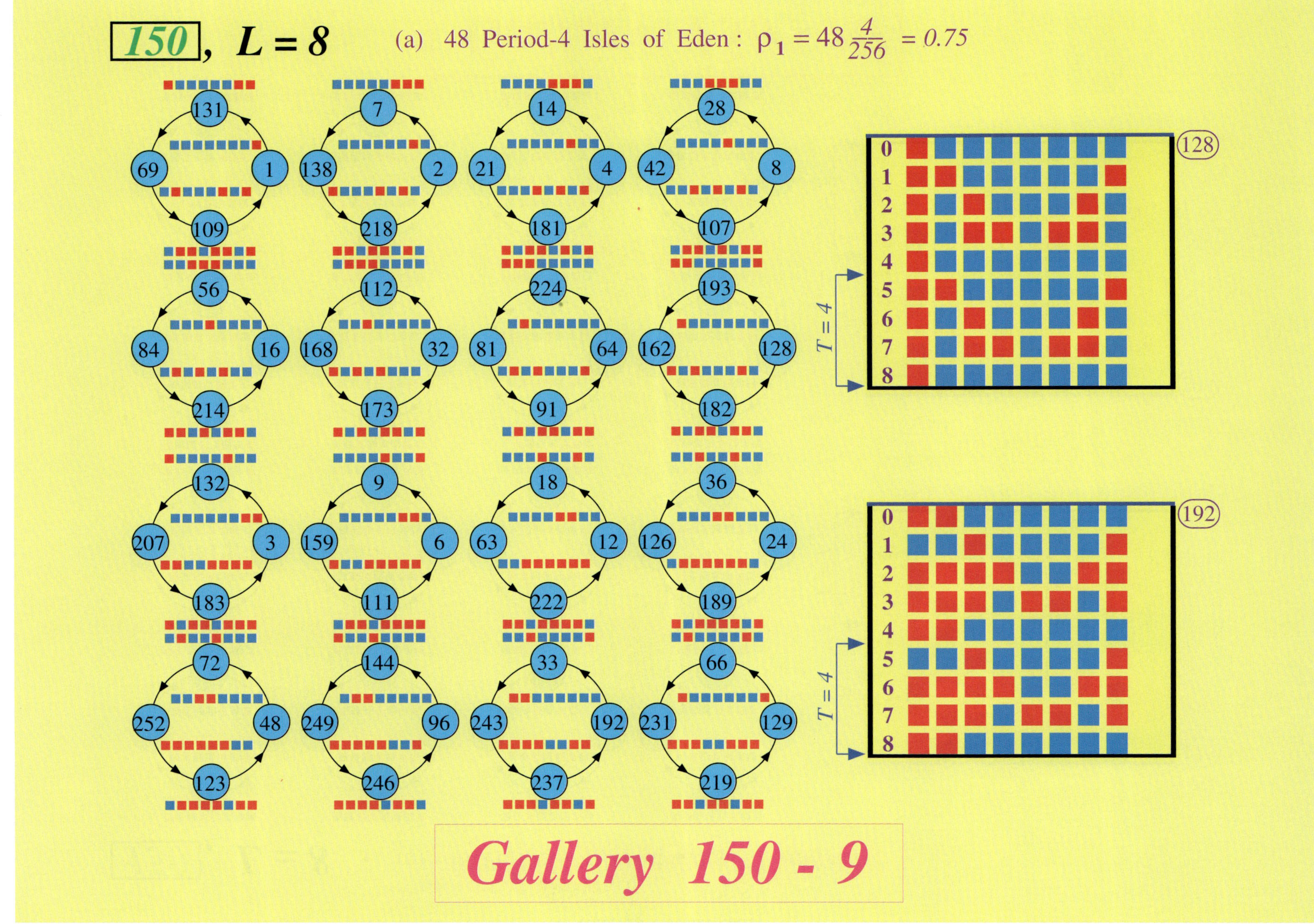

Table 23. (*Continued*)

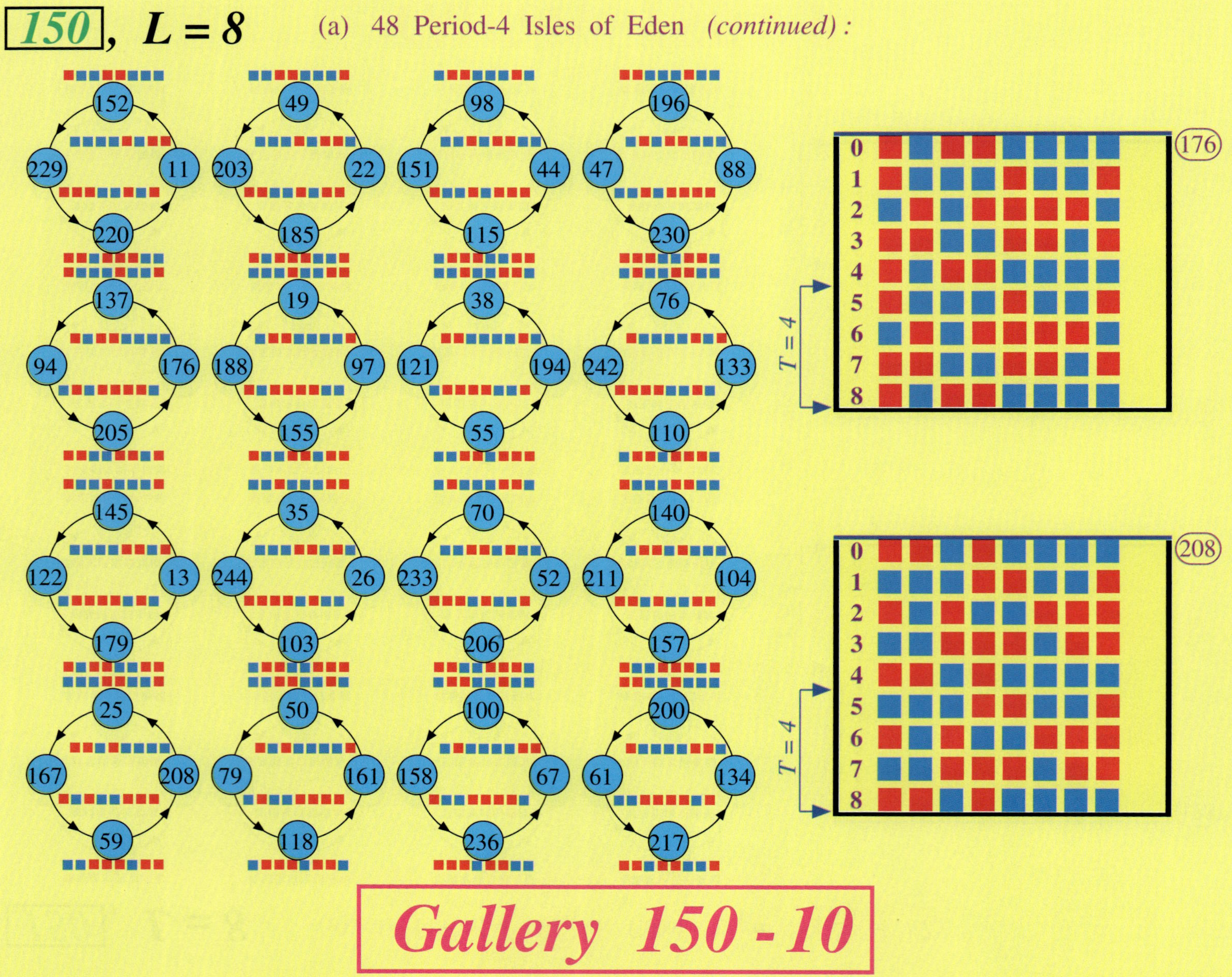

150, *L = 8*   (a) 48 Period-4 Isles of Eden (*continued*) :

Gallery 150 - 10

126

150, **L = 8** (a) 48 Period-4 Isles of Eden *(continued)* :

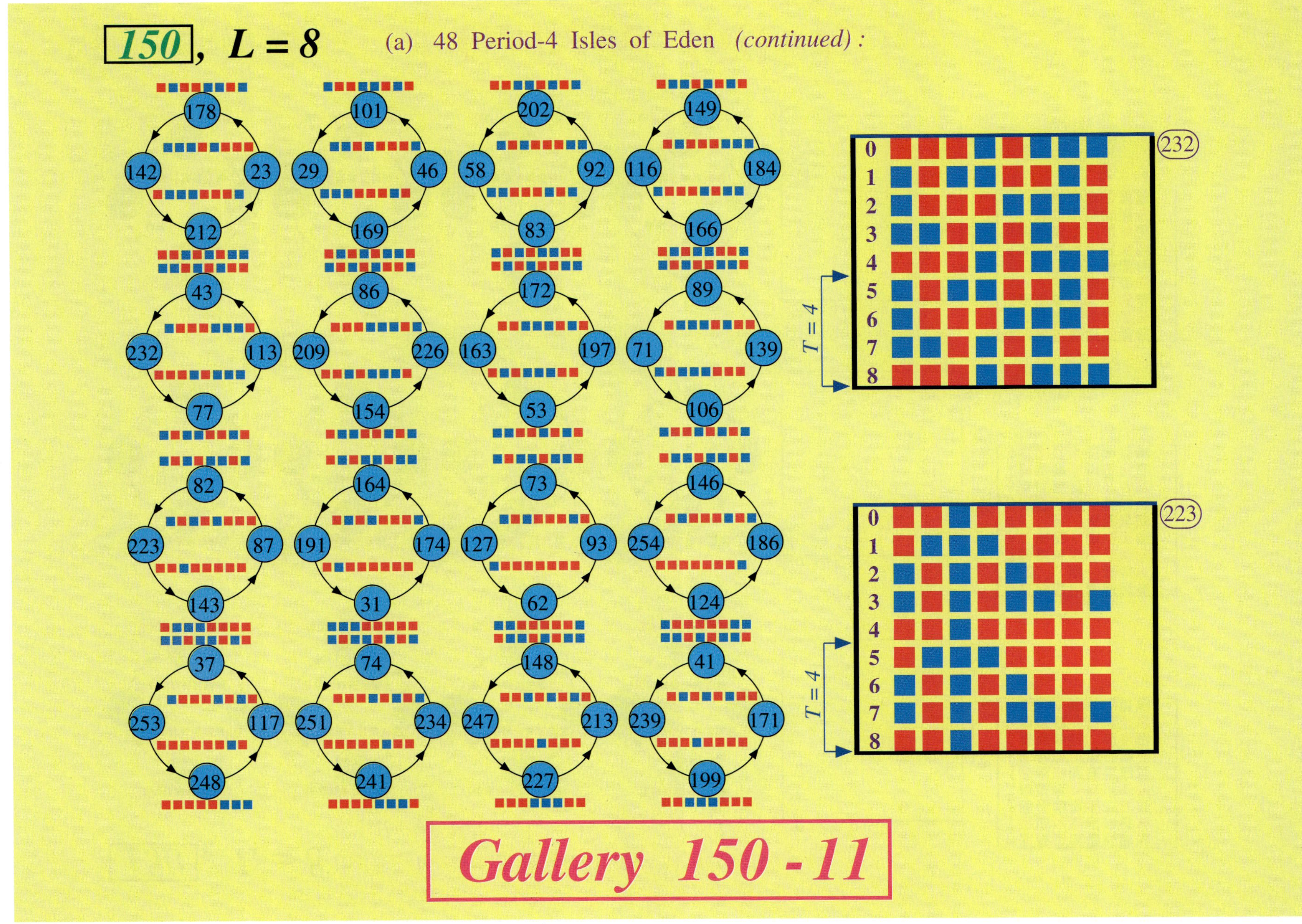

Table 23. (*Continued*)

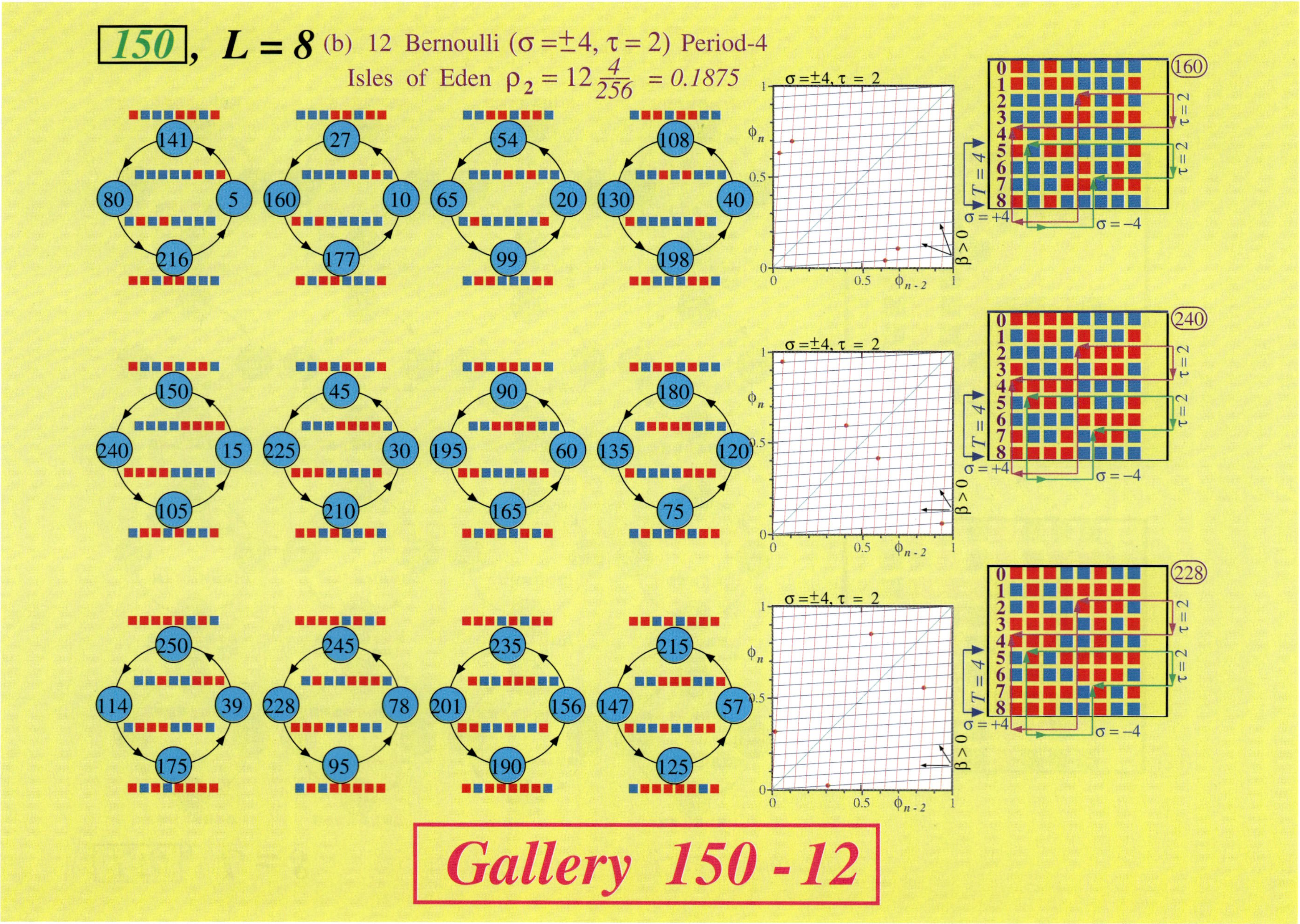

150, $L = 8$ (b) 12 Bernoulli ($\sigma = \pm 4$, $\tau = 2$) Period-4

Isles of Eden $\rho_2 = 12\frac{4}{256} = 0.1875$

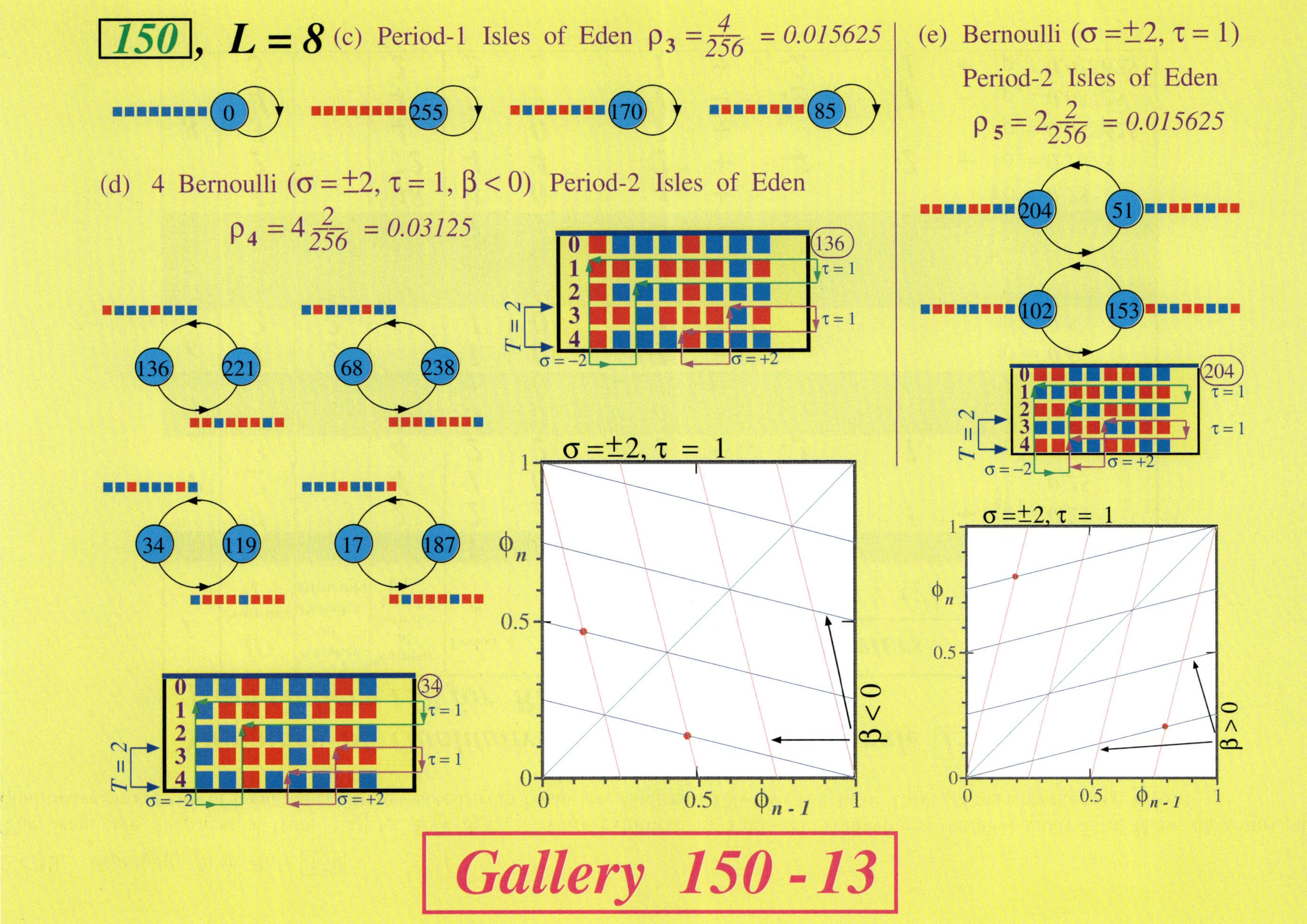
150, L = 8 (c) Period-1 Isles of Eden ρ₃ = 4/256 = 0.015625
(e) Bernoulli (σ = ±2, τ = 1)
Period-2 Isles of Eden
ρ₅ = 2 2/256 = 0.015625
0
255
170
85
204   51
102   153
(d) 4 Bernoulli (σ = ±2, τ = 1, β < 0) Period-2 Isles of Eden
ρ₄ = 4 2/256 = 0.03125
136
136   221
68   238
34   119
17   187
T = 2
0 1 2 3 4
τ = 1
τ = 1
σ = −2
σ = +2
204
T = 2
0 1 2 3 4
τ = 1
τ = 1
σ = −2
σ = +2
34
T = 2
0 1 2 3 4
τ = 1
τ = 1
σ = −2
σ = +2
σ = ±2, τ = 1
φₙ
φₙ₋₁
β < 0
σ = ±2, τ = 1
φₙ
φₙ₋₁
β > 0
Gallery 150 - 13

2.4.10.   *Highlights from Rule* 150

The basin tree diagrams of Rule 150 for $L = 3, 4, \ldots, 8$ are exhibited in Table 23. Following a detailed analysis of these diagrams, the qualitative properties of local rule 150 extracted from basin-tree Galleries 150-1 to 150-13 of Table 23 are summarized below:

Summary of Qualitative properties of local rule 150 extracted from Gallery 150 for Rule 150

L	ID Number i	Number of Period-n attractors	Number of Period-n Isles of Eden	Period n	Bernoulli			Parameters			Robustness coefficient ρ
					σ_1	τ_1	β_1	σ_2	τ_2	β_2	
3	1	2		1	0	1	+				$\rho_1 = 1$
4	1		2	2	2	1	+	-2	1	+	$\rho_1 = 0.25$
	2		4	1	0	1	+				$\rho_2 = 0.25$
	3		4	2	2	1	−	-2	1	−	$\rho_3 = 0.5$
5	1		10	3	0	3	+				$\rho_1 = 0.9375$
	2		2	1	0	1	+				$\rho_2 = 0.0625$
6	1	2		1	0	1	+				$\rho_1 = 0.5$
	2	2		1	0	1	+				$\rho_2 = 0.5$
7	1		2	1	0	1	+				$\rho_1 = 0.015625$
	2		14	7	0	7	+				$\rho_2 = 0.765625$
	3		4	7	1	2	+	-1	2	+	$\rho_3 = 0.21875$
8	1		48	4	0	4	+				$\rho_1 = 0.75$
	2		12	4	4	2	+	-4	2	+	$\rho_2 = 0.1875$
	3		4	1	0	1	+				$\rho_3 = 0.015625$
	4		4	2	2	1	−	-2	1	−	$\rho_4 = 0.03125$
	5		2	2	2	1	+	-2	1	+	$\rho_5 = 0.015625$

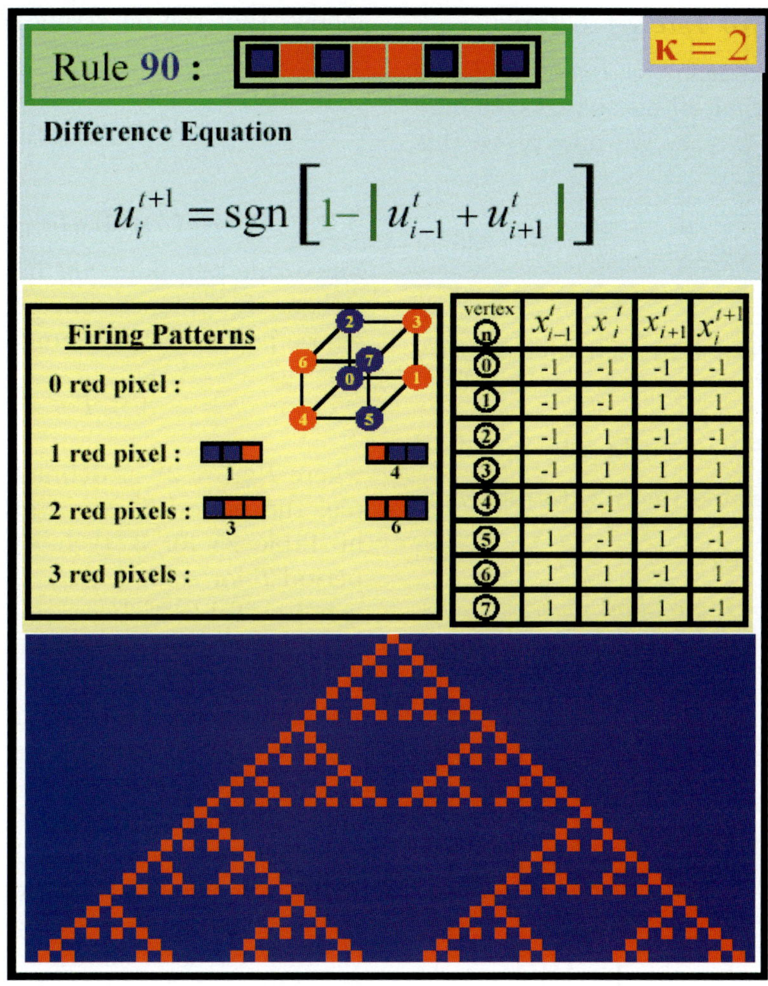

Fig. 2.   Truth table, Boolean cube, Difference Equation, and space-time pattern of local rule $\boxed{90}$.

3. Global Analysis of Local Rule $\boxed{90}$

The *truth table*, *Boolean cube*, and *"Difference Equation"* defining the local rule $\boxed{90}$ along with a space-time pattern (with a single red-pixel initial state) exhibited in Table 5 of [Chua *et al.*, 2003] is reproduced in Fig. 2 for the reader's convenience. For this paper, it is more instructive to recast the Difference Equation defining $\boxed{90}$ into an *equivalent* difference equation involving only a *mod 2 addition* $\oplus$ (defined in Table 24).

Substituting $u_i = 2x_i - 1$ from Eq. (4) of [Chua *et al.*, 2005a] for u_i in the Difference Equation for $\boxed{90}$, we obtain

$$2x_i^{t+1} - 1 = \mathrm{sgn}[1 - |2x_{i-1}^t + 2x_{i+1}^t - 2|]$$
$$= \mathrm{sgn}[1 - 2|x_{i-1}^t + x_{i+1}^t - 1|] \quad (18)$$

Table 24.   Table defining[8] $x_i \oplus x_j \triangleq x_i\ XOR\ x_j$.

$\oplus$	0	1
0	0	1
1	1	0

Simplifying Eq. (18) using Table 24, we obtain the following equivalent Difference Equation:

$$\text{Rule } \boxed{90}\quad \boxed{\begin{aligned} x_i^{t+1} &= (x_{i-1}^t + x_{i+1}^t) \bmod (2)\\ &= x_{i-1}^t \oplus x_{i+1}^t \end{aligned}} \quad (19)$$

---

[8]The *mod 2 operation* $x_i \oplus x_j$ between *two* binary variables is also called an *exclusive OR* operation in *mathematical logic*, and denoted by $x_i \oplus x_j \triangleq x_i\ XOR\ x_j$.

3.1. *Rule* $\boxed{90}$ *has no Isle of Eden*

A cursory glimpse at the basin-tree diagram of rule $\boxed{90}$ in Table 18 reveals that all bit strings converge to an *attractor* for $3 \le L \le 8$. We now prove this property is true for all L.

Theorem 1. *Rule* $\boxed{90}$ *does not have any* isle of *Eden.*

Proof. It follows from Eq. (19) that an arbitrary bit string

$$\boldsymbol{x}^t = (x_0^t \quad x_1^t \quad x_2^t \quad \cdots \quad x_{L-1}^t) \tag{20}$$

at time "t" is *linearly* related (mod 2) to its *image*

$$\boldsymbol{x}^{t+1} = (x_0^{t+1} \quad x_1^{t+1} \quad x_2^{t+1} \quad \cdots \quad x_{L-1}^{t+1}) \tag{21}$$

at time "$t+1$" via an $L \times L$ *circulant matrix* [Davis, 1979] $\boldsymbol{M}(\boxed{90})$, *henceforth called the local time-1 state transition matrix*:

$$\underbrace{\begin{bmatrix} x_0^{t+1} \\ x_1^{t+1} \\ x_2^{t+1} \\ x_3^{t+1} \\ \vdots \\ x_{L-2}^{t+1} \\ x_{L-1}^{t+1} \end{bmatrix}}_{\boldsymbol{x}^{t+1}} = \underbrace{\begin{bmatrix} 0 & 1 & 0 & 0 & 0 & \cdots & 1 \\ 1 & 0 & 1 & 0 & 0 & \cdots & 0 \\ 0 & 1 & 0 & 1 & 0 & \cdots & 0 \\ 0 & 0 & 1 & 0 & 1 & \cdots & 0 \\ \vdots & & & & & \ddots & \vdots \\ 0 & 0 & 0 & 0 & \cdots & 1 & 0 \\ 0 & 0 & 0 & 0 & \cdots & 0 & 1 \\ 1 & 0 & 0 & 0 & \cdots & 1 & 0 \end{bmatrix}}_{\boldsymbol{M}(\boxed{90})} \underbrace{\begin{bmatrix} x_0^t \\ x_1^t \\ x_2^t \\ x_3^t \\ \vdots \\ x_{L-2}^t \\ x_{L-1}^t \end{bmatrix}}_{\boldsymbol{x}^t} \tag{22}$$

where addition is mod (2) sum $\oplus$.

Note the *diagonal* elements of the circulant matrix $\boldsymbol{M}(\boxed{90})$ are all equal to *zero*. Observe also the elements directly below (resp. above) the diagonal of $\boldsymbol{M}(\boxed{90})$ are all equal to *one*. All other elements are zero, except for the top rightmost element, and the bottom leftmost elements, which are equal to *one*, respectively. It follows from this special structure that the *leftmost* column of $\boldsymbol{M}(\boxed{90})$ is equal to the *mod 2 sum* of the remaining L-1 columns. Since the columns of $\boldsymbol{M}(\boxed{90})$ are not *linearly independent*, mod 2, it follows that $\boldsymbol{M}$ does *not* have an inverse. Since the bit string $\boldsymbol{x}^{t+1}$ on the left side of Eq. (19) does not have a *unique preimage*, it follows that the bit string $\boldsymbol{x}^t$ is *not* an *isle of Eden* of $\boxed{90}$.

Since $\boldsymbol{x}^t$ is an arbitrary bit string, it follows that $\boxed{90}$ cannot possess an *isle of Eden* for any L. ■

3.2. *Period of Rule* $\boxed{90}$ *grows with L*

Since rule $\boxed{90}$ does *not* have isles of Eden, all bit strings of $\boxed{90}$ must converge to *period-T attractors* whose *period* "T" is bounded by

$$1 \le T \le T_{\max} \tag{23}$$

where $T_{\max} = 2^L$ as defined in Eq. (6). As an example, the period T of an attractor of $\boxed{90}$ is listed in Table 25 for $3 \le L \le 100$. Observe that the period T for *some* L (e.g. $L = 47, 49, 53$, etc.) is not listed in Table 25 because it is so large that it had exceeded the maximum simulation time allocated. A bit string belonging to one of the many period-T attractors for $3 \le L \le 25$ is given in Table 26. For example, the bit string listed for $L = 3$ corresponds to the third *period-1* attractor (out of 4) listed in Gallery 90-1 of Table 18. The bit string listed for $L = 5$ corresponds to node ⑥ of Gallery 90-3 of Table 18, out of five period-3 attractors. The bit string listed for $L = 6$ corresponds to node ㉚ in the fifth attractor shown on the left of Gallery 90-5. The bit string listed for $L = 7$ corresponds to node (68) in the top left attractor of $\boxed{90}$ shown on the top left of Gallery 90-7.

As examination of Table 25 shows that unlike the period-1 and period-2 local rules listed in Tables 7 and 8, and the period-3 local rule $\boxed{62}$, which have a relatively small period, and independent of L, the period T of rule $\boxed{90}$ can increase at an *exponential* rate as a function of L, as depicted in Fig. 3. Such exponential growth of T as a function of L is a signature of all *complex Bernoulli* rules in Table 11, and *hyper-Bernoulli* rules in Table 12.

In spite of the very large values T of some period-T *attractors* of $\boxed{90}$, these periods are usually many orders of magnitude smaller than the upper bound $T_{\max}$ listed in Table 27 for $3 \le L \le 85$.[9] There exists, however, period T attractors whose period T approaches the upper bound $T_{\max}$. For example, Table 28 shows a period-504 bit string of an *isle of Eden* of rule $\boxed{45}$ for $L = 9$, which is almost as large as $T_{\max} = 2^9 = 512$![10]

---

[9]It would take at least 105, 104, 783, 572 years for a 1 GHz PC to simulate all $T_{\max} = 2^L$ distinct bit strings!

[10]Rule $\boxed{45}$ will be studied in Part VIII where it is proved that *all* bit strings are *isles of Eden* if, and only if L is an *odd* number.

Table 25.   Period *"T"* of attractors of local rule $\boxed{90}$ for $3 \leq L \leq 100$.

L	Attractor	L	Attractor	L	Attractor	L	Attractor	L	Attractor
1		21	$T = 63$	41	$T = 1023$	61		81	
2		22	$T = 62$	42	$T = 126$	62	$T = 62$	82	$T = 2046$
3	$T = 1$	23	$T = 2047$	43	$T = 127$	63	$T = 63$	83	
4	$T = 1$	24	$T = 8$	44	$T = 124$	64	$T = 1$	84	$T = 252$
5	$T = 3$	25	$T = 1023$	45	$T = 4095$	65	$T = 63$	85	$T = 255$
6	$T = 2$	26	$T = 126$	46	$T = 4094$	66	$T = 62$	86	$T = 254$
7	$T = 7$	27	$T = 511$	47		67		87	
8	$T = 1$	28	$T = 28$	48	$T = 16$	68	$T = 60$	88	$T = 248$
9	$T = 7$	29	$T = 16383$	49		69		89	$T = 2047$
10	$T = 6$	30	$T = 30$	50	$T = 2046$	70	$T = 8190$	90	$T = 8190$
11	$T = 31$	31	$T = 31$	51	$T = 255$	71		91	$T = 4095$
12	$T = 4$	32	$T = 1$	52	$T = 252$	72	$T = 56$	92	$T = 8188$
13	$T = 63$	33	$T = 31$	53		73	$T = 511$	93	$T = 1023$
14	$T = 14$	34	$T = 30$	54	$T = 1022$	74	$T = 174762$	94	
15	$T = 15$	35	$T = 4095$	55		75		95	
16	$T = 1$	36	$T = 28$	56	$T = 56$	76	$T = 2044$	96	$T = 32$
17	$T = 15$	37	$T = 87381$	57	$T = 511$	77		97	
18	$T = 14$	38	$T = 1022$	58	$T = 32766$	78	$T = 8190$	98	
19	$T = 511$	39	$T = 4095$	59		79		99	$T = 32767$
20	$T = 12$	40	$T = 24$	60	$T = 60$	80	$T = 48$	100	$T = 4092$

3.3. *Global state-transition formula for rule* $\boxed{90}$

The state transition formula given in Fig. 2 and Eq. (19) for rule $\boxed{90}$ is *local in time* in the sense that it generates from a bit string $\boldsymbol{x}^t = (x_0^t \ x_1^t \ x_2^t \ \cdots \ x_{L-1}^t)$ at time *"t"* the next bit string $\boldsymbol{x}^{t+1} = (x_0^{t+1} \ x_1^{t+1} \ x_2^{t+1} \ \cdots \ x_{L-1}^{t+1})$ at time *"t + 1"*. Our next theorem gives an explicit formula which is *global in time* in the sense that it generates a bit string $\boldsymbol{x}_0^n = (x_0^n \ x_1^n \ x_2^n \ \cdots \ x_{L-1}^n)$ at any future time $n > t$.

Theorem 2. Global State-Transition Formula for $\boxed{90}$.

Each pixel x_i^n at time $n > t$ is determined from "n + 1" initial pixels $x_{i-n}^0, x_{i-n+2}^0, \ldots, x_{i+n-2}^0, x_{i+n}^0$ at $t = 0$ via the binomial formula.

$$\boxed{x_i^n = \sum_{k=0}^{n} \frac{n!}{k!(n-k)!} \bullet x_{i-n+2k}^0 \quad \mathrm{mod}\ (2)} \quad (24)$$

Proof. Apply *mathematical induction* as follow:

(a) $n = 1$

Applying $n = 1$ in Eq. (24), we obtain[11]

$$x_i^1 = x_{i-1}^0 + x_{i+1}^0 \quad \mathrm{mod}\ (2) \quad (25)$$

which is Eq. (19) for $t = 0$.

---

[11] Recall the factorial notation $0! \overset{\triangle}{=} 1$.

Table 26.   Bit strings for generating a period-T attractor of Rule $\boxed{90}$.

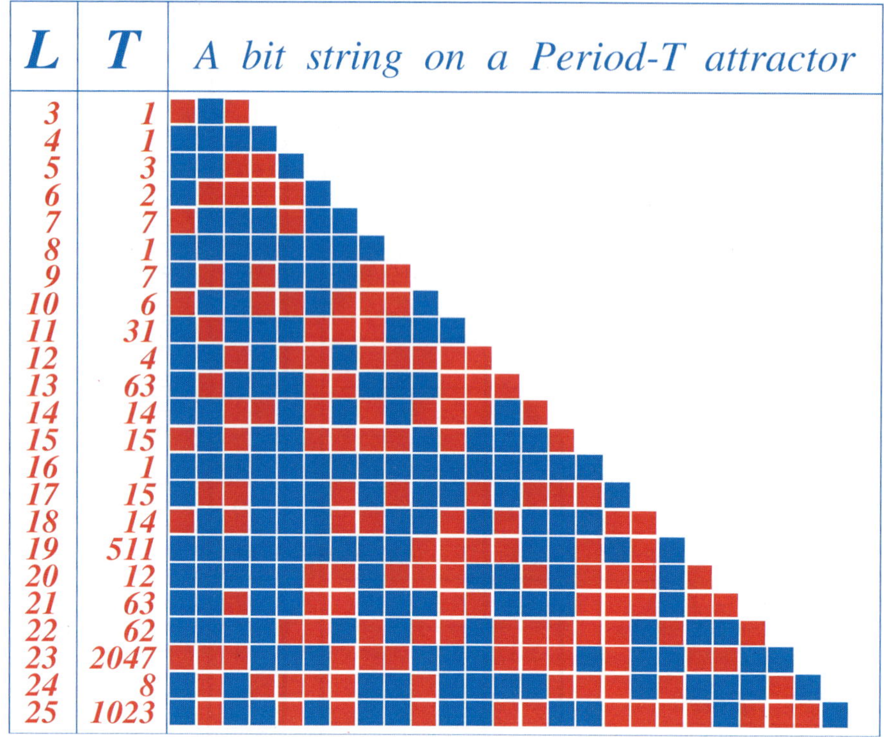

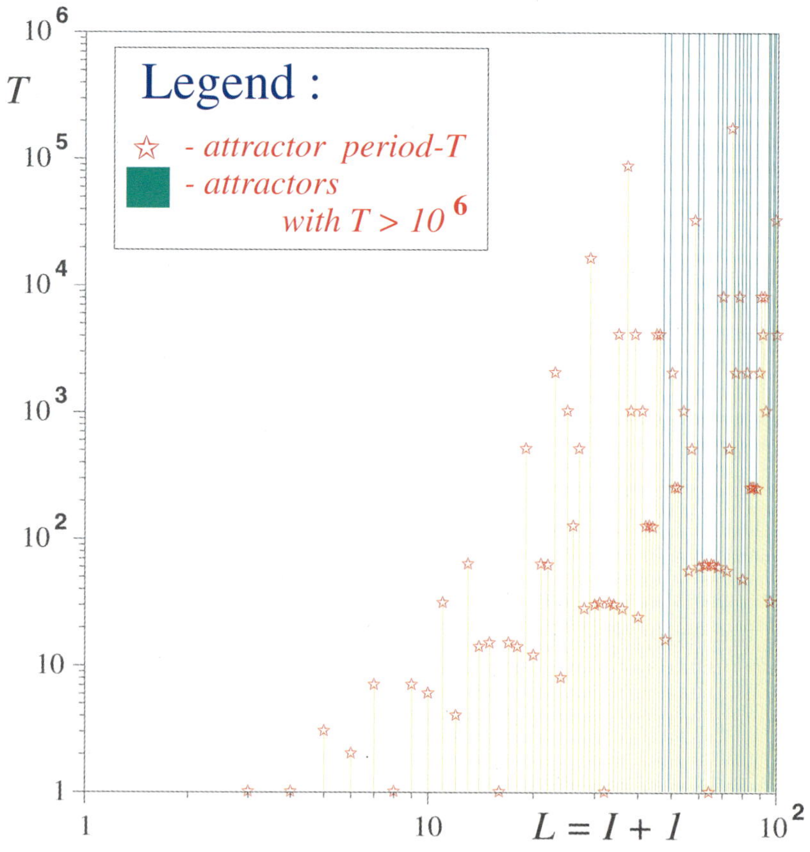

Fig. 3.   Dependence of the period "T" of attractor of rule $\boxed{90}$ as a function of L (in logarithmic scale).

Table 27. The upper bound $T_{\max}$ of the period "T" as function of L for $3 \leq L \leq 85$.

L	$T_{max} = 2^L$
3	8
4	16
5	32
6	64
7	128
8	256
9	512
10	1024
11	2048
12	4096
13	8192
14	16384
15	32768
16	65536
⋮	⋮
85	38685626227668133590597632

(b) Assume Eq. (24) is true for $n = m$ (*induction hypothesis*); namely,

$$x_i^m = \sum_{k=0}^{m} \frac{m!}{k!(m-k)!} \bullet x_{i-m+2k}^0 \quad \text{mod } (2) \quad (26)$$

We must show that incrementing "m" to "$m+1$" in Eq. (26) gives Eq. (24) with $n = m+1$.

Substituting Eq. (26) to Eq. (19), we obtain

$$x_i^{m+1} = x_{i-1}^m + x_{i+1}^m \quad \text{mod } (2)$$

$$= \sum_{k=0}^{m} \frac{m!}{k!(m-k)!} x_{(i-1)-m+2k}^0$$

$$+ \sum_{k=0}^{m} \frac{m!}{k!(m-k)!} x_{(i+1)-m+2k}^0 \quad \text{mod } (2)$$

$$(27)$$

Changing symbol "m" on the right-hand side of Eq. (27) to $m'-1$ gives

$$\sum_{k=0}^{m'-1} \frac{(m'-1)!}{k!(m'-1-k)!} x_{i-m'+2k}^0$$

$$+ \sum_{k=0}^{m'-1} \frac{(m'-1)!}{k!(m'-1-k)!} x_{i-m'+2k+2}^0 \quad \text{mod } (2)$$

$$(28)$$

Changing symbol k in the second summation terms in Eq. (28) to $k'-1$ gives

$$\sum_{k=0}^{m'-1} \frac{(m'-1)!}{k!(m'-1-k)!} x_{i-m'+2k}^0$$

$$+ \sum_{k'=1}^{m'} \frac{(m'-1)!}{(k'-1)!(m'-k')!} x_{i-m'+2k'}^0 \quad (29)$$

Changing the dummy index k' in Eq. (29) back to k, we obtain

$$\left[\sum_{k=0}^{m'-1} \frac{(m'-1)!}{k!(m'-1-k)!} + \sum_{k=1}^{m'} \frac{(m'-1)!}{(k-1)!(m'-k)!} \right]$$

$$\bullet \, x_{i-m'+2k}^0 \quad (30)$$

The terms inside the bracket can be simplified by observing for $k = 1$ to $m'-1$, we have

$$\frac{(m'-1)!}{k!(m'-1-k)!} + \frac{(m'-1)!}{(k-1)!(m'-k)!}$$

$$= \frac{(m'-1)!}{(k-1)!(m'-1-k)!} \left[\frac{1}{k} + \frac{1}{(m'-k)} \right]$$

$$= \frac{(m'-1)!}{(k-1)!(m'-1-k)!} \left[\frac{(m'-k)+k}{k(m'-k)} \right]$$

$$= \frac{m'!}{k!(m'-k)!} \quad (31)$$

Moreover, when $k = 0$ and $k = m'$, Eq. (31) gives the same value as the first term on the left of Eq. (30), and the last term on the right of Eq. (30), respectively. Substituting back $m = m'-1$ in Eq. (31), and making use of Eqs. (27)–(31), we obtain

$$x_i^{m+1} = \sum_{k=0}^{m+1} \frac{(m+1)!}{k!(m+1-k)!} \bullet x_{i-(m+1)+2k}^0$$

$$\text{mod } (2) \quad (32)$$

which is identical to incrementing m in the *induction hypothesis* (26) to $m+1$. ∎

Table 29 gives the global state-transition formula (24) of rule $\boxed{90}$ for $n = 1, 2, 3, 4$ and 5. Observe that the coefficients $\binom{n}{k}$ for each time $n \geq 1$ is identical to the binomial coefficients in the expansion of $(x+y)^n$, as listed in Table 30 for $n = 1, 2, \ldots, 11$. These binomial coefficients are repackaged in Table 31 into the form of a Pascal's triangle where each coefficient under the pyramid is obtained by adding adjacent left and right coefficients above it.

Table 28.   The period of the following isle of Eden for rule $\boxed{45}$ is $T = 504$, which is equal almost to $T_{\max} = 512$.

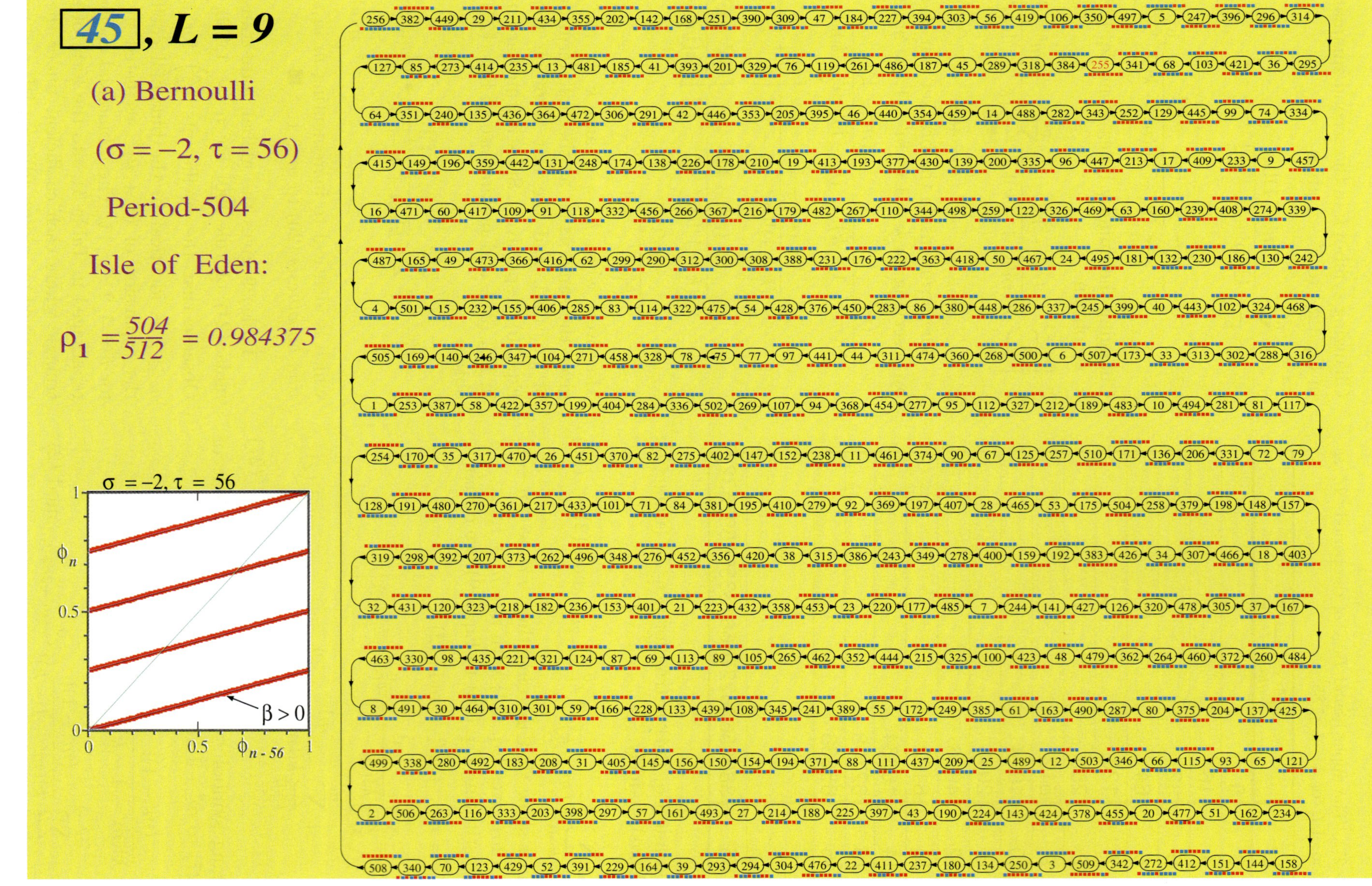

Table 29.   Global state-transition formula for rule $\boxed{90}$ for $1 \leq n \leq 5$.

n	$x_i^n = \sum\limits_{k=0}^{n} \dfrac{n!}{k!(n-k)!} \cdot x_{i-n+2k}^0 \quad \mathbf{mod(2)}$
1	$x_i^1 = x_{i-1}^0 + x_{i+1}^0 \quad \mathbf{mod(2)}$
2	$x_i^2 = x_{i-2}^0 + 2x_i^0 + x_{i+2}^0 \quad \mathbf{mod(2)}$
3	$x_i^3 = x_{i-3}^0 + 3x_{i-1}^0 + 3x_{i+1}^0 + x_{i+3}^0 \quad \mathbf{mod(2)}$
4	$x_i^4 = x_{i-4}^0 + 4x_{i-2}^0 + 6x_i^0 + 4x_{i+2}^0 + x_{i+4}^0 \quad \mathbf{mod(2)}$
5	$x_i^5 = x_{i-5}^0 + 5x_{i-3}^0 + 10x_{i-1}^0 + 10x_{i+1}^0 + 5x_{i+3}^0 + x_{i+5}^0 \quad \mathbf{mod(2)}$
$\cdots$	$\cdots$

Table 30.   Table of $\binom{n}{k} \triangleq n!/k!\,(n-k)!$, $n = 1, 2, \ldots, 11$, $k = 0, 1, 2, \ldots, 11$.

n	k											
	0	1	2	3	4	5	6	7	8	9	10	11
1	1	1										
2	1	2	1									
3	1	3	3	1								
4	1	4	6	4	1							
5	1	5	10	10	5	1						
6	1	6	15	20	15	6	1					
7	1	7	21	35	35	21	7	1				
8	1	8	28	56	70	56	28	8	1			
9	1	9	36	84	126	126	84	36	9	1		
10	1	10	45	120	210	252	210	120	45	10	1	
11	1	11	55	165	330	462	462	330	165	55	11	1

Table 31.   Binomial coefficients $\binom{n}{k}$ repackaged into a Pascal's triangle.

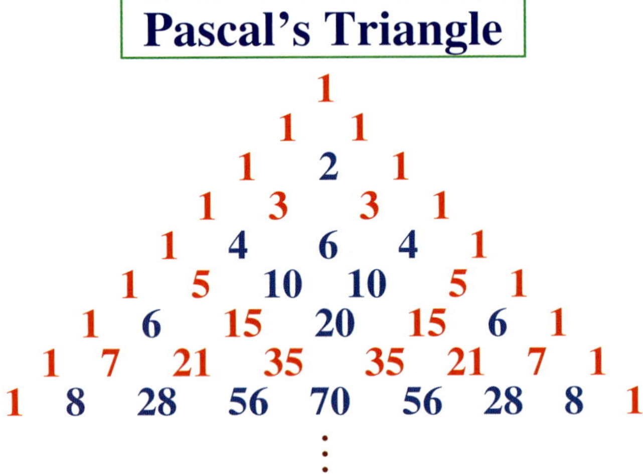

Taking the "*mod 2 equivalent*" of each coefficient in Table 29, we obtain the more compact but equivalent expansion in Table 32 where all nonzero terms correspond to those in Table 29 with "*odd number*" coefficients. The equivalent mod 2 coefficients are repackaged in Table 33. Observe that Table 33 can be obtained from Table 31 by replacing each odd (respectively, even) coefficient in Table 31 by a *one* (respectively, a *zero*). If we fill in the missing slot in each row of the mod 2 Pascal's triangle, we would obtain the pyramidal "fractal" space-time pattern of rule $\boxed{90}$ in Table 34, which is identical

to that shown in the bottom of Fig. 2, where the initial configuration consists of a single red bit at the center, as in [Wolfram, 2002].

Example 1.   Table 35 shows the *space-time pattern* obtained from the *global* state-transition formula of rule $\boxed{90}$ in (a) when the initial configuration consists of a single red bit at the center. The corresponding pattern obtained from the *local* state-transition formula is shown in (b). They are identical, as expected. The minor differences in the graphics and color are due to the differences in the softwares used to generate these patterns.

Example 2.   Table 36 shows the corresponding results when the initial configuration consists of a string of *random* bits.

3.4.   *Periodicity constraints of rule* $\boxed{90}$

Theorem 1 implies that all bit strings of rule $\boxed{90}$ must converge to a period-T attractor, where $T \leq T_{\max} \leq 2^L$. We will prove in this subsection that for finite length $L \triangleq I + 1$, the period T must satisfy certain constraints. Such *periodicity constraints* are useful on many occasions, such as verifying whether certain periodic orbit can exist, or to generate new periodic orbits, etc. The proof of many of these results depends on the following easily

Table 32.   Compact global state-transition formula for rule $\boxed{90}$ for $1 \leq n \leq 5$.

n	$x_i^n = \sum_{k=0}^{n} \dfrac{n!}{k!(n-k)!} \cdot x_{i-n+2k}^0 \quad \mathbf{mod(2)}$
1	$x_i^1 = x_{i-1}^0 + x_{i+1}^0 \quad \mathbf{mod(2)}$
2	$x_i^2 = x_{i-2}^0 \qquad\qquad + x_{i+2}^0 \quad \mathbf{mod(2)}$
3	$x_i^3 = x_{i-3}^0 + x_{i-1}^0 + x_{i+1}^0 + x_{i+3}^0 \quad \mathbf{mod(2)}$
4	$x_i^4 = x_{i-4}^0 \qquad\qquad\qquad + x_{i+4}^0 \quad \mathbf{mod(2)}$
5	$x_i^5 = x_{i-5}^0 + x_{i-3}^0 \qquad\qquad + x_{i+3}^0 + x_{i+5}^0 \quad \mathbf{mod(2)}$
$\cdots$	$\cdots$

Table 33.   Mod 2 binomial coefficients $\binom{n}{k}$ repackaged into a mod 2 Pascal's triangle.

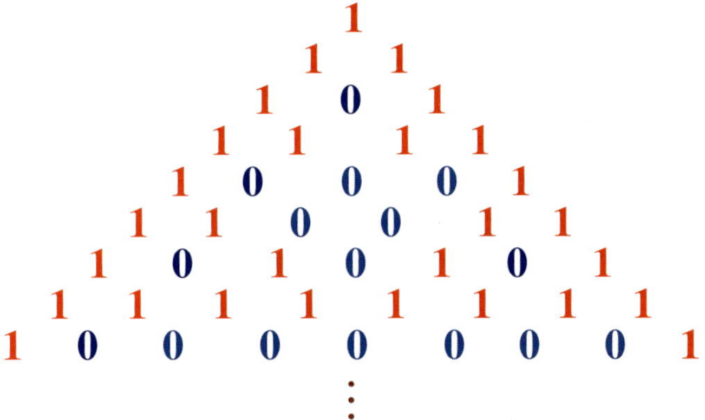

verifiable identities:

Binomial Coefficient Lemma. *If $n = 2^m$, where $m \geq 2$, then the following identities hold:*

(i) $\binom{n-1}{k} = 1 \bmod (2), \quad$ *for $k = 0, 1, 2, \ldots, n-1$*

$$(33)$$

(ii) $\binom{n}{k} = \begin{cases} 0 \bmod (2), & \textit{for } k = 1, 2, \ldots, n-1 \\ 1 \bmod (2), & \textit{for } k = 0, n \end{cases}$

$$(34)$$

(iii) $\binom{n+1}{k} = \begin{cases} 0 \bmod (2), & \textit{for } k = 2, 3, \ldots, n-1 \\ 1 \bmod (2), & \textit{for } k = 0, 1, n, n+1 \end{cases}$

$$(35)$$

where

$$\binom{n}{k} \triangleq \frac{n!}{k!(n-k)!} \tag{36}$$

Theorem 3. Periodicity Condition: $L = 2^m$.
  For $L = 2^m$, $m = 2, 3, 4, \ldots$, rule $\boxed{90}$ has a global period-1 attractor Γ; namely,

$$\boldsymbol{x}(\Gamma) = \underbrace{(0 \quad 0 \quad 0 \quad \cdots \quad 0)}_{L = 2^m} \tag{37}$$

All bit strings not belonging to the attractor Γ converge to Γ in at most 2^{m-1} iterations.

Proof.   Let $n = 2^{m-1}$ in the global state-transition formula (24). It follows from Eqs. (33) and (36) that

$$\frac{n!}{k!(n-k)!} \bmod (2)$$

$$= \begin{cases} 0, & \textit{for } k = 1, 2, \ldots, n-1 \\ 1, & \textit{for } k = 0, n = 2^{m-1} \end{cases} \tag{38}$$

It follows from Eq. (38) and the global state-transition formula (24) that x_i^n contains only two nonzero terms; namely, the leftmost and the

Table 34.   Space-time pattern of the Pascal triangle fractal generated by rule $\boxed{90}$.

n	-8	-7	-6	-5	-4	-3	-2	-1	0	1	2	3	4	5	6	7	8
1	0	0	0	0	0	0	0	1	0	1	0	0	0	0	0	0	0
2	0	0	0	0	0	0	1	0	0	0	1	0	0	0	0	0	0
3	0	0	0	0	0	1	0	1	0	1	0	1	0	0	0	0	0
4	0	0	0	0	1	0	0	0	0	0	0	0	1	0	0	0	0
5	0	0	0	1	0	1	0	0	0	0	0	1	0	1	0	0	0
6	0	0	1	0	0	0	1	0	0	0	1	0	0	0	1	0	0
7	0	1	0	1	0	1	0	1	0	1	0	1	0	1	0	1	0
8	1	0	0	0	0	0	0	0	0	0	0	0	0	0	0	0	1

Table 35.   Space-time pattern of the rule $\boxed{90}$ with red central bit initial configuration: (a) from global state-transition formula; (b) from local state-transition formula.

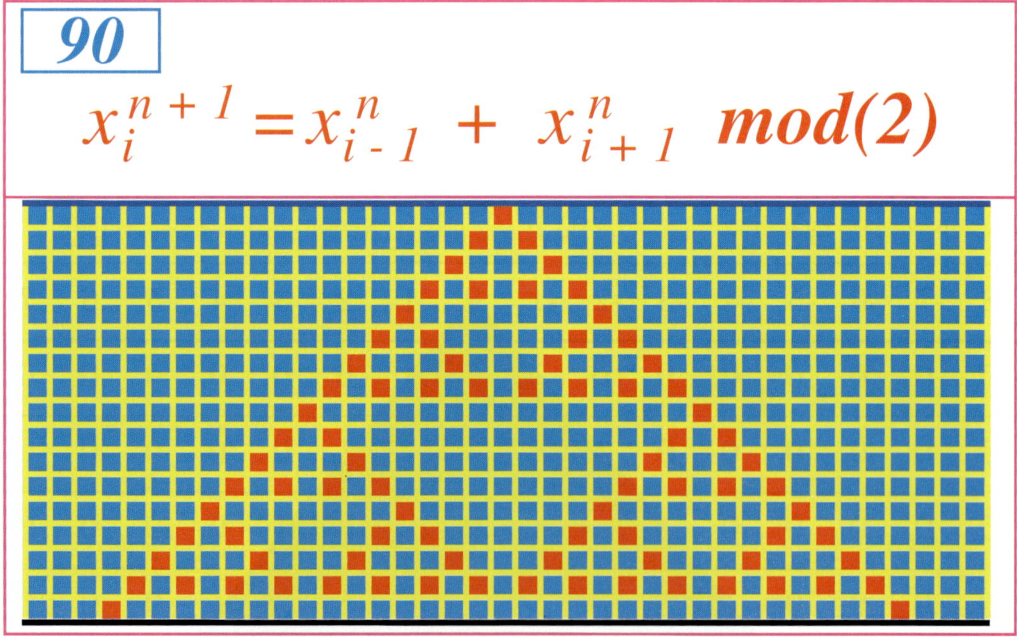

(a)

(b)

Table 36.   Space-time pattern of the rule $\boxed{90}$ obtained from random initial state generated by: (a) global state-transition formula; (b) local state-transition formula.

$$\boxed{90} \quad x_i^{\,n} = \sum_{k=0}^{n} \frac{n!}{k!(n-k)!}\, x_{i-n+2k}^{\,0} \ \ \textit{mod(2)}$$

(a)

$$\boxed{90} \quad x_i^{\,n+1} = x_{i-1}^{\,n} + x_{i+1}^{\,n} \ \ \textit{mod(2)}$$

(b)

rightmost terms. Hence,

$$x_i^n = x_{i-n}^0 + x_{i+n}^0 \quad \text{mod}\ (2) \tag{39}$$

where $n = 2^{m-1}$. Substituting $i = n = 2^{m-1}$ in Eq. (39), we obtain

$$
\begin{aligned}
x_i^n &= x_{n-n}^0 + x_{n+n}^0 \quad \text{mod}\ (2) \\
&= x_0^0 + x_{2n}^0 \quad \text{mod}\ (2) \\
&= x_0^0 + x_{2^m}^0 \quad \text{mod}\ (2) \\
&= x_0^0 + x_L^0 \quad \text{mod}\ (2) \\
&= 2x_0 \quad \text{mod}\ (2) \\
&= 0 \tag{40}
\end{aligned}
$$

because $x_0^0 = x_L$.

Since x_i^0 is arbitrary, it follows that all bit strings must converge to Eq. (7) in at most 2^{m-1} iterations.   ■

Corollary to Theorem 2.
A bit string

$$\mathbf{x^0} = (x_0^0 \ \ x_1^0 \ \ x_2^0 \ \ \cdots \ \ x_{L-1}^0) \tag{41}$$

of length $L = I + 1$ *(under* periodic boundary condition) *is a* period-n attractor *of local rule* $\boxed{90}$ if, and only *if, the* periodicity condition

$$
\boxed{
\begin{aligned}
x_{i\,\text{mod}(L)}^n &= x_i^0 \\
&= \sum_{k=0}^{n} \frac{n!}{k!(n-k)!} \\
&\bullet\, x_{((i-n+2k)\,\text{mod}(L))}^0 \quad \text{mod}\ (2)
\end{aligned}
}
\tag{42}
$$

is satisfied *for all i.*

Proof. Follows directly from *Theorem 2* and the periodic boundary condition.   ■

The *periodicity constraint* equation (42) is applicable to *any* period-n attractor of rule $\boxed{90}$.

The "**mod** (L)" operation attached to the subscript index of x^0 is just a mathematically precise algorithm for implementing the *periodic boundary conditions*. It is also mathematically equivalent to *concatenating* replicas of the L-bit string $x_0\ x_1\ x_2\ \cdots\ x_I$ *ad infinitum*; namely,

$$\underbrace{\cdots x_0 \quad x_1 \quad x_2 \quad \cdots \quad x_I}_{L \text{ bits}} \underbrace{x_0 \quad x_1 \quad x_2 \quad \cdots \quad x_I}_{L \text{ bits}}$$
$$\underbrace{x_0 \quad x_1 \quad x_2 \quad \cdots \quad x_I}_{L \text{ bits}} \cdots \tag{43}$$

where $L = I + 1$.

In the special case where

$$n = 2^m - 1 \tag{44}$$

all binomial coefficients in Eq. (42) are equal to *unity*, in view of the Binomial Coefficient Lemma; namely,

$$\boxed{\frac{n!}{k!(n-k)!}\ \text{mod}\ (2) = 1, \quad k = 0, 1, 2, \ldots, n} \tag{45}$$

Equation (45) is obtained by substituting $n + 1 = 2^m$ from Eq. (44) in place of n in Eq. (33):

$$\binom{(n+1)-1}{k} = 1\ \text{mod}\ (2), \quad k = 0, 1, 2, \ldots, n \tag{46}$$

Substituting Eq. (46) into Eq. (42), we obtain the following simplified periodicity constraint:

$$\text{Valid if}\ n = 2^m - 1 \ \boxed{\begin{aligned} x^n_{i\,\text{mod}(L)} &= x^0_i \\ &= \sum_{k=0}^{2^m-1} x^0_{((i-n+2k)\,\text{mod}(L))}\ \text{mod}\ (2) \\ &\qquad\qquad\qquad\text{for all } i \end{aligned}} \tag{47}$$

If we impose the additional constraint $L = n = 2^m - 1$, then we obtain the following simple method for finding period-($2^m - 1$) attractors:

Theorem 4. Periodicity Condition: $L = 2^m - 1$. *Rule* $\boxed{90}$ *has a* period-n attractor *where* $n = 2^m - 1$ and $L = 2^m - 1$ if, and only if,

$$\text{Valid for}\ \begin{aligned} n &= 2^m - 1 \\ L &= 2^m - 1 \end{aligned}\ \boxed{\sum_{i=0}^{L-1} x^0_i\ \text{mod}\ (2) = 0} \tag{48}$$

Proof.   Let us list all terms from Eq. (47) as follows:

$$\begin{aligned} x^0_{i\,\text{mod}(L)} &= x^0_{((i-n)\,\text{mod}(L))} + x^0_{((i-n+2)\,\text{mod}(L))} \\ &\quad + \cdots + x^0_{((i+n-2)\,\text{mod}(L))} \\ &\quad + x^0_{((i+n)\,\text{mod}(L))}\quad \text{mod}\ (2) \end{aligned} \tag{49}$$

Since "i" is an arbitrary index in Eq. (49), let it be "n". Substituting $i = n$ in Eq. (49), we obtain

$$x^0_0 = x^0_0 + x^0_2 + \cdots + \underbrace{x^0_{n-1}}_{n-1 = 2^m-2 < L} + x^0_{((n+1)\,\text{mod}(L))}$$
$$+ \cdots + x^0_{((2n-2)\,\text{mod}(L))} + x^0_{((2n)\,\text{mod}(L))}$$
$$\text{mod}\ (2) \tag{50}$$

Observe next that for $n = L = 2^m - 1$, we have

$$(2n)\ \text{mod}\ (L) = 0, \quad ((2n-2)\ \text{mod}\ (L)) = L - 2,$$
$$((n+1)\ \text{mod}\ (L)) = 1.$$

Observe also that $L = 2^m - 1$ implies that $L - 2$, $L - 4$, etc. are *odd* numbers. Substituting these mod (L) equivalent indices into Eq. (50), we obtain

$$x^0_0 = x^0_0 + x^0_2 + \cdots + x^0_{L-1} + x^0_1 + x^0_3$$
$$+ \cdots + x^0_{L-2} + x^0_0\quad \text{mod}\ (2) \tag{51}$$

Observe that whereas the first x^0_0 on the right-hand side of Eq. (51) comes from the corresponding first term of Eq. (50), the last x^0_0 of Eq. (51) comes from the last bit $x^0_{((2n)\,\text{mod}(L))} = x^0_0$ of Eq. (50). Rearranging the terms in increasing subscript order in Eq. (51), we obtain

$$x^0_0 = x^0_0 + x^0_0 + x^0_1 + x^0_2 + \cdots + x^0_{L-2}$$
$$+ x^0_{L-1}\ \text{mod}\ (2) \tag{52}$$

Substituting $(x^0_0 + x^0_0)\ \text{mod}\ (2) = 0$ in Eq. (52), we obtain

$$x^0_0 = x^0_1 + x^0_2 + x^0_3 + \cdots + x^0_{L-2}$$
$$+ x^0_{L-1}\ \text{mod}\ (2) \tag{53}$$

By adding the bit x^0_0 to both sides of Eq. (53), we obtain

$$\underbrace{x^0_0 + x^0_0}_{0\,\text{mod}(2)} = x^0_0 + x^0_1 + x^0_2 + \cdots + x^0_{L-2}$$
$$+ x^0_{L-1}\ \text{mod}\ (2) \tag{54}$$

It follows from Eq. (54) that

$$\sum_{i=0}^{L-1} x^0_i\ \text{mod}\ (2) = 0 \tag{55}$$

$\blacksquare$

Our next theorem shows that *the same* periodicity condition in Eq. (45) of Theorem 4 also holds for a *different* $L = 2^m + 1$ but for the same $n = 2^m - 1$.

Theorem 5. Periodicity Condition: $L = 2^m + 1$. *Rule* $\boxed{90}$ *has a* period-n attractor where $n = 2^m - 1$ and $L = 2^m + 1$ if, and only if,

Valid for
$$n = 2^m - 1 \qquad \boxed{\sum_{i=0}^{L-1} x_i^0 \bmod (2) = 0} \qquad (56)$$
$$L = 2^m + 1$$

Proof. Since $n = 2^m - 1$ remains the same as in Theorem 4, Eqs. (47) and (49) remain unchanged. Substituting $i = n$ in Eq. (49), we obtain[12]

$$x_0^0 = x_0^0 + x_2^0 + \cdots + x_{n-1}^0 + \underbrace{x_{n+1}^0}_{n+1 = 2^m < L}$$
$$+ x_{((n+3)\bmod(L))}^0 + \cdots$$
$$+ x_{((2n-2)\bmod(L))}^0$$
$$+ x_{((2n)\bmod(L))}^0 \quad \bmod (2) \qquad (57)$$

Observe next that for $L = 2^m + 1$, we have $L = n + 2$. Hence, unlike in Eq. (51) we must now replace L in mod (L) by $n + 2$ to obtain $(2n \bmod (n+2)) = n - 2$, $((2n-2) \bmod (n+2)) = n - 4, \ldots, ((n+3) \bmod (n+2)) = 1$. Equation (57) now reduces to

$$x_0^0 = x_0^0 + x_2^0 + \cdots + x_{n-1}^0 + x_{n+1}^0 + x_1^0 + x_3^0$$
$$+ \cdots + x_{n-4}^0 + x_{n-2}^0 \bmod (2)$$
$$= x_0^0 + x_1^0 + x_2^0 + x_3^0 + \cdots + x_{n-2}^0 + x_{n-1}^0$$
$$+ x_{n+1}^0 \bmod (2) \qquad (58)$$

Observe that the term x_n^0 is missing from Eq. (58). Replacing n by $n = L - 2$ in Eq. (58) and by choosing $i = L - 2$ in Eq. (47), we obtain

$$x_{L-2}^0 = x_0^0 + x_1^0 + x_2^0 + \cdots + x_{L-4}^0$$
$$+ x_{L-3}^0 + x_{L-1}^0 \bmod (2) \qquad (59)$$

Adding the term $x_{L-2}^0 = x_{I-1}^0$ to both sides of Eq. (59), we obtain

$$\underbrace{x_{L-2}^0 + x_{L-2}^0}_{0\,\bmod(2)} = x_0^0 + x_1^0 + x_2^0 + \cdots + x_{L-4}^0 + x_{L-3}^0$$
$$+ x_{L-2}^0 + x_{L-1}^0 \bmod (2) \qquad (60)$$

Hence, we have

$$\sum_{i=0}^{L-1} x_i^0 \bmod (2) = 0$$

$\blacksquare$

Corollary 1 to Theorems 4 and 5. *The total number of red pixels in the period* $2^m - 1$ *attractors of Theorems 4 and 5, must be an* even *number.*

Corollary 2 to Theorem 3, 4, 5. *Theorems 3–5 hold also for* infinite *bit strings* $(L \to \infty)$.

Recap. Theorems 3–5 give necessary and sufficient conditions for rule $\boxed{90}$ to have the following *period-n attractors*:

Theorem 3
$\quad n = 1 \quad$ and $\quad L = 2^m$
Theorem 4
$\quad n = 2^m - 1 \quad$ and $\quad L = 2^m - 1$
Theorem 5
$\quad n = 2^m - 1 \quad$ and $\quad L = 2^m + 1$

As illustrations of the applications of these *analytically* derived results, let us examine the basin tree diagrams exhibited in Tables 14–23.

Applications of Theorem 3

1. $m = 2$, $L = 2^m = 4$

Gallery 90-2 shows all bit strings converge to the unique global attractor $\textcircled{0}$, as predicted by Theorem 3.

2. $m = 3$, $L = 2^m = 8$

Gallery 90-9 shows all bit strings converge to the global attractor $\textcircled{0}$, as predicted by Theorem 3.

Applications of Theorem 4

$$m = 3, \quad n = 2^m - 1 = 7, \quad L = 2^m - 1 = 7$$

Galleries 90-6, 90-7 and 90-8 show nine period-7 attractors as predicted. Observe the number of red pixels in all attractors is an even number, as predicted. The only other attractor is a *period-1 attractor*, $\textcircled{0}$, which qualifies also as a period-7 attractor, with "0" red pixels, an even number, as predicted.

Applications of Theorem 5

$$m = 2, \quad n = 2^m - 1 = 3, \quad L = 2^m + 1 = 5$$

Gallery 90-3 shows five period-3 attractors, all have only orbits with an even number of red pixels. The only other attractor is a *period-1 attractor*, $\textcircled{0}$, which qualifies also as a period-3 attractor. In this case, there are no red bits, which is an *even number*, as predicted.

---

[12]Observe that unlike Eq. (50) where $n - 1 = 2^m - 2 < L$, we now have $n + 1 = 2^m < 1$.

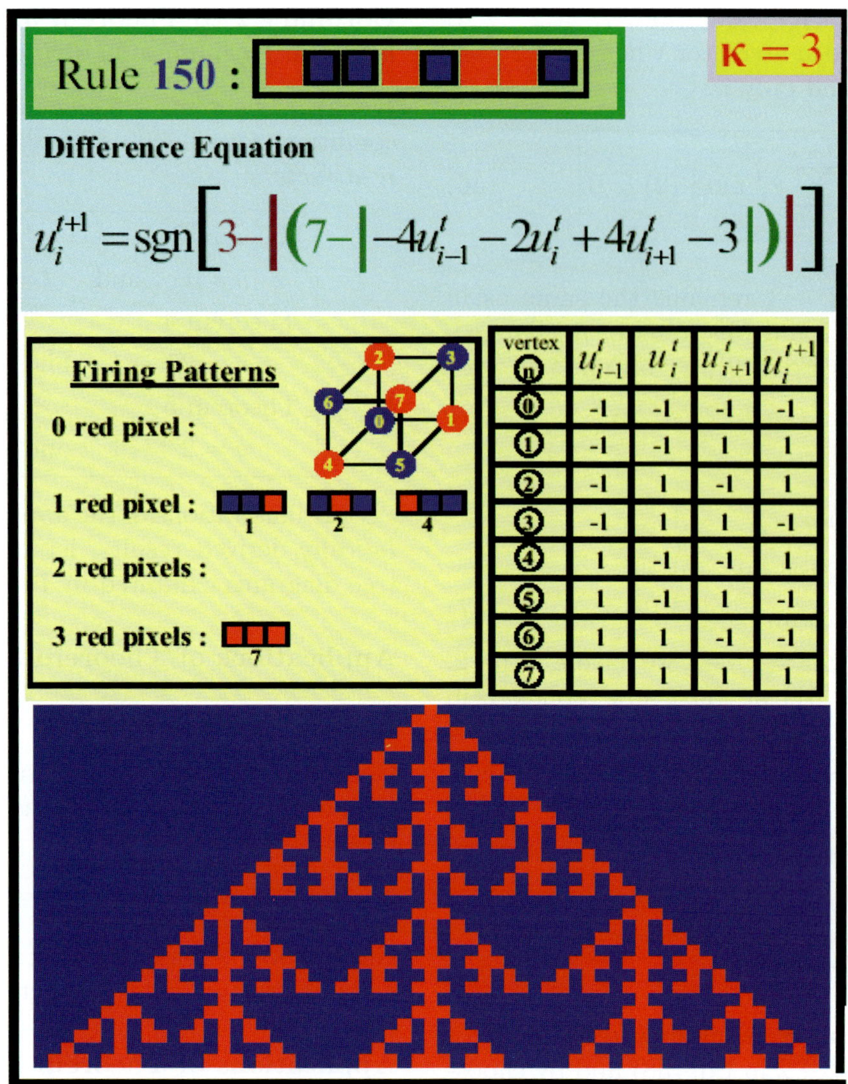

$$u_i^{t+1} = \text{sgn}\left[3 - \left|\left(7 - \left|-4u_{i-1}^t - 2u_i^t + 4u_{i+1}^t - 3\right|\right)\right|\right]$$

vertex (n)	u_{i-1}^t	u_i^t	u_{i+1}^t	u_i^{t+1}
0	-1	-1	-1	-1
1	-1	-1	1	1
2	-1	1	-1	1
3	-1	1	1	-1
4	1	-1	-1	1
5	1	-1	1	-1
6	1	1	-1	-1
7	1	1	1	1

Fig. 4. Truth table, Boolean cube, difference equation, and space-time pattern of local rule $\boxed{150}$.

4. Global Analysis of Local Rules $\boxed{150}$ and $\boxed{105}$

The *truth table*, *Boolean cube*, and "Difference Equation" defining the local rules $\boxed{150}$ and $\boxed{105}$ along with a space-time pattern (with a single red-pixel initial state) exhibited in Table 5 of [Chua *et al.*, 2003] is reproduced in Figs. 4 and 5, respectively, for the readers convenience. For this paper, it is more instructive to recast the Difference Equations defining $\boxed{150}$ and $\boxed{105}$ as follow:

$$\text{Rule } \boxed{150} \quad \boxed{x_i^{t+1} = x_{i-1}^t \oplus x_i^t \oplus x_{i+1}^t} \tag{61}$$

$$\text{Rule } \boxed{105} \quad \boxed{x_i^{t+1} = \overline{x_{i-1}^t \oplus x_i^t \oplus x_{i+1}^t}} \tag{62}$$

where $\oplus$ is the mod 2 sum defined in Table 24, and the bar on top denotes *complementation*. The alert reader will notice that we list $\boxed{150}$ ahead of $\boxed{105}$, counter to our style. This is done to avoid clutter where formula (61) for $\boxed{150}$ is clearly simpler than formula (62) for $\boxed{105}$. Yet, as will be shown in Sec. 4.3, these two rules are related via a global alternating transformation $\widetilde{T}$ where $\boxed{105} = \widetilde{T}(\boxed{150})$, and $\boxed{150} = \widetilde{T}^{-1}(\boxed{105})$, so that it suffices to study only $\boxed{150}$.[13]

---

[13]Unlike the *Vierergruppe* transformations $T^\dagger$, $\overline{T}$ and T^* in [Chua *et al.*, 2004], which are defined for *all* 256 rules, the *alternating transformation* in Sec. 4.2 is defined only for some rules, such as rules $\boxed{150}$ and $\boxed{105}$.

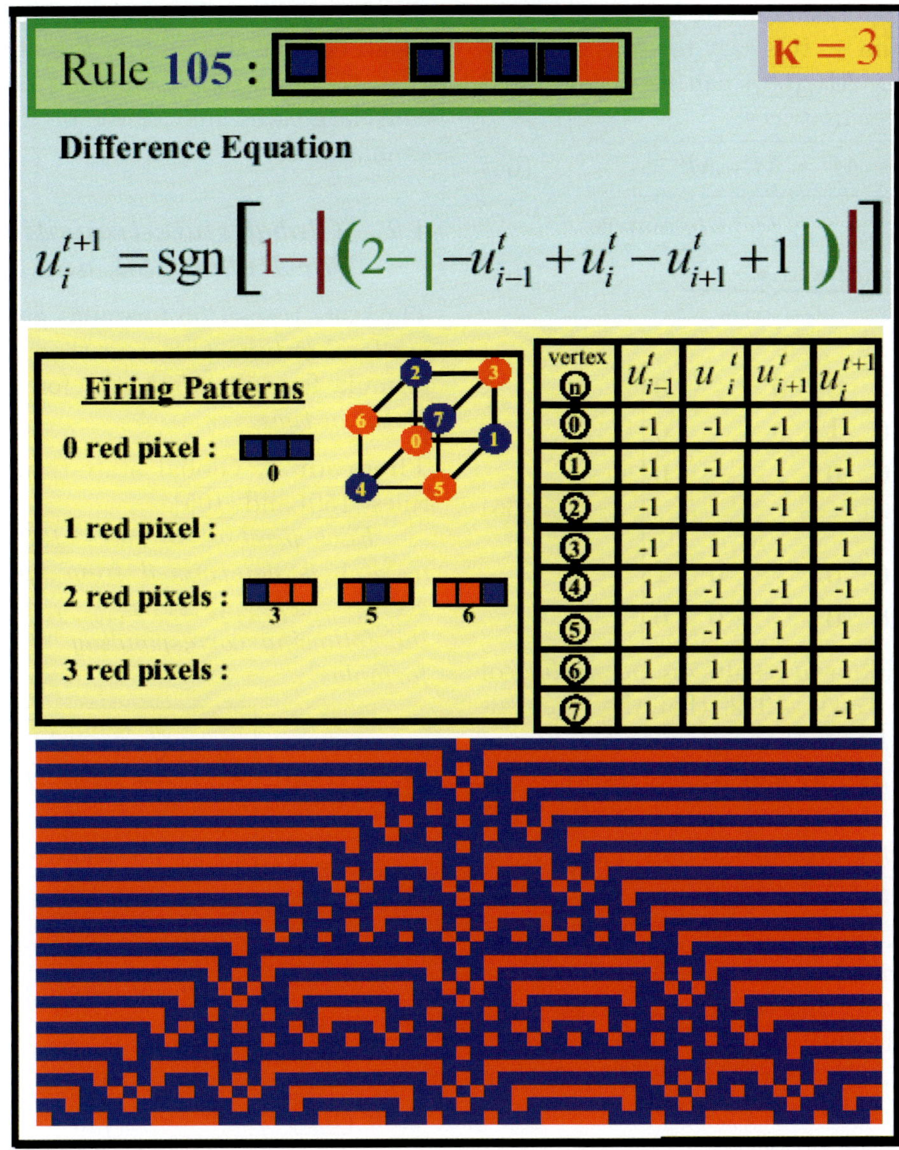

vertex (n)	u_{i-1}^t	u_i^t	u_{i+1}^t	u_i^{t+1}
⓪	-1	-1	-1	1
①	-1	-1	1	-1
②	-1	1	-1	-1
③	-1	1	1	1
④	1	-1	-1	-1
⑤	1	-1	1	1
⑥	1	1	-1	1
⑦	1	1	1	-1

Fig. 5. Truth table, Boolean cube, difference equation, and space-time pattern of local rule $\boxed{105}$.

4.1. Rules $\boxed{150}$ and $\boxed{105}$ are composed of Isles of Eden if L is not divisible by 3

Theorem 6. *Every bit string of rules* $\boxed{150}$ *and* $\boxed{105}$ *is an Isle of Eden if, and only if, $L/3$ is not an Integer.*

Proof. We will present proof only for rule $\boxed{150}$ since rules $\boxed{150}$ and $\boxed{105}$ are globally quasi-equivalent via the *alternating* transformation $\widetilde{T}$ to be defined in Sec. 4.3.

Just like rule $\boxed{90}$, each arbitrary bit string $\boldsymbol{x}^t$ at time t maps into $\boldsymbol{x}^{t+1}$ at time $t+1$ of rule $\boxed{150}$ via an $L \times L$ *circulant* local time-1 state transition matrix; namely,

$$
\underbrace{\begin{bmatrix} x_0^{t+1} \\ x_1^{t+1} \\ x_2^{t+1} \\ \vdots \\ x_{L-2}^{t+1} \\ x_{L-1}^{t+1} \end{bmatrix}}_{\boldsymbol{x}^{t+1}}
=
\underbrace{\begin{bmatrix}
1 & 1 & 0 & 0 & 0 & \cdots & 1 \\
1 & 1 & 1 & 0 & 0 & \cdots & 0 \\
0 & 1 & 1 & 1 & 0 & \cdots & 0 \\
\vdots & & & & \ddots & & \vdots \\
0 & 0 & 0 & \cdots & 1 & 1 & 1 \\
1 & 0 & 0 & \cdots & 0 & 1 & 1
\end{bmatrix}}_{M(\boxed{150})}
\underbrace{\begin{bmatrix} x_0^t \\ x_1^t \\ x_2^t \\ \vdots \\ x_{L-2}^t \\ x_{L-1}^t \end{bmatrix}}_{\boldsymbol{x}^t}
\quad \mathrm{mod}\ (2) \quad (63)
$$

Here, the *addition* operation in the matrix multiplication is *mod (2)* sum $\oplus$. It is easy to verify that the above $L \times L$ matrix $M(\boxed{150})$ can be decomposed into the sum of three matrices

$$\boxed{M(\boxed{150}) = M^0 + M + M^{L-1}} \tag{64}$$

where $M^0 = \mathbf{1}$ is an $L \times L$ *identity* matrix,

$$M^{L-1} = \underbrace{M \bullet M \bullet \cdots M}_{L-1 \text{ times}} \tag{65}$$

and

$$M \triangleq \begin{bmatrix} 0 & 1 & 0 & 0 & \cdots & 0 \\ 0 & 0 & 1 & 0 & \cdots & 0 \\ 0 & 0 & 0 & 1 & \cdots & 0 \\ \vdots & & & & & \vdots \\ 0 & 0 & 0 & \cdots & 0 & 1 \\ 1 & 0 & 0 & \cdots & 0 & 0 \end{bmatrix} \tag{66}$$

It follows from well-known results on *eigenvalues* of *circulant* matrices [Davis, 1979] that the eigenvalues of $M(\boxed{150})$ in Eq. (64) are given by the sum of the eigenvalues of M^0, M and M^{L-1}, respectively; namely,

$$\lambda_k = 1 + \exp\left(2\pi i \frac{k}{L}\right) + \exp\left(2\pi i (L-1) \frac{k}{L}\right) \tag{67}$$

where $k = 0, 1, \ldots, L-1$.

It is easily verified that if $L \equiv 0 \mod (3)$, i.e. if L is divisible by 3, then there must exist some $k \in \{1, 2, \ldots, L-1\}$ such that $\lambda_k = 0$. However, if $L \equiv 1, 2 \mod (3)$, i.e. if L is *not* divisible by 3, then for *any* $k \in \{0, 1, \ldots, L-1\}$, $\lambda_k \neq 0$.

It follows from the property

$$\det M = \lambda_0 \cdot \lambda_1 \cdots \lambda_{L-1}$$

that if L is not divisible by 3, then $\det M(\boxed{150}) \neq 0$. This implies that every bit string in this case has a *unique* preimage, and hence is an *isle of Eden*.

On the other hand, if L is divisible by 3, then every bit string has a multiple preimage, implying that it *cannot* be an isle of Eden. Such a bit string must necessarily lie on a basin tree, or on an attractor. ∎

As illustrations of Theorem 6, Tables 37 and 38 exhibit a list of the period "T" of at least one period-T *isle of Eden* for $L \neq 0 \mod (3)$, and at least one period-T *attractor*, if $L \equiv 0 \mod (3)$, of local rule $\boxed{150}$ and $\boxed{105}$, respectively. Those "blank" rectangles in these two tables without any entry imply that the period of either an isle of Eden, or an attractor for the particular L is larger than

the threshold set by our program. Indeed, Table 28 suggests that the period T of this set of L's (e.g. $L = 47, 49, 53, 55, 67, 69$, etc.) could be astronomically large, and may never be found by brute-force simulations.

4.2. *Global state-transition formula for Rules $\boxed{150}$ and $\boxed{105}$*

The state transition formulas given in Figs. 4 and 5 are *local in time*. Our next theorem gives an explicit formula for rule $\boxed{150}$, and for rule $\boxed{105}$, which is *global in time*.

Theorem 7. Global state-transition Formula for rules $\boxed{150}$ and $\boxed{105}$.

Each pixel x_i^n of rules $\boxed{150}$ and $\boxed{105}$ at time $n > 1$ is determined from "$n + 1$" initial pixels x_{i-n}^0, x_{i-n+2}^0, $\ldots$, x_{i+n-2}^0, x_{i+n}^0 at $t = 0$ via the following corresponding "Composite Binomial formulas":

Rule $\boxed{150}$
$$\boxed{\begin{aligned} x_i^n &= \sum_{k=0}^{n} \frac{n!}{k!(n-k)!} \sum_{j=0}^{k} \frac{k!}{j!(k-j)!} \\ &\bullet x_{i-n+2k-j}^0 \mod (2) \end{aligned}} \tag{68}$$

Rule $\boxed{105}$
$$\boxed{\begin{aligned} x_i^n &= \alpha_n + (-1)^n \sum_{k=0}^{n} \frac{n!}{k!(n-k)!} \\ &\bullet \sum_{j=0}^{k} \frac{k!}{j!(k-j)!} \\ &\bullet x_{i-n+2k-j}^0 \mod (2) \\ \text{where} \\ \alpha_n &\triangleq \frac{1}{2}[1 - (-1)^n]. \end{aligned}} \tag{69}$$

Table 39 exhibits the detailed expansion of the global state transition formula for rule $\boxed{150}$ for $n = 1, 2, \ldots, 8$. The corresponding formula in *mod (2) coefficients* are shown in Table 40. These coefficients are repacked into a Pascal-like triangle in Table 41. Table 42 shows the space-time patterns generated from applying the global and local state transition formulas for rule $\boxed{150}$ when the initial configuration consists of a single red center pixel. Table 43 shows corresponding space-time patterns when a *random* initial configuration is used.

The corresponding illustrations for rule $\boxed{105}$ are shown in Tables 44–48, respectively.

Table 37.   Period "T" of Isles of Eden and attractors of local rule $\boxed{150}$.

L	Isle of Eden	Attractor	L	Isle of Eden	Attractor
1			26	T = 42	
2			27		T = 511
3		T = 1	28	T = 28	
4	T = 2		29	T = 16383	
5	T = 3		30		T = 15
6		T = 1	31	T = 31	
7	T = 7		32	T = 16	
8	T = 4		33		T = 31
9		T = 7	34	T = 30	
10	T = 6		35	T = 4095	
11	T = 31		36		T= 14
12		T = 1	37	T = 29127	
13	T = 21		38	T = 1022	
14	T = 14		39		T = 4095
15		T = 15	40	T = 24	
16	T = 8		41	T = 1023	
17	T = 15		42		T = 63
18		T = 7	43	T = 127	
19	T = 511		44	T = 124	
20	T = 12		45		T = 4095
21		T = 63	46	T = 4094	
22	T = 62		47		
23	T = 2047		48		T = 4
24		T = 2	49		
25	T = 1023		50	T = 2046	

Table 37.   (*Continued*)

L	Isle of Eden	Attractor	L	Isle of Eden	Attractor
51		$T = 255$	76	$T = 2044$	
52	$T = 84$		77		
53			78		$T = 4095$
54		$T = 511$	79		
55			80	$T = 48$	
56	$T = 56$		81		
57		$T = 511$	82	$T = 2046$	
58	$T = 32766$		83		
59			84		$T = 126$
60		$T = 30$	85	$T = 255$	
61			86	$T = 254$	
62	$T = 62$		87		
63		$T = 63$	88	$T = 248$	
64	$T = 32$		89	$T = 2047$	
65	$T = 63$		90		$T = 4095$
66		$T = 31$	91	$T = 4095$	
67			92	$T = 8188$	
68	$T = 60$		93		$T = 1023$
69			94		
70	$T = 8190$		95		
71			96		$T = 8$
72		$T = 28$	97		
73	$T = 511$		98		
74	$T = 58254$		99		$T = 32767$
75			100	$T = 4092$	

Table 37. (*Continued*)

L	Isle of Eden	Attractor	L	Isle of Eden	Attractor
101			126		$T = 63$
102		$T = 255$	127	$T = 127$	
103			128	$T = 64$	
104	$T = 168$		129		$T = 127$
105		$T = 4095$	130	$T = 126$	
106			131		
107			132		$T = 62$
108		$T = 1022$	133	$T = 262143$	
109	$T = 262143$		134		
110			135		
111			136	$T = 120$	
112	$T = 112$		137		
113	$T = 16383$		138		
114		$T = 511$	139		
115			140	$T = 16380$	
116	$T = 65532$		141		
117		$T = 4095$	142		
118			143		
119			144		$T = 56$
120		$T = 60$	145	$T = 16383$	
121			146	$T = 1022$	
122			147		
123			148	$T = 116508$	
124	$T = 124$		149		
125			150		

Table 37.   (*Continued*)

L	Isle of Eden	Attractor	L	Isle of Eden	Attractor
151	$T = 32767$		176	$T = 496$	
152	$T = 4088$		177		
153			178	$T = 4094$	
154			179		
155			180		$T = 8190$
156		$T = 8190$	181		
157			182	$T = 8190$	
158			183		
159			184	$T = 16376$	
160	$T = 96$		185	$T = 262143$	
161			186		$T = 1023$
162			187		
163			188		
164	$T = 4092$		189		
165			190		
166			191		
167			192		$T = 16$
168		$T = 252$	193		
169			194		
170	$T = 510$		195		$T = 4095$
171		$T = 511$	196		
172	$T = 508$		197		
173			198		$T = 32767$
174			199		
175			200	$T = 8184$	

Table 37. (*Continued*)

L	Isle of Eden	Attractor	L	Isle of Eden	Attractor
204		$T = 510$	273		$T = 4095$
205	$T = 1023$		280	$T = 32760$	
208	$T = 336$		288		$T = 112$
210		$T = 4095$	290	$T = 32766$	
216		$T = 2044$	292	$T = 2044$	
217	$T = 32767$		296	$T = 233016$	
218	$T = 524286$		302	$T = 65534$	
224	$T = 224$		304	$T = 8176$	
226	$T = 32766$		312		$T = 16380$
228		$T = 1022$	315		$T = 4095$
232	$T = 131064$		320	$T = 192$	
234		$T = 4095$	328	$T = 8184$	
240		$T = 120$	336		$T = 504$
241	$T = 4095$		340	$T = 1020$	
248	$T = 248$		341	$T = 1023$	
252		$T = 126$	342		$T = 511$
254	$T = 254$		344	$T = 1016$	
255		$T = 255$	352	$T = 992$	
256	$T = 128$		356	$T = 8188$	
257	$T = 255$		360		$T = 16380$
258		$T = 127$	364	$T = 16380$	
260	$T = 252$		368	$T = 32752$	
264		$T = 124$	370	$T = 524286$	
266	$T = 524286$		372		$T = 2046$
272	$T = 240$		384		$T = 32$

Table 37. (*Continued*)

L	Isle of Eden	Attractor	L	Isle of Eden	Attractor
390		*T = 4095*	**516**		*T = 254*
396		*T = 65534*	**520**	*T = 504*	
400	*T = 16368*		**528**		*T = 248*
408		*T = 1020*	**532**	*T = 1048572*	
410	*T = 2046*		**544**	*T = 480*	
416	*T = 672*		**546**		*T = 4095*
420		*T = 8190*	**560**	*T = 65520*	
432		*T = 4088*	**565**	*T = 16383*	
434	*T = 65534*		**576**		*T = 224*
436	*T = 1048572*		**580**	*T = 65532*	
448	*T = 448*		**584**	*T = 4088*	
452	*T = 65532*		**585**		*T = 4095*
455	*T = 4095*		**592**	*T = 466032*	
456		*T = 2044*	**604**	*T = 131068*	
464	*T = 262128*		**608**	*T = 16352*	
468		*T = 8190*	**624**		*T = 32760*
480		*T = 240*	**630**		*T = 4095*
482	*T = 8190*		**640**	*T = 384*	
496	*T = 496*		**656**	*T = 16368*	
504		*T = 252*	**672**		*T = 1088*
508	*T = 508*		**680**	*T = 2040*	
510		*T = 255*	**682**	*T = 2046*	
511	*T = 511*		**683**	*T = 2047*	
512	*T = 256*		**684**		*T = 1022*
513		*T = 511*	**688**	*T = 2032*	
514	*T = 510*		**704**	*T = 1984*	

Table 37.   (*Continued*)

L	Isle of Eden	Attractor
712	T = 16376	
720		T = 32760
728	T = 32760	
736	T = 65504	
740	T = 1048572	
744		T = 4092
768		T = 64
780		T = 8190
792		T = 131068
800	T = 32736	
816		T = 2040
819		T = 4095
820	T = 4092	
832	T = 1344	
840		T = 16380
864		T = 8176
868	T = 131068	
896	T = 896	
904	T = 131064	
910	T = 8190	
912		T = 4088
928	T = 524256	
936		T = 16380
960		T = 480
964	T = 16380	
992	T = 992	

Table 38.   Period "*T*" of Isles of Eden and attractors of local rule $\boxed{105}$.

L	Isle of Eden	Attractor	L	Isle of Eden	Attractor
1			26	$T = 42$	
2			27		$T = 1022$
3		$T = 2$	28	$T = 28$	
4	$T = 2$		29	$T = 32766$	
5	$T = 6$		30		$T = 30$
6		$T = 2$	31	$T = 62$	
7	$T = 14$		32	$T = 16$	
8	$T = 4$		33		$T = 62$
9		$T = 14$	34	$T = 30$	
10	$T = 6$		35	$T = 8190$	
11	$T = 62$		36		$T = 28$
12		$T = 2$	37	$T = 58254$	
13	$T = 42$		38	$T = 1022$	
14	$T = 14$		39		$T = 8190$
15		$T = 30$	40	$T = 24$	
16	$T = 8$		41	$T = 2046$	
17	$T = 30$		42		$T = 126$
18		$T = 14$	43	$T = 254$	
19	$T = 1022$		44	$T = 124$	
20	$T = 12$		45		$T = 8190$
21		$T = 126$	46	$T = 4094$	
22	$T = 62$		47		
23	$T = 4094$		48		$T = 8$
24		$T = 4$	49		
25	$T = 2046$		50	$T = 2046$	

Table 38. (*Continued*)

L	Isle of Eden	Attractor	L	Isle of Eden	Attractor
51		*T = 510*	**76**	*T = 2044*	
52	*T = 84*		**77**		
53			**78**		*T = 8190*
54		*T = 1022*	**79**		
55			**80**	*T = 48*	
56	*T = 56*		**81**		
57		*T = 1022*	**82**	*T = 2046*	
58	*T = 32766*		**83**		
59			**84**		*T = 252*
60		*T = 60*	**85**	*T = 510*	
61			**86**	*T = 254*	
62	*T = 62*		**87**		
63		*T = 126*	**88**	*T = 248*	
64	*T = 32*		**89**	*T = 4094*	
65	*T = 126*		**90**		*T = 8190*
66		*T = 62*	**91**	*T = 8190*	
67			**92**	*T = 8188*	
68	*T = 60*		**93**		*T = 2046*
69			**94**		
70	*T = 8190*		**95**		
71			**96**		*T = 16*
72		*T = 56*	**97**		
73	*T = 1022*		**98**		
74	*T = 58254*		**99**		*T = 65534*
75			**100**	*T = 4092*	

Table 38.   (*Continued*)

L	Isle of Eden	Attractor	L	Isle of Eden	Attractor
101			126		$T = 126$
102		$T = 510$	127	$T = 254$	
103			128	$T = 64$	
104	$T = 168$		129		$T = 254$
105		$T = 8190$	130	$T = 126$	
106			131		
107			132		$T = 124$
108		$T = 2044$	133	$T = 524286$	
109	$T = 524286$		134		
110			135		
111			136	$T = 120$	
112	$T = 112$		137		
113	$T = 32766$		138		
114		$T = 1022$	139		
115			140	$T = 16380$	
116	$T = 65532$		141		
117		$T = 8190$	142		
118			143		
119			144		$T = 112$
120		$T = 120$	145	$T = 32766$	
121			146	$T = 1022$	
122			147		
123			148	$T = 116508$	
124	$T = 124$		149		
125			150		

Table 38.   (*Continued*)

L	Isle of Eden	Attractor	L	Isle of Eden	Attractor
151	$T = 65534$		176	$T = 496$	
152	$T = 4088$		177		
153			178	$T = 4094$	
154			179		
155			180		$T = 16380$
156		$T = 16380$	181		
157			182	$T = 8190$	
158			183		
159			184	$T = 16376$	
160	$T = 96$		185	$T = 524286$	
161			186		$T = 2046$
162			187		
163			188		
164	$T = 4092$		189		
165			190		
166			191		
167			192		$T = 32$
168		$T = 504$	193		
169			194		
170	$T = 510$		195		$T = 8190$
171		$T = 1022$	196		
172	$T = 508$		197		
173			198		$T = 65534$
174			199		
175			200	$T = 8184$	

Table 38.   (*Continued*)

L	Isle of Eden	Attractor
204		T = 1020
205	T = 2046	
208	T = 336	
210		T = 8190
216		T = 4088
217	T = 65534	
218	T = 524286	
224	T = 224	
226	T = 32766	
228		T = 2044
232	T = 131064	
234		T = 8190
240		T = 240
241	T = 8190	
248	T = 248	
252		T = 252
254	T = 254	
255		T = 510
256	T = 128	
257	T = 510	
258		T = 254
260	T = 252	
264		T = 248
266	T = 524286	
272	T = 240	

L	Isle of Eden	Attractor
273		T = 8190
280	T = 32760	
288		T = 224
290	T = 32766	
292	T = 2044	
296	T = 233016	
302	T = 65534	
304	T = 8176	
312		T = 32760
315		T = 8190
320	T = 192	
328	T = 8184	
336		T = 1008
340	T = 1020	
341	T = 2046	
342		T = 1022
344	T = 1016	
352	T = 992	
356	T = 8188	
360		T = 32760
364	T = 16380	
368	T = 32752	
370	T = 524286	
372		T = 4096
384		T = 64

Table 38. (*Continued*)

L	Isle of Eden	Attractor
390		T = 8190
396		T = 131068
400	T = 16368	
408		T = 2040
410	T = 2046	
416	T = 672	
420		T = 16380
432		T = 8176
434	T = 65534	
436	T = 1048572	
448	T = 448	
452	T = 65532	
455	T = 8190	
456		T = 4088
464	T = 262128	
468		T = 16380
480		T = 480
482	T = 8190	
496	T = 496	
504		T = 504
508	T = 508	
510		T = 510
511	T = 1022	
512	T = 256	
513		T = 1022
514	T = 510	

L	Isle of Eden	Attractor
516		T = 508
520	T = 504	
528		T = 496
532	T = 1048572	
544	T = 480	
546		T = 8190
560	T = 65520	
565	T = 32766	
576		T = 448
580	T = 65532	
584	T = 4088	
585		T = 8190
592	T = 466032	
604	T = 131068	
608	T = 16352	
624		T = 65520
630		T = 8190
640	T = 384	
656	T = 16368	
672		T = 2016
680	T = 2040	
682	T = 2046	
683	T = 4094	
684		T = 2044
688	T = 2032	
704	T = 1984	

Table 38.   (*Continued*)

L	Isle of Eden	Attractor
712	$T = 16376$	
720		$T = 65520$
728	$T = 32760$	
736	$T = 65504$	
740	$T = 1048572$	
744		$T = 8184$
768		$T = 128$
780		$T = 16380$
792		$T = 262136$
800	$T = 32736$	
816		$T = 4080$
819		$T = 8190$
820	$T = 4092$	
832	$T = 1344$	
840		$T = 32760$
864		$T = 16352$
868	$T = 131068$	
896	$T = 896$	
904	$T = 131064$	
910	$T = 8190$	
912		$T = 8176$
928	$T = 524256$	
936		$T = 32760$
960		$T = 960$
964	$T = 16380$	
992	$T = 992$	

Table 39.   Global state-transition formula for rule $\boxed{150}$ for $1 \leq n \leq 8$.

n	$x_i^n = \sum\limits_{k=0}^{n} \dfrac{n!}{k!(n-k)!} \sum\limits_{j=0}^{k} \dfrac{k!}{j!(k-j)!} \cdot x_{i-n+2k-j}^0 \quad \mathbf{mod(2)}$
1	$x_i^1 = x_{i-1}^0 + x_i^0 + x_{i+1}^0 \quad \mathbf{mod(2)}$
2	$x_i^2 = x_{i-2}^0 + 2x_{i-1}^0 + 3x_i^0 + 2x_{i+1}^0 + x_{i+2}^0 \quad \mathbf{mod(2)}$
3	$x_i^3 = x_{i-3}^0 + 3x_{i-2}^0 + 6x_{i-1}^0 + 7x_i^0 + 6x_{i+1}^0 + 3x_{i+2}^0 + x_{i+3}^0 \quad \mathbf{mod(2)}$
4	$x_i^4 = x_{i-4}^0 + 4x_{i-3}^0 + 10x_{i-2}^0 + 16x_{i-1}^0 + 19x_i^0$ $\quad + 16x_{i+1}^0 + 10x_{i+2}^0 + 4x_{i+3}^0 + x_{i+4}^0 \quad \mathbf{mod(2)}$
5	$x_i^5 = x_{i-5}^0 + 5x_{i-4}^0 + 15x_{i-3}^0 + 30x_{i-2}^0 + 45x_{i-1}^0 + 51x_i^0$ $\quad + 45x_{i+1}^0 + 30x_{i+2}^0 + 15x_{i+3}^0 + 5x_{i+4}^0 + x_{i+5}^0 \quad \mathbf{mod(2)}$
6	$x_i^6 = x_{i-6}^0 + 6x_{i-5}^0 + 21x_{i-4}^0 + 50x_{i-3}^0 + 90x_{i-2}^0 + 126x_{i-1}^0 + 141x_i^0$ $\quad + 126x_{i+1}^0 + 90x_{i+2}^0 + 50x_{i+3}^0 + 21x_{i+4}^0 + 6x_{i+5}^0 + x_{i+6}^0 \quad \mathbf{mod(2)}$
7	$x_i^6 = x_{i-7}^0 + 7x_{i-6}^0 + 28x_{i-5}^0 + 77x_{i-4}^0 + 161x_{i-3}^0$ $\quad + 266x_{i-2}^0 + 357x_{i-1}^0 + 393x_i^0 + 357x_{i+1}^0 + 266x_{i+2}^0$ $\quad + 161x_{i+3}^0 + 77x_{i+4}^0 + 28x_{i+5}^0 + 7x_{i+6}^0 + x_{i+7}^0 \quad \mathbf{mod(2)}$
8	$x_i^6 = x_{i-8}^0 + 8x_{i-7}^0 + 36x_{i-6}^0 + 112x_{i-5}^0 + 266x_{i-4}^0 + 504x_{i-3}^0$ $\quad + 784x_{i-2}^0 + 1016x_{i-1}^0 + 1107x_i^0 + 1016x_{i+1}^0 + 784x_{i+2}^0$ $\quad + 504x_{i+3}^0 + 266x_{i+4}^0 + 112x_{i+5}^0 + 36x_{i+6}^0 + 8x_{i+7}^0 + x_{i+8}^0 \quad \mathbf{mod(2)}$

The formal proof of Eq. (68) is given below. A similar proof involving more messy expressions can be given for Eq. (69). We omit the proof to avoid clutter. A different and more illuminating proof follows as a *Corollary* to the $\boxed{105} \rightleftarrows \boxed{150}$ *Alternating Symmetry Duality* given in the *Appendix*.

Proof of Global State-Transition Formula (68) for $\boxed{150}$. Apply *mathematical induction* as follow:

(a) Applying $n = 1$ in Eq. (68), we obtain

$$x_i^1 = x_{i-1}^0 + x_i^0 + x_{i+1}^0 \ \mathrm{mod}\ (2) \tag{70}$$

which is Eq. (61) for $t = 0$.

(b) Assuming Eq. (68) is true for $n = m$ (*induction hypothesis*), namely,

$$x_i^m = \sum_{k=0}^{m} \frac{m!}{k!(m-k)!} \sum_{j=0}^{k} \frac{k!}{j!(k-j)!}$$
$$\bullet \, x_{i-m+2k-j}^0 \ \mathrm{mod}\ (2) \tag{71}$$

We must show that incrementing "m" to "$m+1$" in Eq. (71) gives Eq. (68) with $n = m + 1$. Rewriting Eq. (70) as a mapping from m to $m + 1$, we have

$$x_i^{m+1} = x_{i-1}^m + x_i^m + x_{i+1}^m \ \mathrm{mod}\ (2)$$

$$= \sum_{k=0}^{m} \frac{m!}{k!(m-k)!} \sum_{j=0}^{k} \frac{k!}{j!(k-j)!}$$
$$\bullet \, x_{(i-1)-m+2k-j}^0 \tag{72}$$

Table 40.   Mod (2) global state-transition formula in terms of *mod (2)* coefficients for rule $\boxed{150}$ for $1 \leq n \leq 8$.

$$x_i^n = \sum_{k=0}^{n} \frac{n!}{k!(n-k)!} \sum_{j=0}^{k} \frac{k!}{j!(k-j)!} \cdot x_{i-n+2k-j}^0 \quad \mathbf{mod(2)}$$

n	
1	$x_i^1 = x_{i-1}^0 + x_i^0 + x_{i+1}^0 \quad \mathbf{mod(2)}$
2	$x_i^2 = x_{i-2}^0 + x_i^0 + x_{i+2}^0 \quad \mathbf{mod(2)}$
3	$x_i^3 = x_{i-3}^0 + x_{i-2}^0 + x_i^0 + x_{i+2}^0 + x_{i+3}^0 \quad \mathbf{mod(2)}$
4	$x_i^4 = x_{i-4}^0 + x_i^0 + x_{i+4}^0 \quad \mathbf{mod(2)}$
5	$x_i^5 = x_{i-5}^0 + x_{i-4}^0 + x_{i-3}^0 + x_{i-1}^0 + x_i^0 + x_{i+1}^0 + x_{i+3}^0 + x_{i+4}^0 + x_{i+5}^0 \quad \mathbf{mod(2)}$
6	$x_i^6 = x_{i-6}^0 + x_{i-4}^0 + x_i^0 + x_{i+4}^0 + x_{i+6}^0 \quad \mathbf{mod(2)}$
7	$x_i^6 = x_{i-7}^0 + x_{i-6}^0 + x_{i-4}^0 + x_{i-3}^0 + x_{i-1}^0 + x_i^0 + x_{i+1}^0 + x_{i+3}^0 + x_{i+4}^0 + x_{i+6}^0 + x_{i+7}^0 \quad \mathbf{mod(2)}$
8	$x_i^6 = x_{i-8}^0 + x_i^0 + x_{i+8}^0 \quad \mathbf{mod(2)}$

Table 41.   Mod (2) coefficients for global-transition formula for $\boxed{150}$ for $1 \leq n \leq 8$.

n	-8	-7	-6	-5	-4	-3	-2	-1	0	1	2	3	4	5	6	7	8
1	0	0	0	0	0	0	0	1	1	1	0	0	0	0	0	0	0
2	0	0	0	0	0	0	1	0	1	0	1	0	0	0	0	0	0
3	0	0	0	0	0	1	1	0	1	0	1	1	0	0	0	0	0
4	0	0	0	0	1	0	0	0	1	0	0	0	1	0	0	0	0
5	0	0	0	1	1	1	0	1	1	1	0	1	1	1	0	0	0
6	0	0	1	0	1	0	0	0	1	0	0	0	1	0	1	0	0
7	0	1	1	0	1	1	0	1	1	1	0	1	1	0	1	1	0
8	1	0	0	0	0	0	0	0	1	0	0	0	0	0	0	0	1

Table 42.   Space-time pattern of $\boxed{150}$ calculated from a single red center pixel via (a) global state-transition formula, and (b) local state-transition formula.

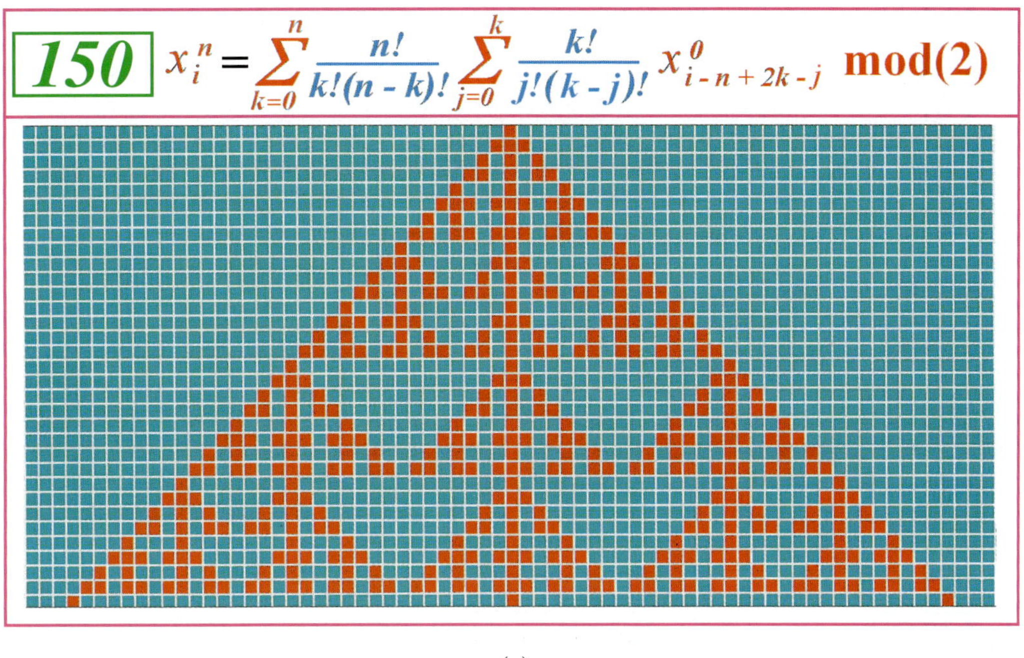

$$\boxed{150}\quad x_i^n = \sum_{k=0}^{n} \frac{n!}{k!(n-k)!} \sum_{j=0}^{k} \frac{k!}{j!(k-j)!}\, x_{i-n+2k-j}^0 \ \mathbf{mod(2)}$$

(a)

$$\boxed{150}\quad x_i^{n+1} = x_{i-1}^n \oplus x_i^n \oplus x_{i+1}^n$$

(b)

Table 43.   Space-time patterns of $\boxed{150}$ from a *random* initial configuration calculated from the global and local state-transition formula, respectively, of rule $\boxed{150}$.

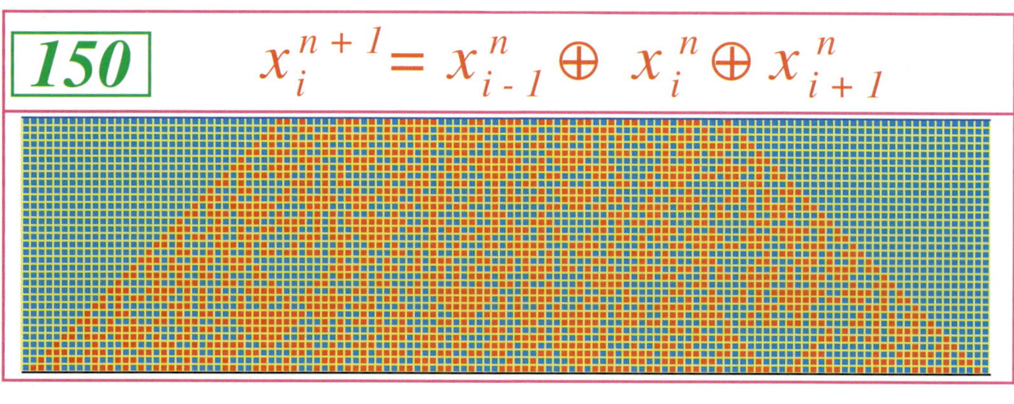

(a)

(b)

$$+ \sum_{k=0}^{m} \frac{m!}{k!(m-k)!} \sum_{j=0}^{k} \frac{k!}{j!(k-j)!}$$

$$\bullet \, x_{i-m+2k-j}^{0} \tag{73}$$

$$+ \sum_{k=0}^{m} \frac{m!}{k!(m-k)!} \sum_{j=0}^{k} \frac{k!}{j!(k-j)!}$$

$$\bullet \, x_{(i+1)-m+2k-j}^{0} \ \mathrm{mod}\,(2) \tag{74}$$

$$+ \sum_{k=0}^{m'+1} \frac{(m'+1)!}{k!(m'+1-k)!} \sum_{j=0}^{k} \frac{k!}{j!(k-j)!}$$

$$\bullet \, x_{i-m'-1+2k-j}^{0} \tag{76}$$

$$+ \sum_{k=0}^{m} \frac{(m)!}{k!(m-k)!} \sum_{j=0}^{k} \frac{k!}{j!(k-j)!}$$

$$\bullet \, x_{i+1-m+2k-j}^{0} \ \mathrm{mod}\,(2) \tag{77}$$

(b.1) Changing symbol "m" in Eq. (73) to $m'+1$ gives

$$x_i^{m+1} = \sum_{k=0}^{m} \frac{m!}{k!(m-k)!} \sum_{j=0}^{k} \frac{k!}{j!(k-j)!}$$

$$\bullet \, x_{i-1-m+2k-j}^{0} \tag{75}$$

Observe that if we substitute $k = m+1$ in Eq. (75) we would get a *zero* because $(m-k)!$ in the denominator gives, in this case, a singularity $(-1)!$ Consequently, the term at $k = m+1$ vanishes and does not affect our derivation. Hence, let us expand the external sum in Eq. (75) from $k = 0$ to $k = m+1$. We omit also in Eq. (76) the prime in m' for convenience.

Table 44. Mod (2) global state-transition formula for rule $\boxed{105}$ for $1 \leq n \leq 8$.

n	$x_i^n = \alpha_n + (-1)^n \sum\limits_{k=0}^{n} \dfrac{n!}{k!(n-k)!} \sum\limits_{j=0}^{k} \dfrac{k!}{j!(k-j)!} \cdot x_{i-n+2k-j}^0 \quad \mathbf{mod(2)}$
1	$x_i^1 = 1 - x_{i-1}^0 - x_i^0 - x_{i+1}^0 \quad \mathbf{mod(2)}$
2	$x_i^2 = x_{i-2}^0 + 2x_{i-1}^0 + 3x_i^0 + 2x_{i+1}^0 + x_{i+2}^0 \quad \mathbf{mod(2)}$
3	$x_i^3 = 1 - x_{i-3}^0 - 3x_{i-2}^0 - 6x_{i-1}^0 - 7x_i^0 - 6x_{i+1}^0 - 3x_{i+2}^0 - x_{i+3}^0 \quad \mathbf{mod(2)}$
4	$x_i^4 = x_{i-4}^0 + 4x_{i-3}^0 + 10x_{i-2}^0 + 16x_{i-1}^0 + 19x_i^0$ $+16x_{i+1}^0 + 10x_{i+2}^0 + 4x_{i+3}^0 + x_{i+4}^0 \quad \mathbf{mod(2)}$
5	$x_i^5 = 1 - x_{i-5}^0 - 5x_{i-4}^0 - 15x_{i-3}^0 - 30x_{i-2}^0 - 45x_{i-1}^0 - 51x_i^0$ $-45x_{i+1}^0 - 30x_{i+2}^0 - 15x_{i+3}^0 - 5x_{i+4}^0 - x_{i+5}^0 \quad \mathbf{mod(2)}$
6	$x_i^6 = x_{i-6}^0 + 6x_{i-5}^0 + 21x_{i-4}^0 + 50x_{i-3}^0 + 90x_{i-2}^0 + 126x_{i-1}^0 + 141x_i^0$ $+126x_{i+1}^0 + 90x_{i+2}^0 + 50x_{i+3}^0 + 21x_{i+4}^0 + 6x_{i+5}^0 + x_{i+6}^0 \quad \mathbf{mod(2)}$
7	$x_i^6 = 1 - x_{i-7}^0 - 7x_{i-6}^0 - 28x_{i-5}^0 - 77x_{i-4}^0 - 161x_{i-3}^0$ $-266x_{i-2}^0 - 357x_{i-1}^0 - 393x_i^0 - 357x_{i+1}^0 - 266x_{i+2}^0$ $-161x_{i+3}^0 - 77x_{i+4}^0 - 28x_{i+5}^0 - 7x_{i+6}^0 - x_{i+7}^0 \quad \mathbf{mod(2)}$
8	$x_i^6 = x_{i-8}^0 + 8x_{i-7}^0 + 36x_{i-6}^0 + 112x_{i-5}^0 + 266x_{i-4}^0 + 504x_{i-3}^0$ $+784x_{i-2}^0 + 1016x_{i-1}^0 + 1107x_i^0 + 1016x_{i+1}^0 + 784x_{i+2}^0$ $+504x_{i+3}^0 + 266x_{i+4}^0 + 112x_{i+5}^0 + 36x_{i+6}^0 + 8x_{i+7}^0 + x_{i+8}^0 \quad \mathbf{mod(2)}$

$$\alpha_n = (1 - (-1)^n)/2$$

Let us combine Eqs. (75) and (76):

$$x_i^{m+1} = \sum_{k=0}^{m+1} \left(\frac{(m)!}{k!(m-k)!} + \frac{(m+1)!}{k!(m+1-k)!} \right)$$

$$\bullet \sum_{j=0}^{k} \frac{k!}{j!(k-j)!} \bullet x_{i-m-1+2k-j}^0$$

$$+ \sum_{k=0}^{m} \frac{(m)!}{k!(m-k)!} \sum_{j=0}^{k} \frac{k!}{j!(k-j)!}$$

$$\bullet \, x_{i+1-m+2k-j}^0 \bmod (2) \qquad (78)$$

The sum of the two binomial coefficients contained between the large parentheses in the first sum is

$$\frac{(m)!}{k!(m-k)!} + \frac{(m+1)!}{k!(m+1-k)!}$$

$$= \frac{(m)!}{k!(m+1-k)!}\big((m+1-k) + (m+1)\big)$$

$$= \frac{(m)!}{k!(m+1-k)!}(2m+2-k)$$

In this case Eq. (78) can be recast as follows:

$$x_i^{m+1} = \sum_{k=0}^{m+1} \frac{\big(2(m+1)-k\big)(m)!}{(k)!(m+1-k)!}$$

$$\bullet \sum_{j=0}^{k} \frac{k!}{j!(k-j)!} \bullet x_{i-m-1+2k-j}^0$$

$$+ \sum_{k=0}^{m} \frac{(m)!}{k!(m-k)!} \sum_{j=0}^{k} \frac{k!}{j!(k-j)!}$$

$$\bullet \, x_{i+1-m+2k-j}^0 \bmod (2) \qquad (79)$$

Separating terms with factors $2(m+1)$ and $(-k)$ we obtain

Table 45.   Mod (2) global state-transition formula in terms of *mod (2)* coefficients for rule $\boxed{105}$ for $1 \le n \le 8$.

n	$x_i^n = \alpha_n + (-1)^n \sum\limits_{k=0}^{n} \dfrac{n!}{k!(n-k)!} \sum\limits_{j=0}^{k} \dfrac{k!}{j!(k-j)!} \cdot x_{i-n+2k-j}^0 \quad \mathbf{mod(2)}$
1	$x_i^1 = 1 - x_{i-1}^0 - x_i^0 - x_{i+1}^0 \quad \mathbf{mod(2)}$
2	$x_i^2 = x_{i-2}^0 + x_i^0 + x_{i+2}^0 \quad \mathbf{mod(2)}$
3	$x_i^3 = 1 - x_{i-3}^0 - x_{i-2}^0 - x_i^0 - x_{i+2}^0 - x_{i+3}^0 \quad \mathbf{mod(2)}$
4	$x_i^4 = x_{i-4}^0 + x_i^0 + x_{i+4}^0 \quad \mathbf{mod(2)}$
5	$x_i^5 = 1 - x_{i-5}^0 - x_{i-4}^0 - x_{i-3}^0 - x_{i-1}^0 - x_i^0 - x_{i+1}^0 - x_{i+3}^0 - x_{i+4}^0 - x_{i+5}^0 \quad \mathbf{mod(2)}$
6	$x_i^6 = x_{i-6}^0 + x_{i-4}^0 + x_i^0 + x_{i+4}^0 + x_{i+6}^0 \quad \mathbf{mod(2)}$
7	$x_i^6 = 1 - x_{i-7}^0 - x_{i-6}^0 - x_{i-4}^0 - x_{i-3}^0 - x_{i-1}^0 - x_i^0 - x_{i+1}^0 - x_{i+3}^0 - x_{i+4}^0 - x_{i+6}^0 - x_{i+7}^0 \quad \mathbf{mod(2)}$
8	$x_i^6 = x_{i-8}^0 + x_i^0 + x_{i+8}^0 \quad \mathbf{mod(2)}$

$$\alpha_n = (1 - (-1)^n)/2$$

Table 46.   Mod (2) coefficients for global-transition formula for $\boxed{105}$ for $1 \le n \le 8$.

n	-8	-7	-6	-5	-4	-3	-2	-1	0	1	2	3	4	5	6	7	8
1	1	1	1	1	1	1	1	0	0	0	1	1	1	1	1	1	1
2	0	0	0	0	0	0	1	0	1	0	1	0	0	0	0	0	0
3	1	1	1	1	1	0	0	1	0	1	0	0	1	1	1	1	1
4	0	0	0	0	1	0	0	0	1	0	0	0	1	0	0	0	0
5	1	1	1	0	0	0	1	0	0	0	1	0	0	0	1	1	1
6	0	0	1	0	1	0	0	0	1	0	0	0	1	0	1	0	0
7	1	0	0	1	0	0	1	0	0	0	1	0	0	1	0	0	1
8	1	0	0	0	0	0	0	0	1	0	0	0	0	0	0	0	1

Table 47.   Space-time pattern of $\boxed{105}$ calculated from a single red center pixel via (a) global state transition function, and (b) local state transition function.

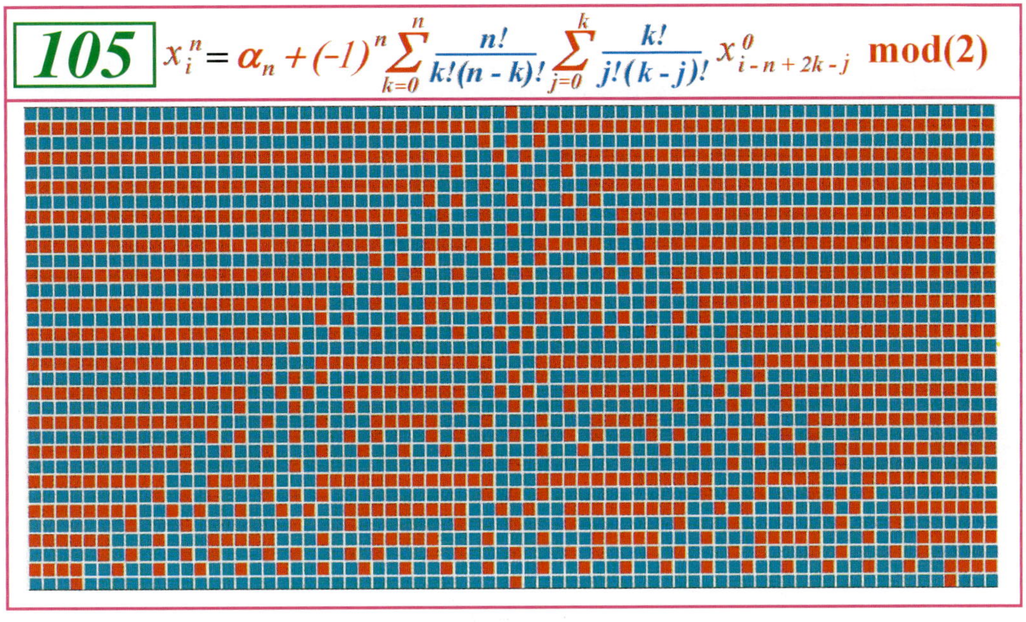

(a)

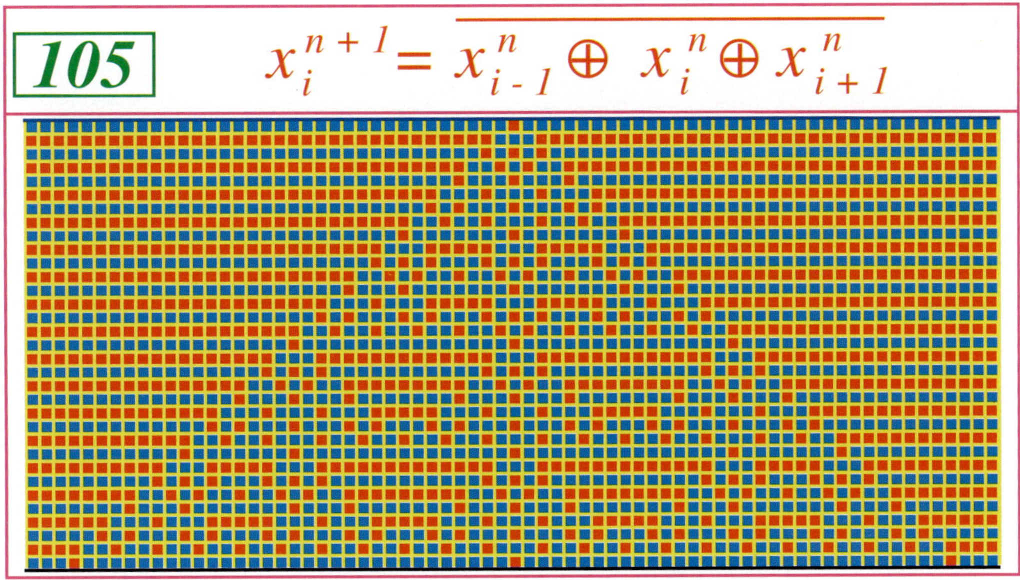

(b)

Table 48.   Space-time patterns of $\boxed{105}$ from a *random* initial configuration calculated from the global and local state transition formula, respectively, of rule $\boxed{105}$.

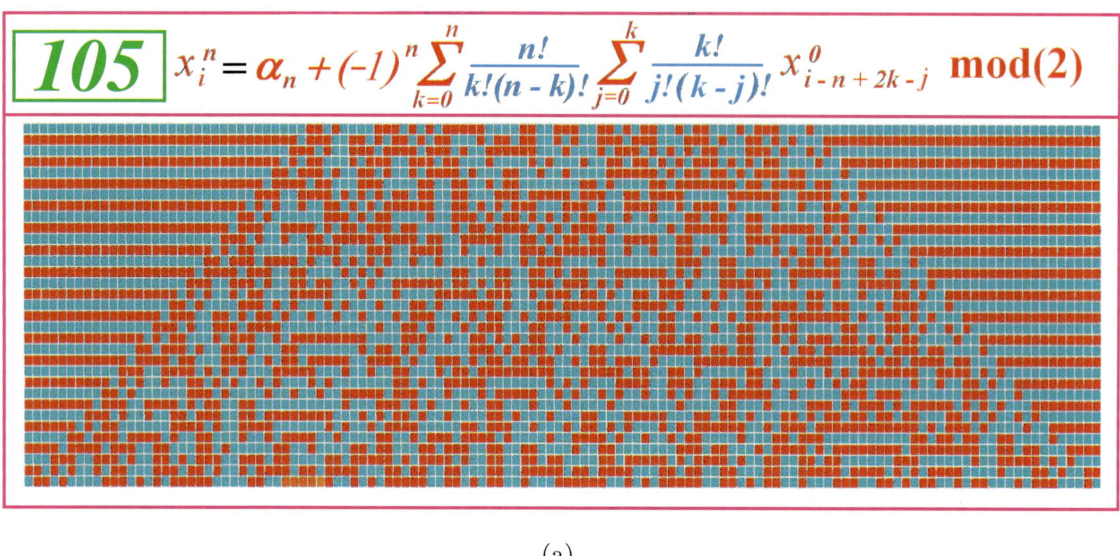

$$\boxed{105}\quad x_i^n = \alpha_n + (-1)^n \sum_{k=0}^{n} \frac{n!}{k!(n-k)!} \sum_{j=0}^{k} \frac{k!}{j!(k-j)!}\, x_{i-n+2k-j}^0 \; \mathbf{mod(2)}$$

(a)

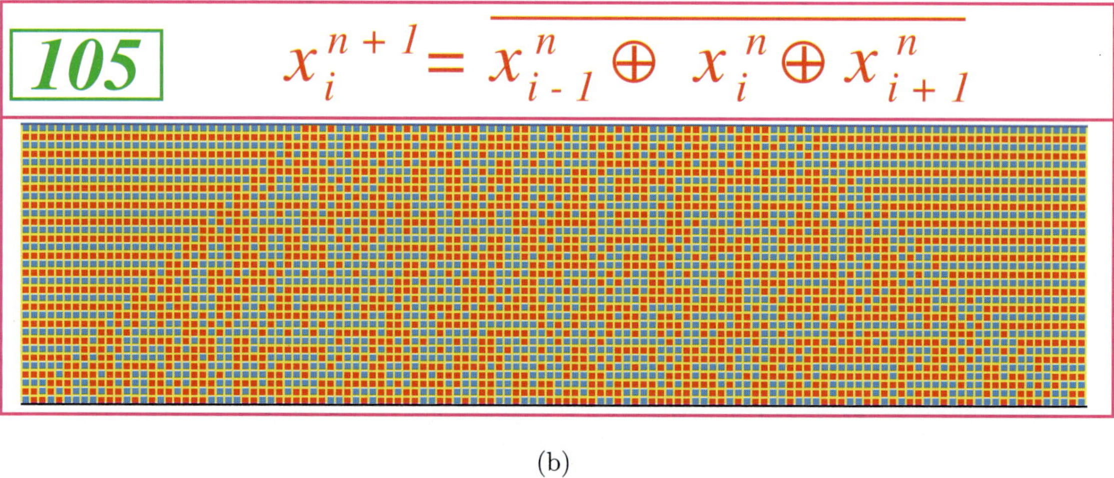

$$\boxed{105}\qquad x_i^{n+1} = \overline{x_{i-1}^n \oplus x_i^n \oplus x_{i+1}^n}$$

(b)

$$x_i^{m+1} = 2 \left(\sum_{k=0}^{m+1} \frac{(m+1)!}{(k)!(m+1-k)!} \sum_{j=0}^{k} \frac{k!}{j!(k-j)!} \bullet x_{i-m-1+2k-j}^0 \right) \tag{80}$$

$$+ \sum_{k=0}^{m+1} \frac{(-k)\bullet(m)!}{(k)!(m+1-k)!} \sum_{j=0}^{k} \frac{k!}{j!(k-j)!} \bullet x_{i-m-1+2k-j}^0 \tag{81}$$

$$+ \sum_{k=0}^{m} \frac{(m)!}{k!(m-k)!} \sum_{j=0}^{k} \frac{k!}{j!(k-j)!} \bullet x_{i+1-m+2k-j}^0 \;\mathrm{mod}\,(2) \tag{82}$$

The first term (80) vanishes under operation **mod (2)** because the multiplier 2 gives rise to all even coefficients. However, let us retain this term until the final computations. The remaining terms are Eqs. (81) and (82).

(b.2) Let us now replace symbol "k" in Eq. (82) by $k' - 1$:

$$x_i^{m+1} = \{\text{Eq. (80)}\} + \left(\sum_{k=0}^{m+1} \frac{(m)!}{(m+1-k)!} \sum_{j=0}^{k} \frac{(-k)}{j!(k-j)!} \bullet x_{i-m-1+2k-j}^0 \right.$$

$$\left. + \sum_{k'=1}^{m+1} \frac{(m)!}{(m-k'+1)!} \sum_{j=0}^{k'-1} \frac{1}{j!(k'-1-j)!} \bullet x_{i-m-1+2k'-j}^0 \right) \bmod (2) \tag{83}$$

Here we cancel, for convenience, the factorials $k!$ and $(k'-1)!$ from both denominator and numerator. Also we transpose the factor $(-k)$ in the first term, see Eq. (81), to the inner sum.

Now by omitting the prime in k' we can combine the two sums as follows:

$$\sum_{k=0}^{m+1} \frac{(m)!}{(m+1-k)!} \sum_{j=0}^{k} \left(\frac{(-k)}{j!(k-j)!} + \frac{1}{j!(k-1-j)!} \right) \bullet x_{i-m-1+2k-j}^0 \bmod (2) \tag{84}$$

The sum of the two binomial coefficients in Eq. (84) gives

$$\frac{-k}{j!(k-j)!} + \frac{1}{j!(k-1-j)!} = \frac{-k+(k-j)}{j!(k-j)!} = \frac{-1}{(j-1)!(k-j)!}.$$

As a result, Eq. (84) transforms to

$$-\sum_{k=0}^{m+1} \frac{(m)!}{k!(m-k+1)!} \sum_{j=0}^{k} \left(\frac{k!}{(j-1)!(k-j)!} \right) \bullet x_{i-(m+1)+2k-j}^0 \bmod (2).$$

Here we reinsert the factorial $k!$ in the denominator and in the numerator again.

(b.3) Let us now replace k by $k' + 1$, and j by $j' + 1$ to obtain

$$-\sum_{k'=-1}^{m} \frac{(m)!}{(k'+1)!(m-k')!} \sum_{j'=-1}^{k'} \frac{(k'+1)!}{(j')!(k'-j')!} \bullet x_{i-(m+1)+1+2k'-j'}^0 \bmod (2).$$

Replacing m on the right by $m' + 1$ we get

$$-\sum_{k'=-1}^{m'+1} \frac{(m'+1)!}{(k')!(m'+1-k')!} \sum_{j'=-1}^{k'} \frac{(k')!}{(j')!(k'-j')!} \bullet x_{i-(m'+1)+2k'-j'}^0 \bmod (2).$$

Observe that at $k' = -1$ and at $j' = -1$, the factorials $(k')!$ and $(j')!$ represented in the denominator are singular, i.e. they are equal to $(-1)!$ Hence, terms with these factorials vanish. Omitting the primes in the symbols m', k' and j' and recalling term (80) we obtain finally

$$x_i^{m+1} = 2 \left(\sum_{k=0}^{m+1} \frac{(m+1)!}{(k)!(m+1-k)!} \sum_{j=0}^{k} \frac{k!}{(j)!(k-j)!} \bullet x_{i-(m+1)+2k-j}^0 \right)$$

$$- \sum_{k=0}^{m+1} \frac{(m+1)!}{(k)!(m+1-k)!} \sum_{j=0}^{k} \frac{(k)!}{(j)!(k-j)!} \bullet x_{i-(m+1)+2k-j}^0 \bmod (2)$$

$$= \sum_{k=0}^{m+1} \frac{(m+1)!}{(k)!(m+1-k)!} \sum_{j=0}^{k} \frac{k!}{(j)!(k-j)!} \bullet x_{i-(m+1)+2k-j}^0 \bmod (2) \tag{85}$$

The result is precisely Eq. (71) with m replaced by $m + 1$, in the *induction hypothesis* in Eq. (71).   ∎

4.3. *Rules* $\boxed{150}$ *and* $\boxed{105}$ *are globally quasi-equivalent*

It follows from the *global state transition formulas* (68) for rule $\boxed{150}$ and (69) for rule $\boxed{105}$ that they are *globally equivalent* in the sense that their *space-time* evolution patterns from *any initial bit string* can be derived from each other via the following *affine transformation*

$$
\underbrace{\begin{bmatrix} x_0^n(\boxed{105}) \\ x_1^n(\boxed{105}) \\ x_2^n(\boxed{105}) \\ \vdots \\ x_{L-1}^n(\boxed{105}) \end{bmatrix}}_{\boldsymbol{x}(\boxed{105})} = \alpha_n \boldsymbol{1} + (-1)^n \underbrace{\begin{bmatrix} 1 & 0 & 0 & \cdots & 0 \\ 0 & 1 & 0 & \cdots & 0 \\ 0 & 0 & 1 & \cdots & 0 \\ \vdots & & & \ddots & \vdots \\ 0 & 0 & 0 & \cdots & 1 \end{bmatrix}}_{\widetilde{\boldsymbol{T}} \triangleq (-1)^n \boldsymbol{1}} \underbrace{\begin{bmatrix} x_0^n(\boxed{150}) \\ x_1^n(\boxed{150}) \\ x_2^n(\boxed{150}) \\ \vdots \\ x_{L-1}^n(\boxed{150}) \end{bmatrix}}_{\boldsymbol{x}(\boxed{150})}
\tag{86}
$$

where

$$
\alpha_n \triangleq \frac{1 - (-1)^n}{2}
\tag{87}
$$

and $\boldsymbol{1}$ denotes the unit (identity) matrix.

The *inverse* affine transformation is given by:

$$
\underbrace{\begin{bmatrix} x_0^n(\boxed{150}) \\ x_1^n(\boxed{150}) \\ x_2^n(\boxed{150}) \\ \vdots \\ x_{L-1}^n(\boxed{150}) \end{bmatrix}}_{\boldsymbol{x}(\boxed{150})} = \alpha_n \boldsymbol{1} + (-1)^n \underbrace{\begin{bmatrix} 1 & 0 & 0 & \cdots & 0 \\ 0 & 1 & 0 & \cdots & 0 \\ 0 & 0 & 1 & \cdots & 0 \\ \vdots & & & \ddots & \vdots \\ 0 & 0 & 0 & \cdots & 1 \end{bmatrix}}_{\widetilde{\boldsymbol{T}} = (-1)^n \boldsymbol{1}} \underbrace{\begin{bmatrix} x_0^n(\boxed{105}) \\ x_1^n(\boxed{105}) \\ x_2^n(\boxed{105}) \\ \vdots \\ x_{L-1}^n(\boxed{105}) \end{bmatrix}}_{\boldsymbol{x}(\boxed{105})}
\tag{88}
$$

Observe that the inverse matrix $\widetilde{\boldsymbol{T}}$ is *itself*.

An analysis of the above global equivalence transformation shows that $\widetilde{\boldsymbol{T}}$ effectively leaves *even-numbered* rows unchanged while complementing the color of all *odd-numbered* rows, for *the same* initial configuration. An example illustrating this transformation is given in Figs. 4 and 5. Consequently, we will henceforth christened $\widetilde{\boldsymbol{T}}$ as an *alternating transformation*. Observe that since $\widetilde{\boldsymbol{T}}$ depends on the *iteration number "n"*, it does *not* defined a *topological conjugacy* between the space-time patterns of the two local rules $\boxed{105}$ and $\boxed{150}$. The global quasi-equivalence transformation $\widetilde{\boldsymbol{T}}$ is therefore *weaker* than the three *topologically conjugate* transformations $\boldsymbol{T}^\dagger$, $\overline{\boldsymbol{T}}$ and $\boldsymbol{T}^*$ from the *Vierergruppe* defined in [Chua *et al.*, 2004]. In particular, the number of *connected* components and the *minimal period* of corresponding attractors of $\boxed{105}$ and $\boxed{150}$ *are not* preserved by $\widetilde{\boldsymbol{T}}$, as is evident from the basin-tree diagrams of rules $\boxed{105}$ and $\boxed{150}$. On the other hand, the alternating transformation $\widetilde{\boldsymbol{T}}$ is *not only a bijection* but it also preserves certain qualitative properties; e.g. *gardens of Eden* maps onto *gardens of Eden, transient* bit strings maps onto *transient* bit strings, some *connected period-T* orbits of $\boxed{105}$ maps onto *two disconnected period-$\frac{T}{2}$* orbits of $\boxed{150}$, etc.

In terms of the *real* variables $u_i^n = 2x_i^n - 1$ defined in Eq. (4) of [Chua *et al.*, 2005a], Eq. (88) assumes the following more compact form:

$$
\underbrace{\begin{bmatrix} u_0^n(\boxed{105}) \\ u_1^n(\boxed{105}) \\ u_2^n(\boxed{105}) \\ \vdots \\ u_{L-1}^n(\boxed{105}) \end{bmatrix}}_{\boldsymbol{u}(\boxed{105})} = (-1)^n \underbrace{\begin{bmatrix} 1 & 0 & 0 & \cdots & 0 \\ 0 & 1 & 0 & \cdots & 0 \\ 0 & 0 & 1 & \cdots & 0 \\ \vdots & & & \ddots & \vdots \\ 0 & 0 & 0 & \cdots & 1 \end{bmatrix}}_{(-1)^n \mathbf{1}} \underbrace{\begin{bmatrix} u_0^n(\boxed{150}) \\ u_1^n(\boxed{150}) \\ u_2^n(\boxed{150}) \\ \vdots \\ u_{L-1}^n(\boxed{150}) \end{bmatrix}}_{\boldsymbol{u}(\boxed{150})}
\tag{89}
$$

where

$$
\widetilde{\boldsymbol{T}} = \begin{bmatrix} (-1)^n & 0 & 0 & \cdots & 0 \\ 0 & (-1)^n & 0 & \cdots & 0 \\ 0 & 0 & (-1)^n & \cdots & 0 \\ & & \vdots & & \\ 0 & 0 & 0 & \cdots & (-1)^n \end{bmatrix}
\tag{90}
$$

is the *alternating* transformation.

5. Concluding Remarks

We cannot overemphasize the usefulness of the *basin tree diagrams* presented in Tables 14–23. Indeed, Theorems 3–5 originated from conjectures suggested by Tables 18, 19 and 23. Space limitation precludes a more detailed analysis of the other tables. They will be the subject of future papers.

We wish to remark that in order to make use of the "empty" spaces in these basin-tree diagrams, we have inserted typical space-time patterns, along with their associated *Bernoulli* $\phi_{n-\tau} \mapsto \phi_n$ *return maps*. These maps typically indicate the associated Bernoulli *velocity* σ, the Bernoulli *return time* τ and the Bernoulli *complementation sign* β. Unlike the elementary Bernoulli σ_τ-shifts studied in Parts IV and VI, where these parameters do *not* depend on L, we now find them to depend crucially on L. Moreover, both $|\sigma|$ and τ can assume arbitrarily large values as $L \to \infty$.

The *period* T associated with each σ and τ of the *complex* and *hyper* Bernoulli rules listed in Tables 11 and 12 is generally equal to $T = \tau L$, if $T_0 \overset{\Delta}{=} \tau L/|\sigma|$ is not an integer, or $T = \tau L/|\sigma|$ if $|\sigma| \geq 2$. The presence of *symmetry* in a period-T bit string can reduce the *minimal* period T further by a factor of "m" for bit strings made of m identical substrings.

For those space-time patterns which do not exhibit a Bernoulli shift with $|\sigma| > 0$, we can generalize our definition of "Bernoulli σ_τ-shift" to include $\sigma = 0$ for all such *period-T* orbits. In this case, the *return map* $\phi_{n-\tau} \mapsto \phi_n$ will consist of points lying on the *diagonal* line $\phi_n = \phi_{n-\tau}$. We usually include such a graph whenever space permits.

We note also that the period "T" of rules $\boxed{90}$, $\boxed{150}$ and $\boxed{105}$ exhibit a *scale free* property as $L \to \infty$. For example, for $L = 2^m$, the *period* of rule $\boxed{90}$ is always equal $T = 1$ with $\underbrace{0\ 0\ \cdots\ 0}_{L\text{ bits}}$ as its global *fixed point* attractor. To its immediate left $(L = 2^m - 1)$ and immediate right $(L = 2^m + 1)$, the period-T orbits have *equal period* $T = 2^m - 1$, at *any scale* $L \to \infty$. To illustrate the *scale-free distribution* of the period "T" of rule $\boxed{90}$, Fig. 6 shows a plot of $\log T$ as a function of $\log L$ of the data listed in Table 25. Observe the six *period-1 red stars* on the horizontal axis $(T = 1)$ are located at $L = 2^m$, $m = 2, 3, 4, 5$; namely, $L = 4, 8, 16, 32, 64$, as predicted by Theorem 3. Observe that all data points from Table 25 lie along straight lines with a slope equal to "*one*".

The distributions of the *period* T of rules $\boxed{150}$ and $\boxed{105}$ are plotted in Figs. 7 and 8, respectively, as a function of the *string length* $L = I + 1$, in base-10 logarithmic scales. The data are extracted from Table 37 for rule $\boxed{150}$, and from Table 38 for rule $\boxed{105}$, respectively. Data points corresponding to *isles of Eden* are shown as *blue dots*. Those

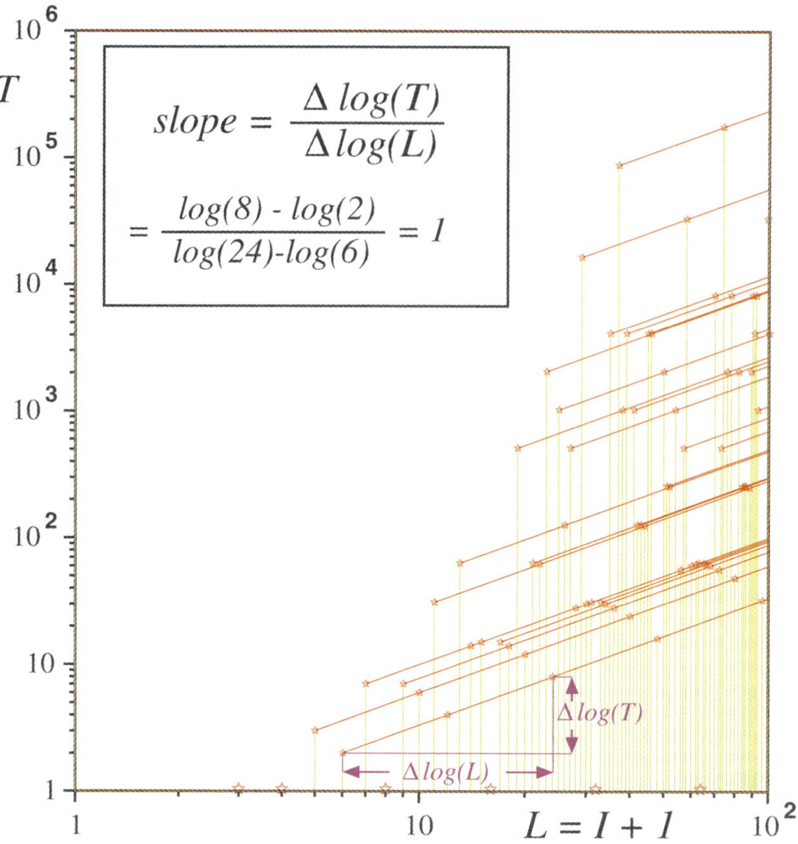

Fig. 6.   Relationship between the *period* T and the *length* $L = I + 1$ of *attractors* of rule $\boxed{90}$ plotted in base-10 logarithmic scales.

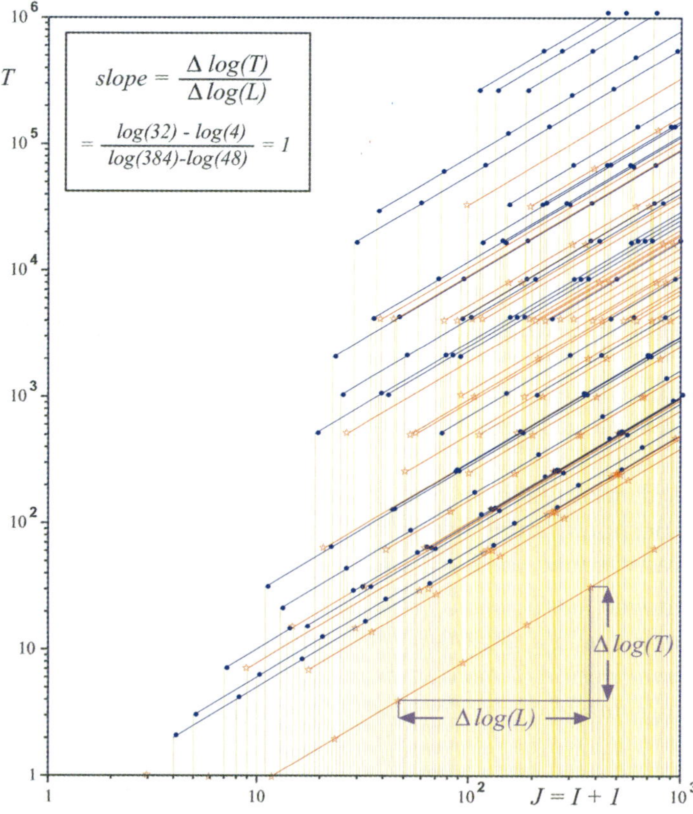

Fig. 7.   Relationship between the *period* T and the *length* $L = I + 1$ of *isles of Eden* (plotted as *blue dots*), and *attractors* (plotted as *red stars*) of rule $\boxed{150}$.

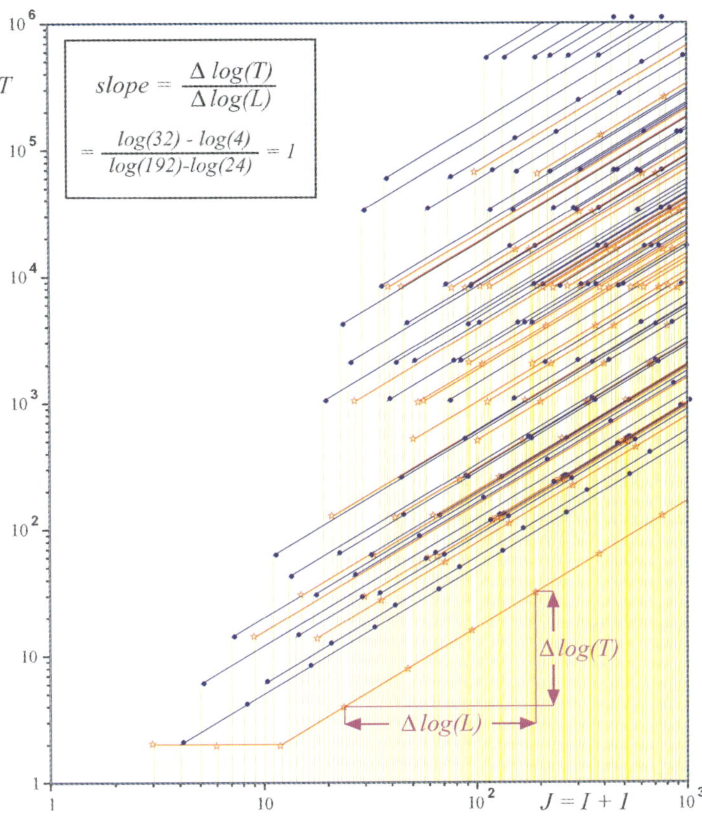

Fig. 8. Relationship between the *period* T and the *length* $L = I + 1$ of *isles of Eden* (plotted as *blue dots*), and attractors (plotted as *red stars*) of rule $\boxed{105}$.

from Table 37 for rule $\boxed{150}$, and from Table 38 for rule $\boxed{105}$, respectively. Data points corresponding to *isles of Eden* are shown as *blue dots*. Those corresponding to *attractors* are shown as *red stars*. Again, the *scale-free* distributions are clearly seen from the parallel straight lines where these data points are located.

Appendix

$\boxed{105} \rightleftarrows \boxed{150}$ *Alternating Symmetry Duality*

The bit strings $\{x_0^n, x_1^n, x_2^n, \ldots, x_I^n\}$ and $\{y_0^n, y_1^n, y_2^n, \ldots, y_I^n\}$ generated respectively by rules $\boxed{150}$ and $\boxed{105}$ from *the same* initial state $\{z_0^0, z_1^0, z_2^0, \ldots, z_I^0\}$ obey the following *alternating symmetry relations*:

$$\boxed{\begin{aligned} y_i^n &= \alpha_n + (-1)^n x_i^n \qquad &\text{(A.1)} \\ x_i^n &= \alpha_n + (-1)^n y_i^n \qquad &\text{(A.2)} \end{aligned}}$$

Proof. Bit string $\{x_0^n, x_1^n, x_2^n, \ldots, x_I^n\}$ evolves under rule $\boxed{150}$ via the formula

$$x_i^{n+1} = x_{i-1}^n + x_i^n + x_{i+1}^n \quad \text{mod} (2) \qquad \text{(A.3)}$$

Bit string $\{y_0^n, y_1^n, y_2^n, \ldots, y_I^n\}$ evolves under rule $\boxed{105}$ via the formula

$$y_i^{n+1} = 1 - (y_{i-1}^n + y_i^n + y_{i+1}^n) \quad \text{mod} (2) \qquad \text{(A.4)}$$

Changing the symbol "y" in Eq. (A.4) into "x" by applying Eq. (A.1) and invoking the identity $(3\alpha_n) \, \text{mod} \, (2) = \alpha_n \, \text{mod} \, (2)$ we obtain

$$\begin{aligned} &\alpha_{n+1} + (-1)^{n+1} x_i^{n+1} \\ &= 1 - (3\alpha_n + (-1)^n x_{i-1}^n + (-1)^n x_i^n \\ &\quad + (-1)^n x_{i+1}^n) \quad \text{mod} (2) \\ &= (1 - \alpha_n) - (-1)^n (x_{i-1}^n + x_i^n + x_{i+1}^n) \quad \text{mod} (2) \end{aligned}$$
$$\text{(A.5)}$$

Consider the following two cases:

(a) Assume n is even in Eq. (A.5).

In this case $\alpha_n = 0$ and $\alpha_{n+1} = 1$. Equation (A.5) reduces to:

$$1 - x_i^{n+1} = 1 - (x_{i-1}^n + x_i^n + x_{i+1}^n) \quad \text{mod} (2)$$
$$\text{(A.6)}$$

Hence,

$$x_i^{n+1} = (x_{i-1}^n + x_i^n + x_{i+1}^n) \quad \text{mod} (2) \qquad \text{(A.7)}$$

(b) Assume n is odd in Eq. (A.5).

In this case $\alpha_n = 1$ and $\alpha_{n+1} = 0$. Equation (A.5) reduces to:

$$x_i^{n+1} = (x_{i-1}^n + x_i^n + x_{i+1}^n) \mod (2) \quad \text{(A.8)}$$

Hence, both Eqs. (A.7) and (A.8) are identical to Eq. (A.3).

Following the same procedure let us change the symbol "x" in Eq. (A.3) into "y" by applying Eq. (A.2) to obtain

$$\begin{aligned}
\alpha_{n+1} &+ (-1)^{n+1} y_i^{n+1} \\
&= (3\alpha_n + (-1)^n y_{i-1}^n + (-1)^n y_i^n \\
&\quad + (-1)^n y_{i+1}^n) \mod (2) \\
&= \alpha_n + (-1)^n (y_{i-1}^n + y_i^n + y_{i+1}^n) \mod (2)
\end{aligned}$$
$$\text{(A.9)}$$

Again, we must consider two cases:

(a) Assume n is even in Eq. (A.9).

In this case $\alpha_n = 0$ and $\alpha_{n+1} = 1$. Equation (A.9) reduces to:

$$1 - y_i^{n+1} = (y_{i-1}^n + y_i^n + y_{i+1}^n) \mod (2)$$
$$\text{(A.10)}$$

Hence,

$$y_i^{n+1} = 1 - (y_{i-1}^n + y_i^n + y_{i+1}^n) \mod (2)$$
$$\text{(A.11)}$$

(b) Assume n is odd in Eq. (A.9).

In this case $\alpha_n = 1$ and $\alpha_{n+1} = 0$. Equation (A.9) reduces to:

$$0 + y_i^{n+1} = 1 - (y_{i-1}^n + y_i^n + y_{i+1}^n) \mod (2)$$
$$\text{(A.12)}$$

Hence, both Eqs. (A.11) and (A.12) are identical to Eq. (A.4).   ∎

Chapter 2

MORE ISLES OF EDEN

This paper presents the basin tree diagrams of all *hyper Bernoulli* σ_τ-shift rules for string lengths $L = 3, 4, \ldots, 8$. These diagrams have revealed many global and time-asymptotic properties that we have subsequently proved to be true for all $L < \infty$. In particular, we have proved that local rule $\boxed{60}$ has no *Isles of Eden* for all L, and that local rules $\boxed{154}$ and $\boxed{45}$ are inhabited by a *dense* set (continuum) of Isles of Eden *if, and only if, L* is an *odd* integer. A novel and powerful graph-theoretic tool, called *Isles-of-Eden digraph*, has been developed and can be used to test the existence of dense Isles of Eden of *any* local rule which satisfies certain constraints, such as rules $\boxed{154}$, $\boxed{45}$, $\boxed{150}$, $\boxed{105}$, as well as *all invariant* local rules, such as rules $\boxed{170}$, $\boxed{240}$, $\boxed{15}$ and $\boxed{85}$, subject to no constraints.

Keywords: Cellular automata; nonlinear dynamics; attractors; Isles of Eden; Bernoulli shift; shift maps; basin tree diagram; Bernoulli velocity; Bernoulli return time; complex Bernoulli shifts; hyper Bernoulli shifts; Binomial series; scale-free phenomena; Rule 45; Rule 60; Rule 90; Rule 105; Rule 150; Rule 154.

1. The Beginning of the End

Recall from Part III [Chua *et al.*, 2004] that among the 256 local rules governing the universe of 1-D cellular automata listed in Table 1, only 88 local rules are *globally-independent*[1] from each other in the sense that the remaining 168 rules can be derived from these 88 rules via one of the three global transformations from the *Vierergruppe* [Chua *et al.*, 2004]; namely, the *Left-Right Transformation* $T^\dagger$, the *Global Complementation* $\overline{T}$, and the *Left-Right Complementation* T^*. The choice of these 88 rules from Table 1 is not unique since each member of an equivalent class is mathematically *identical* to

---

[1]In fact, we can reduce the number of equivalence classes to less than 88 *globally bijective* local rules because the two local rules $\boxed{105}$ and $\boxed{150}$, as well as several other rules exhibiting a similar *alternating symmetry* property, are *globally quasi-equivalent* via the continuous *alternating transformation* $\widetilde{T}$ defined in Eq. (90) of [Chua *et al.*, 2007]. However, $\widetilde{T}$ does *not* apply to the other rules, unlike $T^\dagger, \overline{T}$, and T^*. Moreover, although $\widetilde{T}$ is a *bijection*, it is not a *topological conjugacy*.

Table 1.   List of 256 local rules with their complexity index coded in red ($\kappa = 1$), blue ($\kappa = 2$), and green ($\kappa = 3$), respectively.

0	1	2	3	4	5	6	7	8	9	10	11	12	13	14	15
16	17	18	19	20	21	22	23	24	25	26	27	28	29	30	31
32	33	34	35	36	37	38	39	40	41	42	43	44	45	46	47
48	49	50	51	52	53	54	55	56	57	58	59	60	61	62	63
64	65	66	67	68	69	70	71	72	73	74	75	76	77	78	79
80	81	82	83	84	85	86	87	88	89	90	91	92	93	94	95
96	97	98	99	100	101	102	103	104	105	106	107	108	109	110	111
112	113	114	115	116	117	118	119	120	121	122	123	124	125	126	127
128	129	130	131	132	133	134	135	136	137	138	139	140	141	142	143
144	145	146	147	148	149	150	151	152	153	154	155	156	157	158	159
160	161	162	163	164	165	166	167	168	169	170	171	172	173	174	175
176	177	178	179	180	181	182	183	184	185	186	187	188	189	190	191
192	193	194	195	196	197	198	199	200	201	202	203	204	205	206	207
208	209	210	211	212	213	214	215	216	217	218	219	220	221	222	223
224	225	226	227	228	229	230	231	232	233	234	235	236	237	238	239
240	241	242	243	244	245	246	247	248	249	250	251	252	253	254	255

$\kappa = 1$ (Red) 104 rules

$\kappa = 2$ (Blue) 126 rules

$\kappa = 3$ (Green) 26 rules

the other members of the same equivalent class. Our choice is to simply scan the list in Table 1 consecutively and delete those rules that are equivalent to some preceding rules. These 88 globally-independent local rules are listed in Table 2 (reproduced from [Chua *et al.*, 2007]). Since all our results are stated and derived for the 88 local rules listed in Table 2, the corresponding results for the remaining 168 rules must be inferred from these rules via the appropriate transformation listed in Table 4 of [Chua *et al.*, 2007]. For the readers' convenience, we have identified the appropriate transformation for each of the remaining 168 rules, as shown in Table 3. Observe that some rules in Table 3 can be derived from the 88 rules via *two* distinct global transformations. For example, rule $\boxed{55}$ can be derived from rule $\boxed{19}$ via either $\overline{T}$, or T^*, because $T^{\dagger}\left(\boxed{19}\right) = \boxed{19}$.

The 88 dynamically-independent rules listed in Table 2 can be logically partitioned into six qualitatively distinct classes as shown in Table 4. The steady-state (*attractor* or *Isle of Eden*) behaviors of all 26 *period-1 rules*, and all 13 *period-2 rules*, and the single *period-3 rule*, and all 30 *Bernoulli*

Table 2. The first 88 *globally-independent* local rules among the 256 listed in Table 1.

88 Global Equivalence Classes							
0	1	2	3	4	5	6	7
8	9	10	11	12	13	14	15
18	19	22	23	24	25	26	27
28	29	30	32	33	34	35	36
37	38	40	41	42	43	44	45
46	50	51	54	56	57	58	60
62	72	73	74	76	77	78	90
94	104	105	106	108	110	122	126
128	130	132	134	136	138	140	142
146	150	152	154	156	160	162	164
168	170	172	178	184	200	204	232

σ_{τ}-*shift rules*, have been completely characterized in [Chua *et al.*, 2004; Chua *et al.*, 2005a], and [Chua *et al.*, 2005b].

"*Basin tree diagrams*" of all ten complex Bernoulli-shift rules (listed in Table 11 of [Chua *et al.*, 2007]) for $3 \leq L \leq 8$ are given in [Chua *et al.*, 2007]. In addition, the characteristic steady-state behaviors of three *complex* Bernoulli-shift rules (out of 10); namely, rules $\boxed{90}$, $\boxed{105}$ and $\boxed{150}$, have been analyzed rigorously and presented in [Chua *et al.*, 2007]. Our goal in this paper is to present the *basin tree diagrams* of the eight *Hyper Bernoulli rules* (listed in Table 12 of [Chua *et al.*, 2007]) for $3 \leq L \leq 8$. In addition, we will conduct a rigorous analysis of three *hyper* Bernoulli-shift rules (out of 8); namely, rules $\boxed{60}$, $\boxed{45}$ and $\boxed{154}$, respectively.

2. Basin Tree Diagrams of Eight Hyper Bernoulli Shift Rules

The *basin tree diagram* of a local rule $\boxed{N}$ is defined in [Chua *et al.*, 2007] as the collection of all *period-n* orbits $\Gamma_n\left(\boxed{N}\right)$ for all possible $n = 1, 2, \ldots$, and their associated basin trees $\Im\left(\Gamma_n\left(\boxed{N}\right)\right)$ of an L-bit cellular automata under local rule $\boxed{N}$. The basin tree diagrams for the ten *complex Bernoulli shift* rules $\boxed{18}$, $\boxed{22}$, $\boxed{54}$, $\boxed{73}$, $\boxed{90}$, $\boxed{105}$, $\boxed{122}$, $\boxed{126}$, $\boxed{146}$ and $\boxed{150}$ have been derived and exhibited in Tables 14–23 of [Chua *et al.*, 2007], for length $3 \leq L \leq 8$. The basin tree diagrams for the eight *hyper Bernoulli shift* rules $\boxed{26}$, $\boxed{30}$, $\boxed{41}$, $\boxed{45}$, $\boxed{60}$, $\boxed{106}$, $\boxed{110}$ and $\boxed{154}$ are exhibited in Tables 5–12 below.

2.1. *Highlights from rule* $\boxed{26}$

$$\boxed{\text{Gallery 26-1} : L = 3, n\left(\sum^3\right) = 8}$$

(a) There is a *period-1* attractor; namely, $\{\boxed{0}\}$. It has a robustness coefficient of $\rho_1 = 2/8 = 0.25$.

(b) There is a *period-6 Isle of Eden* with a robustness coefficient of $\rho_2 = 6/8 = 0.75$. The dynamics on this Isle of Eden obeys a Bernoulli σ_{τ}-shift law with $\sigma = -1$, $\tau = 2$ and $\beta > 0$. The period is equal to $T = \tau L/|\sigma| = 6$.

$$\boxed{\text{Gallery 26-2} : L = 4, n\left(\sum^4\right) = 16}$$

Table 3.   Each rule $\boxed{N}$ in the left column is equivalent one of the 88 rules $\boxed{N'}$ listed in Table 2 via the global transformation $(T^{\dagger}, \overline{T}, T^{*})$ indicated in column 2.

N	Equivalent Rule			N	Equivalent Rule		
16	2	$T^{\dagger}$	16	83	27	$T^{\dagger}$	83
17	3	$T^{\dagger}$	17	84	14	$T^{\dagger}$	84
20	6	$T^{\dagger}$	20	85	15	$T^{\dagger}$	85
21	7	$T^{\dagger}$	21	86	30	$T^{\dagger}$	86
31	7	$\overline{T}$	31	87	7	T^{*}	87
39	27	$\overline{T}$	39	88	74	$T^{\dagger}$	88
47	11	$\overline{T}$	47	89	45	T^{*}	89
48	34	$T^{\dagger}$	48	91	19	$\overline{T}$ or T^{*}	91
49	35	$T^{\dagger}$	49	92	78	$T^{\dagger}$	92
52	38	$T^{\dagger}$	52	93	13	T^{*}	93
53	27	T^{*}	53	95	5	$\overline{T}$ or T^{*}	95
55	19	$\overline{T}$ or T^{*}	55	96	40	$T^{\dagger}$	96
59	35	$\overline{T}$	59	97	41	$T^{\dagger}$	97
61	25	T^{*}	61	98	56	$T^{\dagger}$	98
63	3	$\overline{T}$	63	99	57	$T^{\dagger}$	99
64	8	$T^{\dagger}$	64	100	44	$T^{\dagger}$	100
65	9	$T^{\dagger}$	65	101	45	$T^{\dagger}$	101
66	24	$T^{\dagger}$	66	102	60	$T^{\dagger}$	102
67	25	$T^{\dagger}$	67	103	25	$\overline{T}$	103
68	12	$T^{\dagger}$	68	107	41	$\overline{T}$	107
69	13	$T^{\dagger}$	69	109	73	$\overline{T}$ or T^{*}	109
70	28	$T^{\dagger}$	70	111	9	$\overline{T}$	111
71	29	$T^{\dagger}$	71	112	42	$T^{\dagger}$	112
75	45	$\overline{T}$	75	113	43	$T^{\dagger}$	113
79	13	$\overline{T}$	79	114	58	$T^{\dagger}$	114
80	10	$T^{\dagger}$	80	115	35	T^{*}	115
81	11	$T^{\dagger}$	81	116	46	$T^{\dagger}$	116
82	26	$T^{\dagger}$	82	117	11	T^{*}	117

Table 3. (*Continued*)

N	Equivalent Rule	N	Equivalent Rule
118	$62 \mapsto T^{\dagger} \to 118$	163	$58 \mapsto \overline{T} \to 163$
119	$3 \mapsto T^{*} \to 119$	165	$90 \mapsto \overline{T} \text{ or } T^{*} \to 165$
120	$106 \mapsto T^{\dagger} \to 120$	166	$154 \mapsto \overline{T} \to 166$
121	$41 \mapsto T^{*} \to 121$	167	$26 \mapsto \overline{T} \to 167$
123	$33 \mapsto \overline{T} \text{ or } T^{*} \to 123$	169	$106 \mapsto \overline{T} \to 169$
124	$110 \mapsto T^{\dagger} \to 124$	171	$42 \mapsto \overline{T} \to 171$
125	$9 \mapsto T^{*} \to 125$	173	$74 \mapsto \overline{T} \to 173$
127	$1 \mapsto \overline{T} \text{ or } T^{*} \to 127$	174	$138 \mapsto \overline{T} \to 174$
129	$126 \mapsto \overline{T} \text{ or } T^{*} \to 129$	175	$10 \mapsto \overline{T} \to 175$
131	$62 \mapsto \overline{T} \to 131$	176	$162 \mapsto T^{\dagger} \to 176$
133	$94 \mapsto \overline{T} \text{ or } T^{*} \to 133$	177	$58 \mapsto T^{*} \to 177$
135	$30 \mapsto \overline{T} \to 135$	179	$50 \mapsto \overline{T} \text{ or } T^{*} \to 179$
137	$110 \mapsto \overline{T} \to 137$	180	$154 \mapsto T^{*} \to 180$
139	$46 \mapsto \overline{T} \to 139$	181	$26 \mapsto T^{*} \to 181$
141	$78 \mapsto \overline{T} \to 141$	182	$146 \mapsto \overline{T} \text{ or } T^{*} \to 182$
143	$14 \mapsto \overline{T} \to 143$	183	$18 \mapsto \overline{T} \text{ or } T^{*} \to 183$
144	$130 \mapsto T^{\dagger} \to 144$	185	$56 \mapsto T^{*} \to 185$
145	$62 \mapsto T^{*} \to 145$	186	$162 \mapsto \overline{T} \to 186$
147	$54 \mapsto \overline{T} \text{ or } T^{*} \to 147$	187	$34 \mapsto \overline{T} \to 187$
148	$134 \mapsto T^{\dagger} \to 148$	188	$152 \mapsto T^{*} \to 188$
149	$30 \mapsto T^{*} \to 149$	189	$24 \mapsto T^{*} \to 189$
151	$22 \mapsto \overline{T} \text{ or } T^{*} \to 151$	190	$130 \mapsto \overline{T} \to 190$
153	$60 \mapsto T^{*} \to 153$	191	$2 \mapsto \overline{T} \to 191$
155	$38 \mapsto \overline{T} \to 155$	192	$136 \mapsto T^{\dagger} \to 192$
157	$28 \mapsto T^{*} \to 157$	193	$110 \mapsto T^{*} \to 193$
158	$134 \mapsto \overline{T} \to 158$	194	$152 \mapsto T^{\dagger} \to 194$
159	$6 \mapsto \overline{T} \to 159$	195	$60 \mapsto \overline{T} \to 195$
161	$122 \mapsto \overline{T} \text{ or } T^{*} \to 161$	196	$140 \mapsto T^{\dagger} \to 196$

Table 3.   (*Continued*)

N	Equivalent Rule		N	Equivalent Rule
197	78 $\xrightarrow{T^*}$ 197		227	56 $\xrightarrow{\bar{T}}$ 227
198	156 $\xrightarrow{T^\dagger \text{ or } \bar{T}}$ 198		228	172 $\xrightarrow{T^\dagger}$ 228
199	28 $\xrightarrow{\bar{T}}$ 199		229	74 $\xrightarrow{T^*}$ 229
201	108 $\xrightarrow{\bar{T} \text{ or } T^*}$ 201		230	152 $\xrightarrow{\bar{T}}$ 230
202	172 $\xrightarrow{\bar{T}}$ 202		231	24 $\xrightarrow{\bar{T}}$ 231
203	44 $\xrightarrow{\bar{T}}$ 203		233	104 $\xrightarrow{\bar{T} \text{ or } T^*}$ 233
205	76 $\xrightarrow{\bar{T} \text{ or } T^*}$ 205		234	168 $\xrightarrow{\bar{T}}$ 234
206	140 $\xrightarrow{\bar{T}}$ 206		235	40 $\xrightarrow{\bar{T}}$ 235
207	12 $\xrightarrow{\bar{T}}$ 207		236	200 $\xrightarrow{\bar{T} \text{ or } T^*}$ 236
208	138 $\xrightarrow{T^\dagger}$ 208		237	72 $\xrightarrow{\bar{T} \text{ or } T^*}$ 237
209	46 $\xrightarrow{T^*}$ 209		238	136 $\xrightarrow{\bar{T}}$ 238
210	154 $\xrightarrow{T^\dagger}$ 210		239	8 $\xrightarrow{\bar{T}}$ 239
211	38 $\xrightarrow{T^*}$ 211		240	170 $\xrightarrow{T^\dagger \text{ or } T^*}$ 240
212	142 $\xrightarrow{T^\dagger \text{ or } T^*}$ 212		241	42 $\xrightarrow{T^*}$ 241
213	14 $\xrightarrow{T^*}$ 213		242	162 $\xrightarrow{T^*}$ 242
214	134 $\xrightarrow{T^*}$ 214		243	34 $\xrightarrow{T^*}$ 243
215	6 $\xrightarrow{T^*}$ 215		244	138 $\xrightarrow{T^*}$ 244
216	172 $\xrightarrow{T^*}$ 216		245	10 $\xrightarrow{T^*}$ 245
217	44 $\xrightarrow{T^*}$ 217		246	130 $\xrightarrow{T^*}$ 246
218	164 $\xrightarrow{\bar{T} \text{ or } T^*}$ 218		247	2 $\xrightarrow{T^*}$ 247
219	36 $\xrightarrow{\bar{T} \text{ or } T^*}$ 219		248	168 $\xrightarrow{T^*}$ 248
220	140 $\xrightarrow{T^*}$ 220		249	40 $\xrightarrow{T^*}$ 249
221	12 $\xrightarrow{T^*}$ 221		250	160 $\xrightarrow{\bar{T} \text{ or } T^*}$ 250
222	132 $\xrightarrow{\bar{T} \text{ or } T^*}$ 222		251	32 $\xrightarrow{\bar{T} \text{ or } T^*}$ 251
223	4 $\xrightarrow{\bar{T} \text{ or } T^*}$ 223		252	136 $\xrightarrow{T^*}$ 252
224	168 $\xrightarrow{T^\dagger}$ 224		253	8 $\xrightarrow{T^*}$ 253
225	106 $\xrightarrow{T^*}$ 225		254	128 $\xrightarrow{\bar{T} \text{ or } T^*}$ 254
226	184 $\xrightarrow{T^\dagger \text{ or } \bar{T}}$ 226		255	0 $\xrightarrow{\bar{T} \text{ or } T^*}$ 255

Table 4. Steady-state characterization of 88 dynamically- independent local rules.

Topological Classifications of 88 Equivalence Classes

Topologically-distinct Rules	Number
Period-1 Rules	25
Period-2 Rules	13
Period-T Rules $T > 2$	2
Bernoulli σ_τ-Shift Rules	30
Complex Bernoulli-Shift Rules	10
Hyper Bernoulli-Shift Rules	8
Total	**88**

There are 15 bit strings which converge to the *global period-1 attractor* $\{\textcircled{0}\}$. Hence, the *period-1 attractor* $\textcircled{0}$ has maximum robustness coefficient with $\rho_1 = 16/16 = 1$.

$$\boxed{\text{Gallery } 26\text{-}3, 26\text{-}4 : L = 5, n\left(\Sigma^5\right) = 32}$$

(a) There is a *period-1 attractor* $\{\textcircled{0}\}$ with a robustness coefficient $\rho_1 = 2/32 = 0.0625$.

(b) There is a *period-5 Isle of Eden* with a robustness coefficient $\rho_2 = 0.15625$. The dynamics on this Isle of Eden obeys a Bernoulli σ_τ-shift law with $\sigma = -1$, $\tau = 2$ and $\beta > 0$. The period obeys the formula $T = \tau L/|\sigma| = 10$. But symmetry of the pattern reduces T to a minimal period of $T_{\min} = 5$.

(c) There is a *period-20 attractor* with a robustness coefficient $\rho_3 = 25/32 = 0.78125$. The dynamics follows a Bernoulli σ_τ-shift law with $\sigma = -1$, $\tau = 4$ and $\beta > 0$. The period obeys the formula $T = \tau L/|\sigma| = 20$.

$$\boxed{\text{Gallery } 26\text{-}5, 26\text{-}6, 26\text{-}7 : L = 6, n\left(\Sigma^6\right) = 64}$$

(a) There is a *period-1 attractor* $\{\textcircled{0}\}$ with a robustness coefficient $\rho_1 = 4/64 = 0.0625$.

(b) There are three *period-2 attractors* with a combined robustness coefficient $\rho_2 = 3(8/64) = 0.375$. The dynamics on these attractors follow a Bernoulli σ_τ-shift law with $\sigma_1 = 3$, $\tau_1 = 1$ and $\beta > 0$, or $\sigma_2 = -3$, $\tau_2 = 1$ and $\beta > 0$.

(c) There is a *period-6 Isle of Eden* with a robustness coefficient $\rho_3 = 6/64 = 0.09375$. The dynamics on this Isle of Eden obeys a Bernoulli σ_τ-shift law with $\sigma = -1$ and $\tau = 2$ and $\beta > 0$. The period obeys the formula $T = \tau L/|\sigma| = 12$. Symmetry of the pattern reduces T to a minimal period of $T_{\min} = 6$.

(d) There are two *period-3 attractors* with a combined robustness coefficient $\rho_3 = 2(15/64) = 0.46875$. The dynamics on each attractor obeys a Bernoulli σ_τ-shift law with $\sigma = -2$ and $\tau = 1$ and $\beta > 0$. The period obeys the formula $T = \tau L/|\sigma| = 3$.

$$\boxed{\text{Gallery } 26\text{-}8, 26\text{-}9 : L = 7, n\left(\Sigma^7\right) = 128}$$

(a) There is a *period-28 attractor* with a robustness coefficient $\rho_1 = 91/128 = 0.7109375$. The dynamics on this attractor obeys a Bernoulli σ_τ-shift law with $\sigma_1 = -3$, $\tau_1 = 4$ and $\beta > 0$, or $\sigma_2 = 4$, $\tau_2 = 4$ and $\beta > 0$. The period obeys the formula $T = \tau L = 28$ in this case, and not $\tau L/|\sigma|$, because $L/|\sigma| = 7/3$ is not an integer.

(b) There is a second *period-28 attractor* (with a topologically distinct basin tree) with a robustness coefficient $\rho_2 = 35/128 = 0.2734375$.

(c) There is a *period-1 attractor* $\{\textcircled{0}\}$ with a robustness coefficient $\rho_3 = 2/128 = 0.015625$.

$$\boxed{\text{Gallery } 26\text{-}10, 26\text{-}11, 26\text{-}12, 26\text{-}13 : L = 8, \; n\left(\Sigma^8\right) = 256}$$

(a) There are four *period-8 attractors* with a combined robustness coefficient $\rho_1 = 4(16/256) = 0.25$. The dynamics on each attractor obeys a Bernoulli σ_τ-shift law with $\sigma_1 = -4$, $\tau_1 = 4$ and $\beta > 0$, or

$\sigma_2 = 4$, $\tau_2 = 4$ and $\beta > 0$. The period obeys the formula $T = \tau L/|\sigma| = 8$.

(b) There are four *period-8 attractors* (with a topologically distinct basin tree) with a combined robustness coefficient $\rho_2 = 4(20/256) = 0.3125$. The dynamics on each attractor obeys a Bernoulli σ_τ-shift law with $\sigma_1 = -4$, $\tau_1 = 4$ and $\beta > 0$, or $\sigma_2 = 4$, $\tau_2 = 4$ and $\beta > 0$. The period obeys the formula $T = \tau L/|\sigma| = 8$.

(c) There is a *period-16 Isle of Eden* with a robustness coefficient $\rho_3 = 16/256 = 0.0625$. The dynamics on this Isle of Eden obeys a Bernoulli σ_τ-shift law with $\sigma = -1$, $\tau = 2$ and $\beta > 0$. The period obeys the formula $T = \tau L/|\sigma| = 16$.

(d) There is a *period-1 attractor* $\{0\}$ with a robustness coefficient $\rho_4 = 96/256 = 0.375$.

Observe that $\{0\}$ is a *period-1 attractor* of rule $\boxed{26}$ for $3 \leq L \leq 8$. Observe also that rule $\boxed{26}$ has an Isle of Eden for $L = 3, 5, 6$ and 8.

The qualitative properties of local rule $\boxed{26}$ extracted from the above basin-tree Galleries 26-1 to 26-13 are summarized below:

Summary of Qualitative properties of local rule $\boxed{26}$ *extracted from Gallery   26 for Rule* $\boxed{26}$

L	ID Number i	Number of Period-n attractors	Number of Period-n Isles of Eden	Period n	Bernoulli			Parameters			Robustness coefficient ρ
					σ_1	τ_1	β_1	σ_2	τ_2	β_2	
3	1	1		1	0	1	+				$\rho_1 = 0.25$
	2		1	6				-1	2	+	$\rho_2 = 0.75$
4	1	1		1	0	1	+				$\rho_1 = 1$
5	1	1		1	0	1	+				$\rho_1 = 0.0625$
	2		1	5				-1	2	+	$\rho_2 = 0.15625$
	3	1		20				-1	4	+	$\rho_3 = 0.78125$
6	1	1		1	0	1	+				$\rho_1 = 0.0625$
	2	3		2	3	1	+	-3	1	+	$\rho_2 = 0.375$
	3		1	6				-1	2	+	$\rho_3 = 0.09375$
	4	2		3				-2	1	+	$\rho_4 = 0.46875$
7	1	1		28	-3	4	+	4	4	+	$\rho_1 = 0.7109375$
	2	1		28	-3	4	+	4	4	+	$\rho_2 = 0.2734375$
	3	1		1	0	1	+				$\rho_3 = 0.015625$
8	1	4		8	4	4	+	-4	4	+	$\rho_1 = 0.25$
	2	4		8	4	4	+	-4	4	+	$\rho_2 = 0.3125$
	3		1	16				-1	2	+	$\rho_3 = 0.0625$
	4	1		1	0	1	+				$\rho_4 = 0.375$

2.2. *Highlights from rule* $\boxed{30}$

$$\boxed{\text{Gallery } 30\text{-}1 : L = 3, n\left(\textstyle\sum^3\right) = 8}$$

There are seven basin-tree strings, all of which converge to the global *period-1 attractor* $\{0\}$.

Hence, the *period-1 attractor* 0 has maximum robustness coefficient with $\rho_1 = 8/8 = 1$. The dynamics on this attractor obeys a *degenerate Bernoulli σ_τ-shift law* with $\sigma = 0$, $\tau = 1$ and $\beta > 0$.

$$\boxed{\text{Gallery } 30\text{-}2 : L = 4, n\left(\textstyle\sum^4\right) = 16}$$

Basin tree diagrams for Rule 26

26 , $L = 3$

(a) Period-1 Attractor :

$$\rho_1 = \frac{2}{8} = 0.25$$

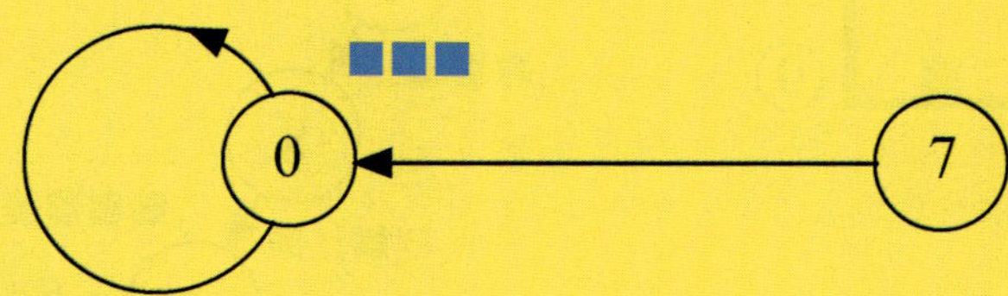

(b) Bernoulli Period-6 Isle of Eden : $\rho_2 = \frac{6}{8} = 0.75$

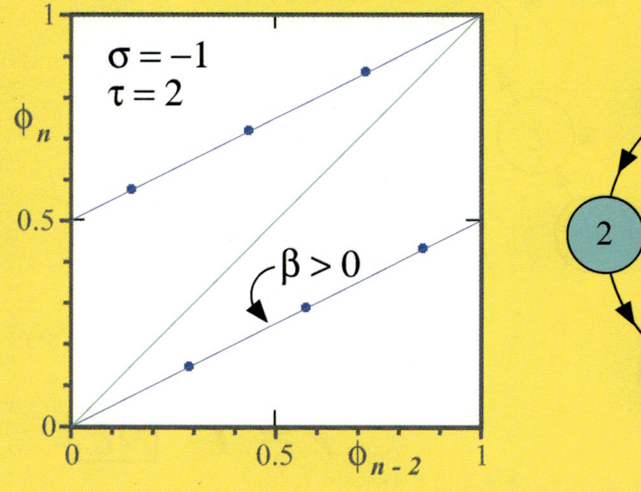

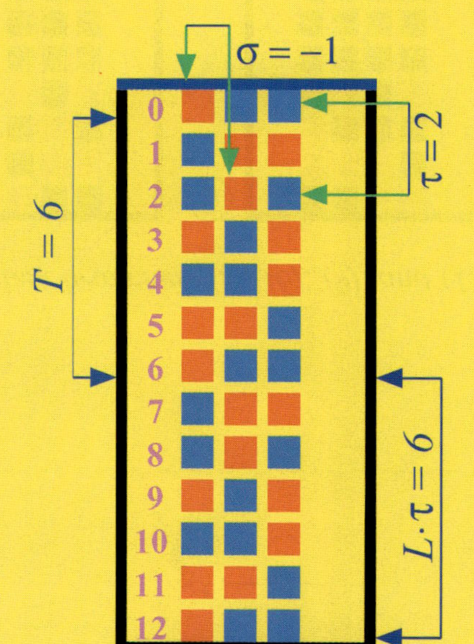

Gallery 26 - 1

Basin tree diagrams for Rule 26

26 , *L = 4*

(a) Period-1 Attractor :

$$\rho_1 = \frac{16}{16} = 1$$

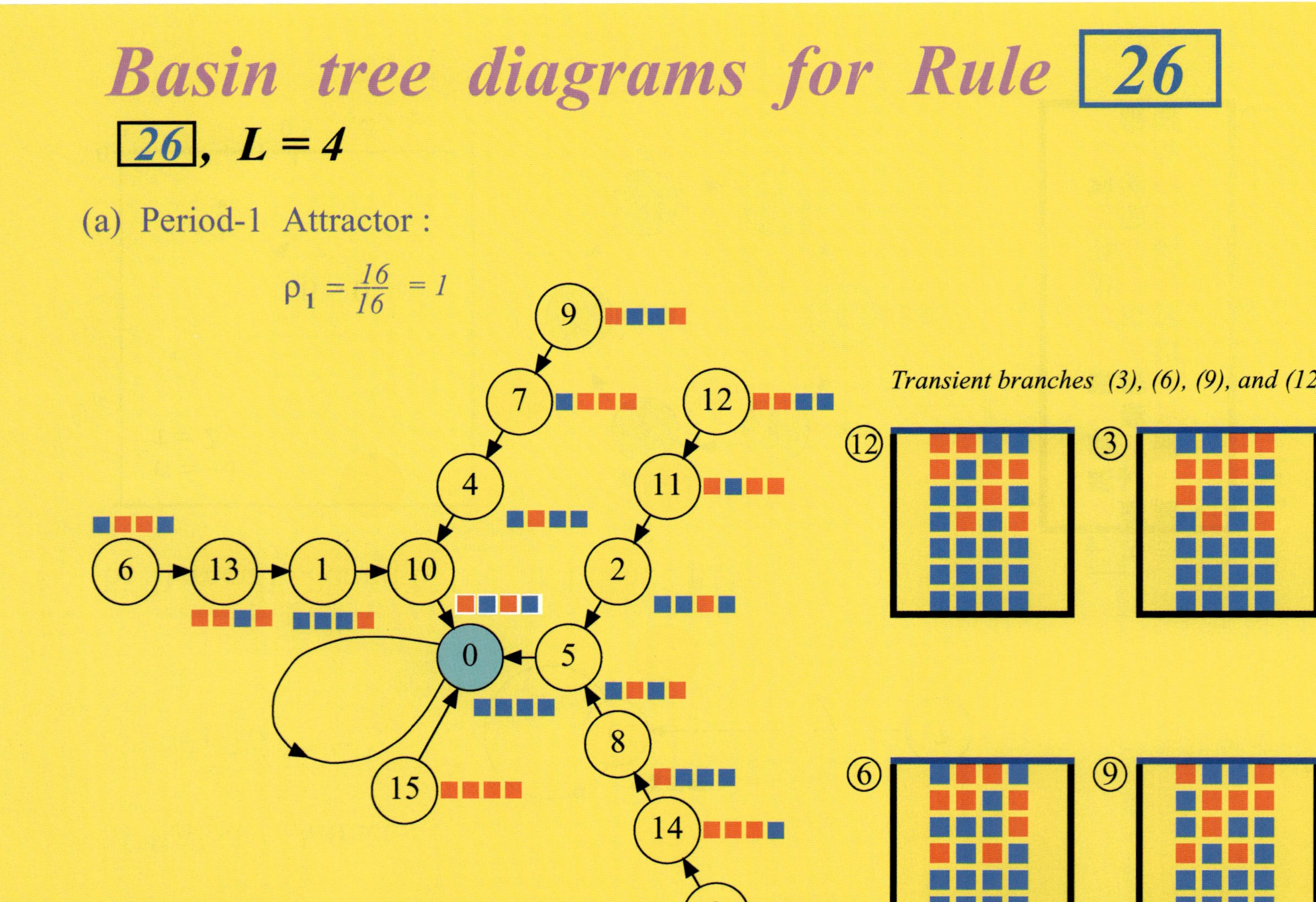

Gallery 26 - 2

Basin tree diagrams for Rule 26

26 , $L = 5$

(a) Period-1 Attractor :

$$\rho_1 = \frac{2}{32} = 0.0625$$

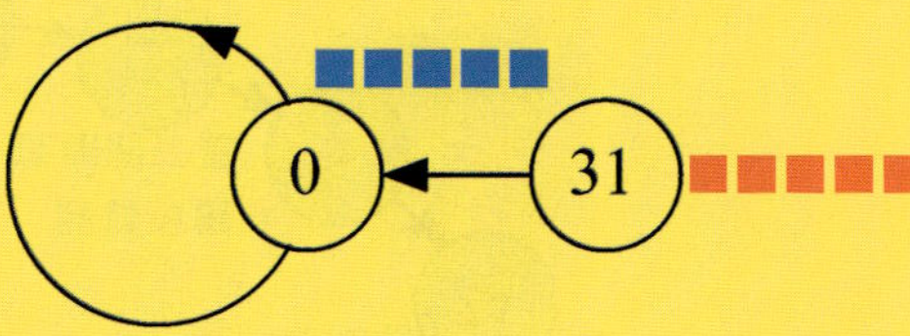

(b) Bernoulli Period-5 Isle of Eden : $\rho_2 = \frac{5}{32} = 0.15625$

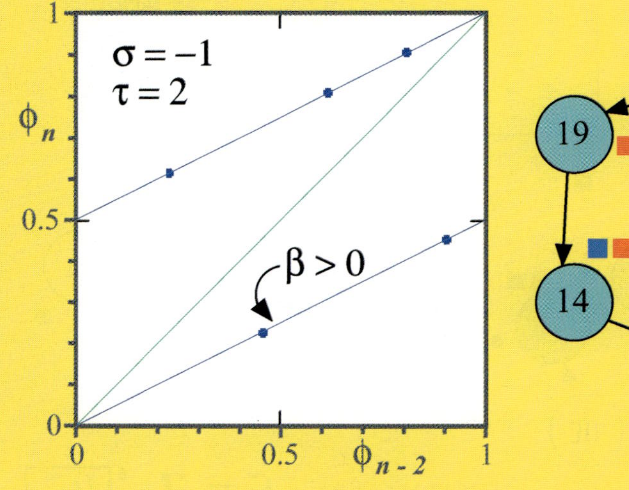

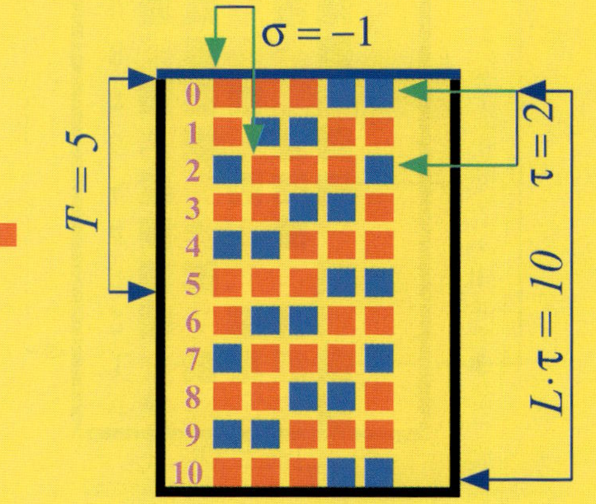

Gallery 26 -3

Table 5. (*Continued*)

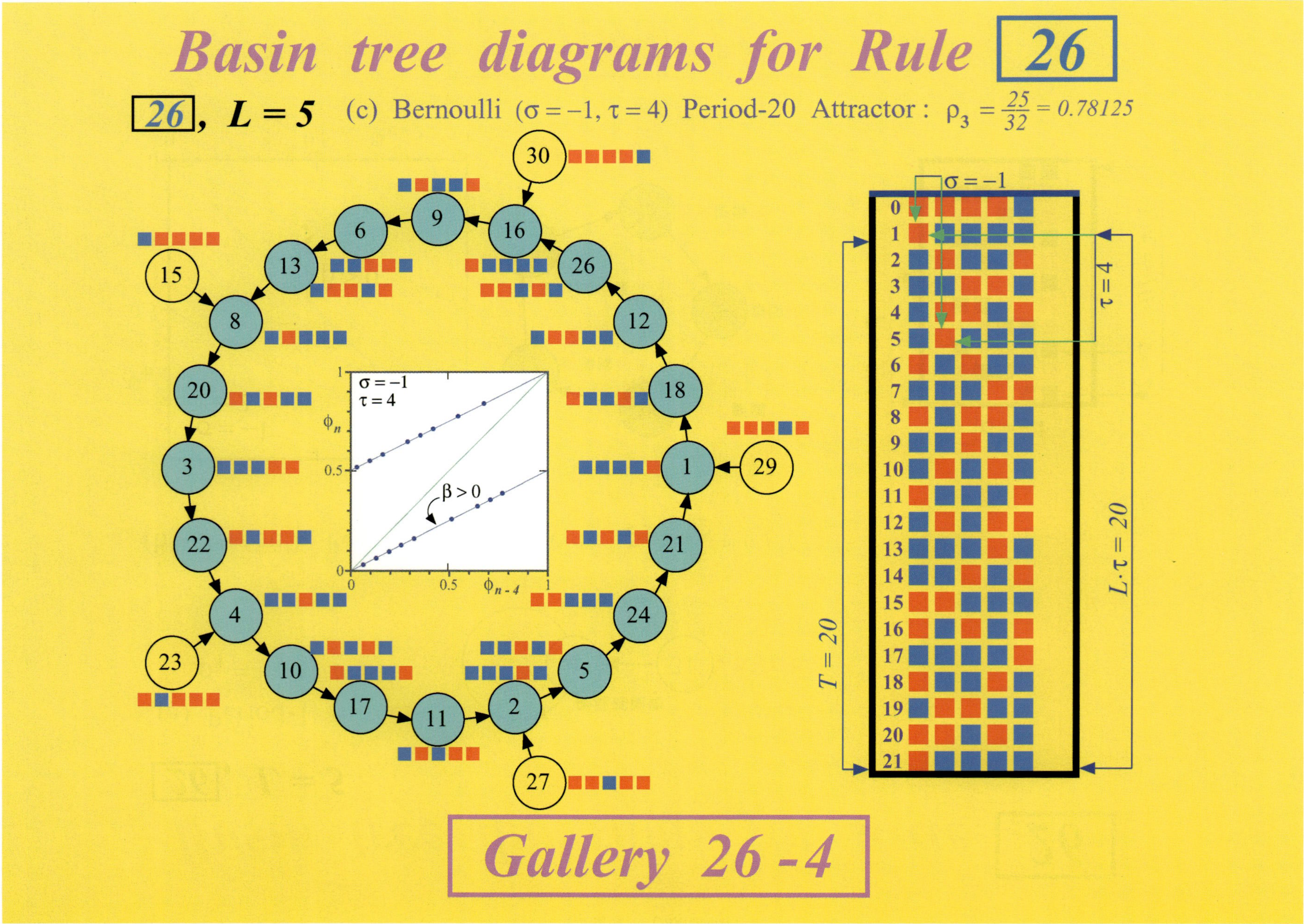
Basin tree diagrams for Rule 26
26, L = 5 (c) Bernoulli (σ = −1, τ = 4) Period-20 Attractor : $\rho_3 = \frac{25}{32} = 0.78125$
30
9
16
6
26
13
15
12
8
20
18
3
1
29
22
21
4
24
23
10
5
17
11
2
27
σ = −1
τ = 4
ϕ_n
0.5
β > 0
ϕ_{n-4}
σ = −1
0
1
2
3
4
5
6
7
8
9
10
11
12
13
14
15
16
17
18
19
20
21
τ = 4
T = 20
L·τ = 20
Gallery 26 -4

Basin tree diagrams for Rule $\boxed{26}$

$\boxed{26}$, $L = 6$ (b) Bernoulli ($\sigma = \pm 3$, $\tau = 1$) Period-2 Attractors :

(a) Bernoulli ($\sigma = 0$, $\tau = 1$)

$$\rho_2 = 3\,\frac{8}{64} = 0.375$$

Period-1 Attractor :

$$\rho_1 = \frac{4}{64} = 0.0625$$

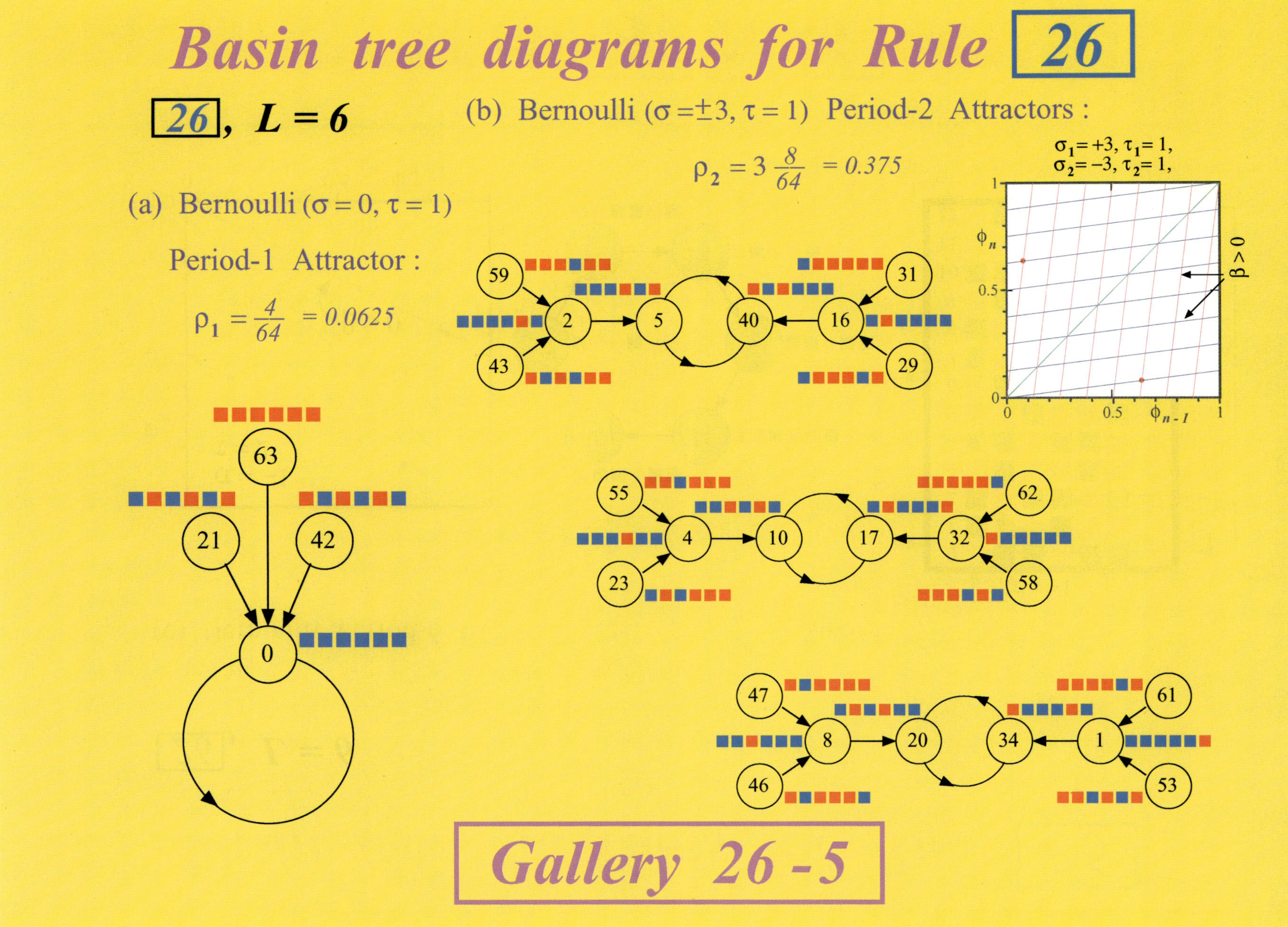

$\boxed{Gallery\ 26 - 5}$

Basin tree diagrams for Rule 26

26 , *L = 6*

(c) Bernoulli Period-6 Isle of Eden : $\rho_3 = \dfrac{6}{64} = 0.09375$

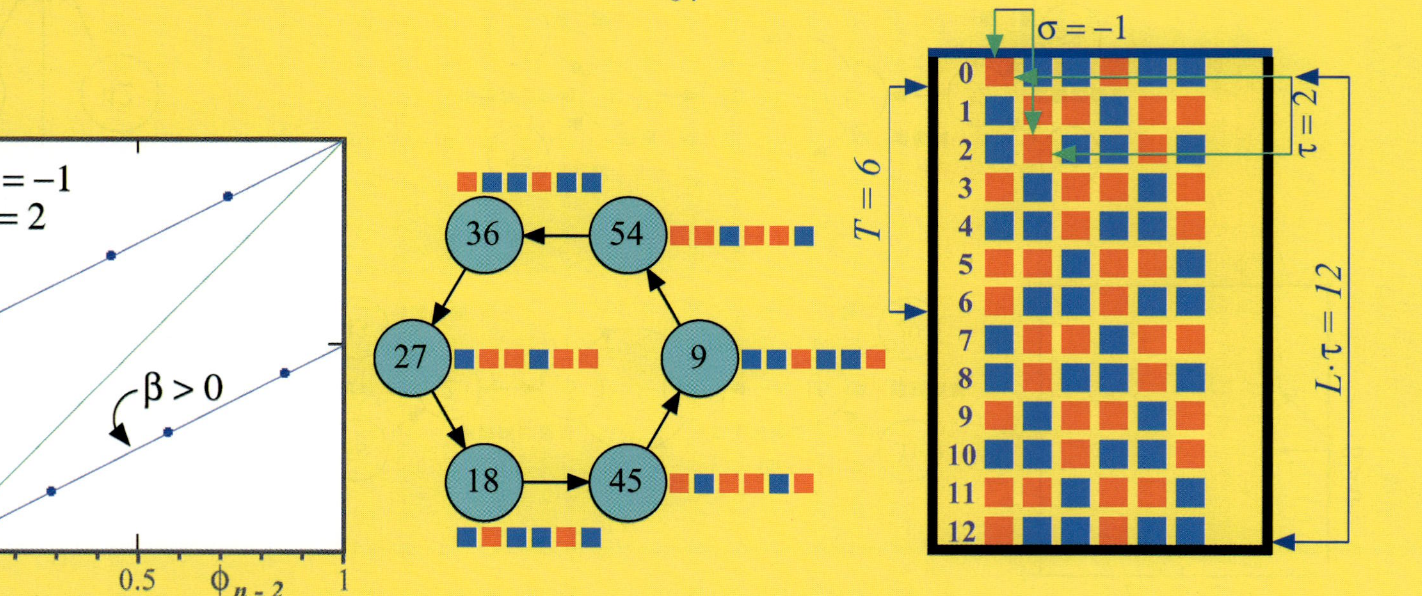

Gallery 26 - 6

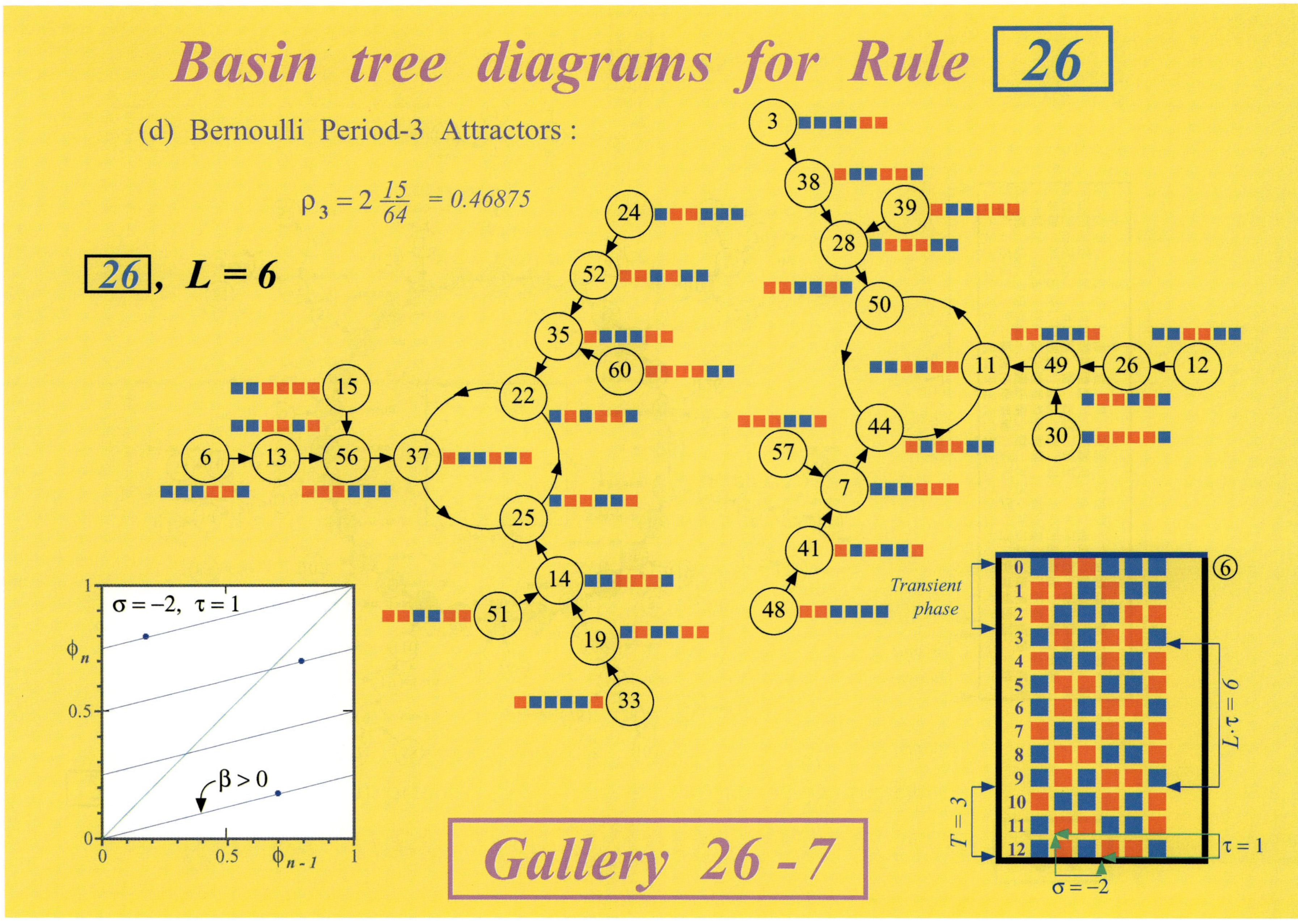

Basin tree diagrams for Rule 26
(d) Bernoulli Period-3 Attractors :
$\rho_3 = 2\,\frac{15}{64} = 0.46875$
26, L = 6
σ = −2, τ = 1
β > 0
Transient phase
T = 3
τ = 1
σ = −2
Gallery 26 - 7

Table 5. (*Continued*)

Basin tree diagrams for Rule $\boxed{26}$

$\boxed{26}$, $\boldsymbol{L = 7}$ (a) Bernoulli ($\sigma = -3$, $\sigma = 4$, $\tau = 4$) Period-28 Attractor :

$$\rho_1 = \frac{91}{128} = 0.7109375$$

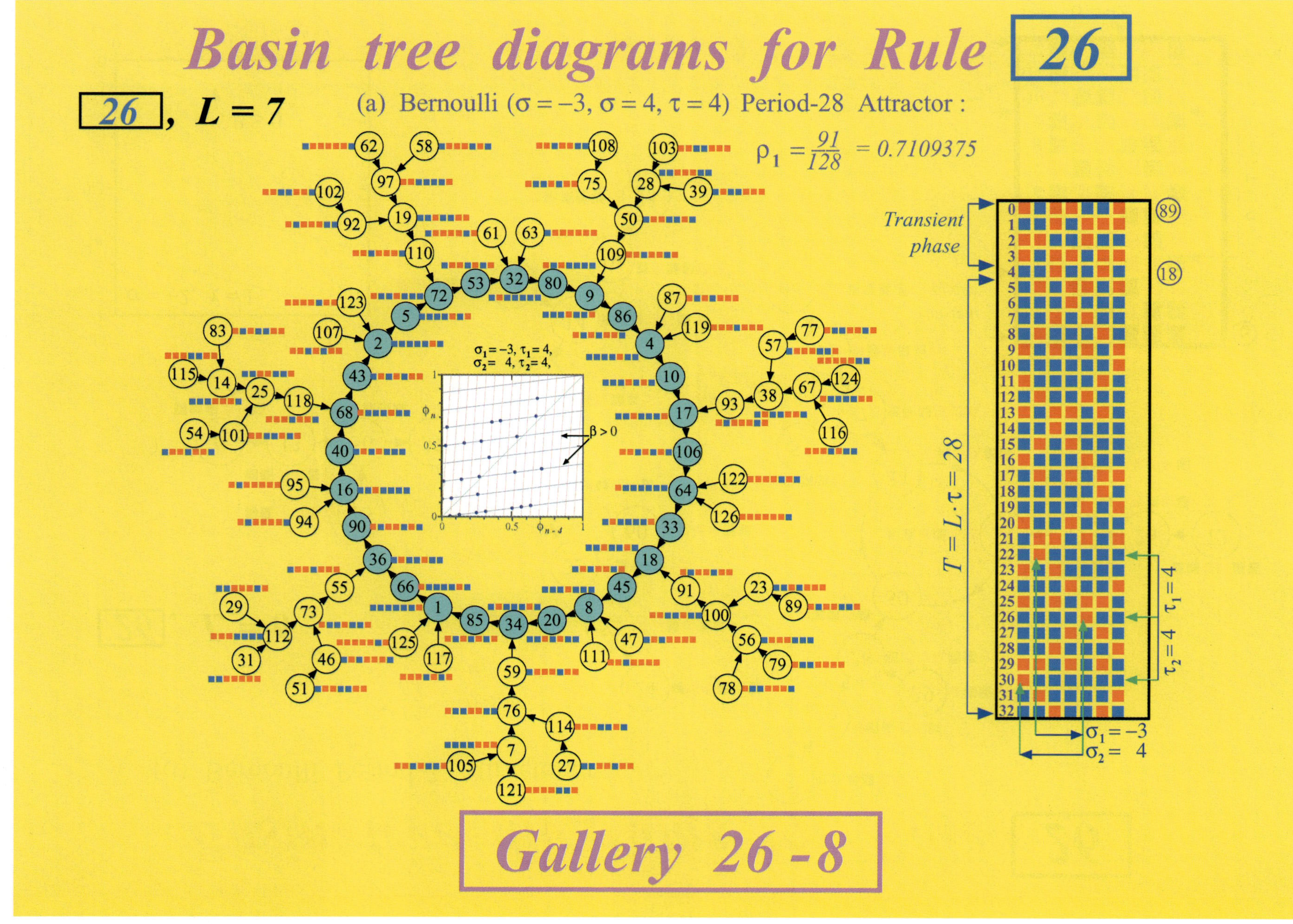

$\boxed{\textit{Gallery 26-8}}$

190

Table 5.   (Continued)

Basin tree diagrams for Rule 26

26, **L = 7** (b) Bernoulli ($\sigma = -3$, $\sigma = 4$, $\tau = 4$) Period-28 Attractor: (c) Period-1 Attractor :

$$\rho_2 = \frac{35}{128} = 0.2734375 \qquad \rho_3 = \frac{2}{128} = 0.015625$$

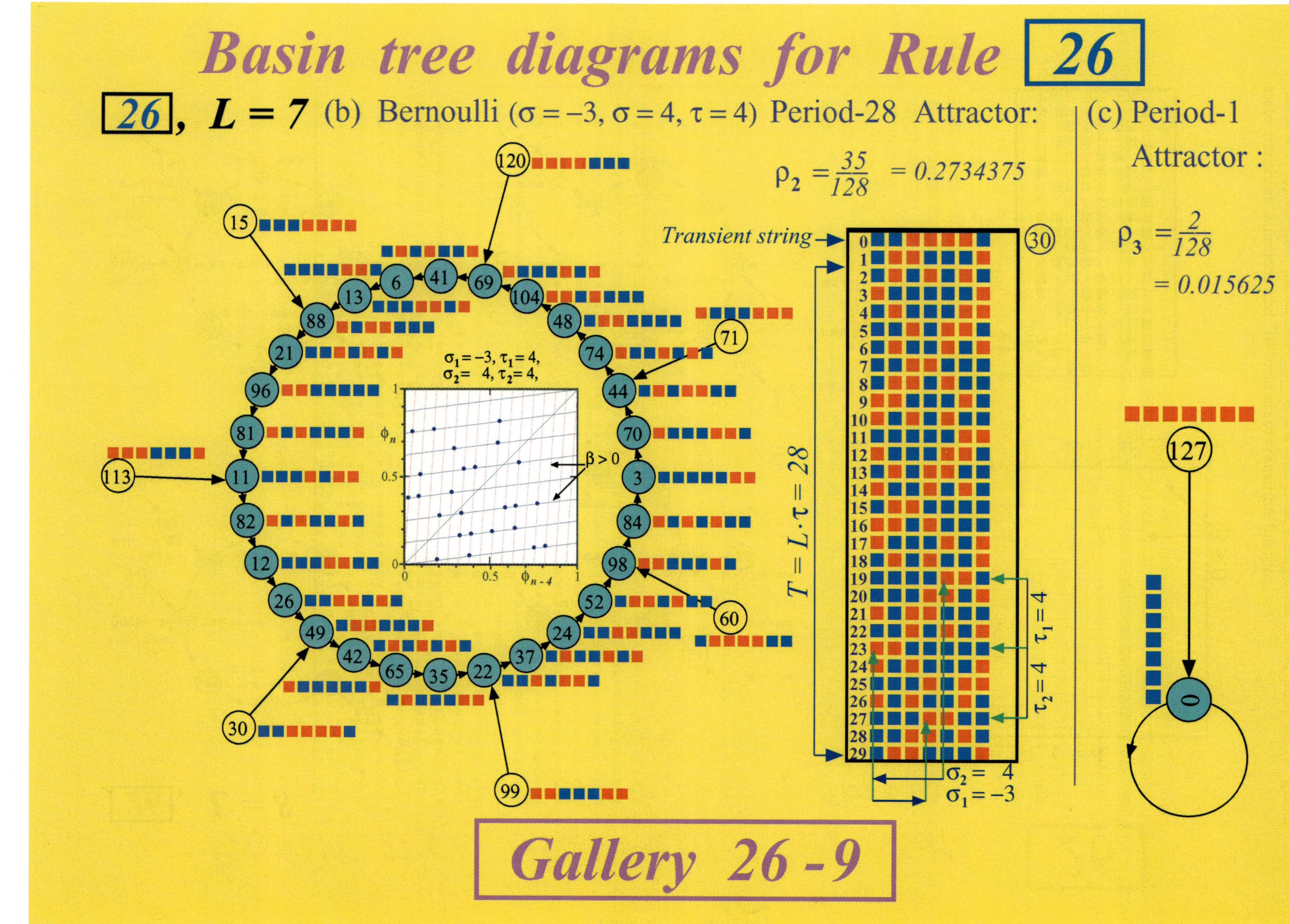

Gallery 26 - 9

191

Basin tree diagrams for Rule 26

26, $L = 8$

(a) Bernoulli ($\sigma = \pm 4$, $\tau = 4$) 4 Period-8 Attractors :

$$\rho_1 = 4\frac{16}{256} = 0.25$$

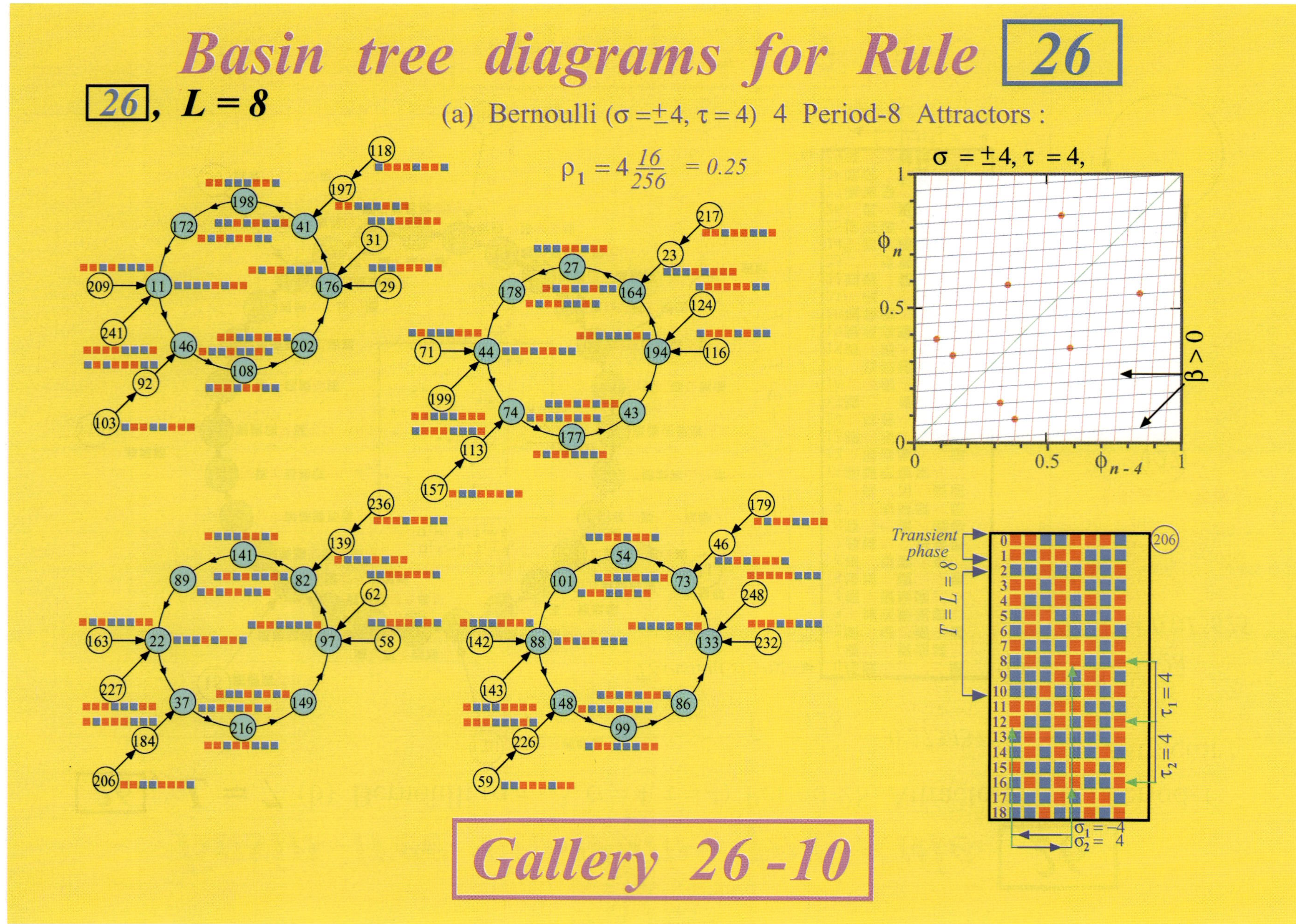

Gallery 26 -10

Basin tree diagrams for Rule 26

26, *L* = 8   (b) Bernoulli ($\sigma = \pm 4$, $\tau = 4$) 4 Period-8 Attractors :

$$\rho_2 = 4\frac{20}{256} = 0.3125$$

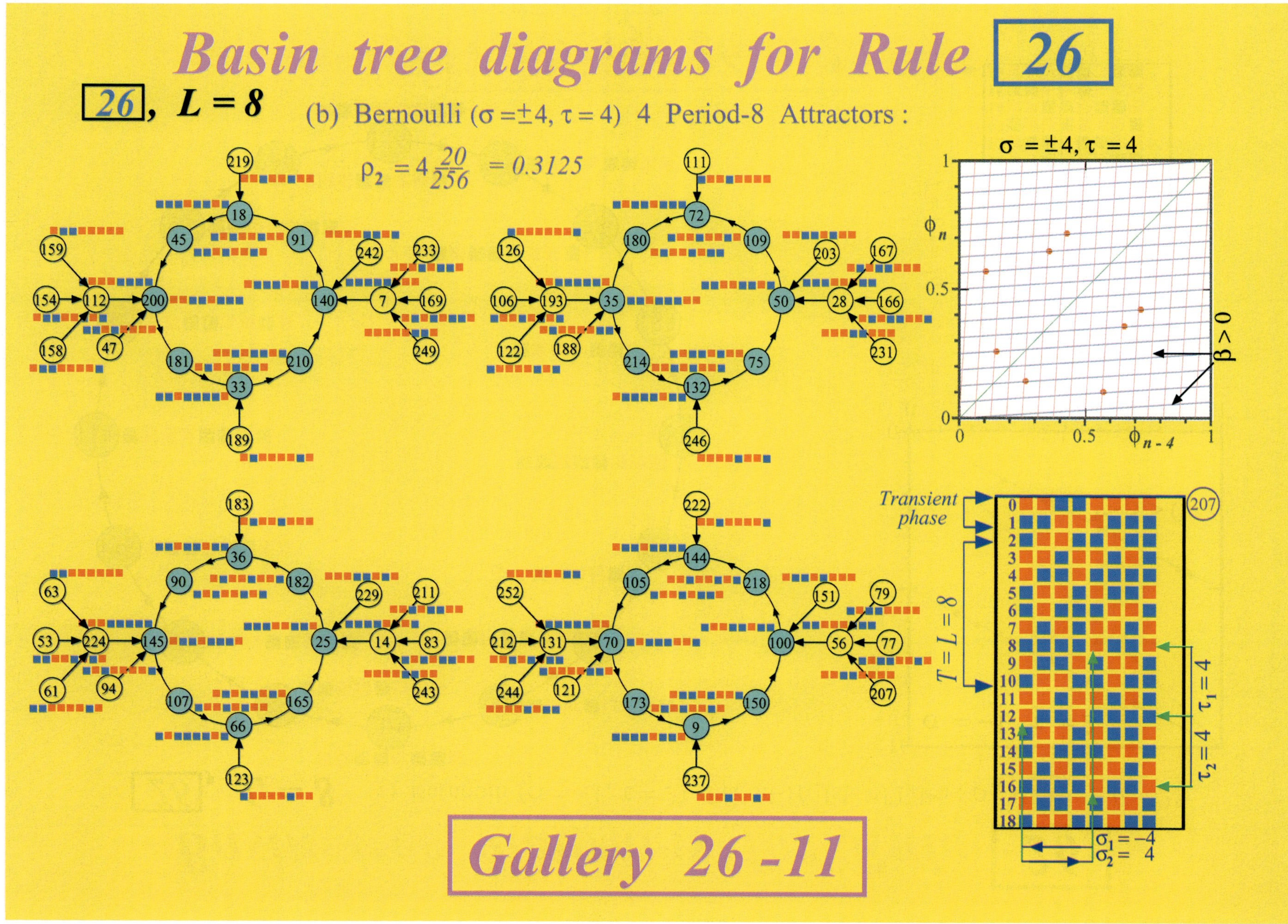

Gallery 26 -11

Table 5. (*Continued*)

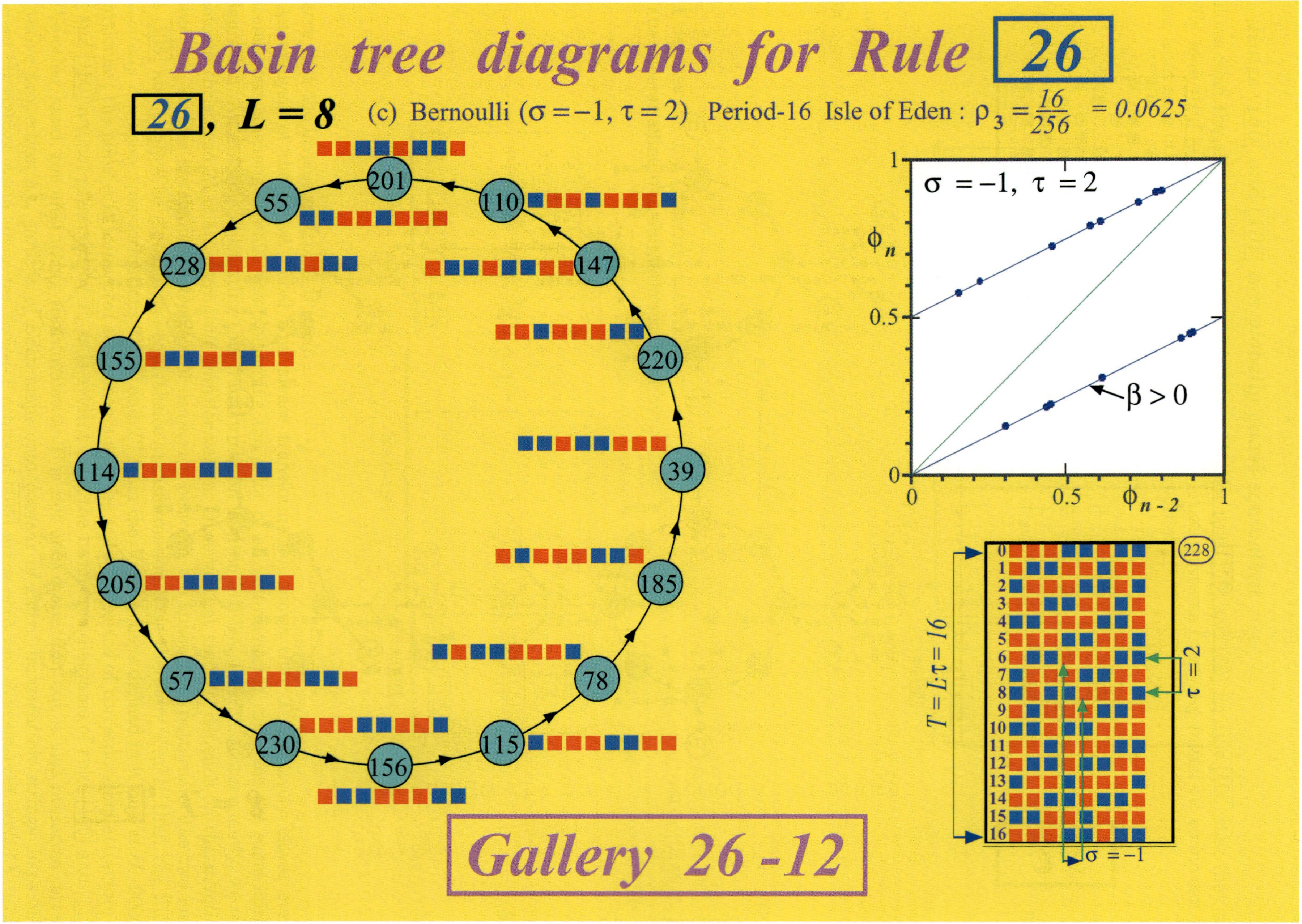

Basin tree diagrams for Rule 26

26, **L = 8**   (c) Bernoulli ($\sigma = -1$, $\tau = 2$)   Period-16   Isle of Eden : $\rho_3 = \frac{16}{256} = 0.0625$

Gallery 26 -12

Table 5. (*Continued*)

Basin tree diagrams for Rule 26

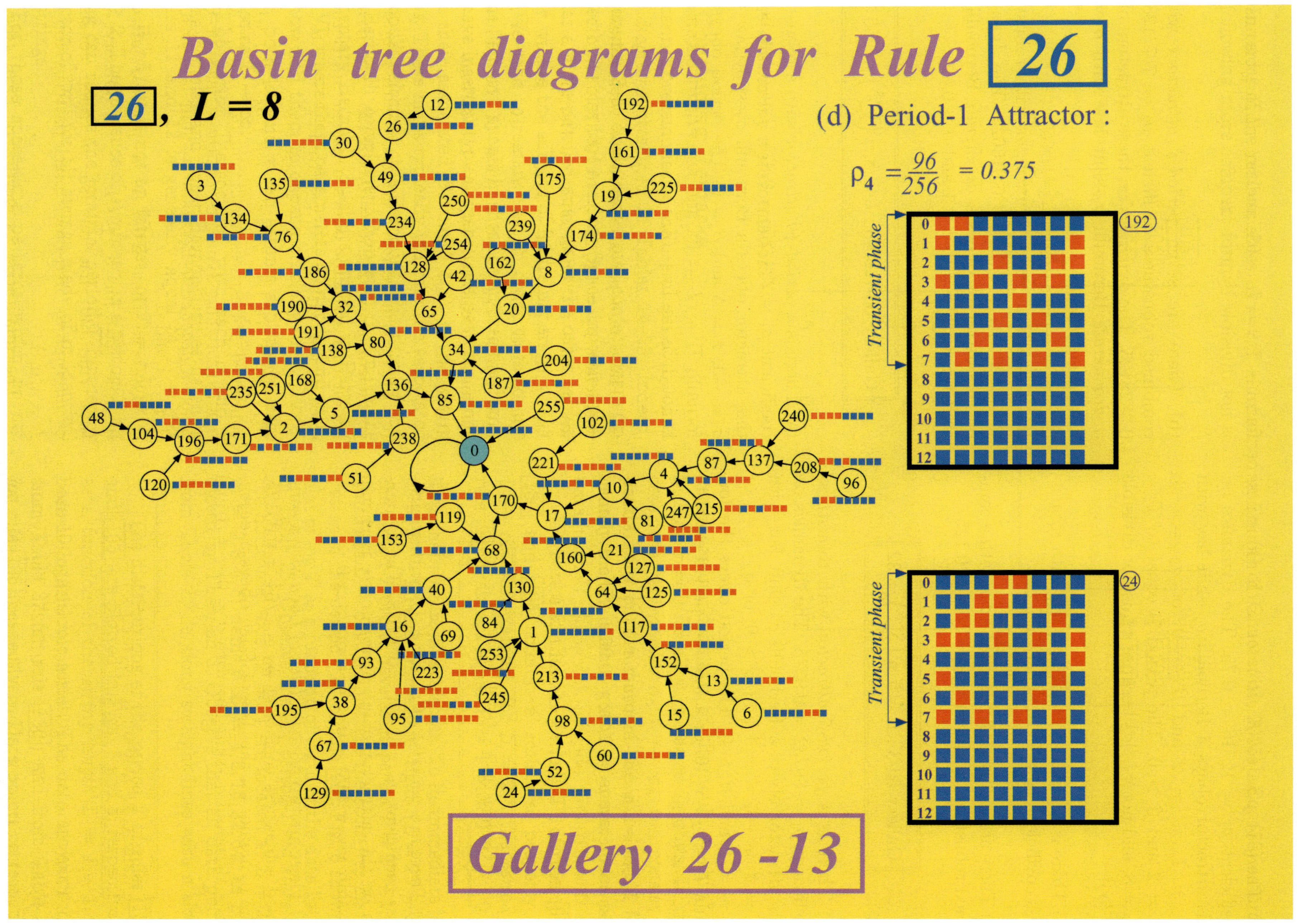

195

(a) There is a *period-8 attractor* with robustness coefficient $\rho_1 = 12/16 = 0.75$. The dynamics on this attractor obeys a Bernoulli σ_τ-shift law with $\sigma_1 = 1$, $\tau_1 = 2$ and $\beta > 0$, or $\sigma_2 = -3$, $\tau_2 = 2$ and $\beta > 0$. The period is equal to $T = \tau L = 8$, and not $\tau L/|\sigma|$ because τL is not divisible by $|\sigma_2|$.

(b) There is a *period-1 attractor* $\{\boxed{0}\}$ with robustness coefficient $\rho_2 = 2/16 = 0.125$.

(c) There are two *period-1 Isles of Eden* with a combined robustness coefficient $\rho_3 = 2/16 = 0.125$.

$$\boxed{\text{Gallery } 30\text{-}3 : L = 5, n\left(\Sigma^5\right) = 32}$$

(a) There is a *period-5 attractor* with robustness coefficient $\rho_1 = 30/32 = 0.9375$. The dynamics on this attractor obeys a Bernoulli σ_τ-shift law with $\sigma_1 = 2$ or -3, $\tau_1 = 1$ and $\beta > 0$; or $\sigma_2 = -1$ or 4, $\tau_2 = 2$ and $\beta > 0$. The period obeys the formula $T = \tau_2 L = 10$, or a *minimal* period equal to $T_{\min} = \tau_1 L = 5$.

(b) There is a *period-1 attractor* $\{\boxed{0}\}$ with robustness coefficient $\rho_2 = 2/32 = 0.0625$. The dynamics on this attractor obeys a *degenerate* Bernoulli σ_τ-shift law with $\sigma = 0$, $\tau = 1$ and $\beta > 0$.

$$\boxed{\text{Gallery } 30\text{-}4 : L = 6, n\left(\Sigma^6\right) = 64}$$

(a) There is a *period-1 attractor* $\{\boxed{0}\}$ with robustness coefficient $\rho_1 = 62/64 = 0.96875$. The dynamics on this attractor obeys a *degenerate* Bernoulli σ_τ-shift law with $\sigma = 0$, $\tau = 1$ and $\beta > 0$.

(b) There are two *period-1 Isles of Eden* with a combined robustness coefficient $\rho_2 = 2/64 = 0.03125$. The dynamics on this Isle of Eden obeys a *degenerate* Bernoulli σ_τ-shift law with $\sigma = 0$, $\tau = 1$ and $\beta > 0$.

$$\boxed{\text{Gallery } 30\text{-}5, 30\text{-}6 : L = 7, n\left(\Sigma^7\right) = 128}$$

(a) There is a *period-63 attractor* with robustness coefficient $\rho_1 = 77/128 = 0.6015625$. The dynamics on this attractor obeys a Bernoulli σ_τ-shift law with $\sigma = 2$, $\tau = 9$ and $\beta > 0$. The period obeys the

formula $T = \tau L = 63$, and not $\tau L/|\sigma|$ because 63 is not divisible by $|\sigma| = 2$.

(b) There are seven *period-4 attractors* with a combined robustness coefficient $\rho_2 = 7(7/128) = 0.3828125$. The dynamics on these attractors obeys a *degenerate* Bernoulli σ_τ-shift law with $\sigma = 0$, $\tau = 4$ and $\beta > 0$.

(c) There is a *period-1 attractor* with robustness coefficient $\rho_3 = 2/128 = 0.015625$. The dynamics on this attractor obeys a *degenerate* Bernoulli σ_τ-shift law with $\sigma = 0$, $\tau = 1$ and $\beta > 0$.

$$\boxed{\text{Gallery } 30\text{-}7, 30\text{-}8 : L = 8, n\left(\Sigma^8\right) = 256}$$

(a) There is a *period-40 attractor* with a robustness coefficient $\rho_1 = 224/256 = 0.875$. The dynamics on this attractor obeys a Bernoulli σ_τ-shift law with $\sigma = -3$, $\tau = 5$ and $\beta > 0$. The period obeys the formula $T = \tau L = 40$, and not $\tau L/|\sigma|$ because 40 is not divisible by $|-3|$.

(b) There is a *period-8 attractor* with robustness coefficient $\rho_2 = 28/256 = 0.109375$. The dynamics on this attractor obeys a Bernoulli σ_τ-shift law with $\sigma_1 = 1$ or -3, $\tau = 2$ and $\beta > 0$. The period obeys $T = \tau L = 16$, with a minimal period $T_{\min} = 8$ in view of the additional symmetry exhibited by the bit strings on this attractor.

(c) There is a *period-1 attractor* with robustness coefficient $\rho_3 = 2/256 = 0.0078125$. The dynamics on this attractor obeys a *degenerate* Bernoulli σ_τ-shift law with $\sigma = 0$, $\tau = 1$ and $\beta > 0$.

(d) There are two *period-1 Isles of Eden* with a combined robustness coefficient $\rho_4 = 2/256 = 0.0078125$. The dynamics on these Isles of Eden obeys a *degenerate* Bernoulli σ_τ-shift law with $\sigma = 0$, $\tau = 1$ and $\beta > 0$.

Observe that $\{\boxed{0}\}$ is a *period-1 attractor* of rule $\boxed{30}$ for $3 \leq L \leq 8$. Observe also that rule $\boxed{30}$ has an *Isle of Eden* consisting of *alternating* "red-blue" pixel patterns for even $L = 4, 6, 8$.

The qualitative properties of local rule $\boxed{30}$ extracted from the above basin-tree Galleries 30-1

to 30-8 are summarized below:

Summary of Qualitative properties of local rule $\boxed{30}$ extracted from Gallery 30 for Rule $\boxed{30}$

L	ID Number i	Number of Period-n attractors	Number of Period-n Isles of Eden	Period n	Bernoulli			Parameters			Robustness coefficient ρ
					σ_1	τ_1	β_1	σ_2	τ_2	β_2	
3	1	1		1	0	1	+				$\rho_1 = 1$
4	1	1		8	1	2	+	-3	2	+	$\rho_1 = 0.75$
	2	1		1	0	1	+				$\rho_2 = 0.125$
	3		2	1	0	1	+				$\rho_3 = 0.125$
5	1	1		5	2, -3	1	+	-1, 4	2	+	$\rho_1 = 0.9375$
	2	1		1	0	1	+				$\rho_2 = 0.0625$
6	1	1		1	0	1	+				$\rho_1 = 0.96875$
	2		2	1	0	1	+				$\rho_2 = 0.03125$
7	1	1		63	2	9	+				$\rho_1 = 0.6015625$
	2	7		4	0	4	+				$\rho_2 = 0.3828125$
	3	1		1	0	1	+				$\rho_3 = 0.015625$
8	1	1		40				-3	5	+	$\rho_1 = 0.875$
	2	1		8	1	2	+	-3	2	+	$\rho_2 = 0.109375$
	3	1		1	0	1	+				$\rho_3 = 0.0078125$
	4		2	1	0	1	+				$\rho_4 = 0.0078125$

2.3. Highlights from rule $\boxed{41}$

$$\boxed{\text{Gallery 41-1}: L = 3, n\left(\Sigma^3\right) = 8}$$

(a) There is a *period-2 attractor* with a robustness coefficient $\rho_1 = 5/8 = 0.625$. The dynamics on this attractor obeys a *degenerate* Bernoulli σ_τ-shift law with $\sigma_1 = 1, 2$ or 3, $\tau_1 = 1$ and $\beta_1 < 0$, or $\sigma_2 = 1, 2$ or 3, $\tau_2 = 2$ and $\beta_2 > 0$.

(b) There is a *period-3 Isle of Eden* with robustness coefficient $\rho_2 = 3/8 = 0.375$. The dynamics on this Isle of Eden obeys a Bernoulli σ_τ-shift law with $\sigma = 1$, $\tau = 1$ and $\beta > 0$. The period obeys the formula $T = \tau L/|\sigma| = 3$.

$$\boxed{\text{Gallery 41-2}: L = 4, n\left(\Sigma^4\right) = 16}$$

(a) There is a *period-2 Isle of Eden* with a robustness coefficient $\rho_1 = 2/16 = 0.125$. The dynamics on this Isle of Eden obeys a *degenerate* Bernoulli σ_τ-shift law with $\sigma_1 = 1, 2, 3$ or 4, $\tau_1 = 1$, $\beta_1 < 0$, or $\sigma_2 = 1, 2, 3$ or 4, $\tau_2 = 2$, $\beta_2 > 0$.

(b) There is a *period-2 Isle of Eden* with a robustness coefficient $\rho_2 = 2/16 = 0.125$. The dynamics on this Isle of Eden obeys a Bernoulli σ_τ-shift law with $\sigma = 1$ or -1, $\tau = 1$ and $\beta > 0$.

(c) There are two *period-2 attractors* with a robustness coefficient robustness coefficient $\rho_3 = 2(6/16) = 0.75$. The dynamics on these attractors obey a σ_τ-shift law with $\sigma = 2$, or -2, $\tau = 1$ and $\beta > 0$. The period obeys the formula $T = \tau L/|\sigma| = 2$.

$$\boxed{\text{Gallery 41-3, 41-4}: L = 5, n\left(\Sigma^5\right) = 32}$$

(a) There is a *period-2 Isle of Eden* with robustness coefficient $\rho_1 = 2/32 = 0.0625$. The dynamics on this Isle of Eden obeys a *degenerate* Bernoulli σ_τ-shift law with $\sigma_1 = 1, 2, 3$ or 4, $\tau_1 = 1$ and $\beta_1 < 0$, or $\sigma_2 = 1, 2, 3$ or 4, $\tau_2 = 2$ and $\beta_2 > 0$.

Table 6.   Basin tree diagrams for rule 30 .

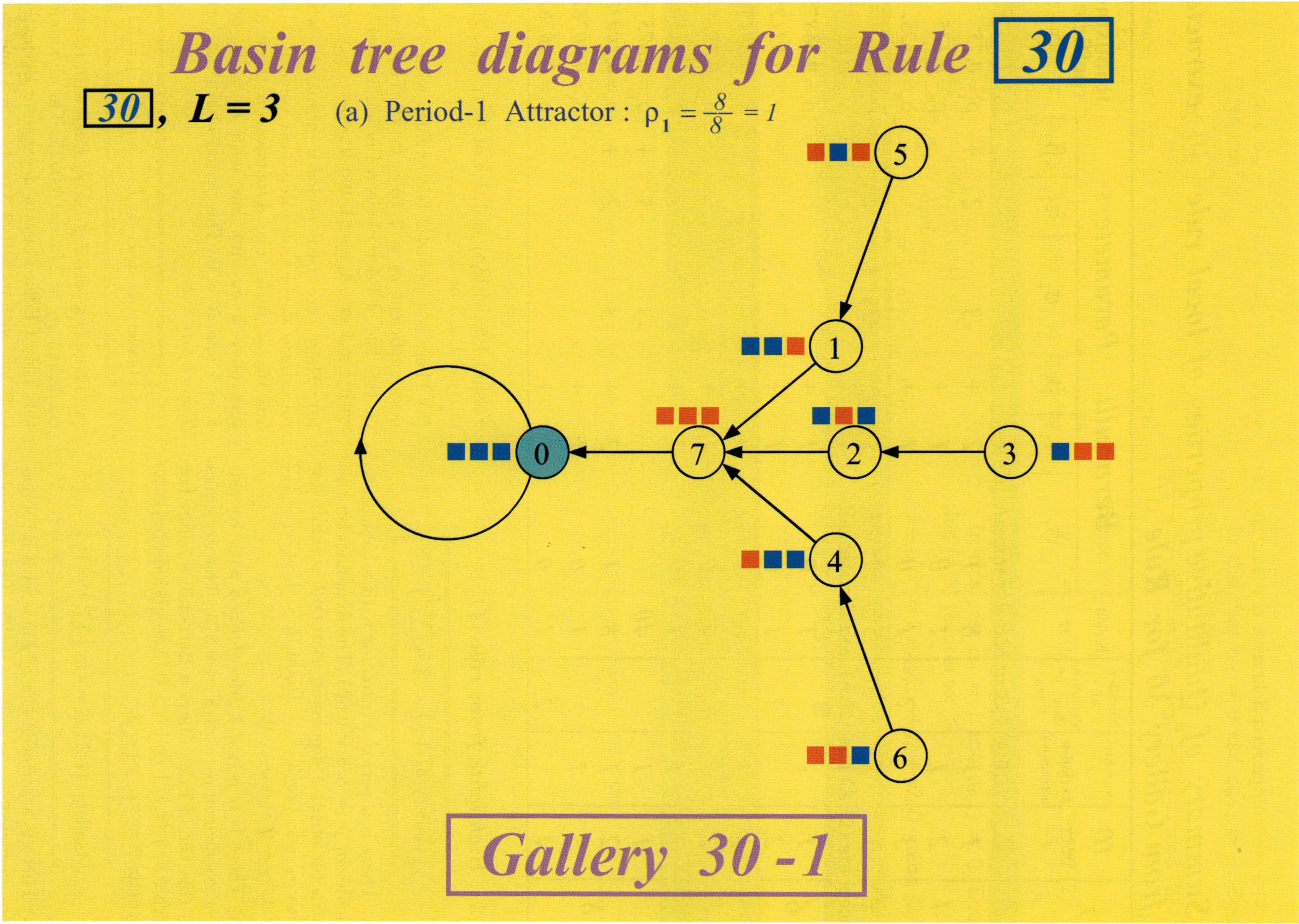

Table 6. (*Continued*)

Basin tree diagrams for Rule 30
30, L = 4
(a) Bernoulli : (σ = +1, σ = −3, τ = 2) Period-8 Attractor : ρ1 = 12/16 = 0.75
σ = +1, σ = −3, τ = 2
ϕn
ϕn-2
β > 0
Transient phase
T = 8
τ = 2
σ = +1
σ = −3
(b) Period-1 Attractor :
ρ2 = 2/16 = 0.125
(c) Period-1 Isles of Eden : ρ3 = 2/16 = 0.125
Gallery 30 - 2

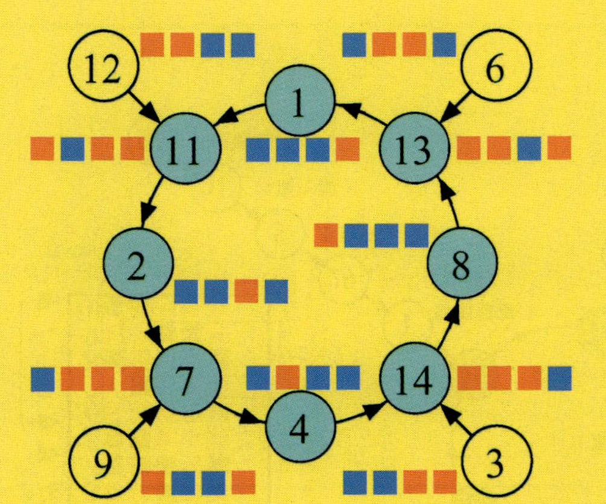

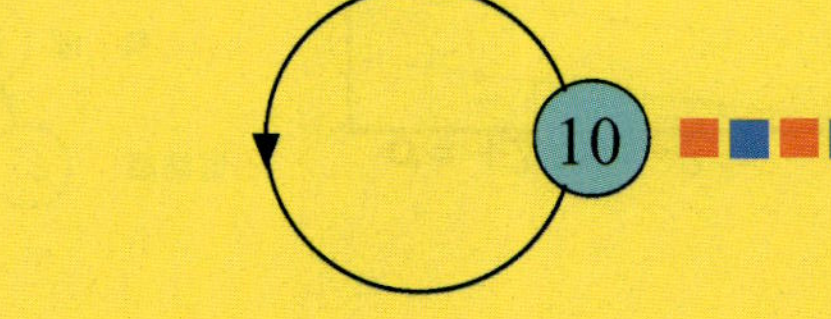

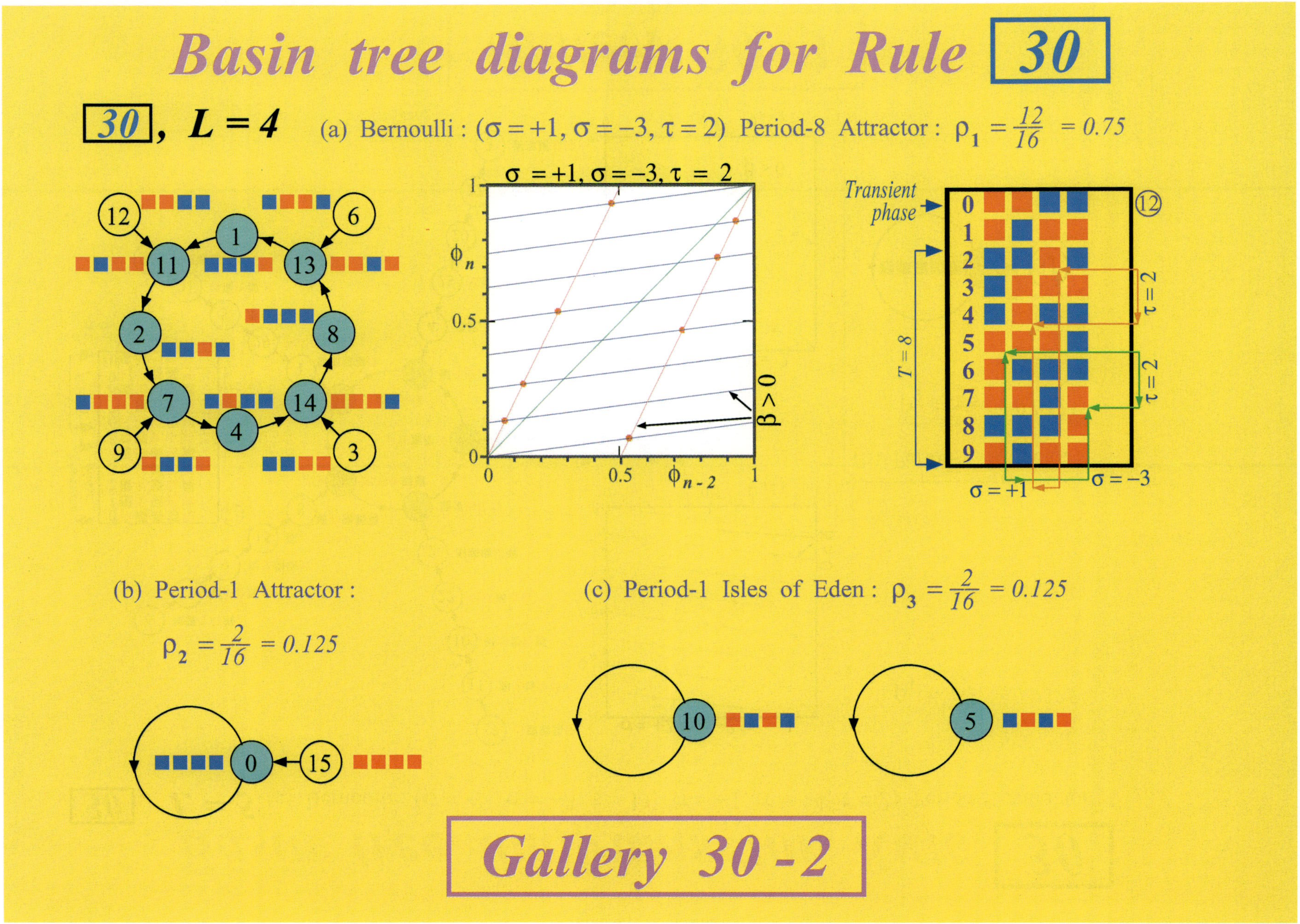

Basin tree diagrams for Rule $\boxed{30}$

$\boxed{30}$, $L = 5$ (a) Bernoulli : $(\sigma = +2, \sigma = -3, \tau = 1)$, $(\sigma = -1, \sigma = +4, \tau = 2)$ Period-5 Attractor :

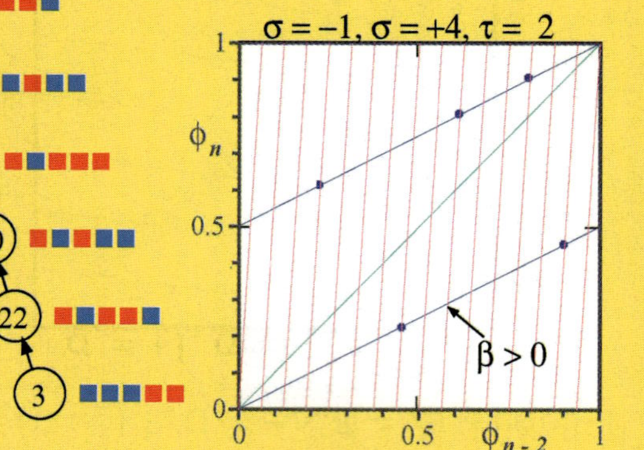

$$\rho_1 = \frac{30}{32} = 0.9375$$

(b) Period-1 Attractor :

$$\rho_2 = \frac{2}{32} = 0.0625$$

Gallery 30-3

Basin tree diagrams for Rule 30

30, *L* = 6

(a) Period-1 Attractor : $\rho_1 = \frac{62}{64} = 0.96875$

(b) Period-1 Isles of Eden :

$$\rho_2 = \frac{2}{64} = 0.03125$$

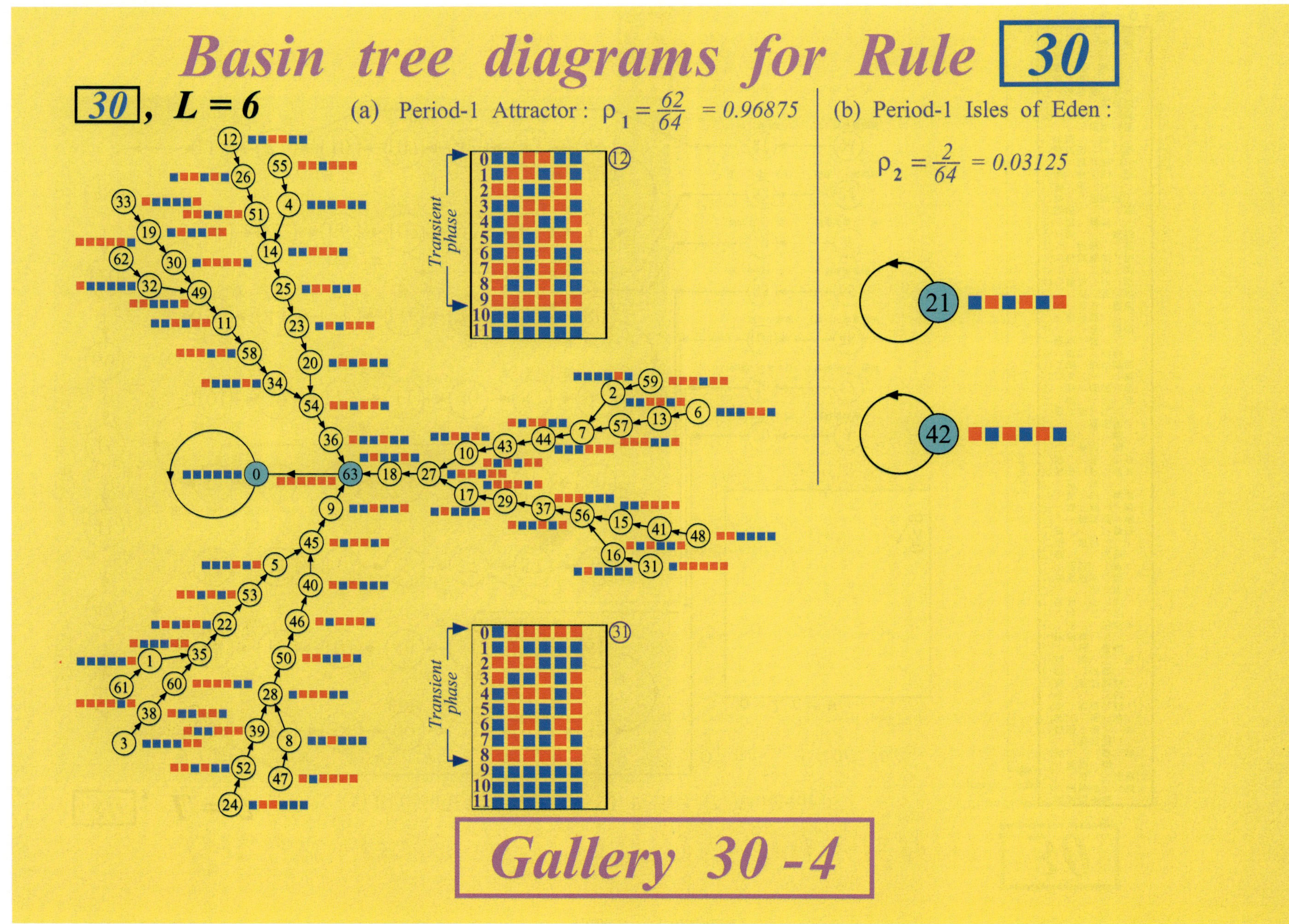

Gallery 30 -4

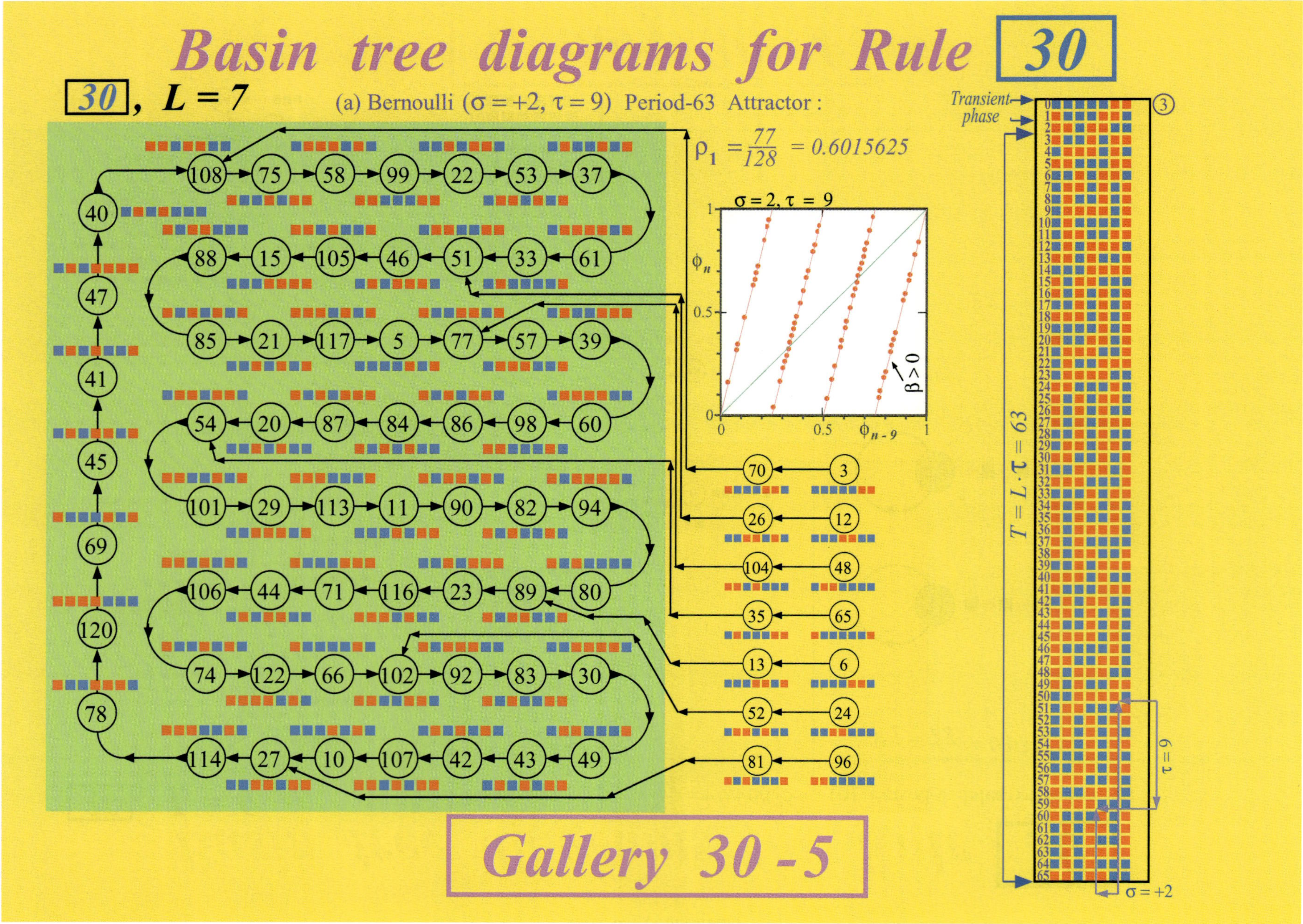

Table 6.   (*Continued*)

Basin tree diagrams for Rule 30

30, $L = 7$ (a) Bernoulli ($\sigma = +2$, $\tau = 9$) Period-63 Attractor :

$$\rho_1 = \frac{77}{128} = 0.6015625$$

Gallery 30 - 5

202

Basin tree diagrams for Rule $\boxed{30}$

$\boxed{30}$, $L = 7$ (b)   Bernoulli ($\sigma = 0, \tau = 4$) Period-4 Attractors : $\rho_2 = 7\frac{7}{128} = 0.3828125$

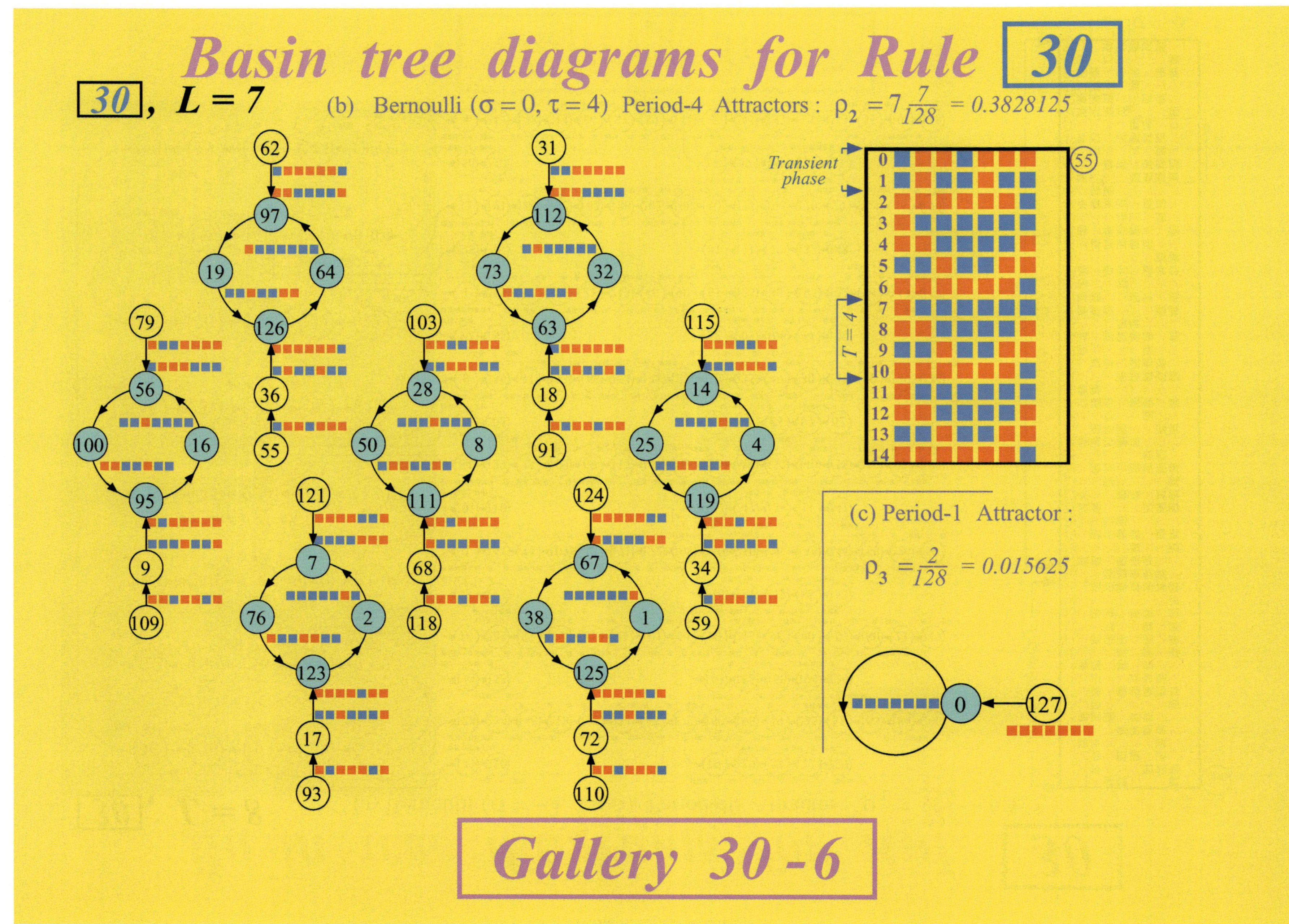

$\boxed{\textit{Gallery 30 - 6}}$

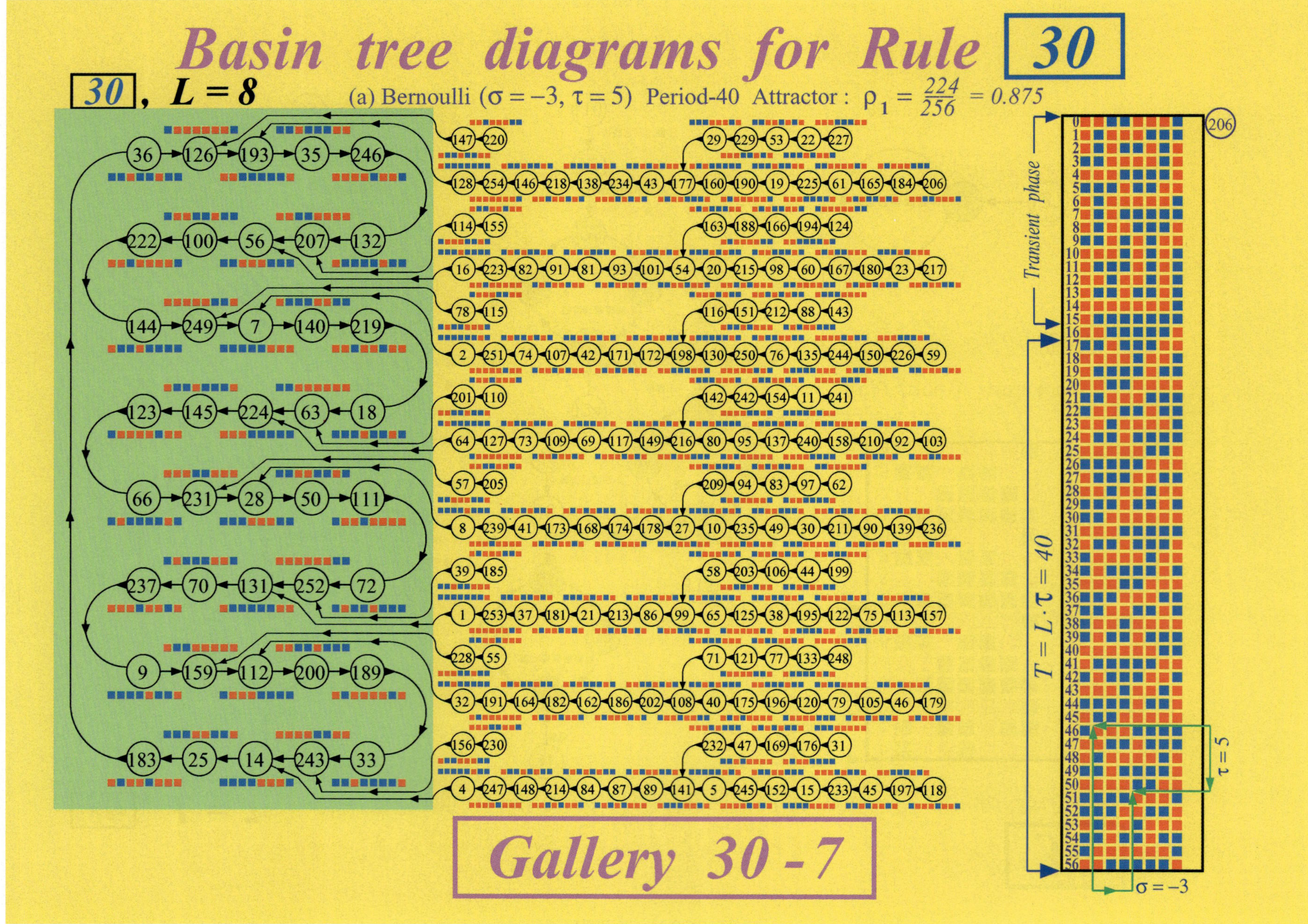

Basin tree diagrams for Rule 30
30, L = 8
(a) Bernoulli ($\sigma = -3$, $\tau = 5$) Period-40 Attractor : $\rho_1 = \frac{224}{256} = 0.875$
Gallery 30 - 7
$T = L \cdot \tau = 40$
Transient phase
$\varsigma = 1$
$\sigma = -3$
$\tau = 5$

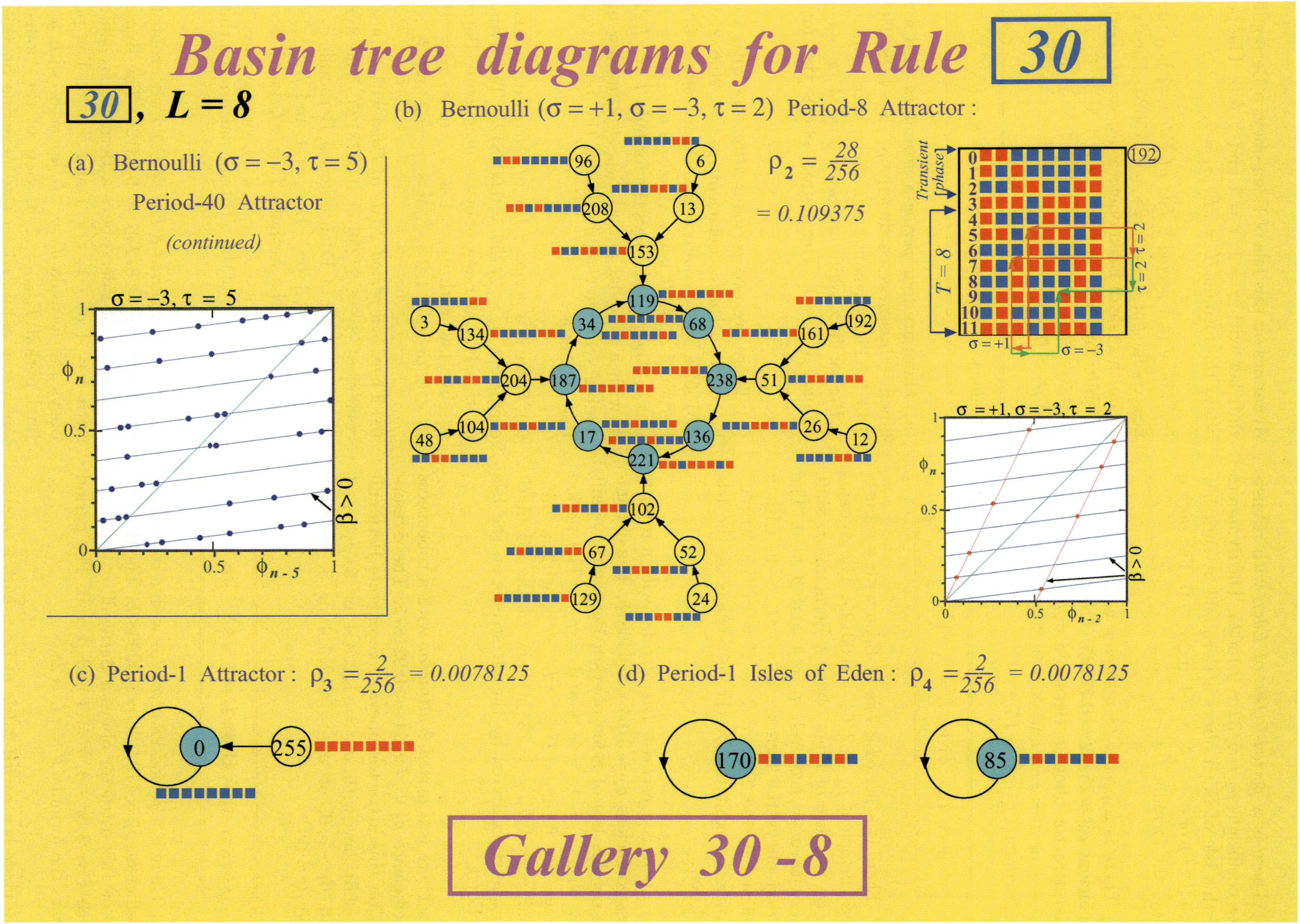

Basin tree diagrams for Rule $\boxed{30}$

$\boxed{30}$, $L = 8$

(a) Bernoulli ($\sigma = -3$, $\tau = 5$)

Period-40 Attractor

(continued)

(b) Bernoulli ($\sigma = +1$, $\sigma = -3$, $\tau = 2$) Period-8 Attractor :

$$\rho_2 = \frac{28}{256} = 0.109375$$

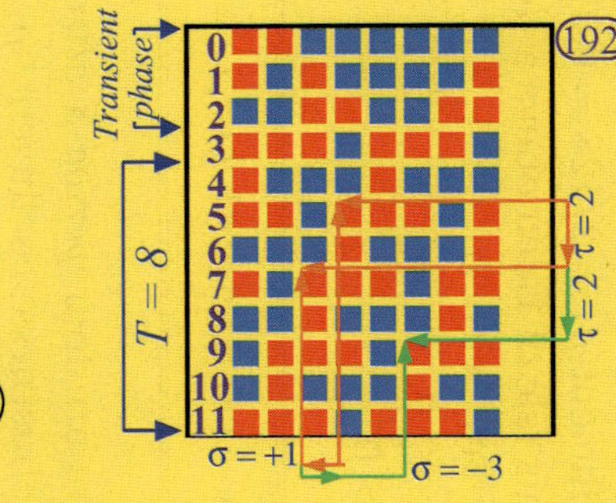

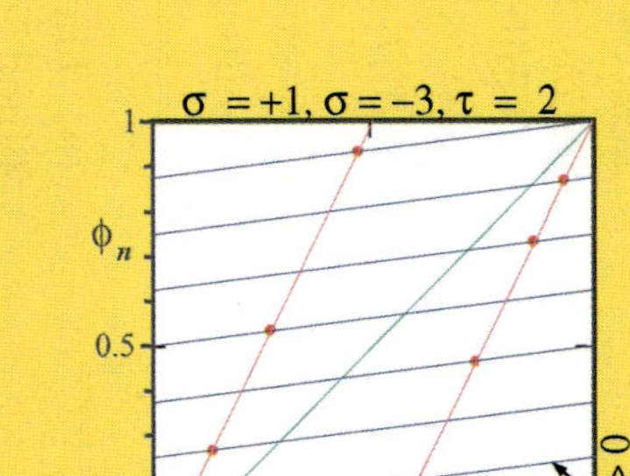

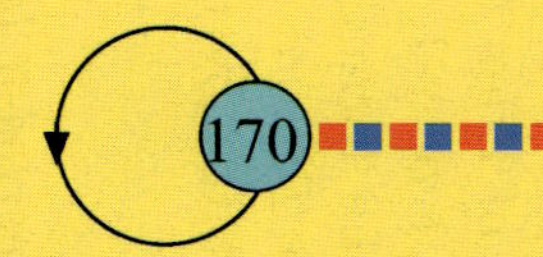

(c) Period-1 Attractor : $\rho_3 = \frac{2}{256} = 0.0078125$

(d) Period-1 Isles of Eden : $\rho_4 = \frac{2}{256} = 0.0078125$

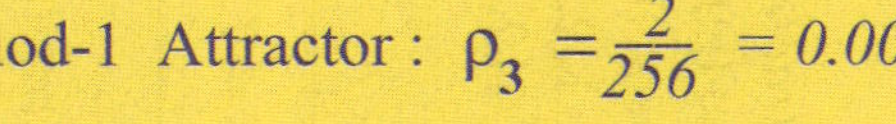

Gallery 30 - 8

(b) There is a *period-5 Isle of Eden* with robustness coefficient $\rho_2 = 5/32 = 0.15625$. The dynamics on this Isle of Eden obeys a Bernoulli σ_τ-shift law with $\sigma = 1$ or $\sigma = -4$, $\tau = 1$ and $\beta > 0$. The period obeys the formula $T = \tau L = 5$.

(c) There is a *period-15 attractor* with robustness coefficient $\rho_3 = 25/32 = 0.78125$. The dynamics on this attractor obeys a Bernoulli σ_τ-shift law with $\sigma = -1$ or 4, $\tau = 3$ and $\beta > 0$. The period obeys the formula $T = \tau L = 15$.

$$\boxed{\text{Gallery } 41\text{-}5, 41\text{-}6 : L = 6, n\left(\Sigma^6\right) = 64}$$

(a) There is a *period-2 Isle of Eden* with a robustness coefficient $\rho_1 = 2/64 = 0.03125$. The dynamics on this Isle of Eden obeys a Bernoulli σ_τ-shift law with $\sigma = 1$ or -1, $\tau = 1$ and $\beta > 0$. The period obeys the formula $T = \tau L = 6$, with a minimal period $T_{\min} = 2$ in view of the "triple" symmetry exhibited by the bit strings on the Isle of Eden.

(b) There is a *period-3 Isle of Eden* with a robustness coefficient $\rho_2 = 3/64 = 0.046875$. The dynamics on this Isle of Eden obeys a Bernoulli σ_τ-shift law with $\sigma = 1$, $\tau = 1$ and $\beta > 0$. The period obeys the formula $T = \tau L = 6$, with a minimal period $T_{\min} = 3$ in view of the "double" symmetry exhibited by the bit strings on the Isle of Eden.

(c) There is a *period-2 attractor* with a robustness coefficients $\rho_3 = 59/64 = 0.921875$. The dynamics on this attractor obeys a *degenerate* Bernoulli σ_τ-shift law with $\sigma_1 = 1, 2, 3, 4$ or 5, $\tau_1 = 1$ and $\beta_1 < 0$, or $\sigma_2 = 1, 2, 3, 4$ or 5, $\tau_2 = 2$ and $\beta_2 > 0$.

$$\boxed{\text{Gallery } 41\text{-}7, 41\text{-}8 : L = 7, n\left(\Sigma^7\right) = 128}$$

(a) There is a *period-28 attractor* with a robustness coefficient $\rho_1 = 119/128 = 0.9296875$. The dynamics on this attractor obeys a Bernoulli σ_τ-shift law with $\sigma = 3$ or -4, $\tau = 4$ and $\beta > 0$. The period obeys the formula $T = \tau L = 28$.

(b) There is a *period-7 Isle of Eden* with a robustness coefficient $\rho_2 = 7/128 = 0.0546875$. The

dynamics on this Isle of Eden obeys a Bernoulli σ_τ-shift law with $\sigma = 1$, $\tau = 1$ and $\beta > 0$. The period obeys the formula $T = \tau L/|\sigma| = 7$.

(c) There is a *period-2 Isle of Eden* with a robustness coefficient $\rho_3 = 2/128 = 0.015625$. The dynamics on this Isle of Eden obeys a *degenerate* Bernoulli σ_τ-shift law with $\sigma_1 = 1, 2, 3, 4, 5$ or 6, $\tau_1 = 1$ and $\beta_1 < 0$, or $\sigma_2 = 1, 2, 3, 4, 5$ or 6, $\tau_2 = 2$ and $\beta_2 > 0$.

$$\boxed{\text{Gallery } 41\text{-}9, 41\text{-}10, 41\text{-}11 : L = 8, n\left(\Sigma^8\right) = 256}$$

(a) There are four *period-8 attractors* with a combined robustness coefficient $\rho_1 = 4(28/256) = 0.4375$. The dynamics on these attractors obeys a Bernoulli σ_τ-shift law with $\sigma = 4$ or -4, $\tau = 4$ and $\beta > 0$. The period obeys the formula $T = \tau L/|\sigma| = 8$.

(b) There are two *period-2 attractors* with a combined robustness coefficient $\rho_2 = 2(66/256) = 0.515625$. The dynamics on these attractors obeys a Bernoulli σ_τ-shift law with $\sigma = 2$ or -2, $\tau = 1$ and $\beta > 0$. The period obeys the formula $T = \tau L/|\sigma| = 4$, with a minimal period $T_{\min} = 2$ in view of the double symmetry exhibited by the bit strings on this attractor.

(c) There is a *period-2 Isle of Eden* with robustness coefficient $\rho_3 = 2/256 = 0.0078125$. The dynamics on this Isle of Eden obeys a *degenerate* Bernoulli σ_τ-shift law with $\sigma_1 = 1, 2, \ldots$, or 8, $\tau_1 = 1$ and $\beta_1 < 0$, or $\sigma_2 = 1, 2, \ldots$, or 8, $\tau_2 = 2$ and $\beta_2 > 0$.

(d) There is a *period-2 Isle of Eden* with robustness coefficient $\rho_4 = 2/256 = 0.0078125$. The dynamics of this Isle of Eden obeys a Bernoulli σ_τ-shift law with $\sigma = 1$ or -1, $\tau = 1$ and $\beta > 0$. The period obeys the formula $T = \tau L/|\sigma| = 8$, with a minimal period $T_{\min} = 2$ in view of the "four-fold" symmetry of the bit string patterns on this Isle of Eden.

(e) There is a *period-8 Isle of Eden* with robustness coefficient $\rho_5 = 8/256 = 0.03125$. The dynamics on this Isle of Eden obeys a Bernoulli σ_τ-shift law with $\sigma = 1$ or -1, $\tau = 1$ and $\beta > 0$. The period obeys the formula $T = \tau L/|\sigma| = 8$, with a minimal period $T_{\min} = 2$ in view of the "four-fold" symmetry exhibited by the bit string on this Isle of Eden.

The qualitative properties of local rule $\boxed{41}$ extracted from the above basin-tree Galleries 41-1 to 41-11 are summarized below:

Summary of Qualitative properties of local rule $\boxed{41}$ extracted from Gallery 41 for Rule $\boxed{41}$

L	ID Number i	Number of Period-n attractors	Number of Period-n Isles of Eden	Period n	Bernoulli			Parameters			Robustness coefficient ρ
					σ_1	τ_1	β_1	σ_2	τ_2	β_2	
3	1	1		2	1, 2, 3	1	−	1, 2, 3	1	+	$\rho_1 = 0.625$
	2		1	3	1	1	+				$\rho_2 = 0.375$
4	1		1	2	1,2,3,4	1	−	1,2,3,4	1	+	$\rho_1 = 0.125$
	2		1	2	1	1	+	-1	1	+	$\rho_2 = 0.125$
	3	2		2	2	1	+	-2	1	+	$\rho_3 = 0.75$
5	1		1	2	1,2,...,5	1	−	1,2,...,5	1	+	$\rho_1 = 0.0625$
	2		1	5	1	1	+	-4	1	+	$\rho_2 = 0.15625$
	3	1		15	-1	3	+	4	3	+	$\rho_3 = 0.78125$
6	1		1	2	1	1	+	-1	1	+	$\rho_1 = 0.03125$
	2		1	3	1	1	+				$\rho_2 = 0.046875$
	3	1		2	1,2,...,6	1	−	1,2,...,6	1	+	$\rho_3 = 0.921875$
7	1	1		28	3	4	+	-4	4	+	$\rho_1 = 0.9296875$
	2		1	7	1	1	+				$\rho_2 = 0.0546875$
	3		1	2	1,2,...,7	1	−	1,2,...,7	1	+	$\rho_3 = 0.015625$
8	1	4		8	4	4	+	-4	4	+	$\rho_1 = 0.4375$
	2	2		2	2	1	+	-2	1	+	$\rho_2 = 0.515625$
	3		1	2	1,2,...,8	1	−	1,2,...,8	1	+	$\rho_3 = 0.0078125$
	4		1	2	1	1	+	-1	1	+	$\rho_4 = 0.0078125$
	5		1	8	1	1	+				$\rho_5 = 0.03125$

2.4.   *Highlights from rule $\boxed{45}$*

$$\boxed{\text{Gallery } 45\text{-}1 : L = 3, n\left(\Sigma^3\right) = 8}$$

(a) There are three *period-1 Isles of Eden* with a combined robustness coefficient $\rho_1 = 3/8 = 0.375$. The dynamics on these Isles of Eden obey a *degenerate* Bernoulli σ_τ-shift law with $\sigma = 3$, $\tau = 1$ and $\beta > 0$.

(b) There is a *period-2 Isle of Eden* with a robustness coefficient $\rho_2 = 2/8 = 0.25$. The dynamics on this Isle of Eden obeys a *degenerate* Bernoulli σ_τ-shift law with $\sigma_1 = 1, 2$ or 3, $\tau_1 = 1$ and $\beta_1 < 0$, or $\sigma_2 = 1, 2$ or 3, $\tau_2 = 2$ and $\beta_2 > 0$.

(c) There is a *period-3 Isle of Eden* with a robustness coefficient $\rho_3 = 3/8 = 0.375$. The dynamics on this Isle of Eden obeys a Bernoulli σ_τ-shift law with $\sigma = 1$, $\tau = 1$ and $\beta > 0$. The period obeys the formula $T = \tau L/|\sigma| = 3$.

$$\boxed{\text{Gallery } 45\text{-}2 : L = 4, n\left(\Sigma^4\right) = 16}$$

(a) There is a *period-2 attractor* with a robustness coefficient $\rho_1 = 16/16 = 1$. The dynamics on this attractor obeys a *degenerate* Bernoulli σ_τ-shift law with $\sigma_1 = 1, 2, 3$, or 4, $\tau_1 = 1$ and $\beta_1 < 0$, or $\sigma_2 = 1, 2, 3$ or 4, $\tau_2 = 2$ and $\beta_2 > 0$.

Table 7.   Basin tree diagrams for rule $\boxed{41}$.

Basin tree diagrams for Rule $\boxed{41}$

$\boxed{41}$, $L = 3$

(a) Period-2 Attractor :

$$\rho_1 = \frac{5}{8} = 0.625$$

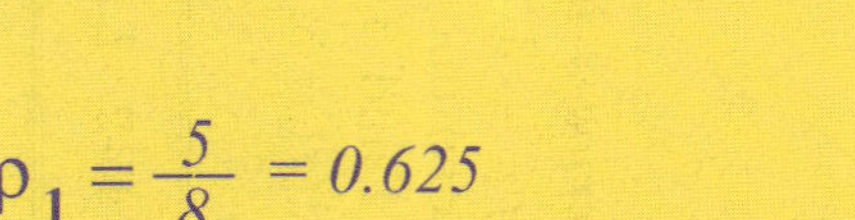

(b) Bernoulli ($\sigma = 1, \tau = 1$)

Period-3 Isle of Eden : $\rho_2 = \frac{3}{8} = 0.375$

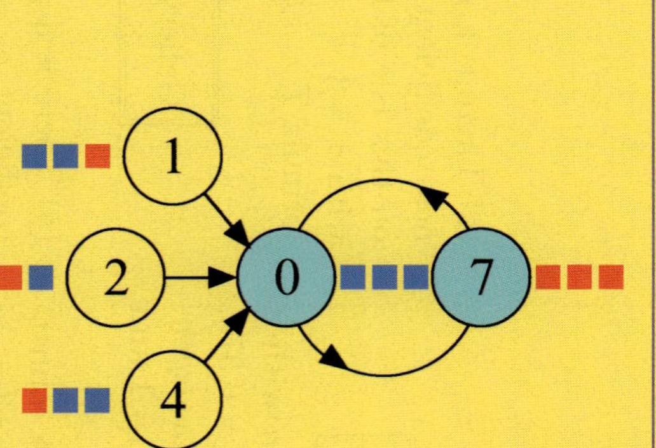

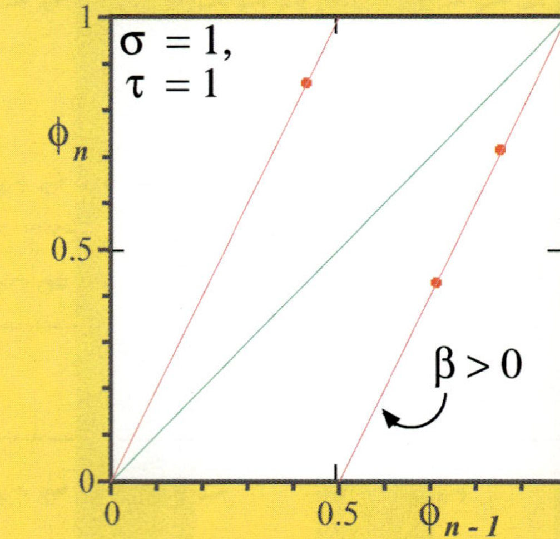

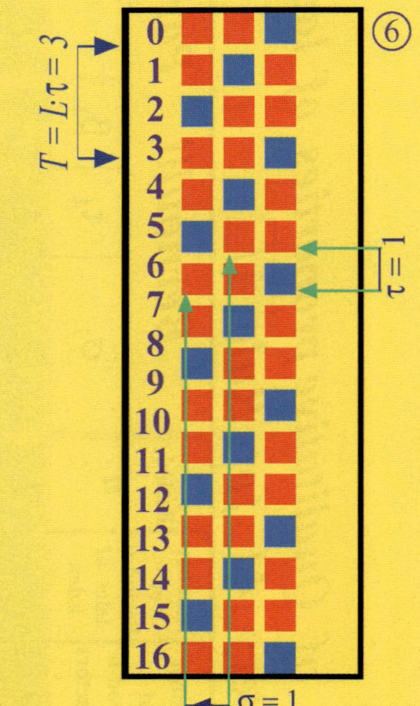

Gallery 41 -1

Table 7. (*Continued*)

Basin tree diagrams for Rule 41
41, L = 4
(a) Period-2 Isle of Eden :
$\rho_1 = \frac{2}{16} = 0.125$
0
15
(b) Bernoulli ($\sigma = \pm 1, \tau = 1$)
Period-2 Isle of Eden :
$\rho_2 = \frac{2}{16} = 0.125$
10
5
(c) Bernoulli ($\sigma = \pm 2, \tau = 1$) Period-2 Attractors :
$\rho_3 = 2\frac{6}{16} = 0.75$
7
12
8
2
3
13
11
6
4
1
9
14
$\sigma = \pm 1$, $\tau = 1$
ϕ_n
ϕ_{n-1}
$\beta > 0$
T = 2
$\tau = 1$
$\sigma = +1$
$\sigma = -1$
10
$\sigma = \pm 2$, $\tau = 1$
2
$\sigma = -2$
$\sigma = +2$
Gallery 41 - 2

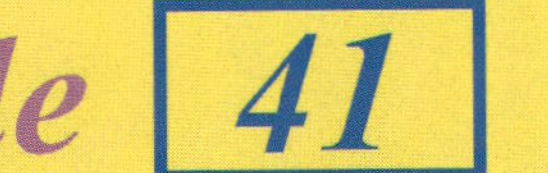

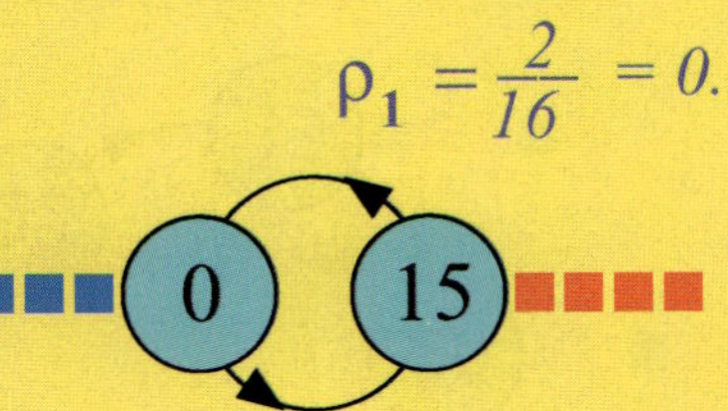

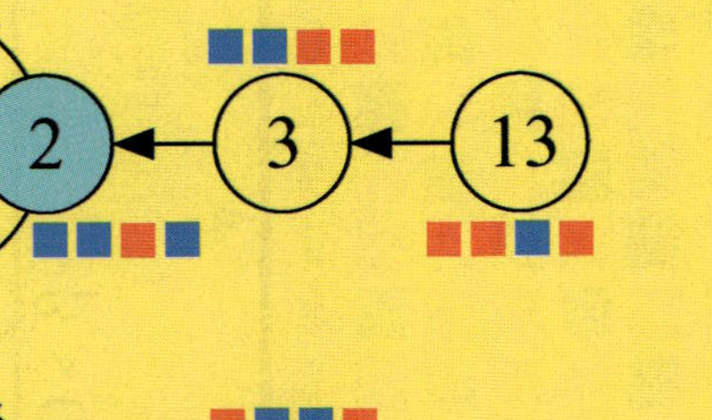

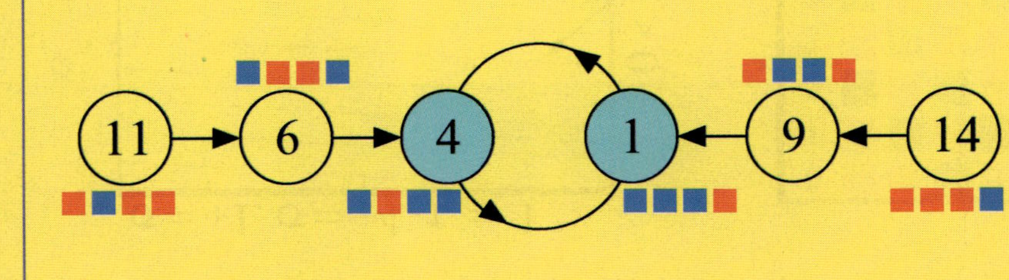

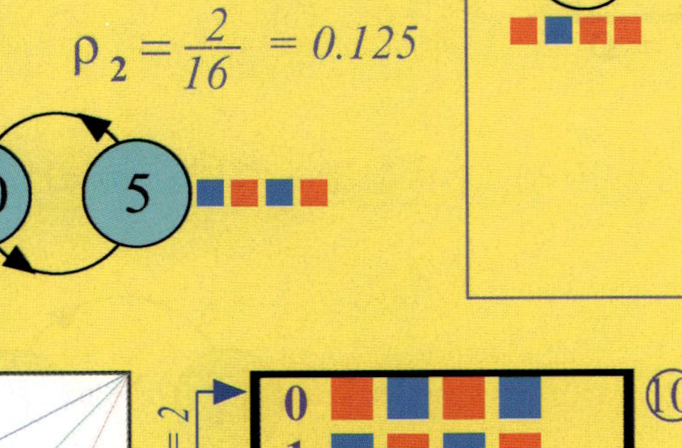

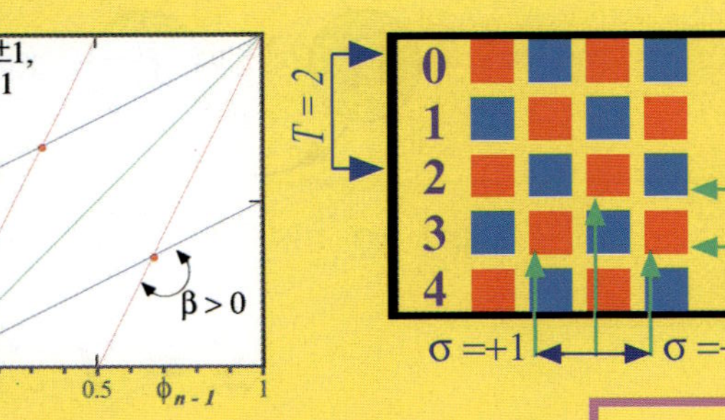

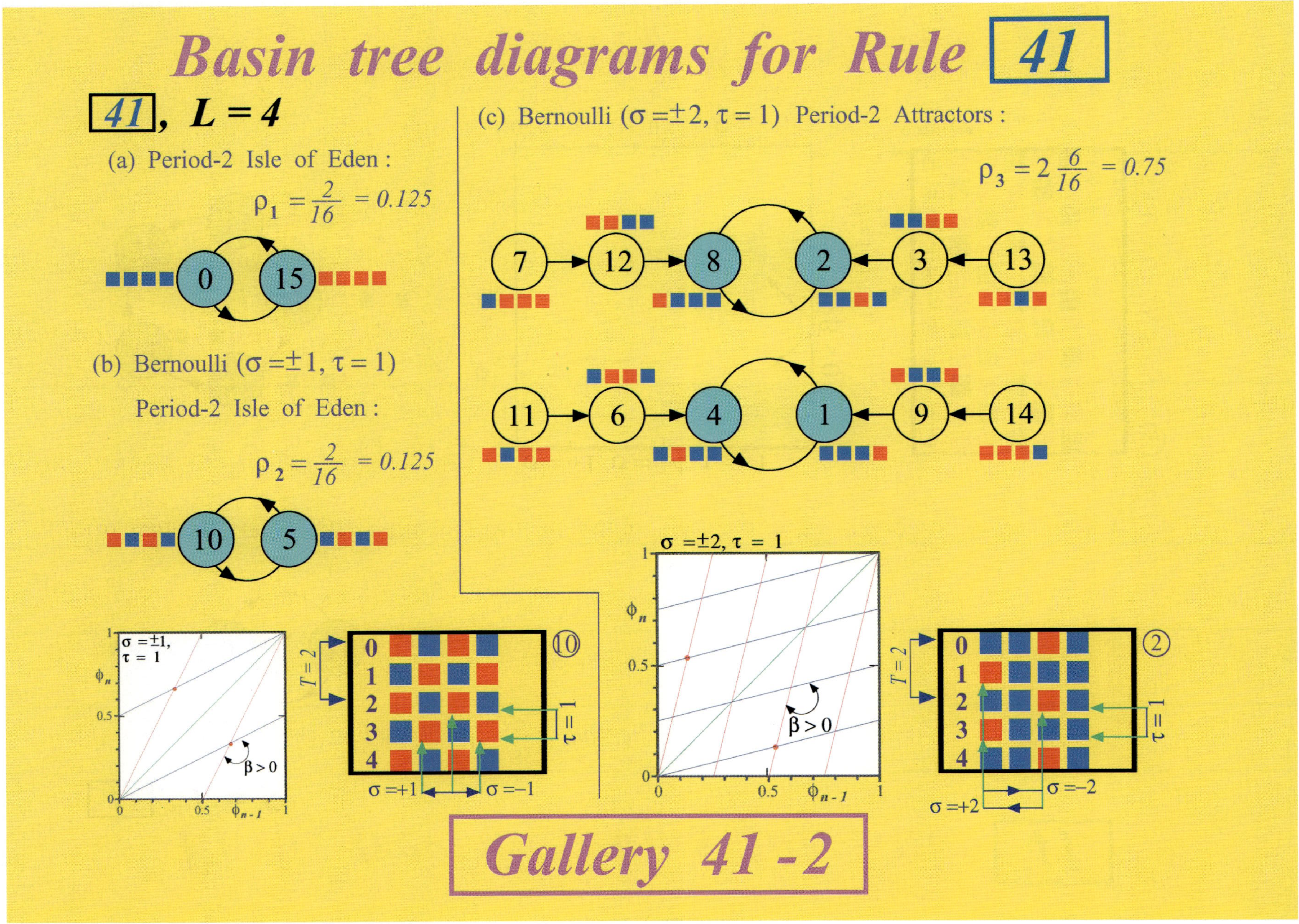

Basin tree diagrams for Rule $\boxed{41}$

$\boxed{41}$, **$L = 5$**

(a) Period-2 Isle of Eden : $\rho_1 = \dfrac{2}{32} = 0.0625$

(b) Bernoulli ($\sigma = +1$, $\sigma = -4$, $\tau = 1$) Period-5 Isle of Eden : $\rho_2 = \dfrac{5}{32} = 0.15625$

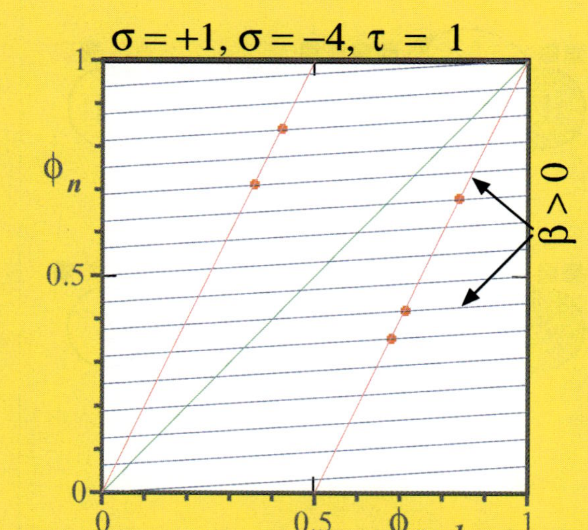

$\boxed{\textit{Gallery 41 -3}}$

Basin tree diagrams for Rule $\boxed{41}$

$\boxed{41}$, **L = 5**   (c) Bernoulli ($\sigma = -1$, $\sigma = +4$, $\tau = 3$) Period-15 Attractor : $\rho_3 = \frac{25}{32} = 0.78125$

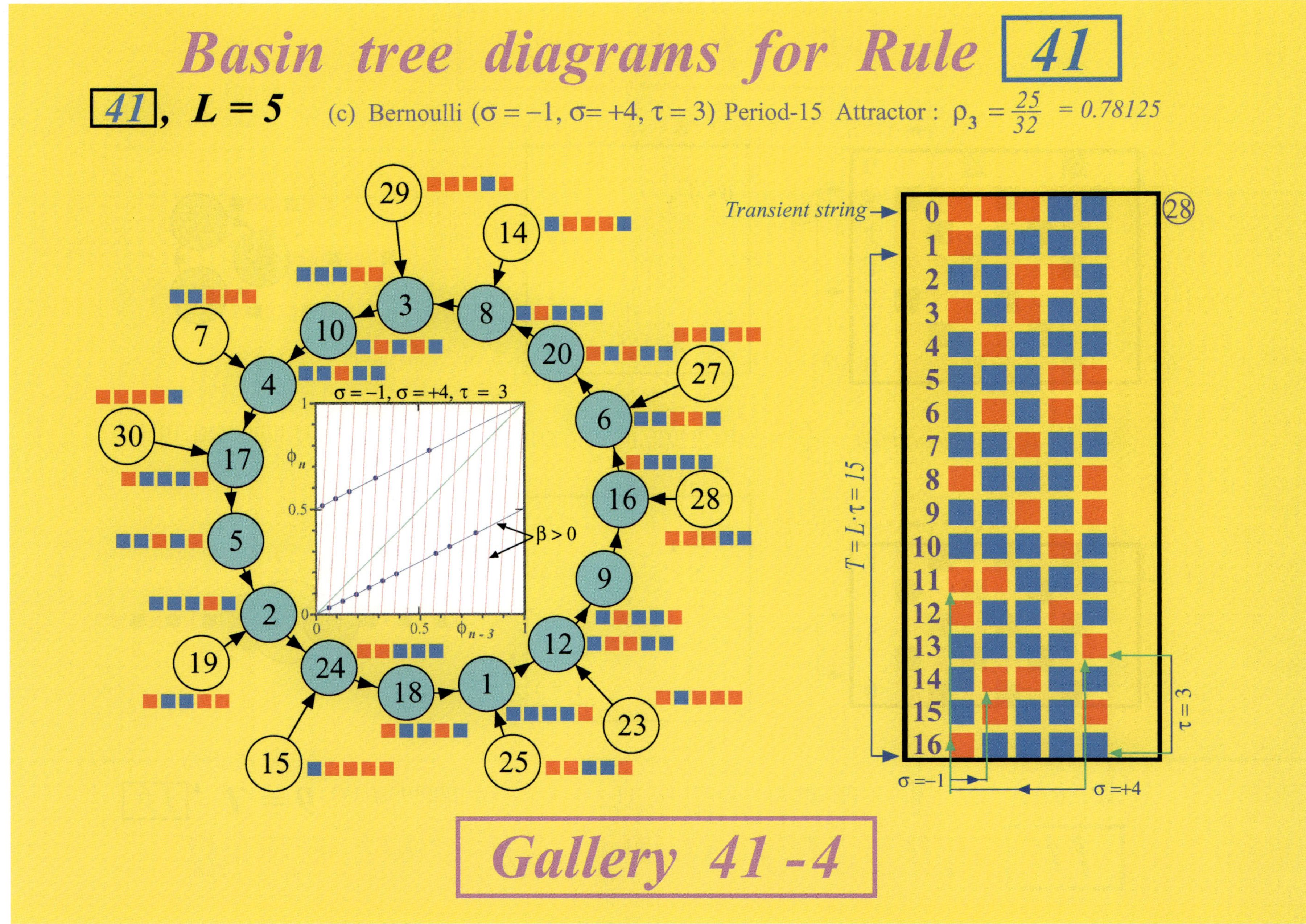

$$\boxed{\textit{Gallery 41 - 4}}$$

Table 7. (*Continued*)

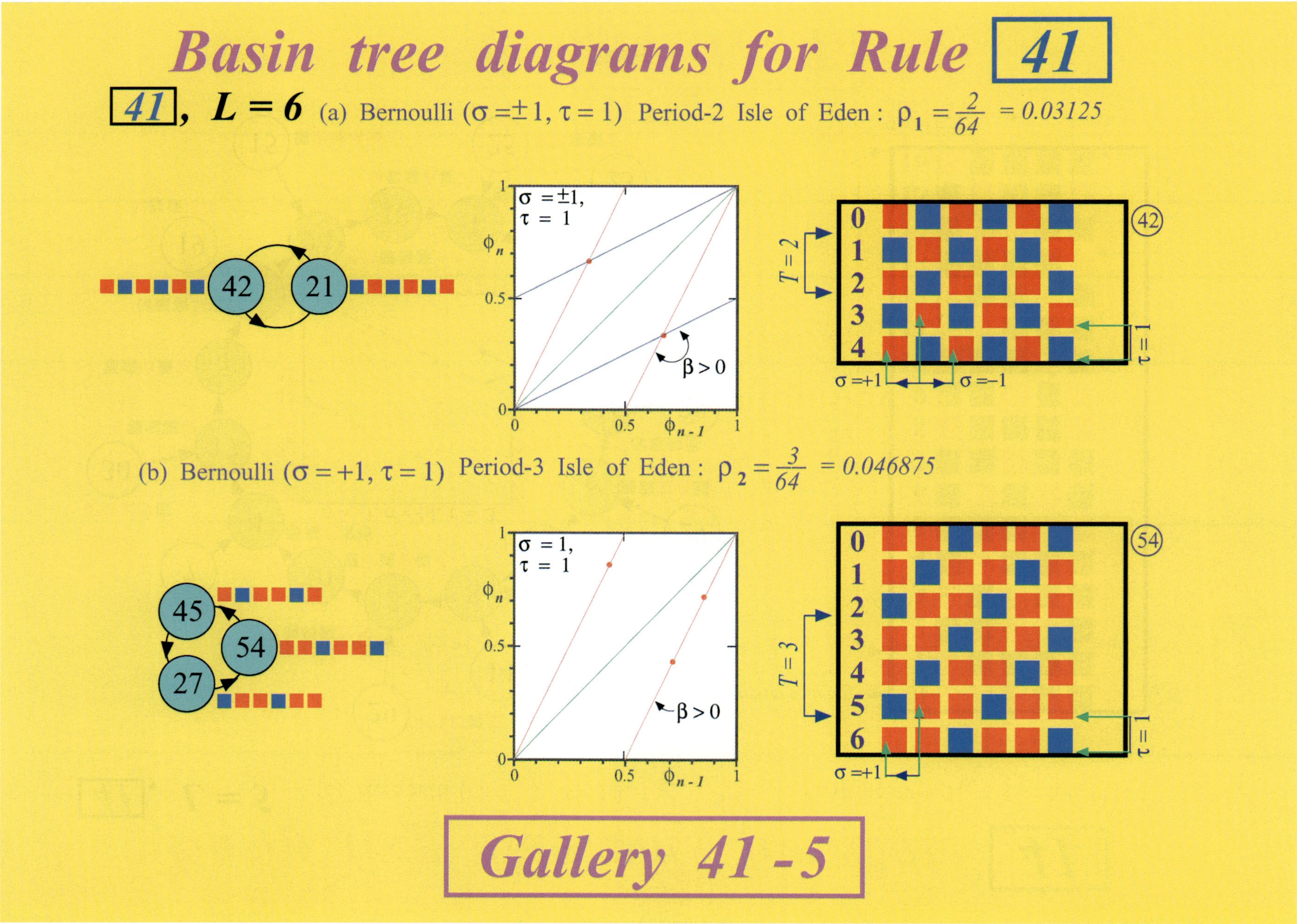

Basin tree diagrams for Rule 41
41, L = 6 (a) Bernoulli (σ = ±1, τ = 1) Period-2 Isle of Eden : $\rho_1 = \frac{2}{64} = 0.03125$
σ = ±1, τ = 1
ϕ_n
ϕ_{n-1}
β > 0
T = 2
σ = +1
σ = −1
τ = 1
(b) Bernoulli (σ = +1, τ = 1) Period-3 Isle of Eden : $\rho_2 = \frac{3}{64} = 0.046875$
σ = 1, τ = 1
T = 3
Gallery 41 - 5

Basin tree diagrams for Rule 41

41 , $L = 6$

(c) Period-2 Attractor :

$$\rho_3 = \frac{59}{64} = 0.921875$$

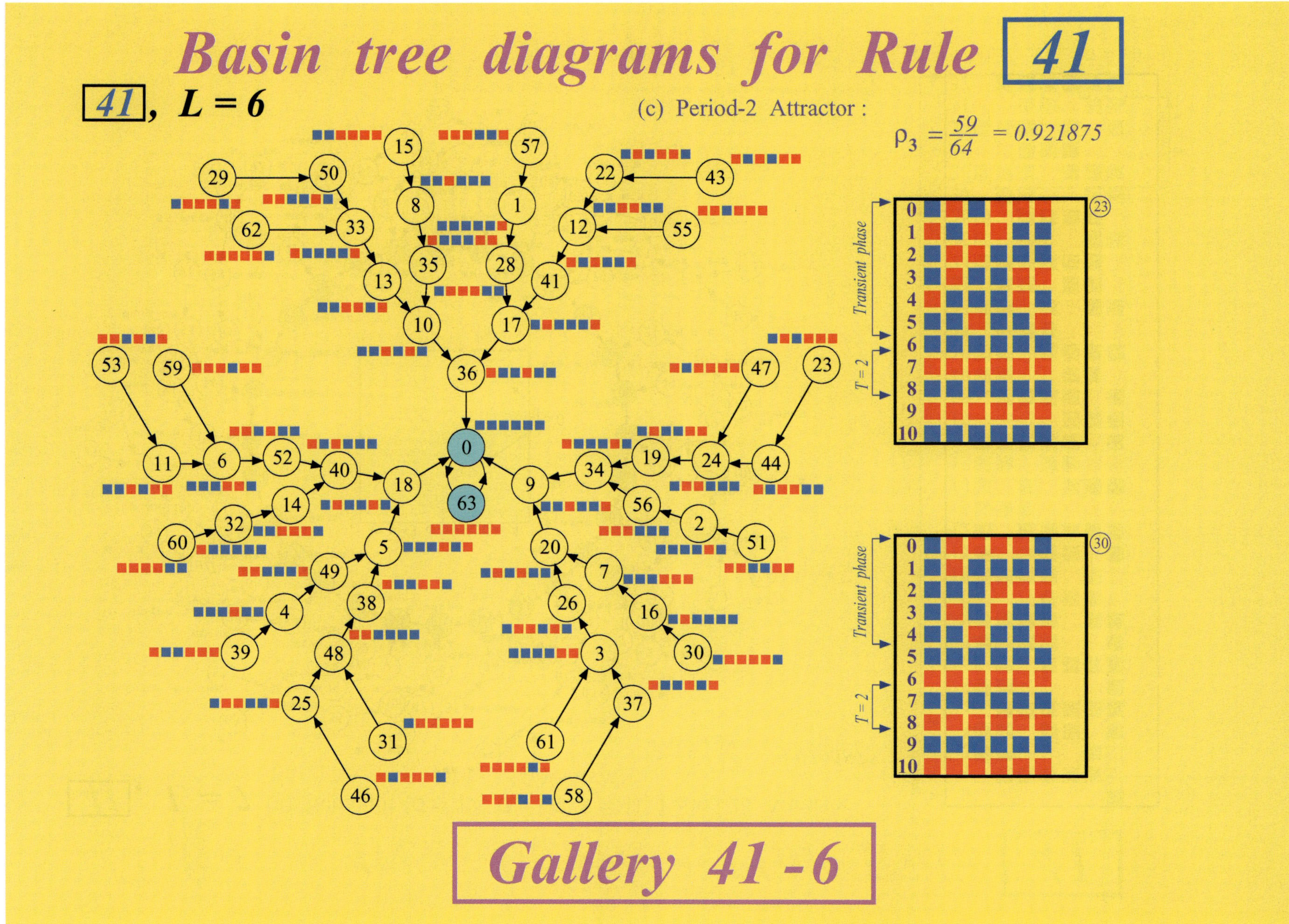

Gallery 41 - 6

Table 7.   (Continued)

Basin tree diagrams for Rule 41

41, **L = 7** (a) Bernoulli ($\sigma = +3$, $\sigma = -4$, $\tau = 4$) Period-28 Attractor :
$$\rho_1 = \frac{119}{128} = 0.9296875$$

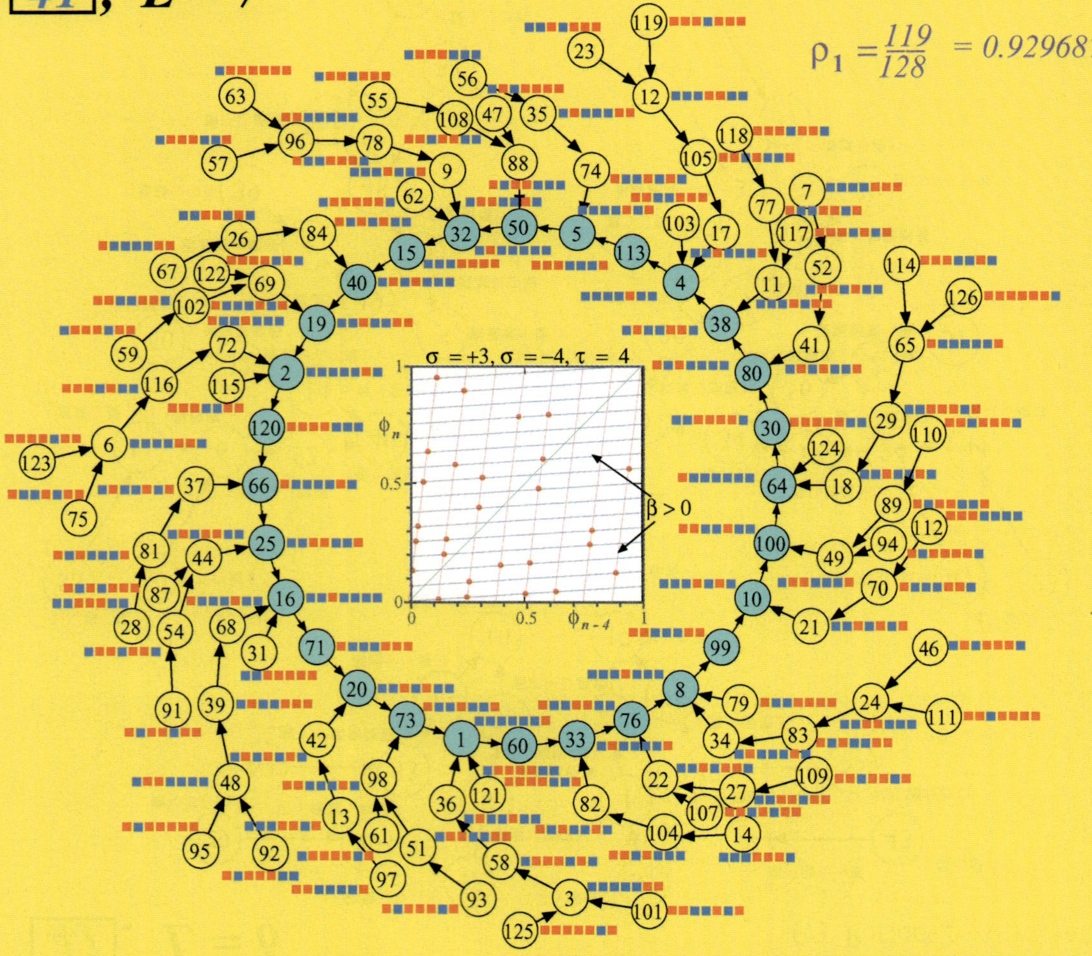

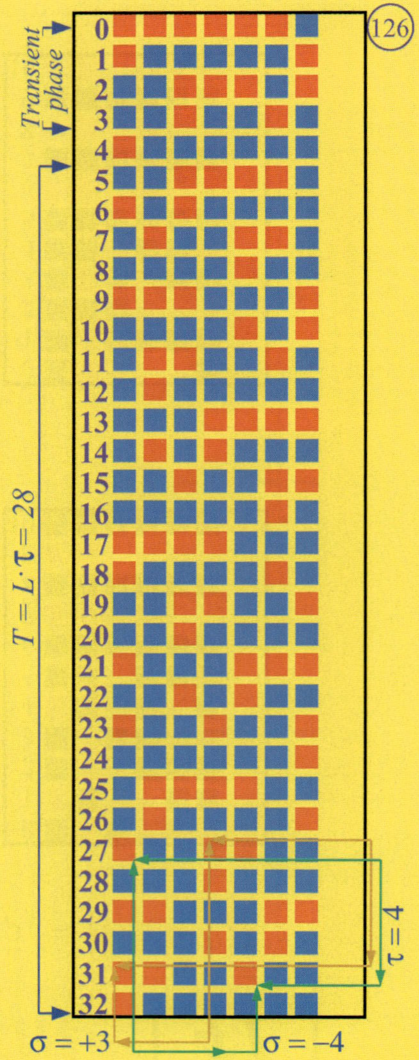

Gallery 41 - 7

Basin tree diagrams for Rule $\boxed{41}$

$\boxed{41}$, $L = 7$ (b) Bernoulli ($\sigma = +1$, $\tau = 1$) Period-7 Isle of Eden : $\rho_2 = \frac{7}{128} = 0.0546875$

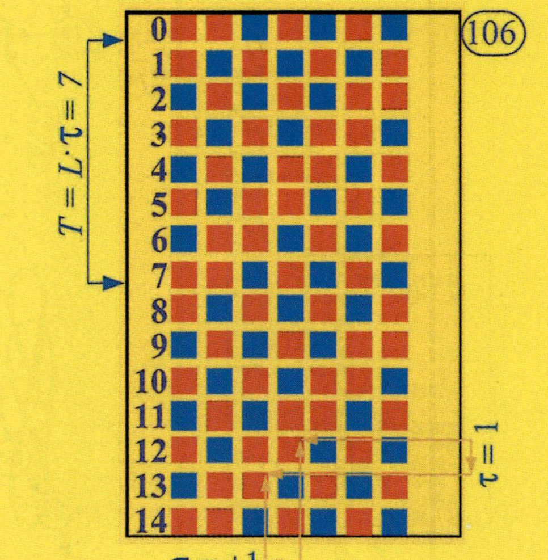

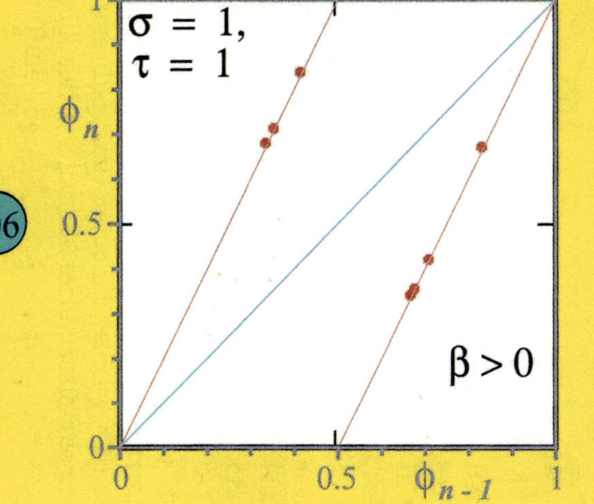

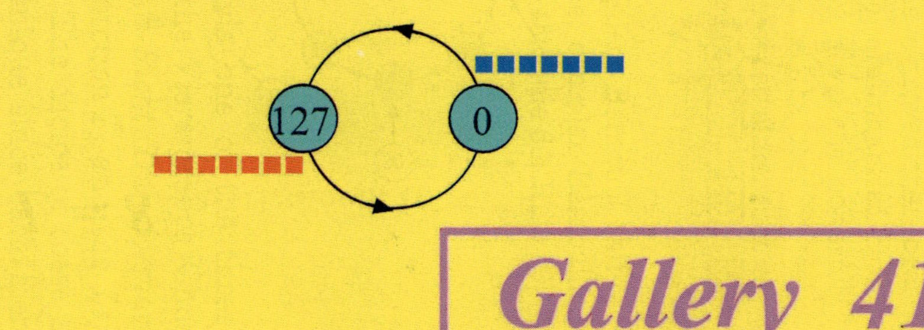

(c) Period-2 Isle of Eden : $\rho_3 = \frac{2}{128} = 0.015625$

$\boxed{\textit{Gallery 41 -8}}$

Basin tree diagrams for Rule 41

41, *L* = 8   (a) Bernoulli ($\sigma = \pm 4$, $\tau = 4$) Period-8 Attractors :

$$\rho_1 = 4\frac{28}{256} = 0.4375$$

Gallery 41 - 9

Basin tree diagrams for Rule 41

41 , L = 8 (b) Bernoulli ($\sigma = +2$, $\sigma = -2$, $\tau = 1$) Period-2 Attractors : $\rho_2 = 2\frac{66}{256} = 0.515625$

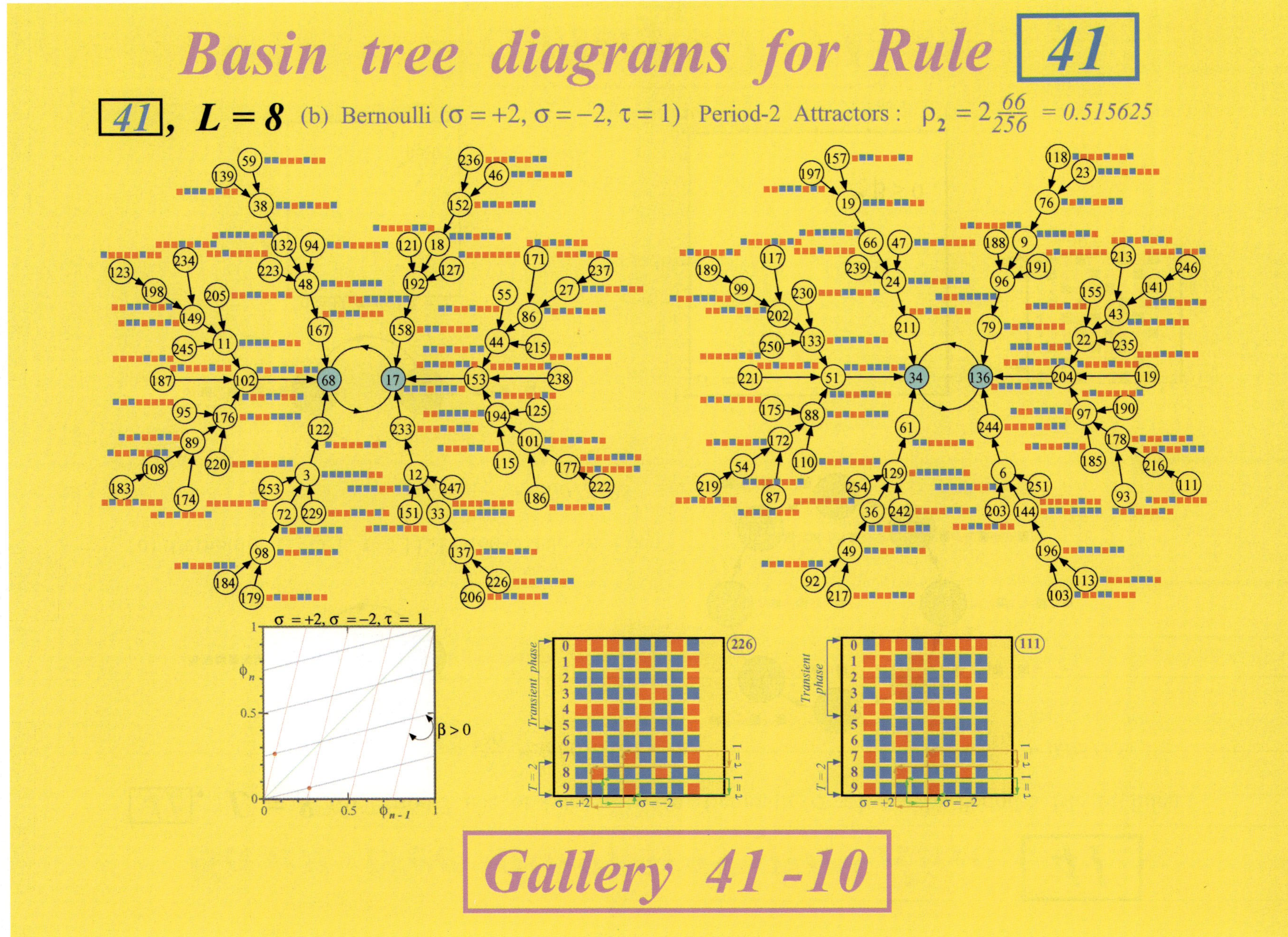

Gallery 41 -10

Table 7.   (*Continued*)

Basin tree diagrams for Rule $\boxed{41}$

$\boxed{41}$, $L = 8$ (c) Period-2 Isle of Eden :   (e) Bernoulli ($\sigma = 1, \tau = 1$) Period-8 Isle of Eden :

$$\rho_3 = \frac{2}{256} = 0.0078125$$

$$\rho_5 = \frac{8}{256} = 0.03125$$

(d) Bernoulli ($\sigma = \pm 1, \tau = 1$) Period-2 Isle of Eden :

$$\rho_4 = \frac{2}{256} = 0.0078125$$

$\boxed{\text{Gallery } 41\text{ -}11}$

$$\boxed{\text{Gallery } 45\text{-}3 : L = 5,\, n\left(\Sigma^5\right) = 32}$$

(a) There is a *period-30 Isle of Eden* with a robustness coefficient $\rho_1 = 30/32 = 0.9375$. The dynamics on this Isle of Eden obeys a Bernoulli σ_τ-shift law with $\sigma = -1$, $\tau = 6$ and $\beta > 0$. The period obeys the formula $T = \tau L/|\sigma| = 30$.

(b) There is a *period-2 Isle of Eden* with a robustness coefficient $\rho_2 = 2/32 = 0.0675$. The dynamics on this Isle of Eden obeys a *degenerate* Bernoulli σ_τ-shift law with $\sigma_1 = 1, 2, 3$ or 4, $\tau_1 = 1$ and $\beta_1 < 0$, or $\sigma_2 = 1, 2, 3$ or 4, $\tau_2 = 2$ and $\beta_2 > 0$.

$$\boxed{\text{Gallery } 45\text{-}4,\, 45\text{-}5 : L = 6,\, n\left(\Sigma^6\right) = 64}$$

(a) There is a *period-18 attractor* with a robustness coefficient $\rho_1 = 54/64 = 0.84375$. The dynamics on this attractor obeys a Bernoulli σ_τ-shift law with $\sigma = 1$, $\tau = 3$ and $\beta > 0$. The period obeys the formula $T = \tau L/|\sigma| = 18$.

(b) There is a *period-2 attractor* with a robustness coefficient $\rho_2 = 4/64 = 0.0625$. The dynamics on this attractor obeys a *degenerate* Bernoulli σ_τ-shift law with $\sigma_1 = 1, 2, \ldots$, or 6, $\tau_1 = 1$, and $\beta_1 < 0$, or $\sigma_2 = 1, 2, \ldots$, or 6, $\tau_2 = 2$ and $\beta_2 > 0$.

(c) There are three *period-1 Isles of Eden* with a combined robustness coefficient $\rho_3 = 3/64 = 0.046875$. The dynamics on this Isle of Eden obeys a *degenerate* Bernoulli σ_τ-shift law with $\sigma = 0$ or 6, $\tau = 1$ and $\beta > 0$.

(d) There is a period-3 Isle of Eden with a robustness coefficient $\rho_4 = 3/64 = 0.046875$. The dynamics on this Isle of Eden obeys a Bernoulli σ_τ-shift law with $\sigma = 1$, $\tau = 1$ and $\beta > 0$. The period obeys the formula $T = \tau L/|\sigma| = 6$, with a minimal period $T_{\min} = 3$ in view of the "two-fold" symmetry exhibited by the bit strings on this Isle of Eden.

$$\boxed{\text{Gallery } 45\text{-}6,\, 45\text{-}7 : L = 7,\, n\left(\Sigma^7\right) = 128}$$

(a) There is a *period-126 Isle of Eden* with a robustness coefficient $\rho_1 = 126/128 = 0.984375$. The dynamics on this Isle of Eden obeys a Bernoulli σ_τ-shift law with $\sigma = 3$ or -4, $\tau = 18$ and $\beta > 0$. The

period obeys the formula $T = \tau L = 126$, and not $\tau L/|\sigma|$ because 126 is not divisible by 3 or 4.

(b) There is a *period-2 Isle of Eden* with a robustness coefficient $\rho_2 = 2/128 = 0.015625$. The dynamics on this Isle of Eden obeys a *degenerate* Bernoulli σ_τ-shift law with $\sigma_1 = 1, 2, \ldots$, or 7, $\tau_1 = 1$ and $\beta_1 < 0$, or $\sigma_2 = 1, 2, \ldots$, or 7, $\tau_2 = 2$ and $\beta_2 > 0$.

$$\boxed{\text{Gallery } 45\text{-}8,\, 45\text{-}9,\, 45\text{-}10,\, 45\text{-}11 : L = 8,\; n\left(\Sigma^8\right) = 256}$$

(a) There are two *period-24 Isles of Eden* with a combined robustness coefficient $\rho_1 = 2(24/256) = 0.1875$. The dynamics on these Isles of Eden obey a Bernoulli σ_τ-shift law with $\sigma = 2$, $\tau = 6$ and $\beta > 0$. The period obeys the formula $T = \tau L/|\sigma| = 24$.

(b) There is a *period-32 Isle of Eden* with a robustness coefficient $\rho_2 = 32/256 = 0.125$. The dynamics on this Isle of Eden obeys a Bernoulli σ_τ-shift law with $\sigma = 1$, $\tau = 4$ and $\beta > 0$. The period obeys the formula $T = \tau L/|\sigma| = 32$.

(c) There is a *period-16 Isle of Eden* with a robustness coefficient $\rho_3 = 16/256 = 0.0625$. The dynamics on this Isle of Eden obeys a Bernoulli σ_τ-shift law with $\sigma = 3$, $\tau = 2$ and $\beta > 0$. The period obeys the formula $T = \tau L = 16$.

(d) There are two *period-4 Isles of Eden* with a robustness coefficient $\rho_4 = 2(4/256) = 0.03125$. The dynamics on these Isles of Eden obeys a Bernoulli σ_τ-shift law with $\sigma = 2$, $\tau = 1$ and $\beta > 0$. The period obeys the formula $T = \tau L/|\sigma| = 4$.

(e) There is a *period-2 attractor* with a robustness coefficient $\rho_5 = 152/256 = 0.59375$. The dynamics on this attractor obeys a *degenerate* Bernoulli σ_τ-shift law with $\sigma_1 = 1, 2, \ldots$, or 8, $\tau_1 = 1$ and $\beta_1 < 0$, or $\sigma_2 = 1, 2, \ldots$, or 8, $\tau_2 = 2$ and $\beta_2 > 0$.

Observe that *all* bit strings of rule $\boxed{45}$ are *Isles of Eden* for *odd* $L = 3, 5$ and 7, whereas *there is at least one attractor for even* $L = 4, 6$ and 8. We will prove in Sec. 5 below that this surprising property is true for all L!

The qualitative properties of local rule $\boxed{45}$ extracted from the above basin-tree Galleries 45-1

to 45-11 are summarized below:

Summary of Qualitative properties of local rule $\boxed{45}$ extracted from Gallery 45 for Rule $\boxed{45}$

L	ID Number i	Number of Period-n attractors	Number of Period-n Isles of Eden	Period n	Bernoulli Parameters						Robustness coefficient ρ
					σ_1	τ_1	β_1	σ_2	τ_2	β_2	
3	1		3	1	0	1	+	3	1	+	$\rho_1 = 0.375$
	2		1	2	1, 2, 3	1	−	1, 2, 3	1	+	$\rho_2 = 0.25$
	3		1	3	1	1	+				$\rho_3 = 0.375$
4	1	1		2	1,2,3,4	1	−	1,2,3,4	1	+	$\rho_1 = 1$
5	1		1	30				-1	6	+	$\rho_1 = 0.9375$
	2		1	2	1,2,...,5	1	−	1,2,...,5	1	+	$\rho_2 = 0.0675$
6	1	1		18	1	3	+				$\rho_1 = 0.84375$
	2	1		2	1,2,...,6	1	−	1,2,...,6	1	+	$\rho_2 = 0.0625$
	3		3	1	0	1	+	6	1	+	$\rho_3 = 0.046875$
	4		1	3	1	1	+				$\rho_4 = 0.046875$
7	1		1	126	3	18	+	-4	18	+	$\rho_1 = 0.984375$
	2		1	2	1,2,...,7	1	−	1,2,...,7	1	+	$\rho_2 = 0.015625$
8	1		2	24	2	6	+				$\rho_1 = 0.1875$
	2		1	32	1	4	+				$\rho_2 = 0.125$
	3		1	16	3	2	+				$\rho_3 = 0.0625$
	4		2	4	2	1	+				$\rho_4 = 0.03125$
	5	1		2	1,2,...,8	1	−	1,2,...,8	1	+	$\rho_5 = 0.59375$

2.5. *Highlights from rule $\boxed{60}$*

$$\boxed{\text{Gallery 60-1}: L = 3, n\left(\textstyle\sum^3\right) = 8}$$

(a) There is a *period-3 attractor* with a robustness coefficient $\rho_1 = 6/8 = 0.75$. The dynamics on this attractor obeys a Bernoulli σ_τ-shift law with $\sigma = 1$, $\tau = 1$ and $\beta > 0$.

(b) There is a period-1 attractor $\{\boxed{0}\}$ with a robustness coefficient $\rho_2 = 2/8 = 0.25$. The dynamics on this attractor obeys a *degenerate* Bernoulli σ_τ-shift law with $\sigma = 0$, $\tau = 1$ and $\beta > 0$.

$$\boxed{\text{Gallery 60-2}: L = 4, n\left(\textstyle\sum^4\right) = 16}$$

There is a *period-1 attractor* $\{\boxed{0}\}$ with a robustness coefficient $\rho_1 = 16/16 = 1$. The dynamics on this attractor obeys a *degenerate* Bernoulli σ_τ-shift law with $\sigma = 0$, $\tau = 1$ and $\beta > 0$.

$$\boxed{\text{Gallery 60-3}: L = 5, n\left(\textstyle\sum^5\right) = 32}$$

(a) There is a *period-15 attractor* with a robustness coefficient $\rho_1 = 30/32 = 0.9375$. The dynamics on this attractor obeys a Bernoulli σ_τ-shift law with $\sigma = 1$, $\tau = 3$ and $\beta > 0$. The period obeys the formula $T = \tau L/|\sigma| = 15$.

(b) There is a *period-1 attractor* $\{\boxed{0}\}$ with a robustness coefficient $\rho_2 = 2/32 = 0.0625$. The dynamics on this attractor obeys a *degenerate* Bernoulli σ_τ-shift law with $\sigma = 0$, $\tau = 1$ and $\beta > 0$.

$$\boxed{\text{Gallery 60-4, 60-5}: L = 6, n\left(\textstyle\sum^6\right) = 64}$$

(a) There are two *period-6 attractors* with a combined robustness coefficients $\rho_1 = 2(24/64) = 0.75$.

Table 8. Basin tree diagrams for rule [45].

Basin tree diagrams for Rule 45
45, L = 3
(a) Period-1 Isles of Eden :
ρ1 = 3/8 = 0.375
4
2
1
(b) Period-2 Isle of Eden : ρ2 = 2/8 = 0.25
7
0
(c) Bernoulli (σ = +1, τ = 1)
Period-3 Isle of Eden :
ρ3 = 3/8 = 0.375
3
5
6
⑥
0
1
2
3
4
5
6
τ = 1
T = L·τ = 3
σ = +1
σ = +1, τ = 1
1
ϕn
0.5
0
0
0.5
ϕn - 1
1
β > 0
Gallery 45 -1

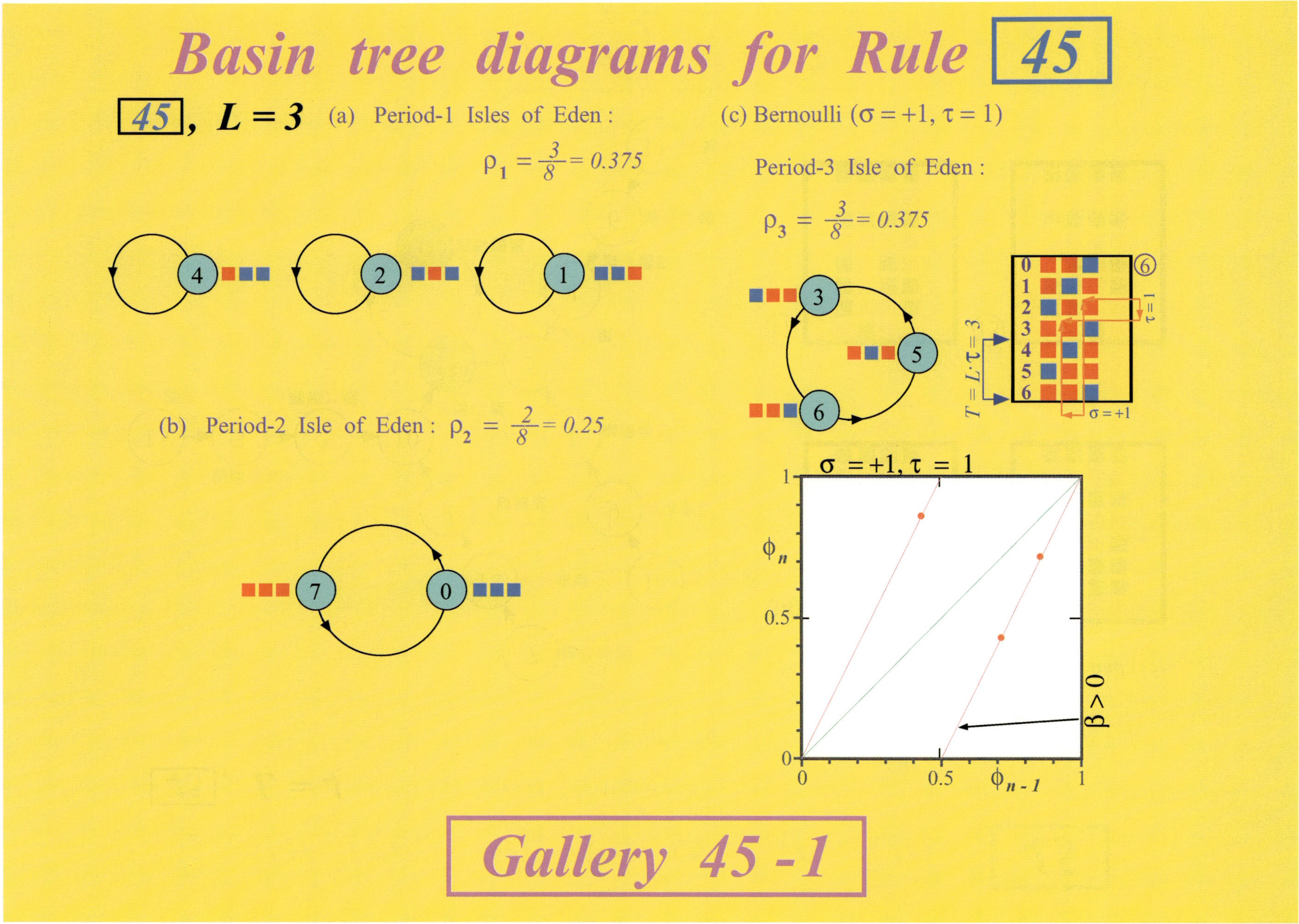

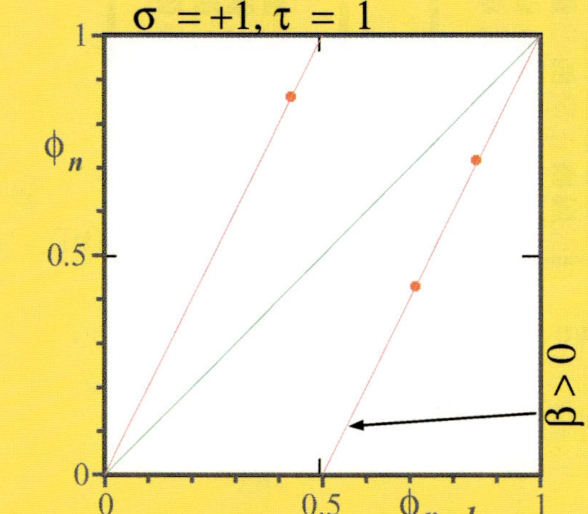

Table 8. (*Continued*)

Basin tree diagrams for Rule 45

45, *L = 4*   (a) Period-2 Attractor :

$$\rho_1 = 1$$

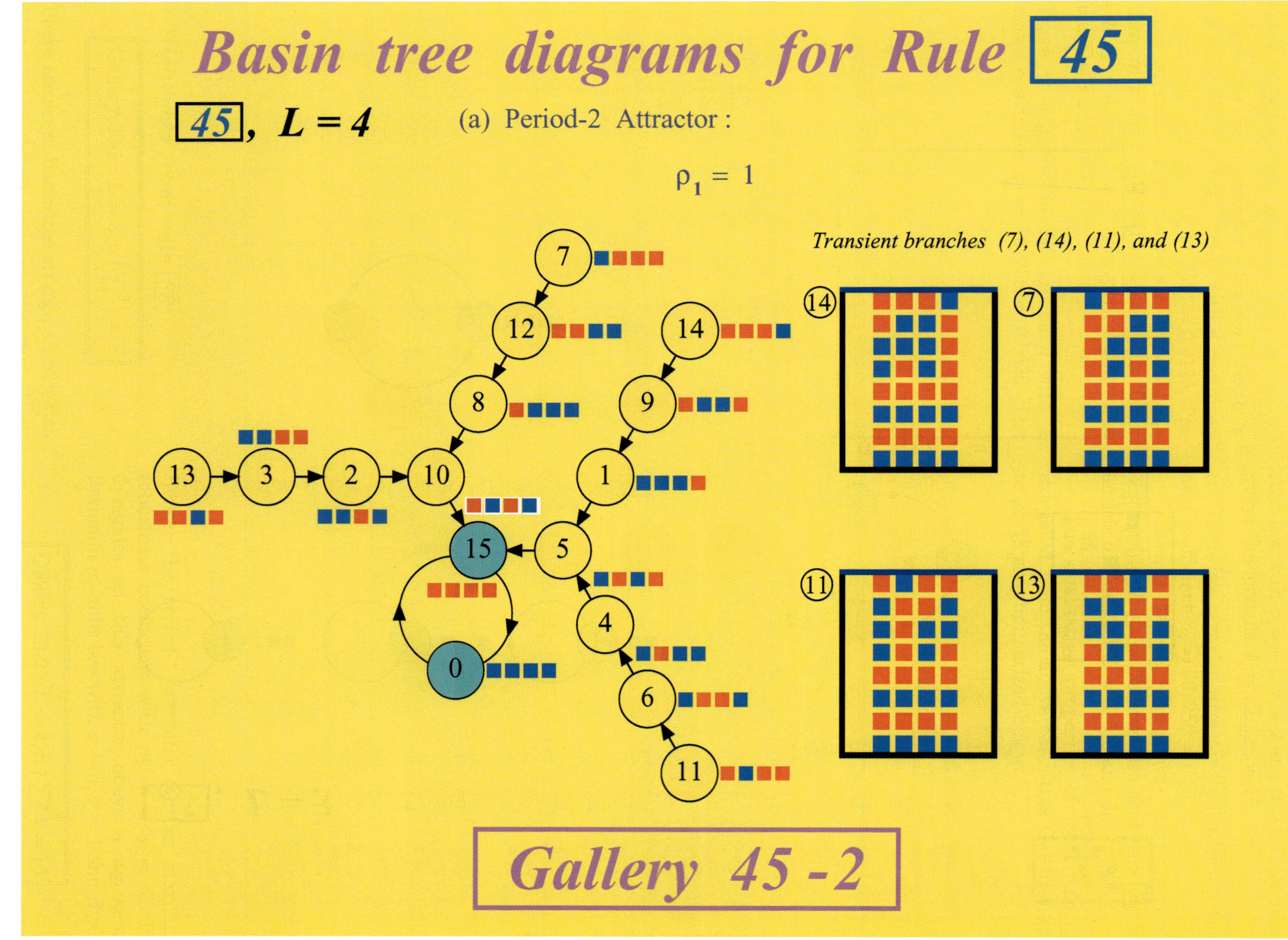

Gallery 45 - 2

Table 8. (*Continued*)

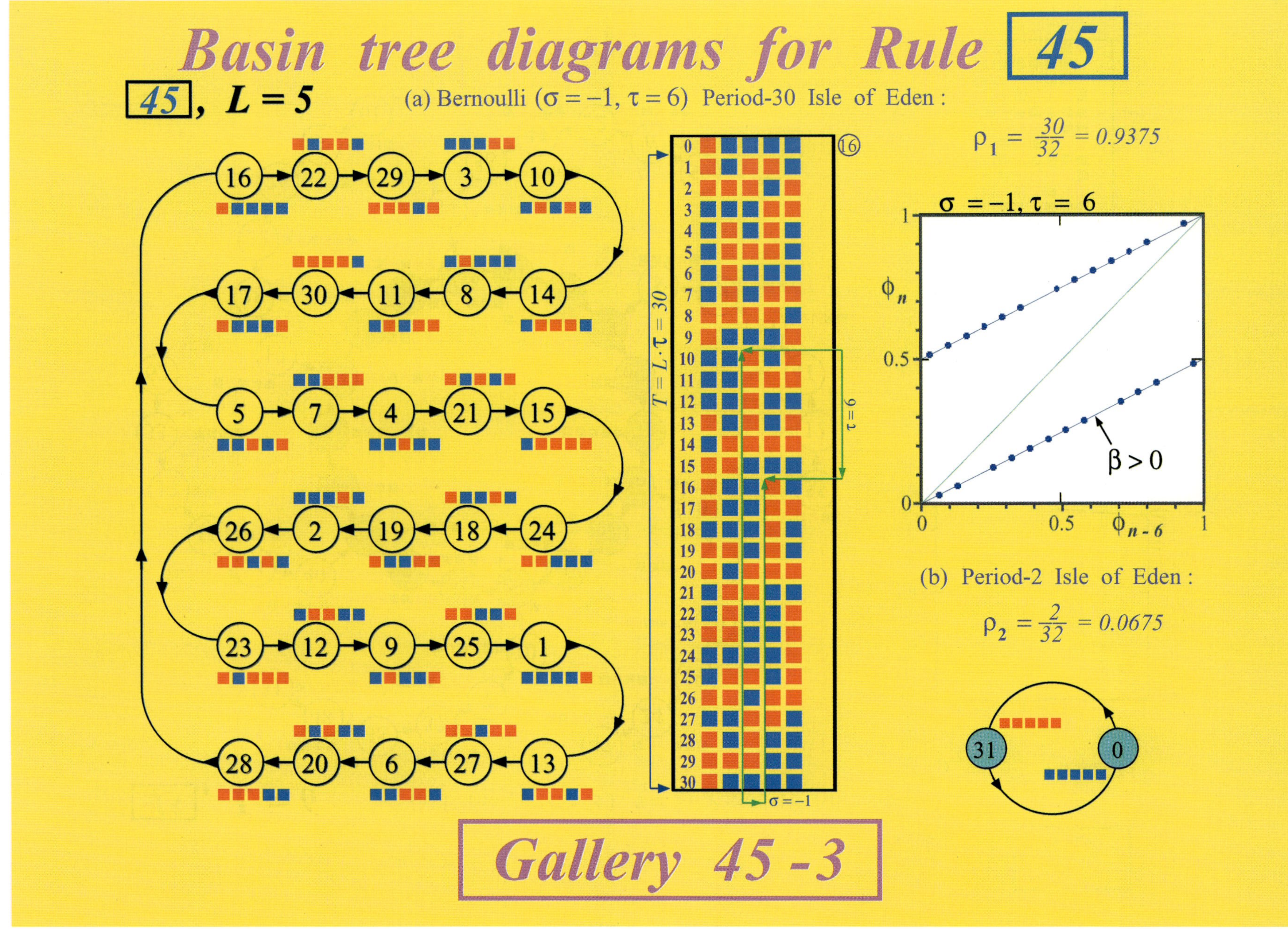

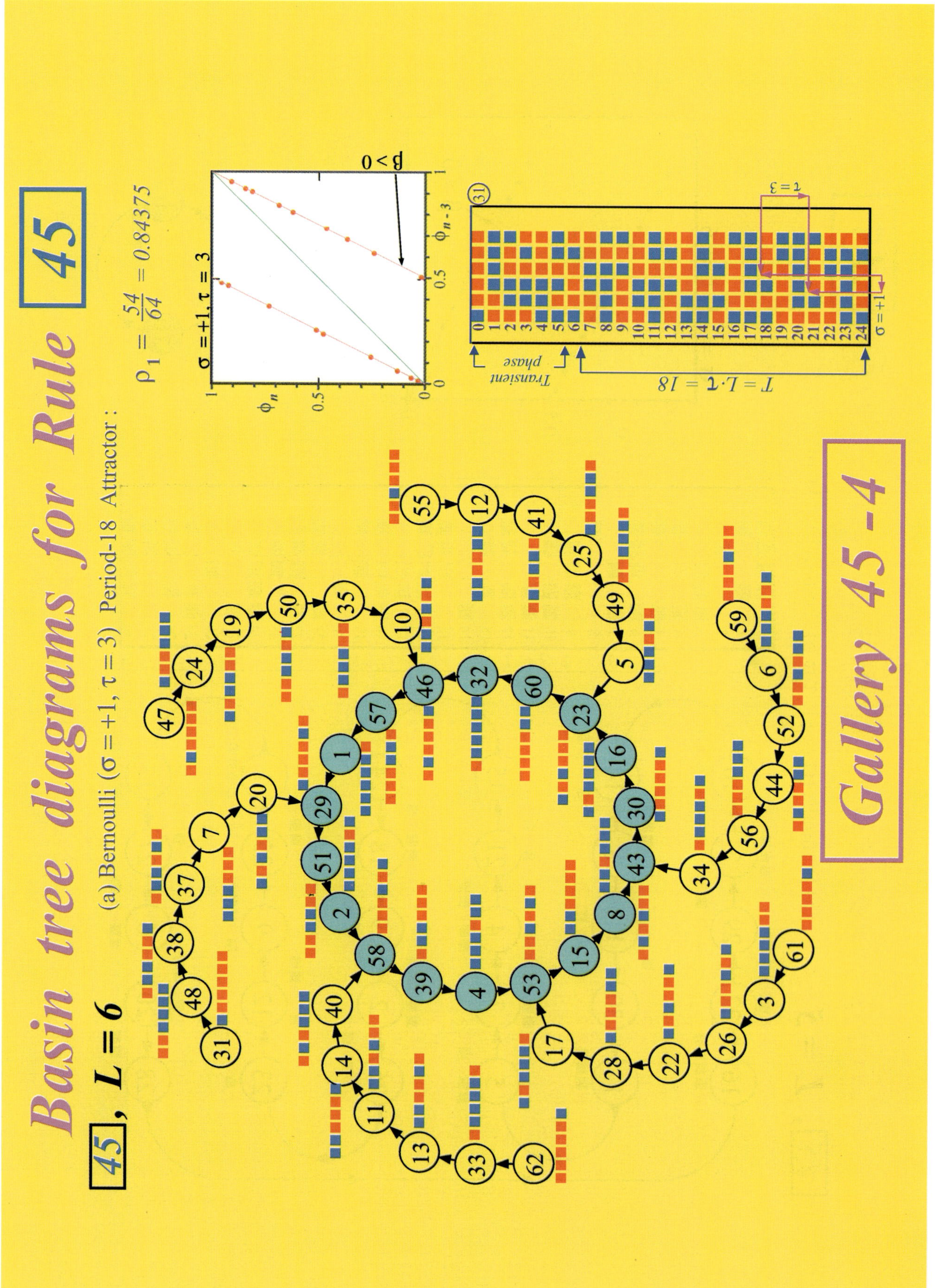

Table 8.   (*Continued*)

Basin tree diagrams for Rule 45

45 , **L = 6**   (a) Bernoulli ($\sigma = +1$, $\tau = 3$) Period-18 Attractor :

$$\rho_1 = \frac{54}{64} = 0.84375$$

Gallery 45 - 4

Table 8. (Continued)

Basin tree diagrams for Rule 45
45, L = 6
(b) Period-2 Attractor :
ρ2 = 4/64 = 0.0625
0
63
42
21
(c) Period-1 Isles of Eden : ρ3 = 3/64 = 0.046875
36
18
9
(d) Bernoulli (σ = +1, τ = 1) Period-3 Isle of Eden :
ρ4 = 3/64 = 0.046875
27
45
54
0
1
2
3
4
5
6
T = 3
τ = 1
σ = +1
σ = +1, τ = 1
φn
φn-1
0
0.5
1
β > 0
Gallery 45 -5

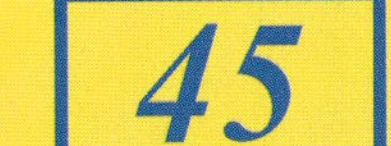

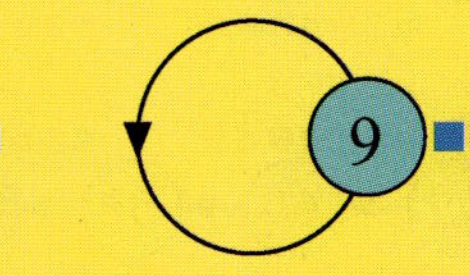

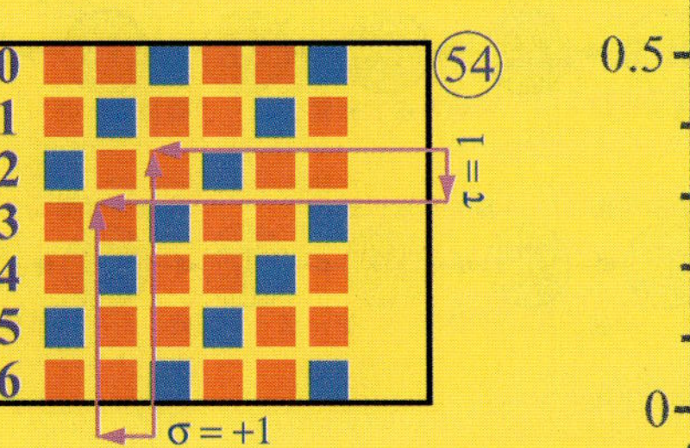

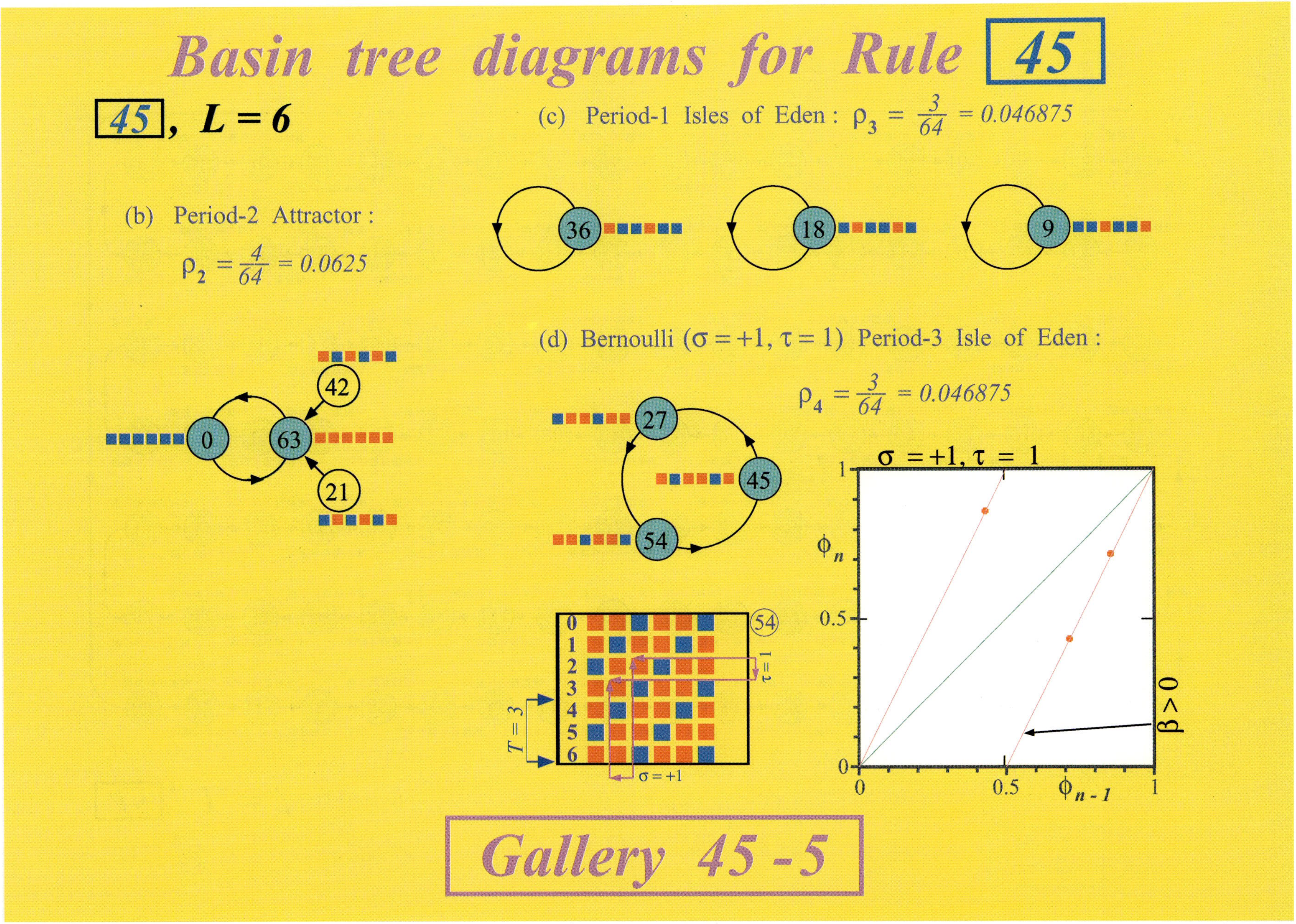

Table 8. (*Continued*)

Basin tree diagrams for Rule 45

45 , **L = 7** (a) Bernoulli ($\sigma = +3$, $\sigma = -4$, $\tau = 18$) Period-126 Isle of Eden :

$$\rho_1 = \frac{126}{128} = 0.984375$$

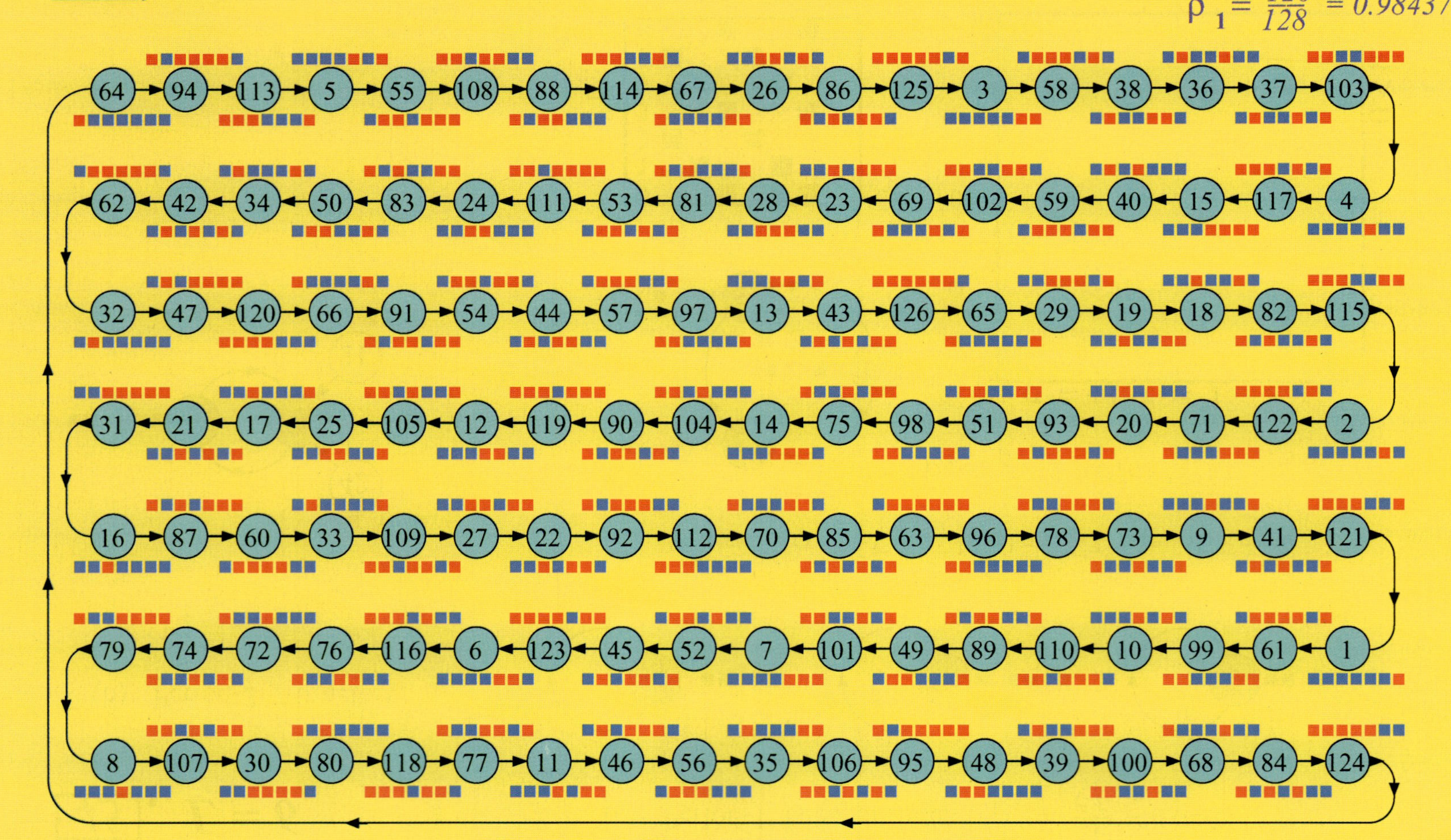

Gallery 45 - 6

Basin tree diagrams for Rule 45

45 , **L = 7** (a) Bernoulli ($\sigma = +3$, $\sigma = -4$, $\tau = 18$) Period-126 Isle of Eden *(continued)* :

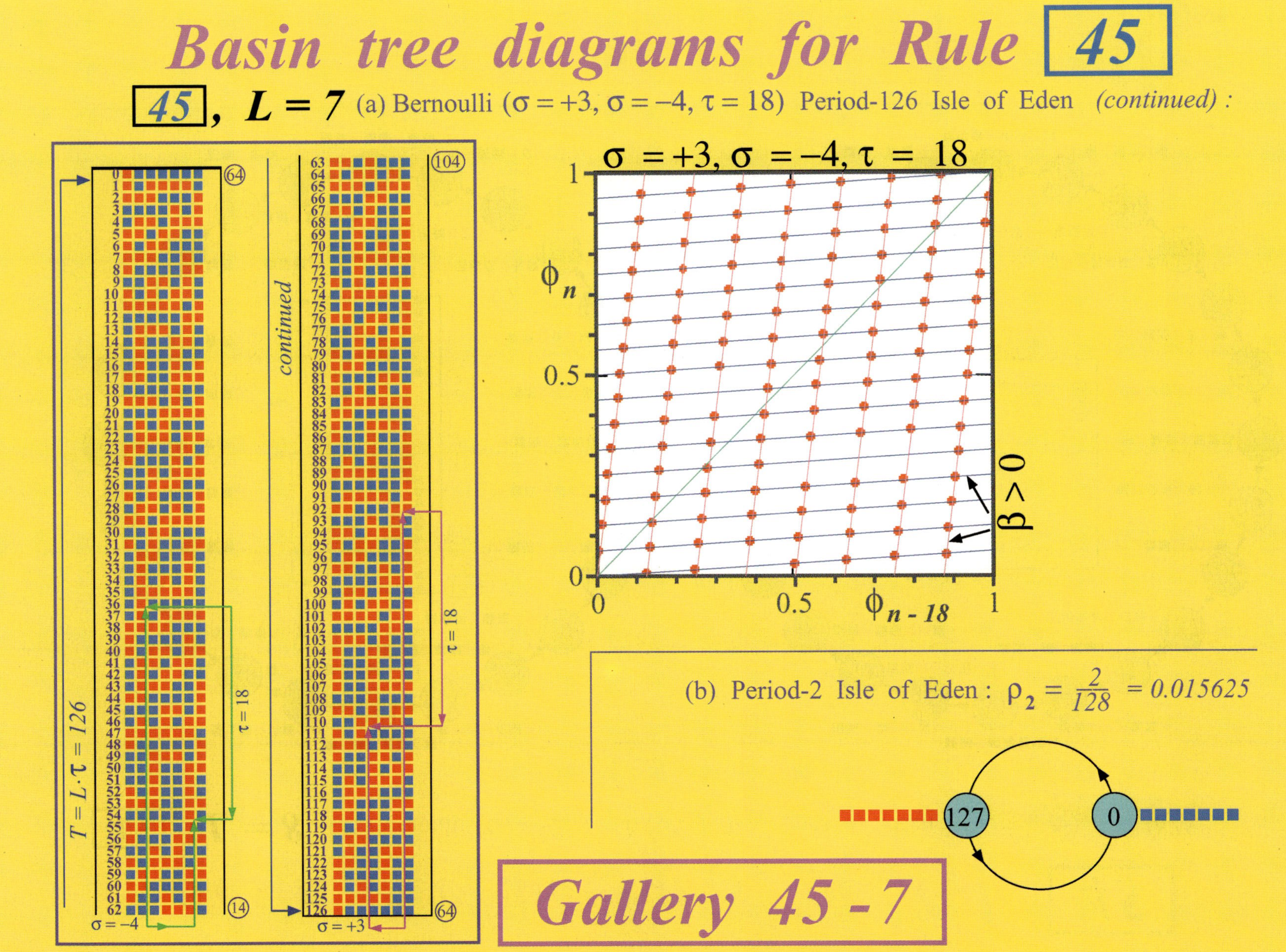

Gallery 45 - 7

Basin tree diagrams for Rule $\boxed{45}$

$\boxed{45}$, $L = 8$ (a) Bernoulli ($\sigma = +2$, $\tau = 6$) Period-24 Isles of Eden : $\rho_1 = 2\frac{24}{256} = 0.1875$

Gallery 45 - 8

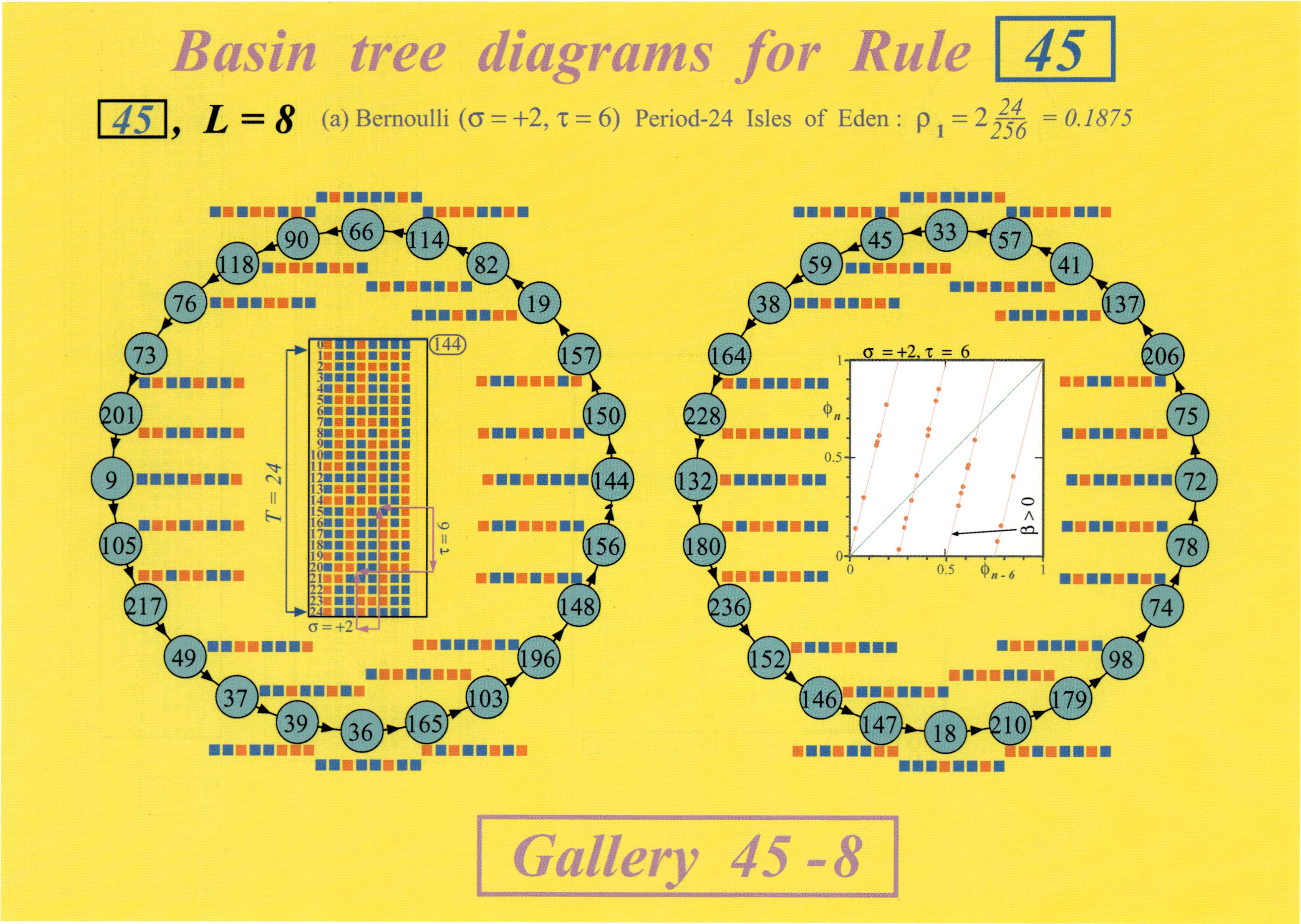

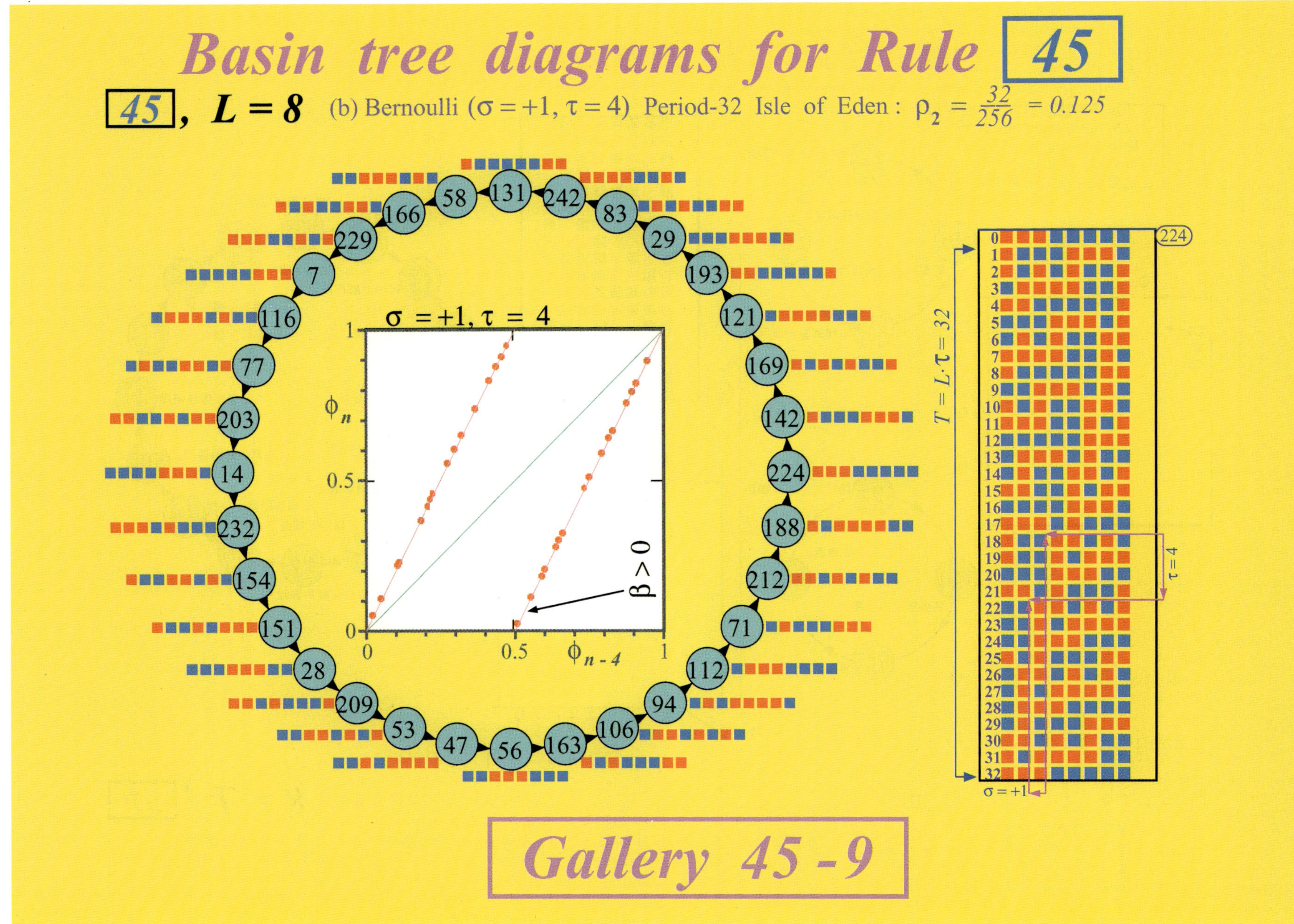

Table 8. (Continued)

Basin tree diagrams for Rule 45

45, $L = 8$ (b) Bernoulli ($\sigma = +1$, $\tau = 4$) Period-32 Isle of Eden : $\rho_2 = \frac{32}{256} = 0.125$

Gallery 45 - 9

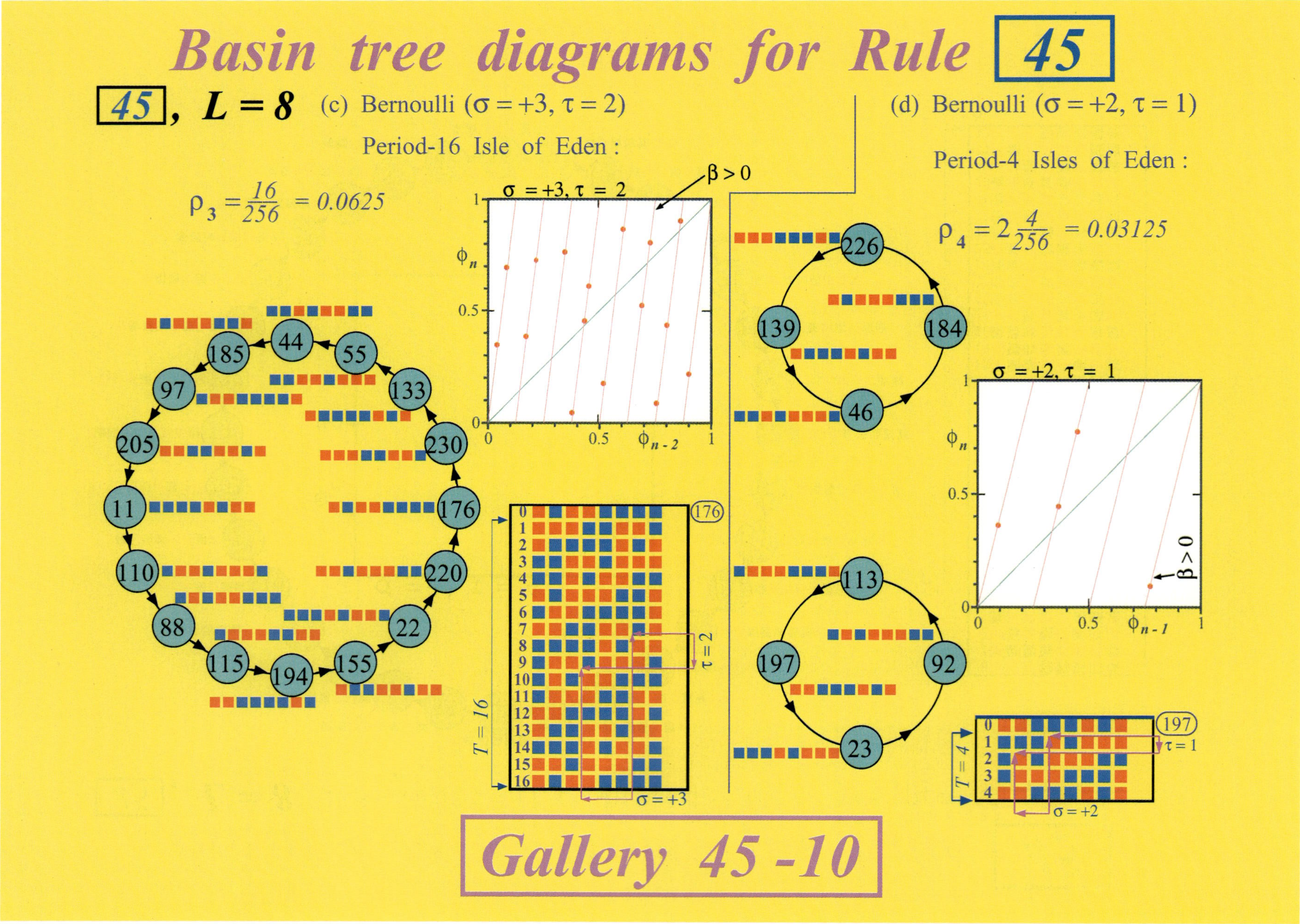

Basin tree diagrams for Rule 45
45, L = 8
(c) Bernoulli (σ = +3, τ = 2)
Period-16 Isle of Eden :
ρ3 = 16/256 = 0.0625
σ = +3, τ = 2
β > 0
φn
φn-2
T = 16
τ = 2
σ = +3
176
(d) Bernoulli (σ = +2, τ = 1)
Period-4 Isles of Eden :
ρ4 = 2 4/256 = 0.03125
σ = +2, τ = 1
φn
φn-1
β > 0
T = 4
τ = 1
σ = +2
197
Gallery 45 -10

Basin tree diagrams for Rule 45

45, **L = 8**

(e) Period-2 Attractor :

$$\rho_5 = \frac{152}{256} = 0.59375$$

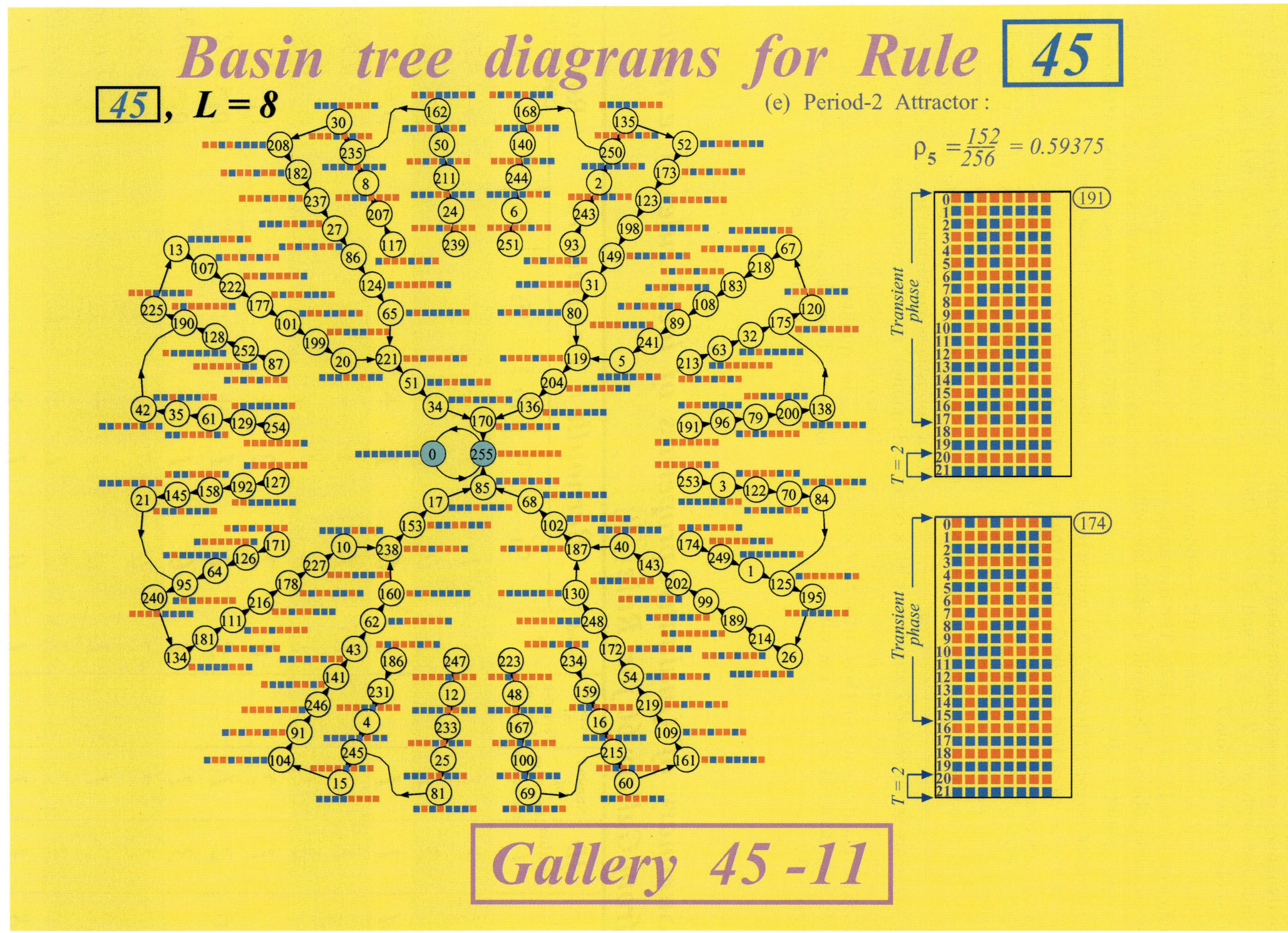

Gallery 45 -11

The dynamics on these attractors obey a Bernoulli σ_τ-shift law with $\sigma = 2$, $\tau = 2$ and $\beta > 0$. The period obeys the formula $T = \tau L/|\sigma| = 6$.

(b) There is a *period-3 attractor* with a robustness coefficient $\rho_2 = 12/64 = 0.1875$. The dynamics on this attractor obeys a Bernoulli σ_τ-shift law with $\sigma = 1$, $\tau = 1$ and $\beta > 0$.

(c) There is a *period-1 attractor* $\{\boxed{0}\}$ with a robustness coefficient $\rho_3 = 4/64 = 0.0625$. The dynamics on this attractor obeys a *degenerate* Bernoulli σ_τ-shift law with $\sigma = 0$, $\tau = 1$ and $\beta > 0$.

$$\boxed{\text{Gallery } 60\text{-}6, 60\text{-}7 : L = 7, n\left(\textstyle\sum^7\right) = 128}$$

(a) There are seven *period-7 attractors* with a combined robustness coefficients $\rho_1 = 7(14/128) = 0.765625$. The dynamics on these attractors obeys a *degenerate* Bernoulli σ_τ-shift law with $\sigma = 0$, $\tau = 7$ and $\beta > 0$.

(b) There is a *period-1 attractor* $\{\boxed{0}\}$ with a robustness coefficient $\rho_2 = 2(1/128) = 0.015625$. The dynamics on this attractor obeys a *degenerate* Bernoulli σ_τ-shift law with $\sigma = 0$, $\tau = 1$ and $\beta > 0$.

(c) There are two *period-7 attractors* with a combined robustness coefficient $\rho_3 = 2(14/128) = 0.21875$. The dynamics on these attractors obeys a Bernoulli σ_τ-shift law with $\sigma = 2$, or $\sigma = -3$, $\tau = 1$ and $\beta > 0$.

$$\boxed{\text{Gallery } 60\text{-}8 : L = 8, n\left(\textstyle\sum^8\right) = 256}$$

(a) There is a *period-1 global attractor* $\{\boxed{0}\}$ with a robustness coefficient $\rho_1 = 256/256 = 1$. The dynamics on this attractor obeys a *degenerate* Bernoulli σ_τ-shift law with $\sigma = 0$, $\tau = 1$ and $\beta > 0$.

The qualitative properties of local rule $\boxed{60}$ extracted from the above basin-tree Galleries 60-1 to 60-8 are summarized below:

Summary of Qualitative properties of local rule $\boxed{60}$ extracted from Gallery 60 for Rule $\boxed{60}$

L	ID Number i	Number of Period-n attractors	Number of Period-n Isles of Eden	Period n	σ_1	τ_1	β_1	σ_2	τ_2	β_2	Robustness coefficient ρ
3	1	1		3	1	1	+				$\rho_1 = 0.75$
	2	1		1	0	1	+				$\rho_2 = 0.25$
4	1	1		1	0	1	+				$\rho_1 = 1$
5	1	1		15	1	3	+				$\rho_1 = 0.9375$
	2	1		1	0	1	+				$\rho_2 = 0.0625$
6	1	2		6	2	2	+				$\rho_1 = 0.75$
	2	1		3	1	1	+				$\rho_2 = 0.1875$
	3	1		1	0	1	+				$\rho_3 = 0.0625$
7	1	7		7	0	7	+				$\rho_1 = 0.765625$
	2	1		1	0	1	+				$\rho_2 = 0.015625$
	3	2		7	2	1	+	-3	1	+	$\rho_3 = 0.21875$
8	1	1		1	0	1	+				$\rho_1 = 1$

Basin tree diagrams for Rule $\boxed{60}$

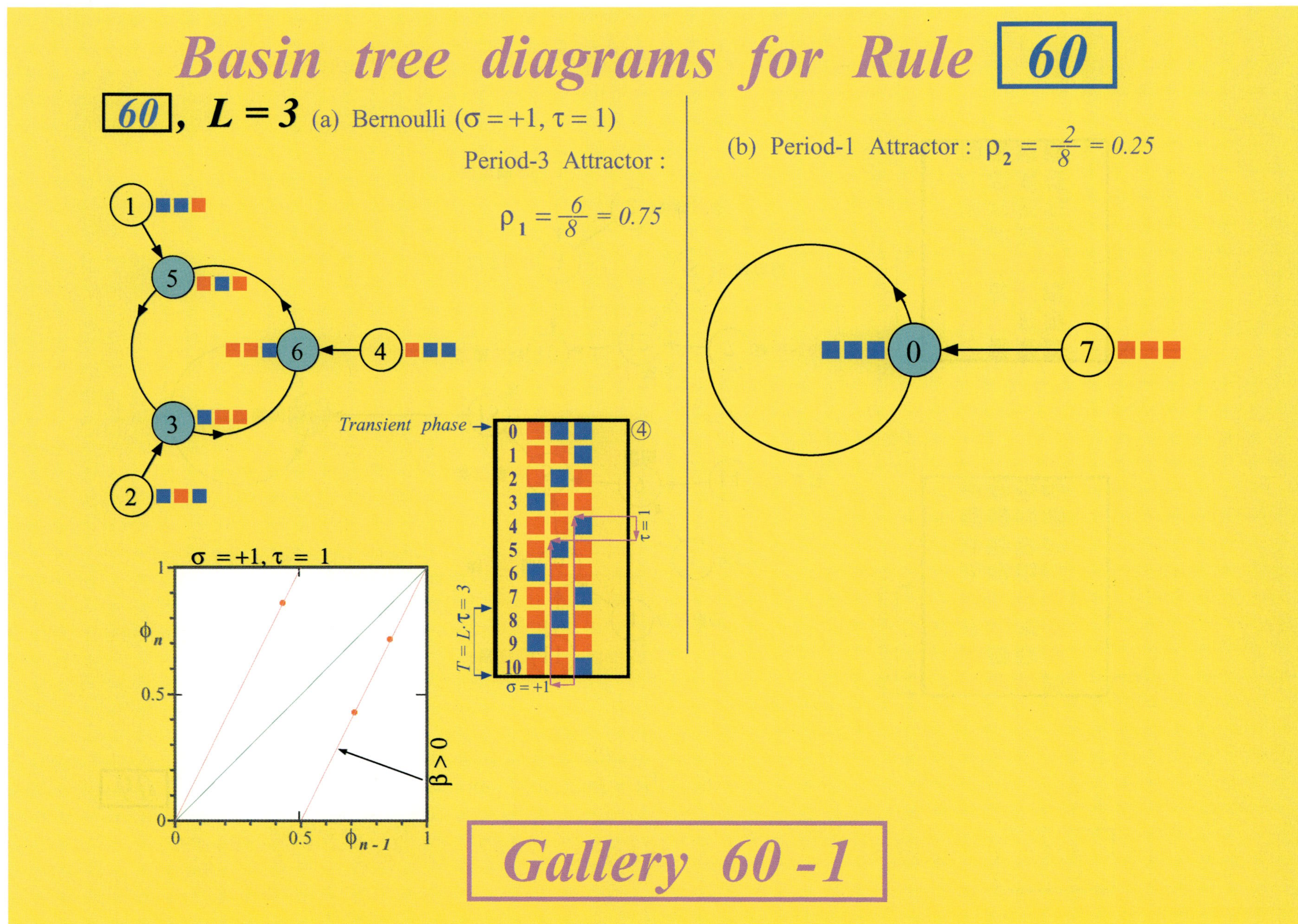

Table 9. (*Continued*)

Basin tree diagrams for Rule 60

60 , $L = 4$ (a) Period-1 Attractor : $\rho_1 = \frac{16}{16} = 1$

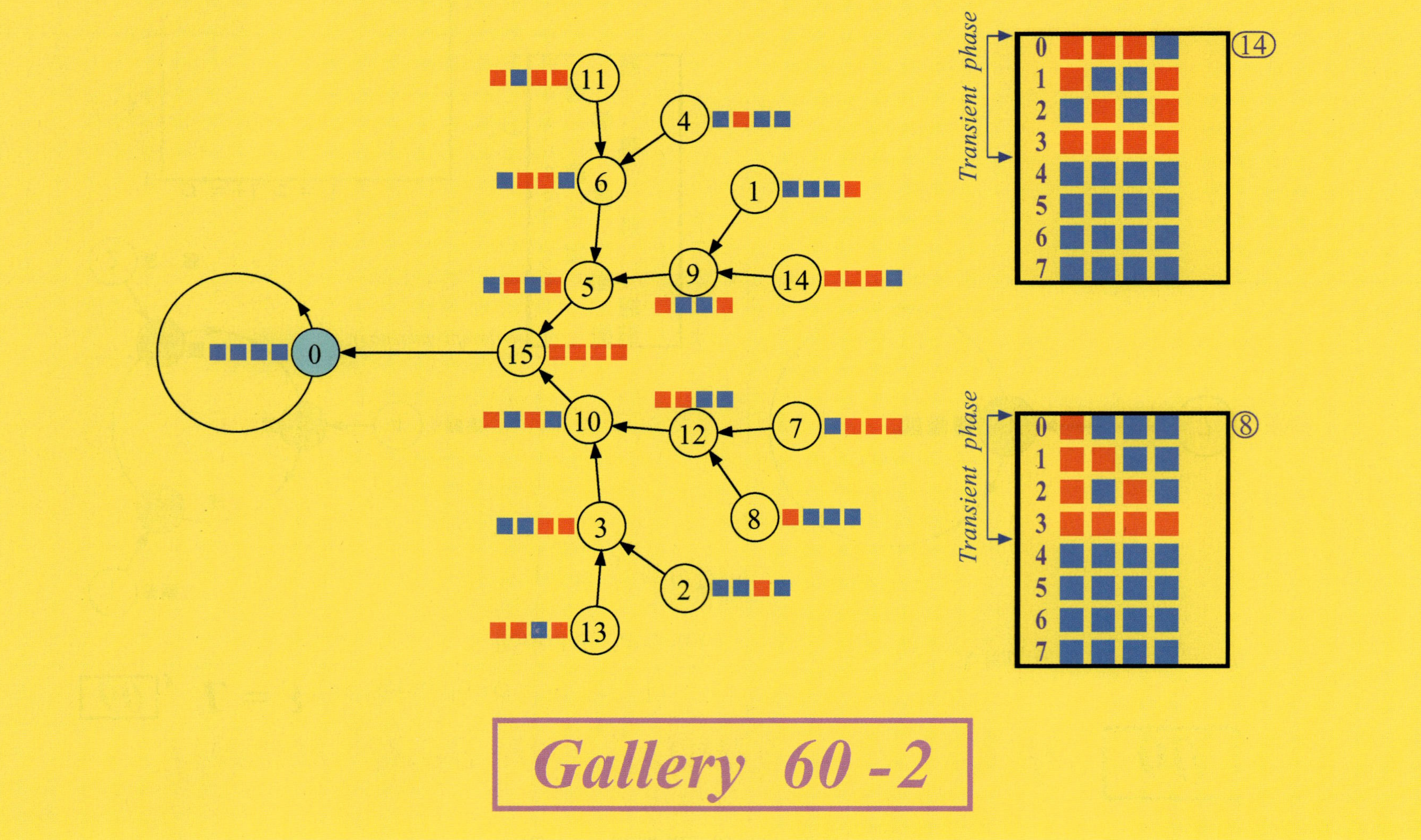

Gallery 60 - 2

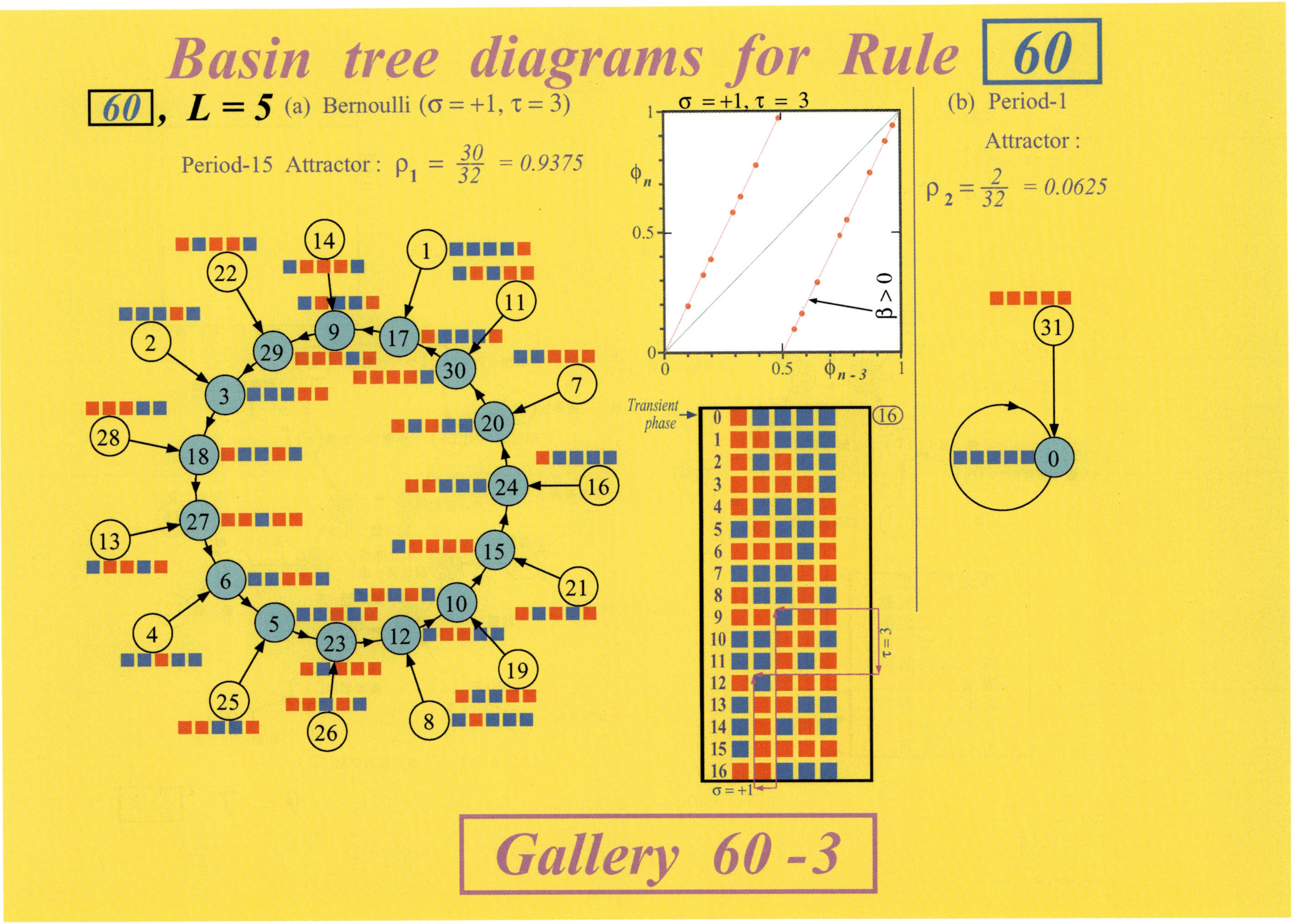

Table 9. (*Continued*)

Basin tree diagrams for Rule $\boxed{60}$

$\boxed{60}$, **$L = 5$** (a) Bernoulli ($\sigma = +1$, $\tau = 3$)

Period-15 Attractor : $\rho_1 = \frac{30}{32} = 0.9375$

(b) Period-1

Attractor :

$\rho_2 = \frac{2}{32} = 0.0625$

$\boxed{\textit{Gallery 60 -3}}$

Table 9. (*Continued*)

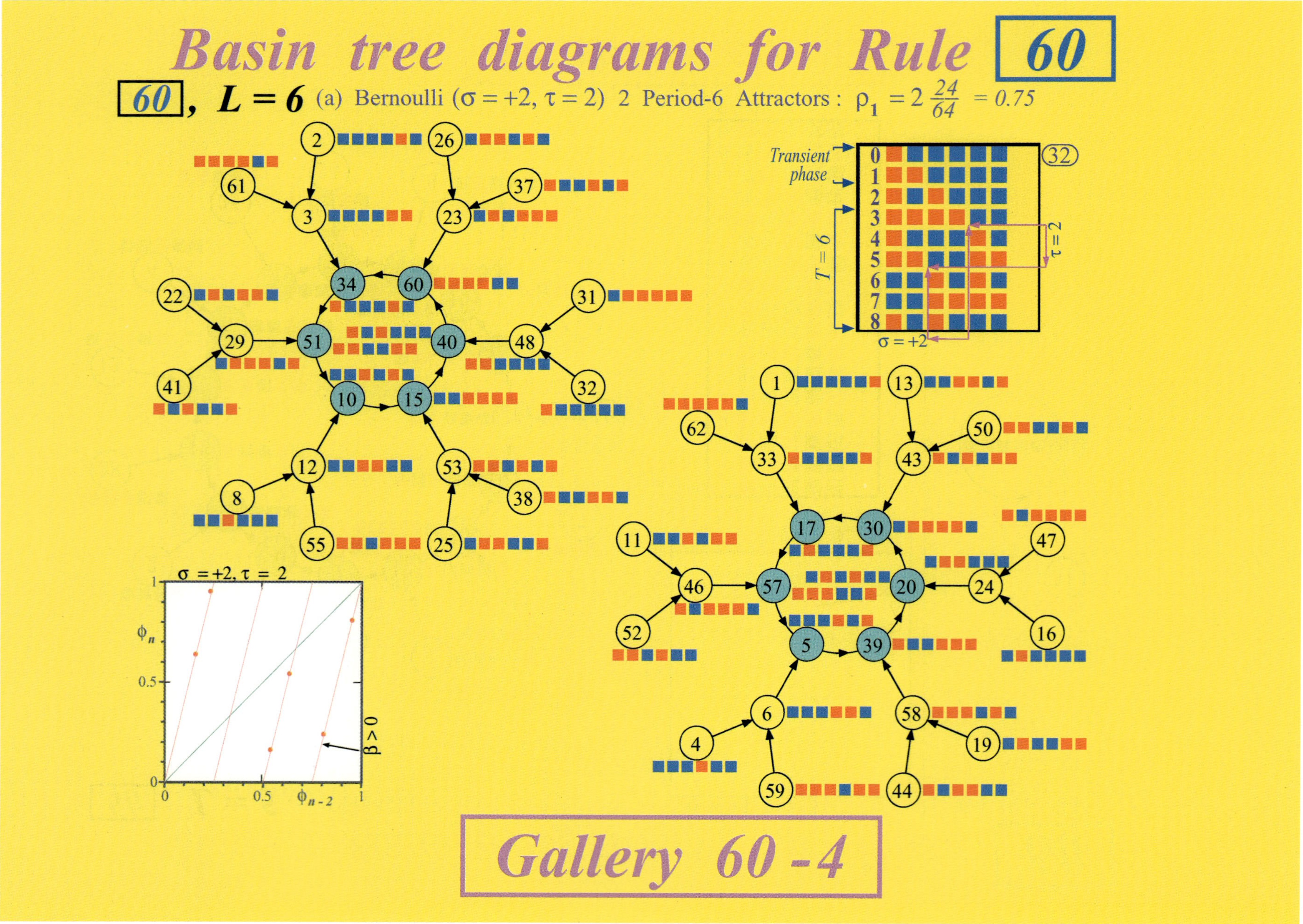

Basin tree diagrams for Rule 60
60, L = 6 (a) Bernoulli (σ = +2, τ = 2) 2 Period-6 Attractors : ρ1 = 2 24/64 = 0.75
Transient phase
T = 6
τ = 2
σ = +2
32
σ = +2, τ = 2
φn
φn-2
β > 0
Gallery 60 -4

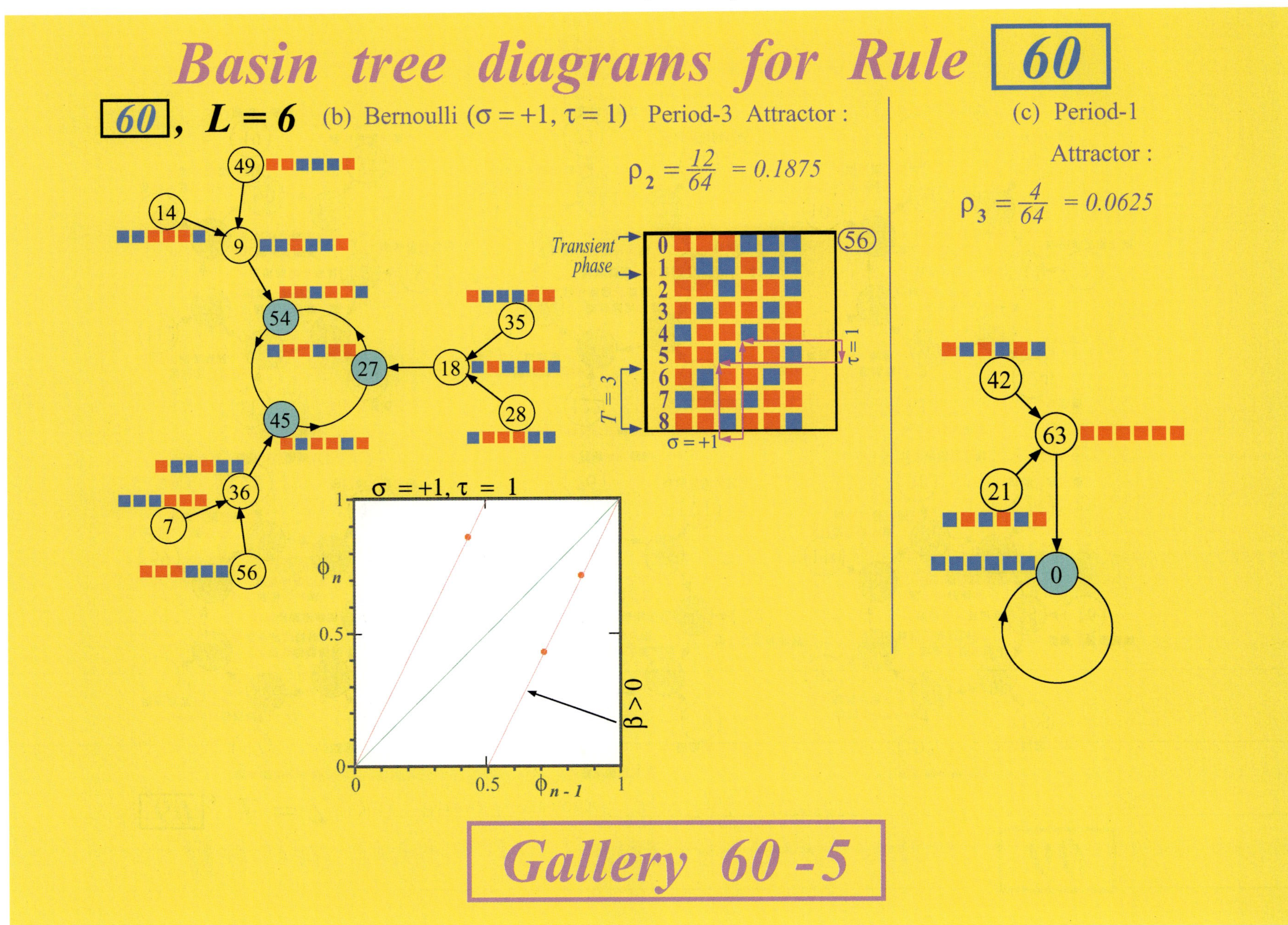

Basin tree diagrams for Rule   60

60 , L = 6   (b) Bernoulli (σ = +1, τ = 1) Period-3 Attractor :

(c) Period-1

Attractor :

$\rho_2 = \frac{12}{64} = 0.1875$

$\rho_3 = \frac{4}{64} = 0.0625$

49
14
9
54
27
45
18
35
28
7
36
56

Transient phase
0
1
2
3
4
5
6
7
8
T = 3
σ = +1
I = 1
56

σ = +1, τ = 1
φ_n
0.5
0
0.5
φ_{n-1}
β > 0

42
63
21
0

Gallery 60 - 5

Table 9.   (*Continued*)

Basin tree diagrams for Rule $\boxed{60}$

$\boxed{60}$, **L = 7**   (a)   Period-7 Attractors : $\rho_1 = 7\frac{14}{128} = 0.765625$

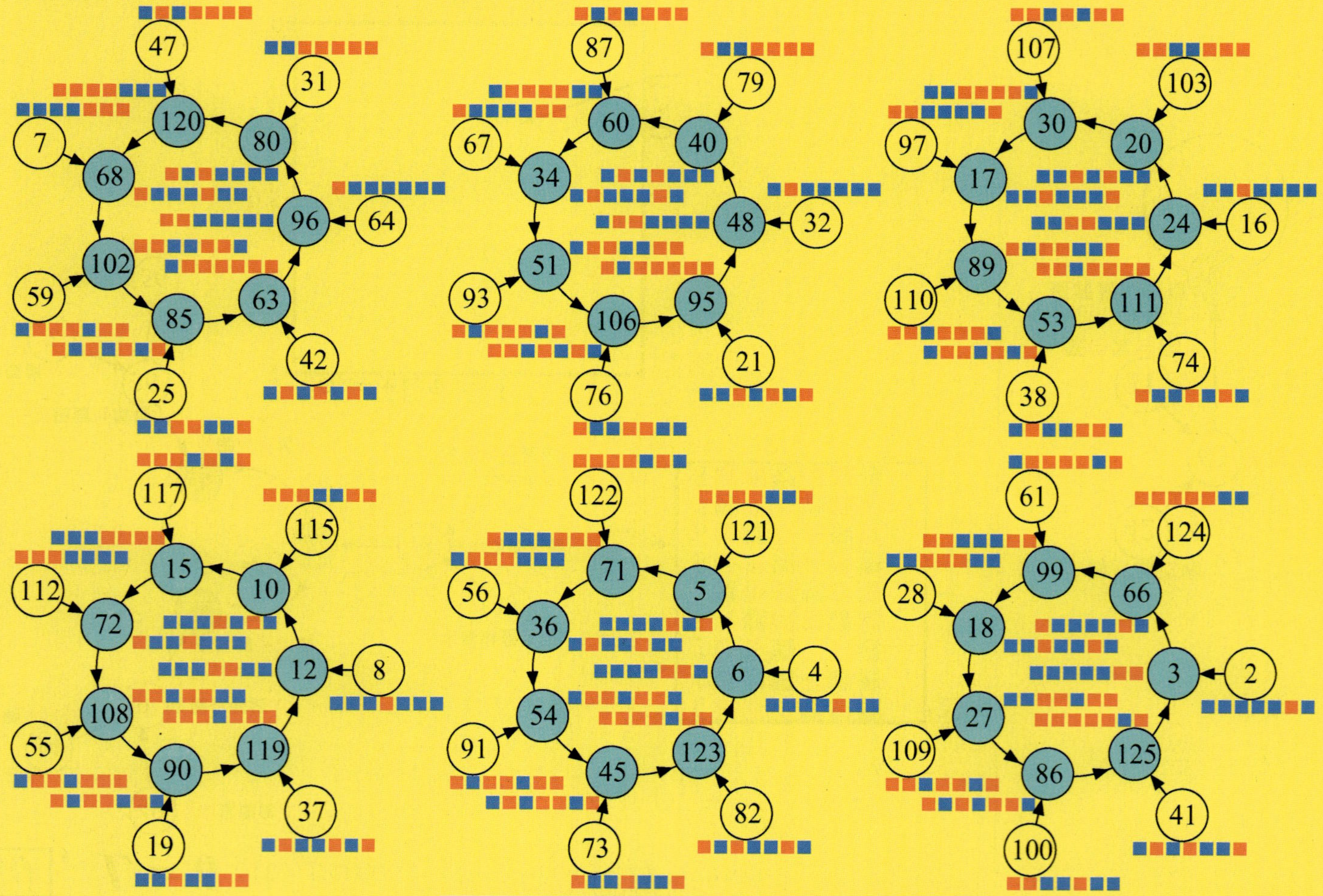

$$\boxed{Gallery\ 60\text{-}6}$$

Basin tree diagrams for Rule 60

60, **L = 7**

(c) Bernoulli ($\sigma = +2$, $\sigma = -3$, $\tau = 1$) Period-7 Attractors :

(a) 7th Period-7 Attractor (*continued*) :

(b) Period-1 Attractor
$$\rho_2 = \frac{2}{128} = 0.015625$$

$$\rho_3 = 2\frac{14}{128} = 0.21875$$

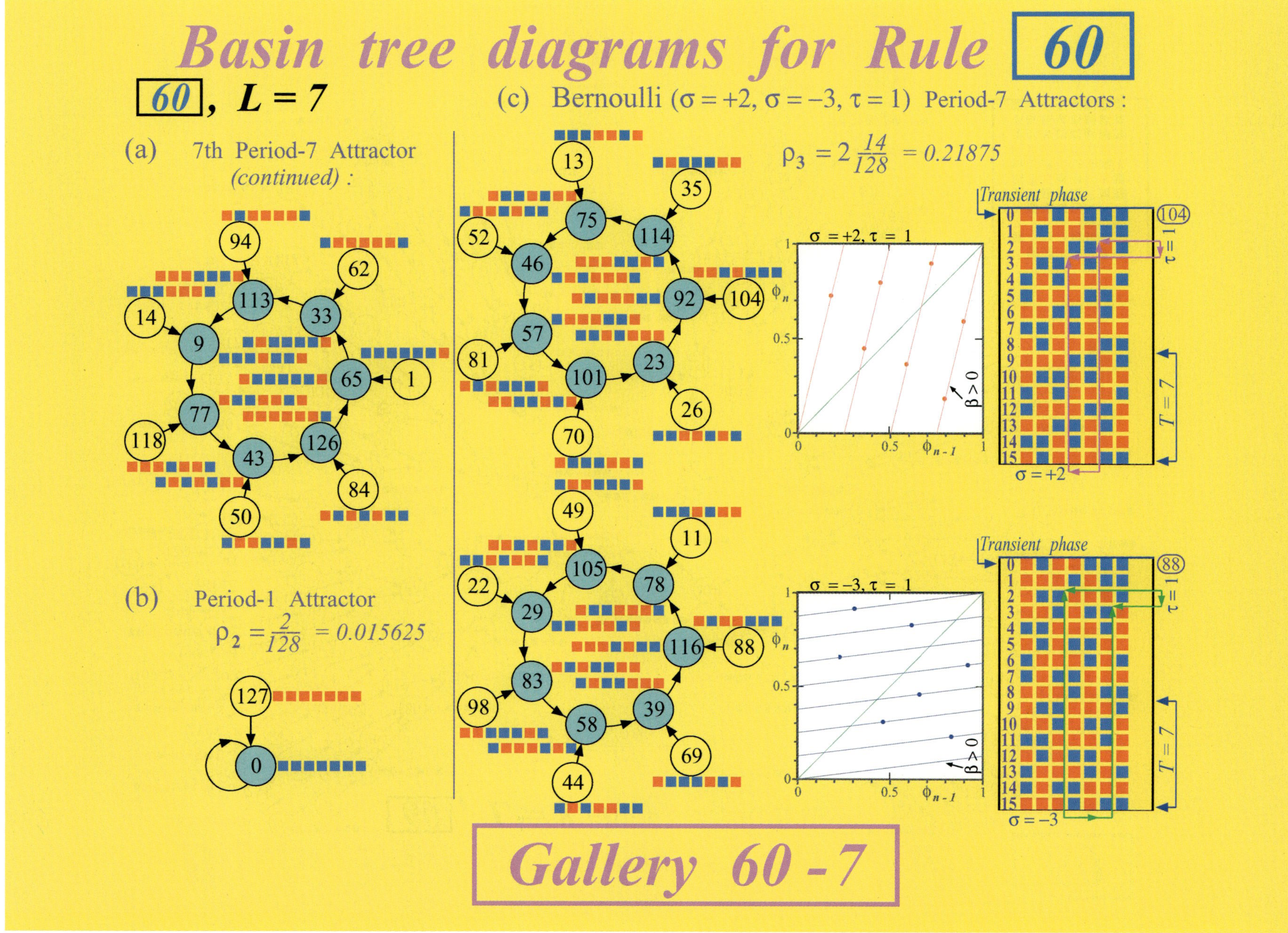

Gallery 60 - 7

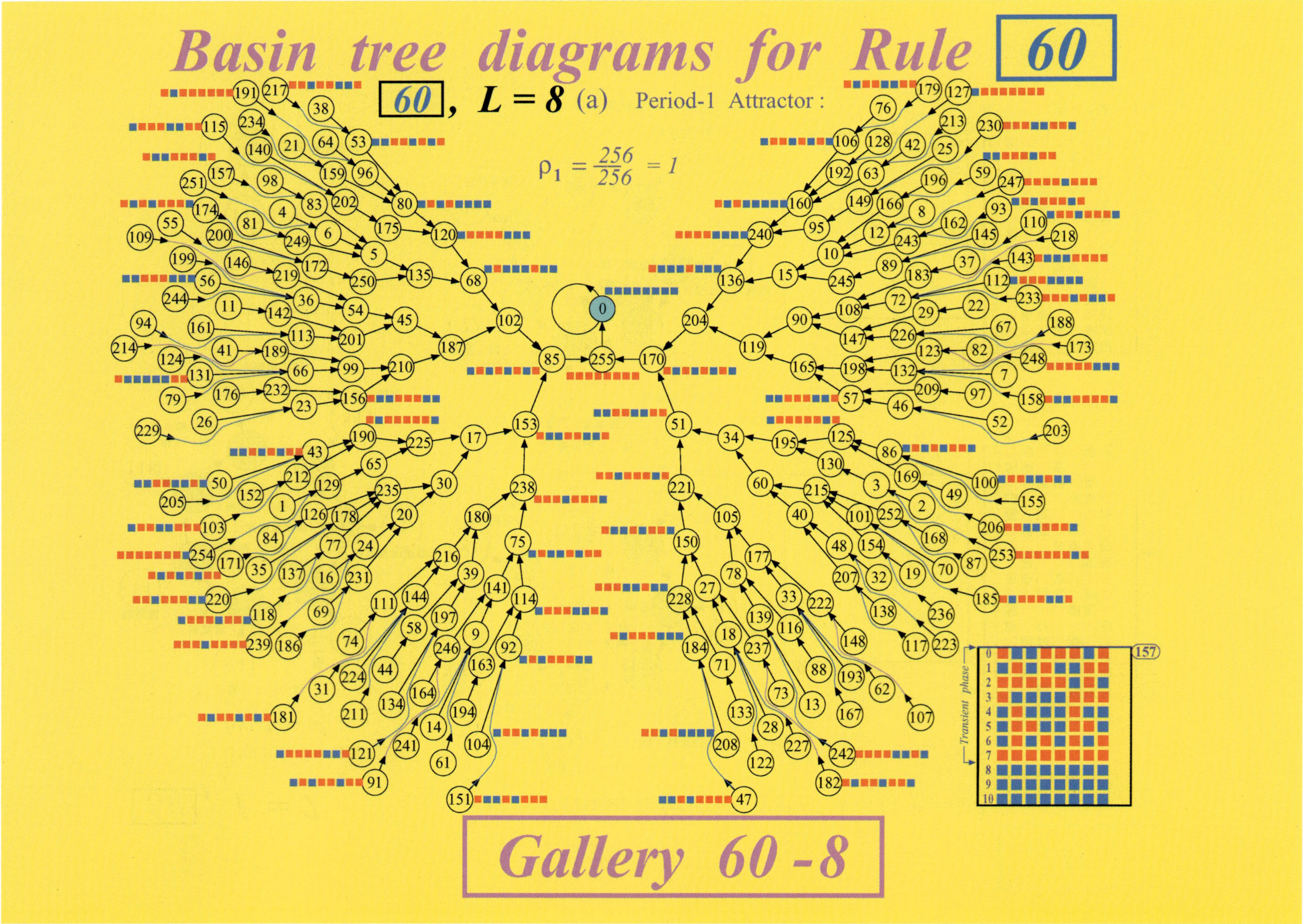

Basin tree diagrams for Rule $\boxed{60}$

$\boxed{60}$, $L = 8$ (a)   Period-1 Attractor:

$$\rho_1 = \frac{256}{256} = 1$$

Gallery 60 -8

2.6. *Highlights from rule* $\boxed{106}$

$$\boxed{\text{Gallery } 106\text{-}1 : L = 3,\, n\left(\sum\nolimits^3\right) = 8}$$

(a) There is a *period-1 attractor* $\{\boxed{0}\}$ with a robustness coefficient $\rho_1 = 5/8 = 0.625$. The dynamics on this attractor obeys a *degenerate* Bernoulli σ_τ-shift law with $\sigma = 0$, $\tau = 1$ and $\beta > 0$.

(b) There is a *period-3 Isle of Eden* with a robustness coefficient $\rho_2 = 3/8 = 0.375$. The dynamics on this attractor obeys a Bernoulli σ_τ-shift law with $\sigma = 1$, or $\sigma = -4$, $\tau = 1$ and $\beta > 0$. The period obeys the formula $T = \tau L/|\sigma| = 3$.

$$\boxed{\text{Gallery } 106\text{-}2 : L = 4,\, n\left(\sum\nolimits^4\right) = 16}$$

(a) There is a *period-2 Isle of Eden* with a robustness coefficient $\rho_1 = 2/16 = 0.125$. The dynamics on this Isle of Eden obeys a Bernoulli σ_τ-shift law with $\sigma = 1$, or $\sigma = -1$, $\tau = 1$ and $\beta > 0$.

(b) There is a *period-4 Isle of Eden* with a robustness coefficient $\rho_2 = 4/16 = 0.25$. The dynamics on this Isle of Eden obeys a Bernoulli σ_τ-shift law with $\sigma = 1$, $\tau = 1$ and $\beta > 0$. The period obeys the formula $T = \tau L/|\sigma| = 4$.

(c) There are two *period-2 attractors* with a combined robustness coefficient $\rho_3 = 2(4/16) = 0.5$. The dynamics on these attractors obey a Bernoulli σ_τ-shift law with $\sigma = 2$, or -2 $\tau = 1$ and $\beta > 0$. The period obeys the formula $T = \tau L/|\sigma| = 2$.

(d) There is a *period-1 attractor* $\{\boxed{0}\}$ with a robustness coefficient $\rho_4 = 2/16 = 0.125$. The dynamics on this attractor obeys a *degenerate* Bernoulli σ_τ-shift law with $\sigma = 0$, $\tau = 1$ and $\beta > 0$.

$$\boxed{\text{Gallery } 106\text{-}3 : L = 5,\, n\left(\sum\nolimits^5\right) = 32}$$

(a) There is a *period-1 attractor* $\{\boxed{0}\}$ with a robustness coefficient $\rho_1 = 2/32 = 0.0625$. The dynamics on this attractor obeys a *degenerate* Bernoulli σ_τ-shift law with $\sigma = 0$, $\tau = 1$ and $\beta > 0$.

(b) There is *a period-15 attractor* with a robustness coefficient $\rho_2 = 20/32 = 0.625$. The dynamics on

this attractor obeys a Bernoulli σ_τ-shift law with $\sigma = -1$, $\tau = 3$ and $\beta > 0$. The period obeys the formula $T = \tau L/|\sigma| = 15$.

(c) There are two *period-5 Isles of Eden* with a combined robustness coefficient $\rho_3 = 2(5/32) = 0.3125$. The dynamics on these Isles of Eden obey a Bernoulli σ_τ-shift law with $\sigma = 1$, $\tau = 1$ and $\beta > 0$. The period obeys the formula $T = \tau L/|\sigma| = 5$.

$$\boxed{\text{Gallery } 106\text{-}4 : L = 6,\, n\left(\sum\nolimits^6\right) = 64}$$

(a) There is a *period-1 attractor* with a robustness coefficient $\rho_1 = 47/64 = 0.734375$. The dynamics on this attractor obeys a *degenerate* Bernoulli σ_τ-shift law with $\sigma = 0$, $\tau = 1$ and $\beta > 0$.

(b) There are two *period-6 Isles of Eden* with a combined robustness coefficient $\rho_2 = 2(6/64) = 0.1875$. The dynamics on these Isles of Eden obeys a Bernoulli σ_τ-shift law with $\sigma = 1$, $\tau = 1$ and $\beta > 0$. The period obeys the formula $T = \tau L/|\sigma| = 6$.

(c) There is a *period-3 Isle of Eden* with a robustness coefficient $\rho_3 = 3/64 = 0.046875$. The dynamics on this Isle of Eden obeys a Bernoulli σ_τ-shift law with $\sigma = 1$, $\tau = 1$ and $\beta > 0$. The period obeys the formula $T = \tau L/|\sigma| = 6$ with a minimal period $T_{\min} = 3$.

(d) There is a *period-2 Isle of Eden* with a robustness coefficient $\rho_4 = 2/64 = 0.03125$. The dynamics on this Isle of Eden obeys a *degenerate* Bernoulli σ_τ-shift law with $\sigma = 0$, $\tau = 1$ and $\beta > 0$.

$$\boxed{\text{Gallery } 106\text{-}5,\, 106\text{-}6 : L = 7,\, n\left(\sum\nolimits^7\right) = 128}$$

(a) There is a *period-49 attractor* with a robustness coefficient $\rho_1 = 98/128 = 0.765625$. The dynamics on this attractor obeys a Bernoulli σ_τ-shift law with $\sigma = 1$, $\tau = 7$ and $\beta > 0$. The period obeys the formula $T = \tau L/|\sigma| = 49$.

(b) There are four *period-7 Isles of Eden* with a combined robustness coefficient $\rho_2 = 4(7/128) = 0.21875$. The dynamics on these Isles of Eden obeys a Bernoulli σ_τ-shift law with $\sigma = 1$, $\tau = 1$ and $\beta > 0$. The period obeys the formula $T = \tau L/|\sigma| = 7$.

(c) There is a *period-1 attractor* $\{\boxed{0}\}$ with a robustness coefficient $\rho_3 = 2/128 = 0.015625$.

242   *A Nonlinear Dynamics Perspective of Wolfram's New Kind of Science*

The dynamics on this attractor obeys a *degenerate* Bernoulli σ_τ-shift law with $\sigma = 0$, $\tau = 1$ and $\beta > 0$.

$$\boxed{\text{Gallery } 106\text{-}7, 106\text{-}8, 106\text{-}9 : L = 8, n\left(\textstyle\sum^8\right) = 256}$$

(a) There are eight *period-15 attractors* with a combined robustness coefficient $\rho_1 = 8(20/256) = 0.625$. The dynamics on these attractors obey a *degenerate* Bernoulli σ_τ-shift law with $\sigma = 0$, $\tau = 15$ and $\beta > 0$.

(b) There is a *period-1 attractor* $\{\boxed{0}\}$ with a robustness coefficient $\rho_2 = 2/256 = 0.0078125$. The dynamics on the attractor obeys a *degenerate* Bernoulli σ_τ-shift law with $\sigma = 0$, $\tau = 1$ and $\beta > 0$.

(c) There is a *period-2 Isle* of Eden with a robustness coefficient $\rho_3 = 2/256 = 0.0078125$. The dynamics on this Isle of Eden obeys a Bernoulli σ_τ-shift law with $\sigma = 1$, or -1, $\tau = 1$ and $\beta > 0$.

(d) There are two *period-2 attractors* with a combined robustness coefficient $\rho_4 = 2(24/256) = 0.1875$. The dynamics on these attractors obeys a Bernoulli σ_τ-shift law with $\sigma = 2$, or -2, $\tau = 1$ and $\beta > 0$. The period obeys the formula $T = \tau L/|\sigma| = 4$ with a minimal $T_{\min} = 2$.

(e) There are five *period-8 Isles of Eden* with a combined robustness coefficient $\rho_5 = 5(8/256) = 0.15625$. The dynamics on these Isles of Eden obey a Bernoulli σ_τ-shift law with $\sigma = 1$, $\tau = 1$ and $\beta > 0$. The period obeys the formula $T = \tau L/|\sigma| = 8$.

(f) There is a *period-4 Isle of Eden* with a robustness coefficient $\rho_6 = 4/256 = 0.015625$. The dynamics on this Isle of Eden obeys a Bernoulli σ_τ-shift law with $\sigma = 1$, $\tau = 1$ and $\beta > 0$. The period obeys the formula $T = \tau L/|\sigma| = 8$, with a minimum period $T = 4$.

The qualitative properties of local rule $\boxed{106}$ extracted from the above basin-tree Galleries 106-1 to 106-9 are summarized below:

Summary of Qualitative properties of local rule $\boxed{106}$ extracted from Gallery 106 for Rule $\boxed{106}$

L	ID Number i	Number of Period-n attractors	Number of Period-n Isles of Eden	Period n	σ_1	τ_1	β_1	σ_2	τ_2	β_2	Robustness coefficient ρ
3	1	1		1	0	1	+				$\rho_1 = 0.625$
	2		1	3	1	1	+				$\rho_2 = 0.375$
4	1		1	2	1	1	+	-1	1	+	$\rho_1 = 0.125$
	2		1	4	1	1	+				$\rho_2 = 0.25$
	3	2		2	2	1	+	-2	1	+	$\rho_3 = 0.5$
	4	1		1	0	1	+				$\rho_4 = 0.125$
5	1	1		1	0	1	+				$\rho_1 = 0.0625$
	2	1		15				-1	3	+	$\rho_2 = 0.625$
	3		2	5	1	1	+	-4	1	+	$\rho_3 = 0.3125$
6	1	1		1	0	1	+				$\rho_1 = 0.734375$
	2		2	6	1	1	+				$\rho_2 = 0.1875$
	3		1	3	1	1	+				$\rho_3 = 0.046875$
	4		1	2	1	1	+	-1	1	+	$\rho_4 = 0.03125$
7	1	1		49	1	7	+				$\rho_1 = 0.765625$
	2		4	7	1	1	+				$\rho_2 = 0.21875$
	3	1		1	0	1	+				$\rho_3 = 0.015625$
8	1	8		15	0	15	+				$\rho_1 = 0.625$
	2	1		1	0	1	+				$\rho_2 = 0.0078125$
	3		1	2	1	1	+	-1	1	+	$\rho_3 = 0.0078125$
	4	2		2	2	1	+	-2	1	+	$\rho_4 = 0.1875$
	5		5	8	1	1	+				$\rho_5 = 0.15625$
	6		1	4	1	1	+				$\rho_6 = 0.015625$

Basin tree diagrams for Rule 106

106 , $L = 3$ (a) Period-1 Attractor :   (b) Bernoulli ($\sigma = +1, \tau = 1$)

$$\rho_1 = \frac{5}{8} = 0.625$$

Period-3 Isle of Eden :

$$\rho_2 = \frac{3}{8} = 0.375$$

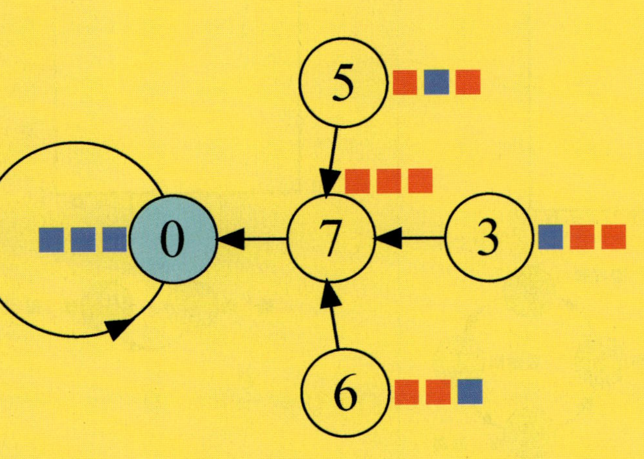

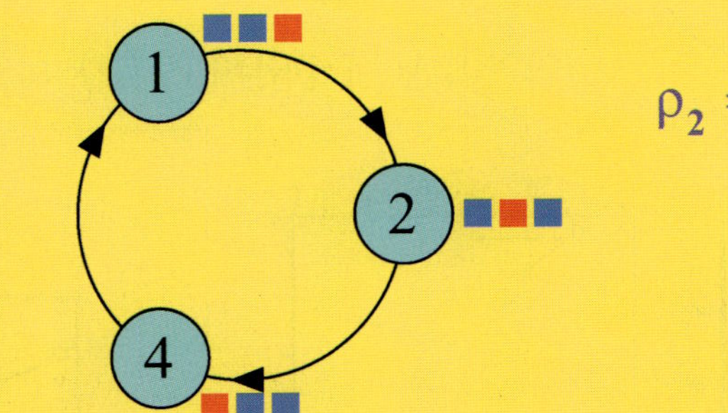

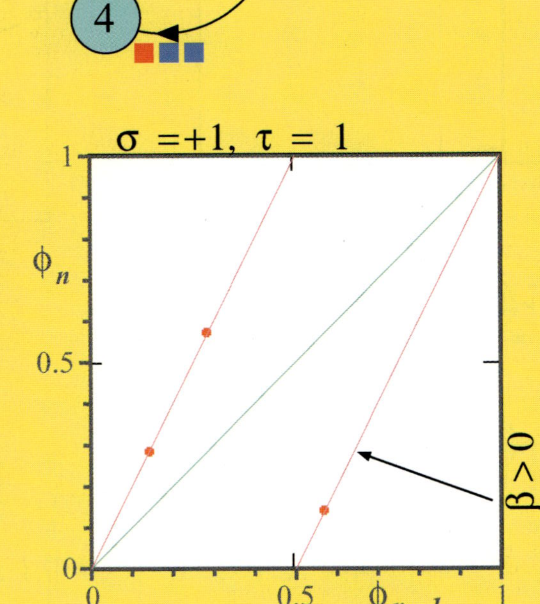

Gallery 106 - 1

Basin tree diagrams for Rule $\boxed{106}$

$\boxed{106}$, $L = 4$

(b) Bernoulli ($\sigma = +1$, $\tau = 1$)

(c) Bernoulli ($\sigma = \pm 2$, $\tau = 1$)

(a) Bernoulli ($\sigma = \pm 1$, $\tau = 1$)

Period-2 Isle of Eden :

$$\rho_1 = \frac{2}{16} = 0.125$$

Period-4 Isle of Eden :

$$\rho_2 = \frac{4}{16} = 0.25$$

Period-2 Attractors : $\rho_3 = 2\frac{4}{16} = 0.5$

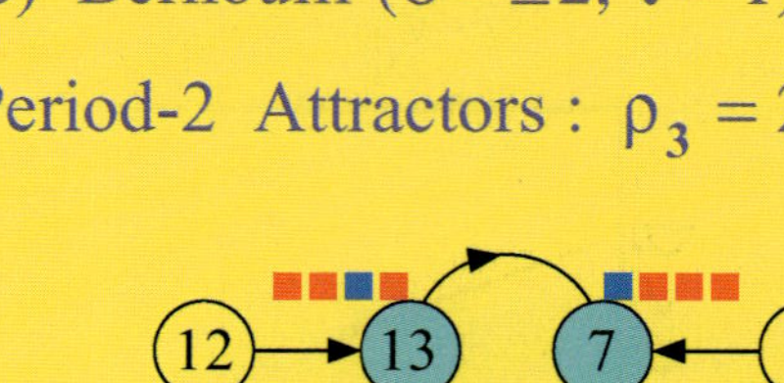
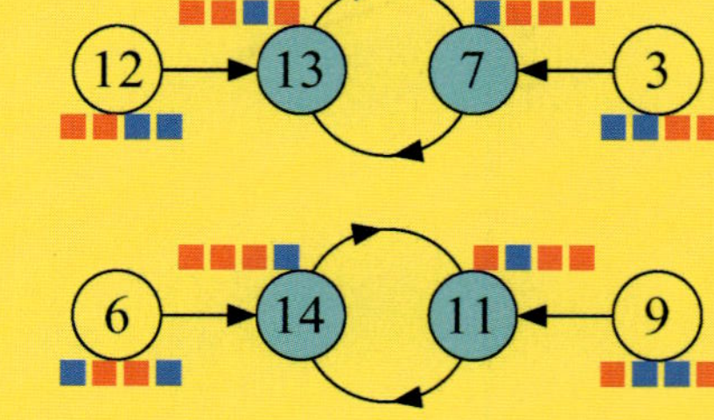
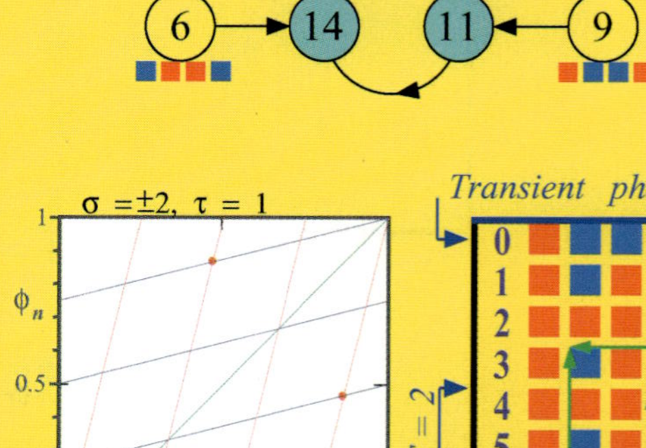
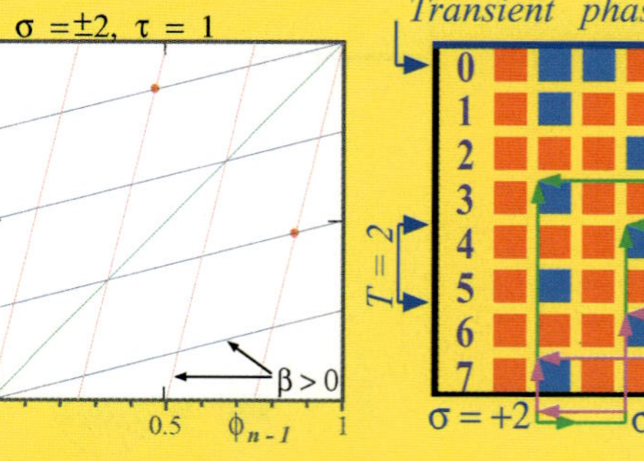
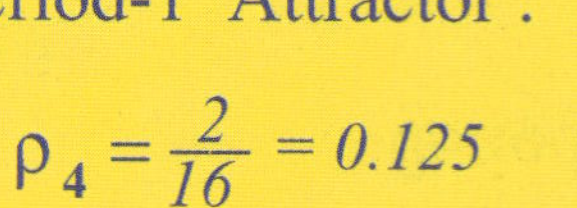
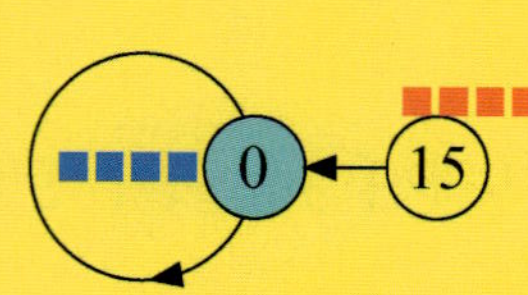

(d) Period-1 Attractor :

$$\rho_4 = \frac{2}{16} = 0.125$$

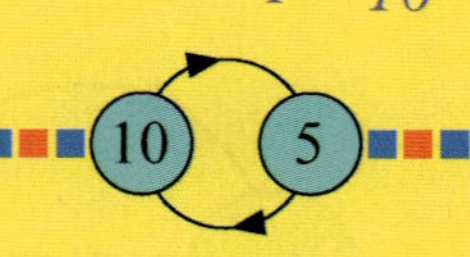

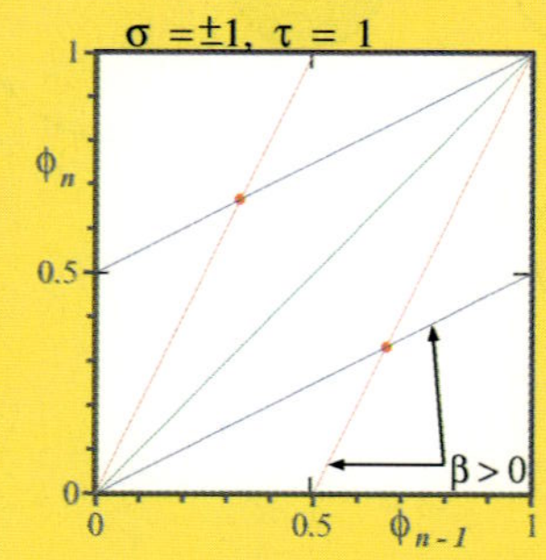
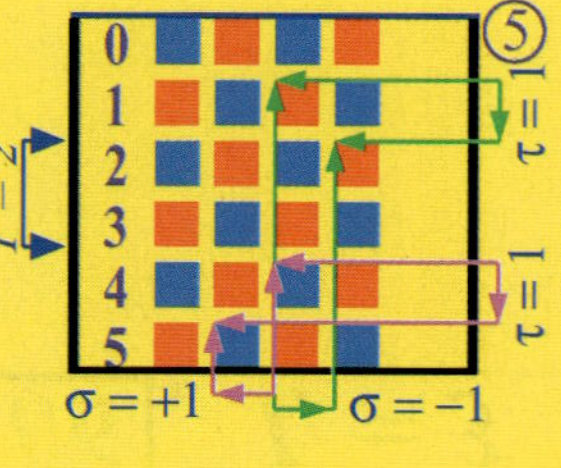

Gallery 106 - 2

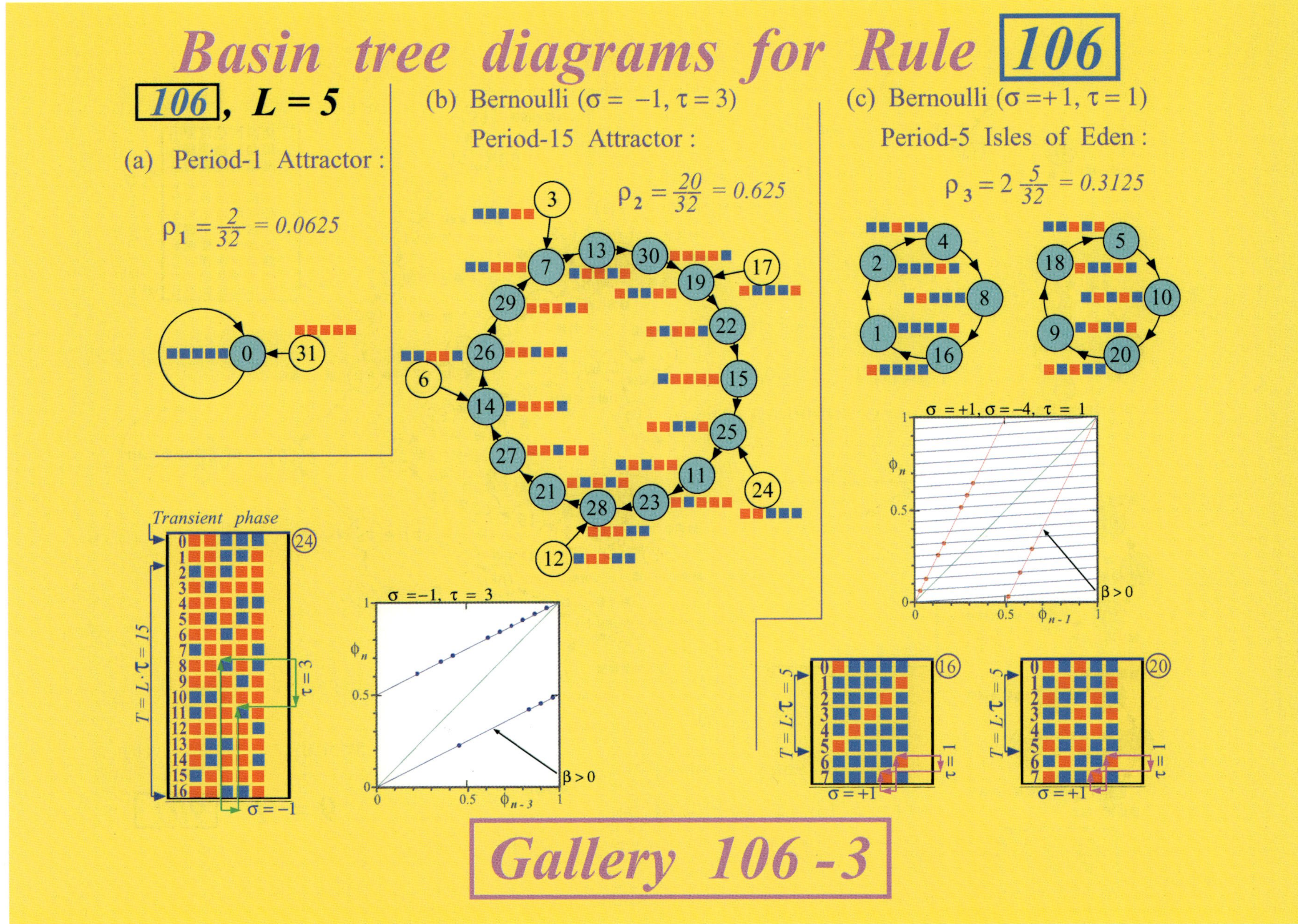

Basin tree diagrams for Rule 106
106, L = 5
(a) Period-1 Attractor :
ρ₁ = 2/32 = 0.0625
(b) Bernoulli (σ = −1, τ = 3)
Period-15 Attractor :
ρ₂ = 20/32 = 0.625
(c) Bernoulli (σ = +1, τ = 1)
Period-5 Isles of Eden :
ρ₃ = 2 5/32 = 0.3125
Transient phase
σ = −1, τ = 3
σ = +1, σ = −4, τ = 1
Gallery 106 -3

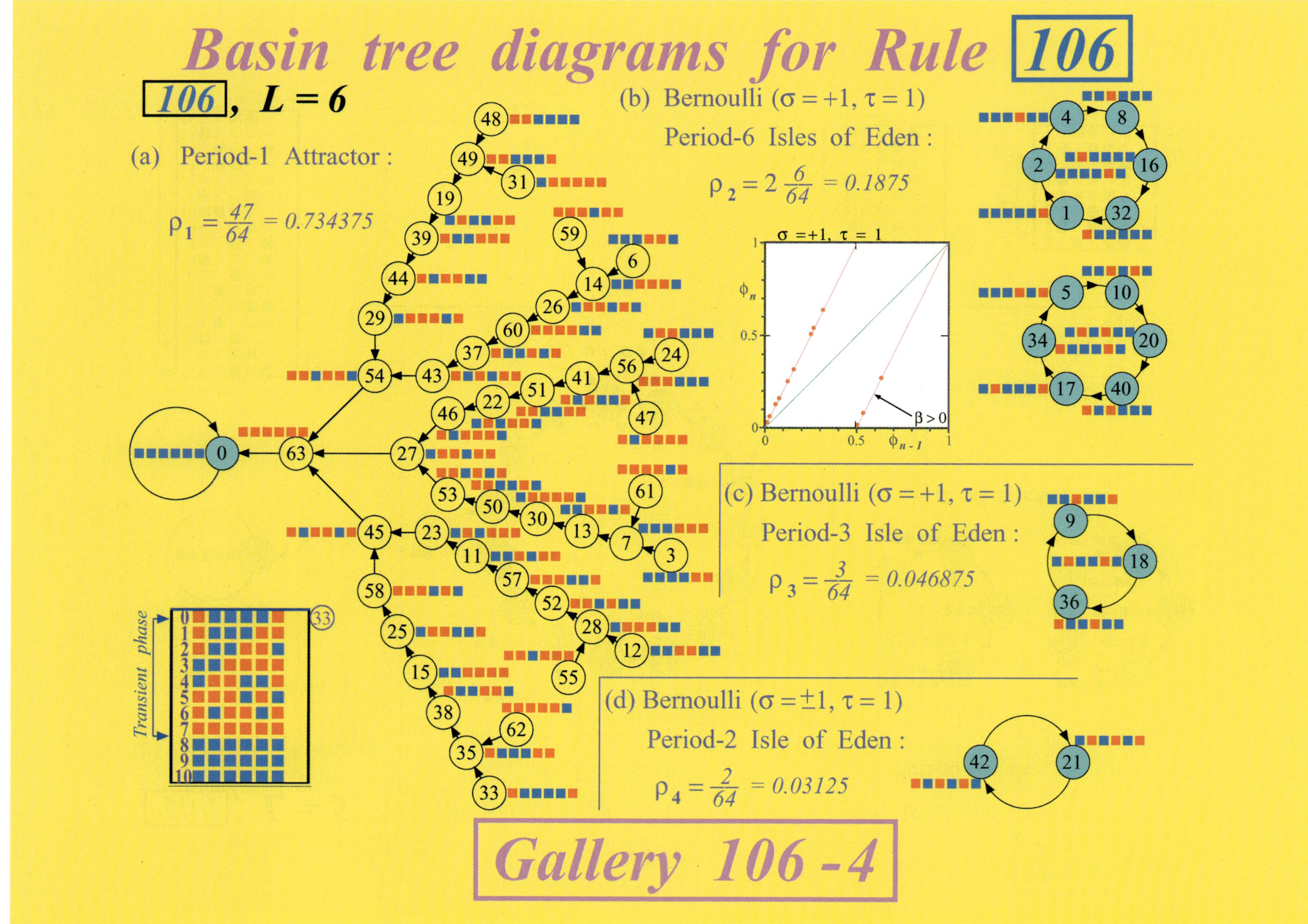

Basin tree diagrams for Rule 106

106, L = 6

(a) Period-1 Attractor :

$\rho_1 = \frac{47}{64} = 0.734375$

(b) Bernoulli ($\sigma = +1, \tau = 1$)
Period-6 Isles of Eden :
$\rho_2 = 2\frac{6}{64} = 0.1875$

$\sigma = +1, \tau = 1$
$\beta > 0$
ϕ_n
ϕ_{n-1}

(c) Bernoulli ($\sigma = +1, \tau = 1$)
Period-3 Isle of Eden :
$\rho_3 = \frac{3}{64} = 0.046875$

(d) Bernoulli ($\sigma = \pm 1, \tau = 1$)
Period-2 Isle of Eden :
$\rho_4 = \frac{2}{64} = 0.03125$

Transient phase

Gallery 106-4

Basin tree diagrams for Rule 106

106, $L = 7$ (a) Bernoulli ($\sigma = +1$, $\tau = 7$) Period-49 Attractor :

$$\rho_1 = \frac{98}{128} = 0.765625$$

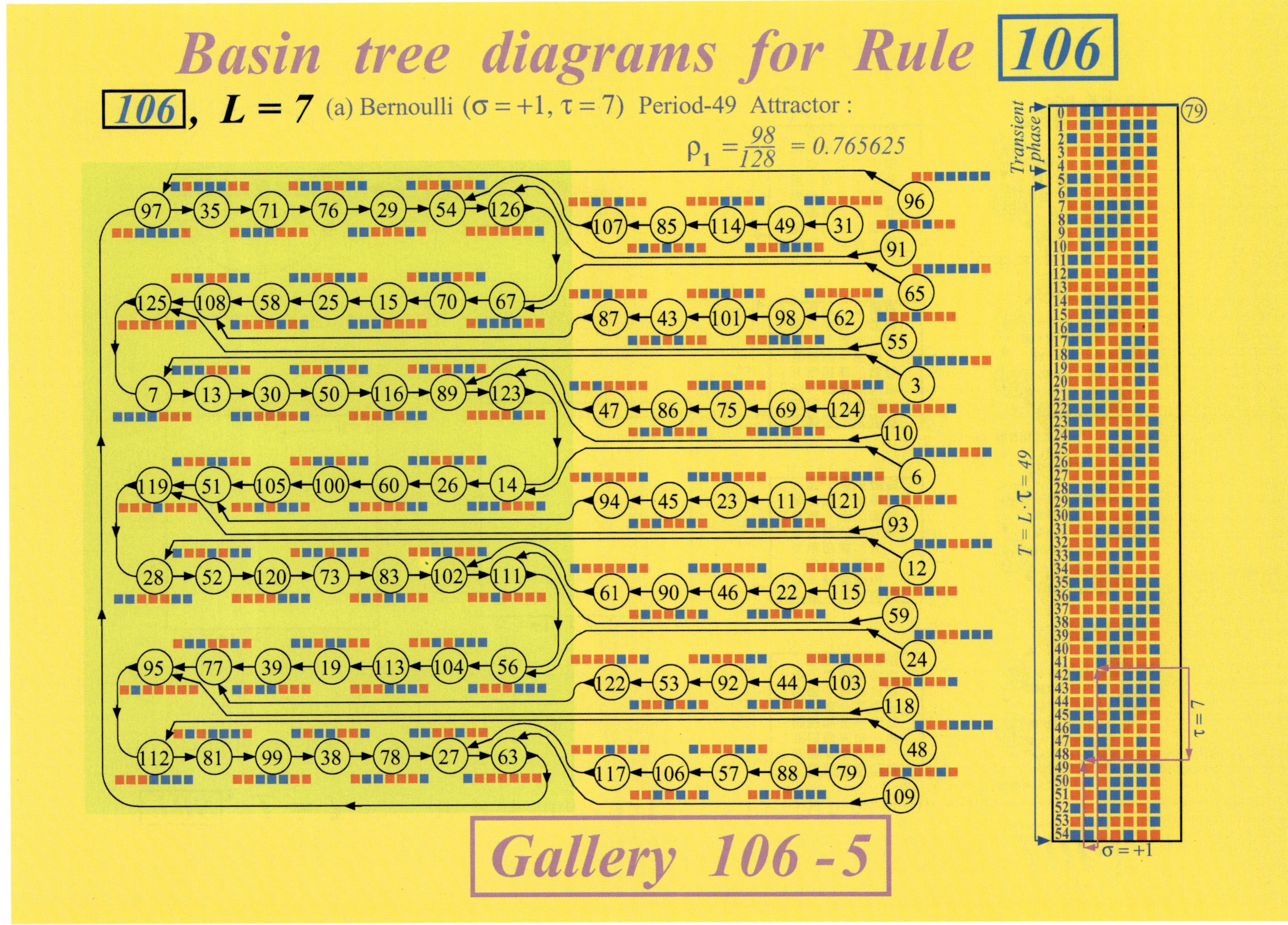

Gallery 106 - 5

Basin tree diagrams for Rule $\boxed{106}$

$\boxed{106}$, **L = 7**   (b) Bernoulli ($\sigma = +1, \tau = 1$) Period-7 Isles of Eden :

(a) Bernulli ($\sigma = +1, \tau = 7$) $\rho_2 = 4 \frac{7}{128} = 0.21875$

Period-49 Attractor

(continued)

(c) Period-1 Attractor : $\rho_3 = \frac{2}{128} = 0.015625$

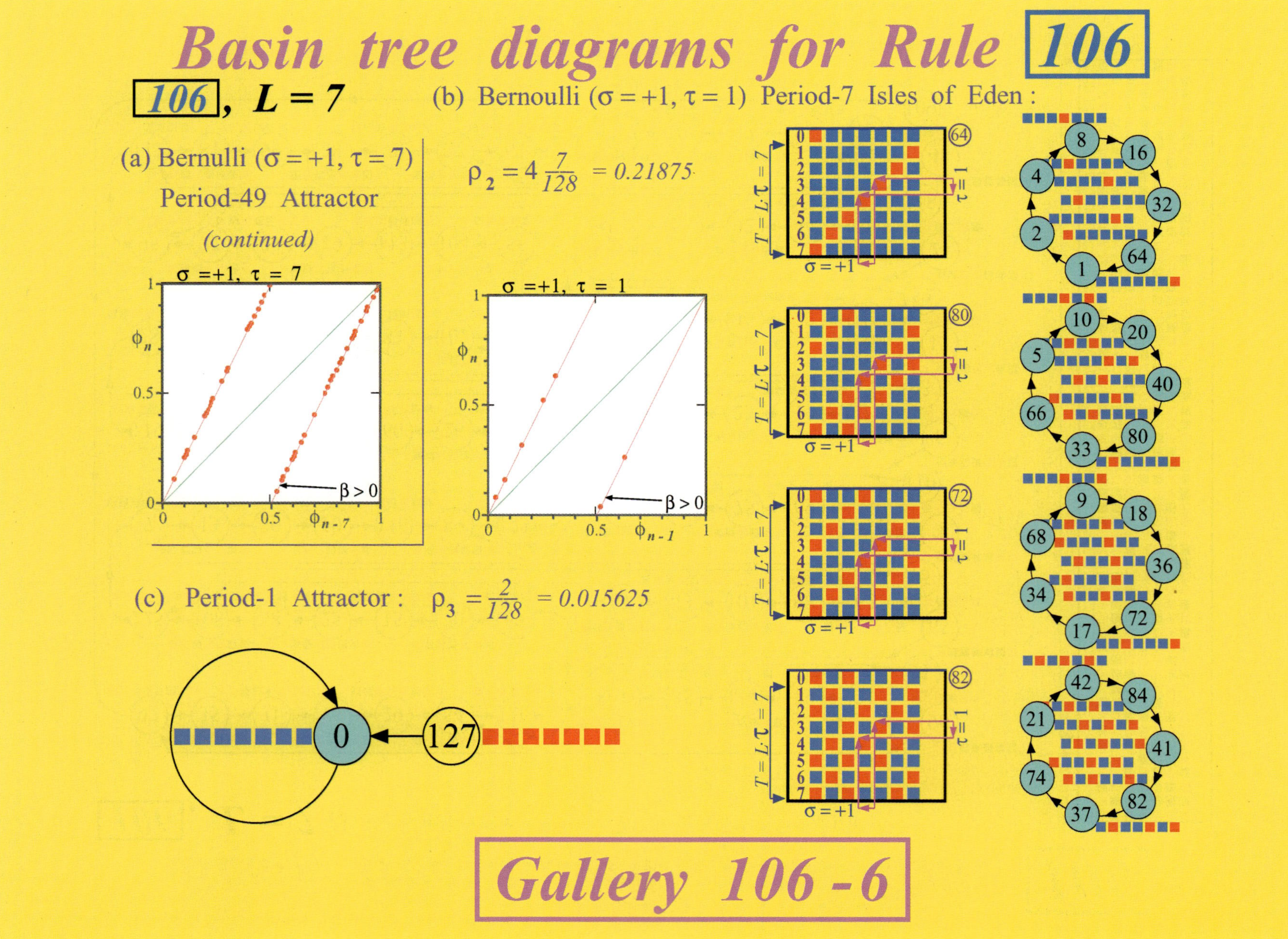

Gallery 106 - 6

Basin tree diagrams for Rule 106

106, $L = 8$ (a) 8 Bernoulli ($\sigma = 0, \tau = 15$) Period-15 Attractors : $\rho_1 = 8\frac{20}{256} = 0.625$

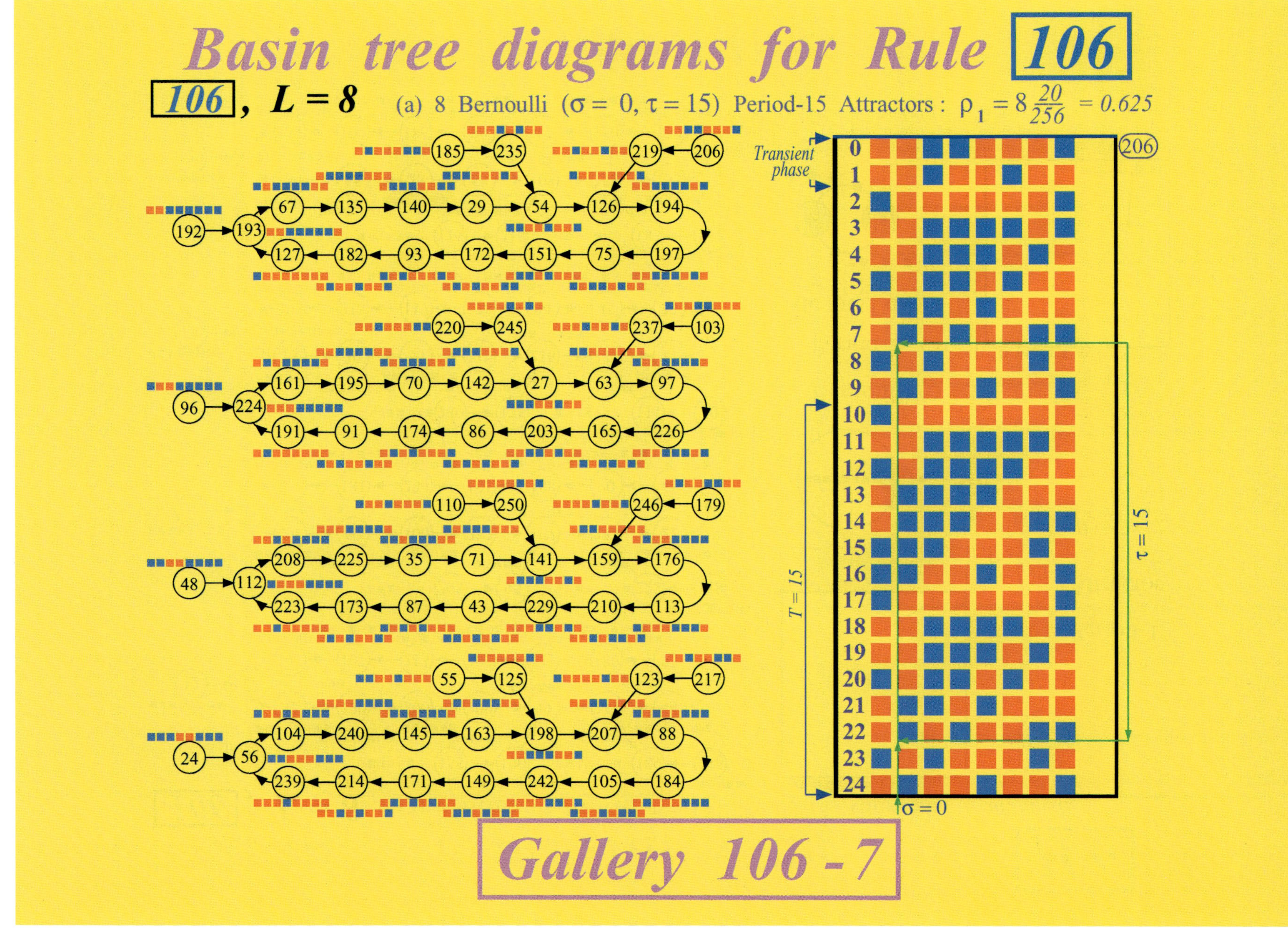

Gallery 106 - 7

Basin tree diagrams for Rule 106

$\boxed{106}$, $\boldsymbol{L = 8}$ (a) 8 Bernoulli ($\sigma = 0$, $\tau = 15$) Period-15 Attractors : *(continued)*

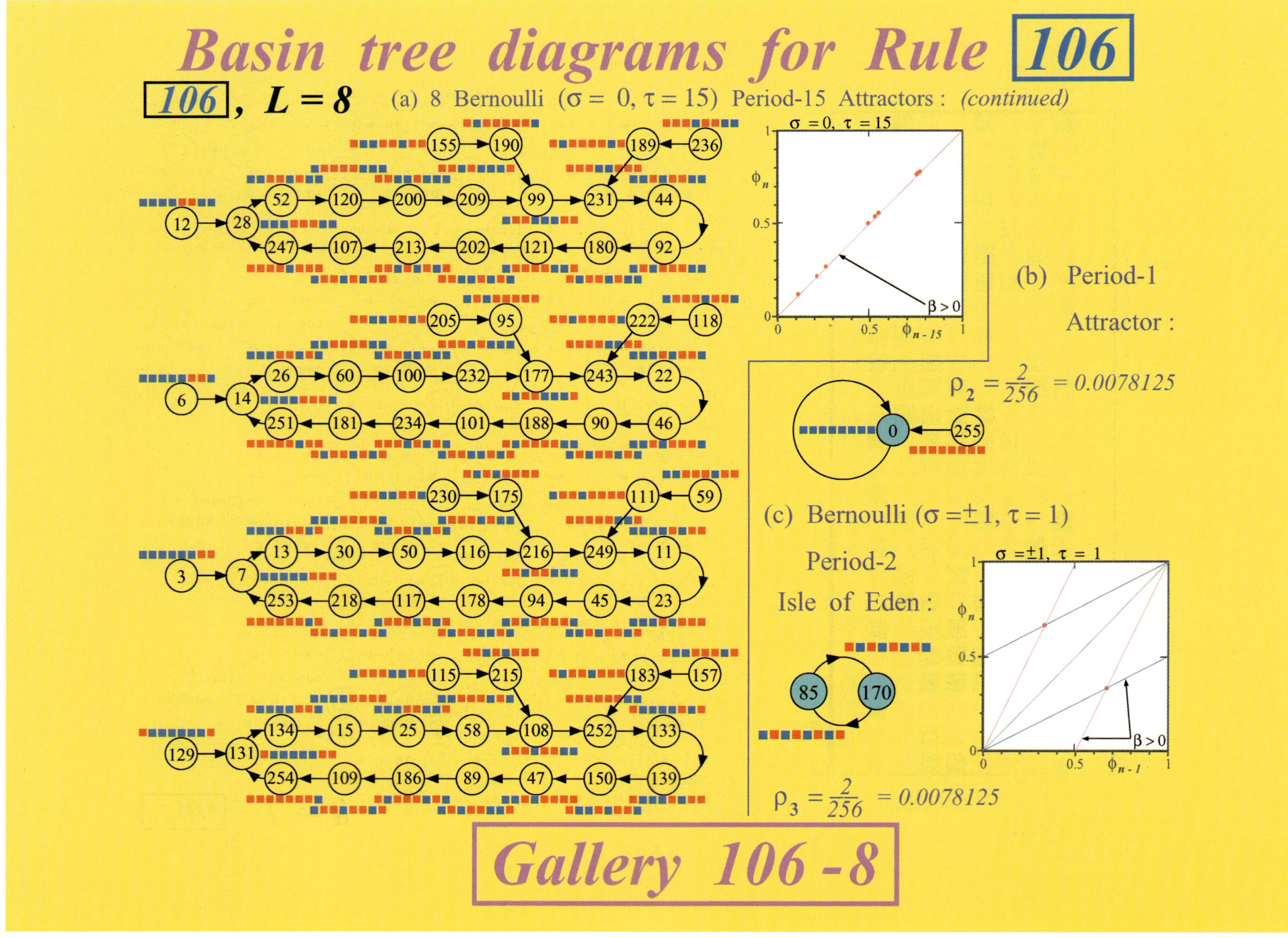

Gallery 106 - 8

Table 10. (Continued)

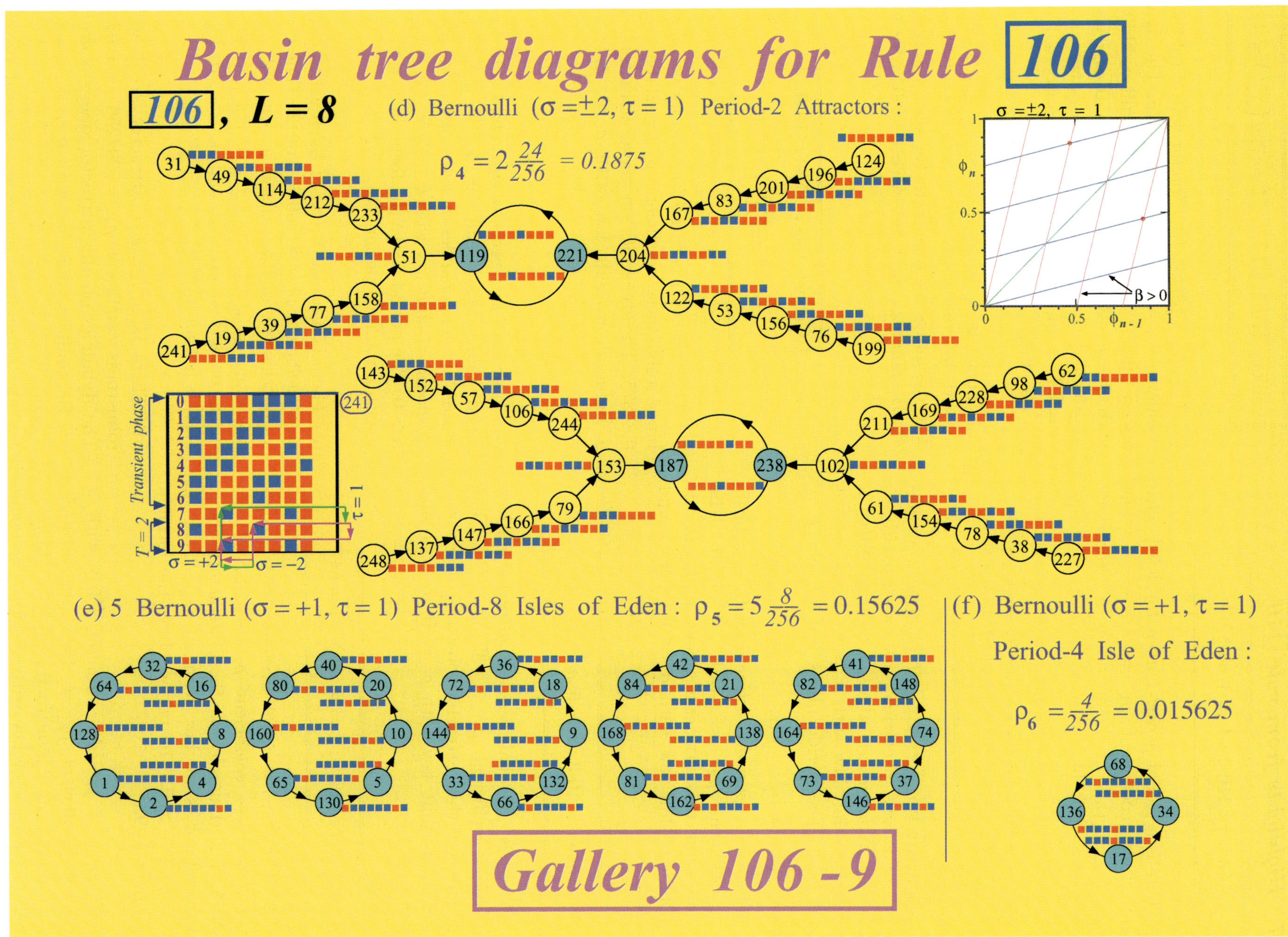

2.7. *Highlights from rule* $\boxed{110}$

$$\boxed{\text{Gallery } 110\text{-}1 : L = 3, n\left(\textstyle\sum^3\right) = 8}$$

(a) There is a global *period-1 attractor* $\{\boxed{0}\}$ with a robustness coefficient $\rho_1 = 8/8 = 1$. The dynamics on this attractor obeys a *degenerate* Bernoulli σ_τ-shift law with $\sigma = 0$, $\tau = 1$ and $\beta > 0$.

$$\boxed{\text{Gallery } 110\text{-}2 : L = 4, n\left(\textstyle\sum^4\right) = 16}$$

(a) There are two *period-2 attractors* with a combined robustness coefficient $\rho_1 = 2(6/16) = 0.75$. The dynamics on these attractors obeys a Bernoulli σ_τ-shift law with $\sigma = 2$, or -2, $\tau = 1$ and $\beta > 0$. The period obeys the formula $T = \tau L/|\sigma| = 2$.

(b) There is a *period-1 attractor* with a robustness coefficient $\rho_2 = 4/16 = 0.25$. The dynamics on this attractor obeys a *degenerate* Bernoulli σ_τ-shift law with $\sigma = 0$, $\tau = 1$ and $\beta > 0$.

$$\boxed{\text{Gallery } 110\text{-}3 : L = 5, n\left(\textstyle\sum^5\right) = 32}$$

(a) There is a global *period-1 attractor* with a robustness coefficient $\rho_1 = 32/32 = 1$. The dynamics on this attractor obeys a *degenerate* Bernoulli σ_τ-shift law with $\sigma = 0$, $\tau = 1$ and $\beta > 0$.

$$\boxed{\text{Gallery } 110\text{-}4 : L = 6, n\left(\textstyle\sum^6\right) = 64}$$

(a) There are two *period-9 attractors* with a combined robustness coefficient $\rho_1 = 2(27/64) = 0.84375$. The dynamics on these attractors obeys a Bernoulli σ_τ-shift law with $\sigma = -2$, or 4, $\tau = 3$ and $\beta > 0$. The period obeys the formula $T = \tau L/|\sigma| = 9$.

(b) There is a *period-1 attractor* $\{\boxed{0}\}$ with a robustness coefficient $\rho_2 = 10/64 = 0.15625$. The dynamics on this attractor obeys a *degenerate* Bernoulli σ_τ-shift law with $\sigma = 0$, $\tau = 1$ and $\beta > 0$.

$$\boxed{\boxed{\text{Gallery } 110\text{-}5,\ 110\text{-}6\ : L = 7, n\left(\textstyle\sum^7\right) = 128}}$$

(a) There is a *period-14 attractor* with a robustness coefficient $\rho_1 = 119/128 = 0.9296875$. The dynamics on this attractor obeys a Bernoulli σ_τ-shift law with $\sigma_1 = -3$, $\tau_1 = 2$ and $\beta_1 > 0$ or $\sigma_2 = 1$, $\tau_2 = 4$ and $\beta_2 > 0$. The period obeys the formula $T = \tau L = 14$, and not $\tau L/|\sigma|$ because $14/|-3|$ is not an integer.

(b) There is a *period-1 attractor* $\{\boxed{0}\}$ with a robustness coefficient $\rho_2 = 9/128 = 0.0703125$. The dynamics on this attractor obeys a *degenerate* Bernoulli σ_τ-shift law with $\sigma = 0$, $\tau = 1$ and $\beta > 0$.

$$\boxed{\begin{array}{c}\text{Gallery } 110\text{-}7, 110\text{-}8, 110\text{-}9 : \\[4pt] L = 8, n\left(\textstyle\sum^8\right) = 256\end{array}}$$

(a) There are two *period-16 attractors* with a combined robustness coefficient $\rho_1 = 2(108/256) = 0.84375$. The dynamics on these attractors obeys a Bernoulli σ_τ-shift law with $\sigma = 2$, $\tau = 4$ and $\beta > 0$. The period obeys the formula $T = \tau L/|\sigma| = 16$.

(b) There are two *period-2 attractors* with a combined robustness coefficient $\rho_2 = 2(6/256) = 0.046875$. The dynamics on these attractors obeys a Bernoulli σ_τ-shift law with $\sigma = 2$, or -2, $\tau = 1$ and $\beta > 0$. The period obeys the formula $T = \tau L/|\sigma| = 4$ with a minimal period $T_{\min} = 2$.

(c) There is a *period-8 Isle of Eden* with a robustness coefficient $\rho_3 = 8/256 = 0.03125$. The dynamics on this Isle of Eden obeys a Bernoulli σ_τ-shift law with $\sigma = 3$, $\tau = 1$ and $\beta > 0$. The period obeys the formula $T = \tau L = 8$, and not $\tau L/|\sigma|$ because $8/|3|$ is not an integer.

(d) There is a *period-1 attractor* $\{\boxed{0}\}$ with a robustness coefficient $\rho_4 = 20/256 = 0.078125$. The dynamics on this attractor obeys a *degenerate* Bernoulli σ_τ-shift law with $\sigma = 0$, $\tau = 1$ and $\beta > 0$.

The qualitative properties of local rule $\boxed{110}$ extracted from the above basin-tree Galleries 110-1

to 110-9 are summarized below:

Summary of Qualitative properties of local rule $\boxed{110}$ extracted from Gallery 110 for Rule $\boxed{110}$

L	ID Number i	Number of Period-n attractors	Number of Period-n Isles of Eden	Period n	Bernoulli Parameters						Robustness coefficient ρ
					σ_1	τ_1	β_1	σ_2	τ_2	β_2	
3	1	1		1	0	1	+				$\rho_1 = 1$
4	1	2		2	2	1	+	-2	1	+	$\rho_1 = 0.75$
	2	1		1	0	1	+				$\rho_2 = 0.25$
5	1	1		1	0	1	+				$\rho_1 = 1$
6	1	2		9	-2	3	+	4	3	+	$\rho_1 = 0.84375$
	2	1		1	0	1	+				$\rho_2 = 0.15625$
7	1	1		14	-3	2	+	1	4	+	$\rho_1 = 0.9296875$
	2	1		1	0	1	+				$\rho_2 = 0.0703125$
8	1	2		16	2	4	+				$\rho_1 = 0.84375$
	2	2		2	2	1	+	-2	1	+	$\rho_2 = 0.046875$
	3		1	8	3	1	+				$\rho_3 = 0.03125$
	4	1		1	0	1	+				$\rho_4 = 0.078125$

2.8. Highlights from rule $\boxed{154}$

$$\boxed{\text{Gallery } 154\text{-}1 : L = 3, n\left(\sum\nolimits^3\right) = 8}$$

(a) There are two *period-1 Isles of Eden* with a combined robustness coefficient $\rho_1 = 2(1/8) = 0.25$. The dynamics on these Isles of Eden obeys a *degenerate* Bernoulli σ_τ-shift law with $\sigma = 0$, $\tau = 1$ and $\beta > 0$.

(b) There is a *period-6 Isle of Eden* with a robustness coefficient $\rho_2 = 6/8 = 0.75$. The dynamics on this Isle of Eden obeys a Bernoulli σ_τ-shift law with $\sigma = -1$, or 2, $\tau = 2$ and $\beta > 0$. The period obeys the formula $T = \tau L/|\sigma| = 6$.

$$\boxed{\text{Gallery } 154\text{-}2 : L = 4, n\left(\sum\nolimits^4\right) = 16}$$

(a) There is a *period-1 Isle of Eden* $\{\textcircled{15}\}$ with a robustness coefficient $\rho_1 = 1/16 = 0.0625$. The dynamics on this Isle of Eden obeys a *degenerate* Bernoulli σ_τ-shift law with $\sigma = 0$, $\tau = 1$ and $\beta > 0$.

(b) There is a *period-1 attractor* $\{\textcircled{0}\}$ with a robustness coefficient $\rho_2 = 7/16 = 0.4375$. The dynamics on this attractor obeys a *degenerate* Bernoulli σ_τ-shift law with $\sigma = 0$, $\tau = 1$ and $\beta > 0$.

(c) There are two *period-4 Isles of Eden* with a combined robustness coefficient $\rho_3 = 2(4/16) = 0.5$. The dynamics on these Isles of Eden obeys a Bernoulli σ_τ-shift law with $\sigma = 2$, or -2, $\tau = 2$ and $\beta > 0$. The period obeys the formula $T = \tau L/|\sigma| = 4$.

$$\boxed{\text{Gallery } 154\text{-}3, 154\text{-}4 : L = 5, n\left(\sum\nolimits^5\right) = 32}$$

Basin tree diagrams for Rule 110

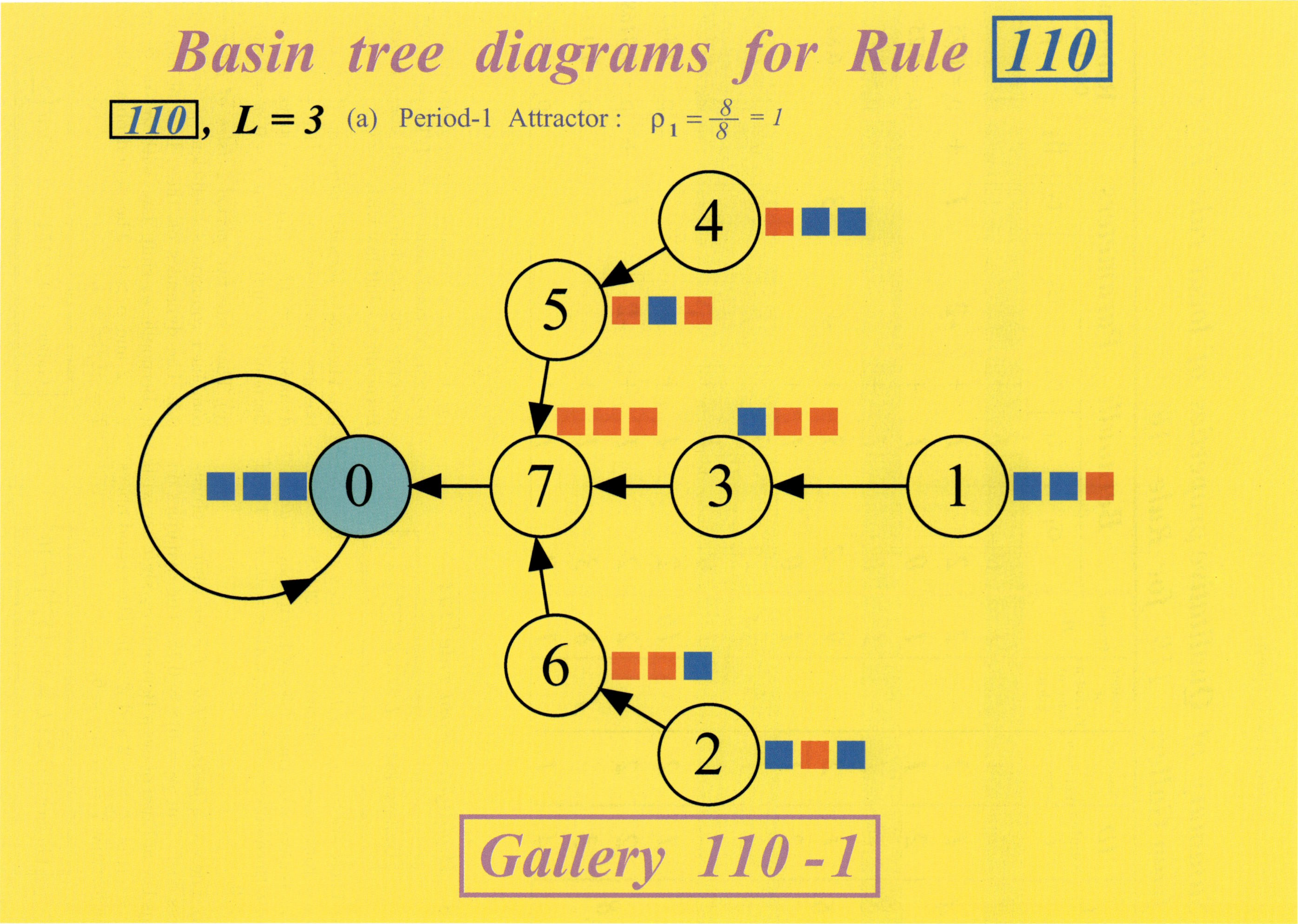

110, **L = 3** (a) Period-1 Attractor : $\rho_1 = \frac{8}{8} = 1$

Basin tree diagrams for Rule $\boxed{110}$

$\boxed{110}$, **L = 4** (a) Bernoulli ($\sigma = \pm 2$, $\tau = 1$) Period-2 Attractors : $\rho_1 = 2\frac{6}{16} = 0.75$

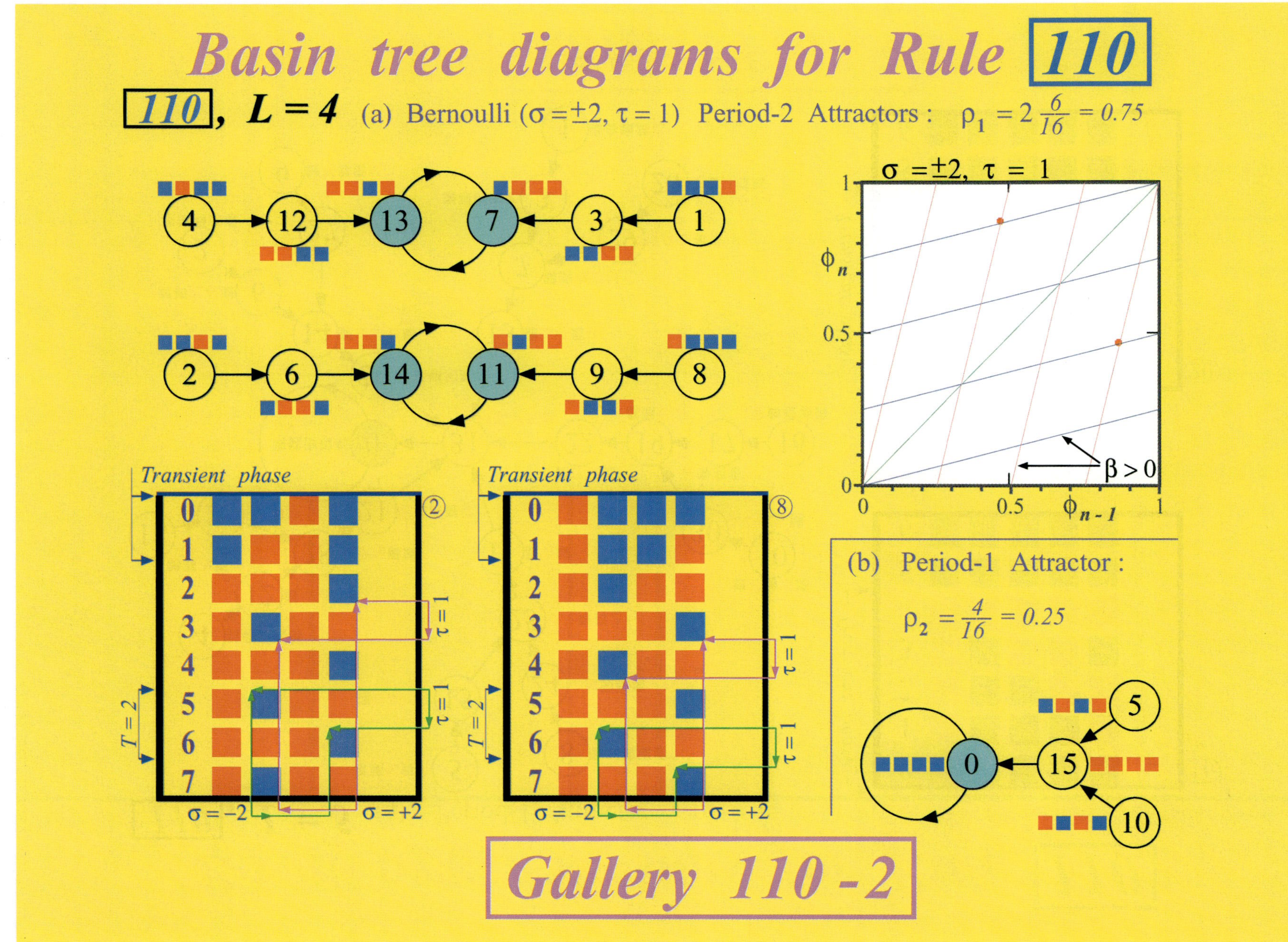

$\boxed{\textit{Gallery 110-2}}$

Table 11. (*Continued*)

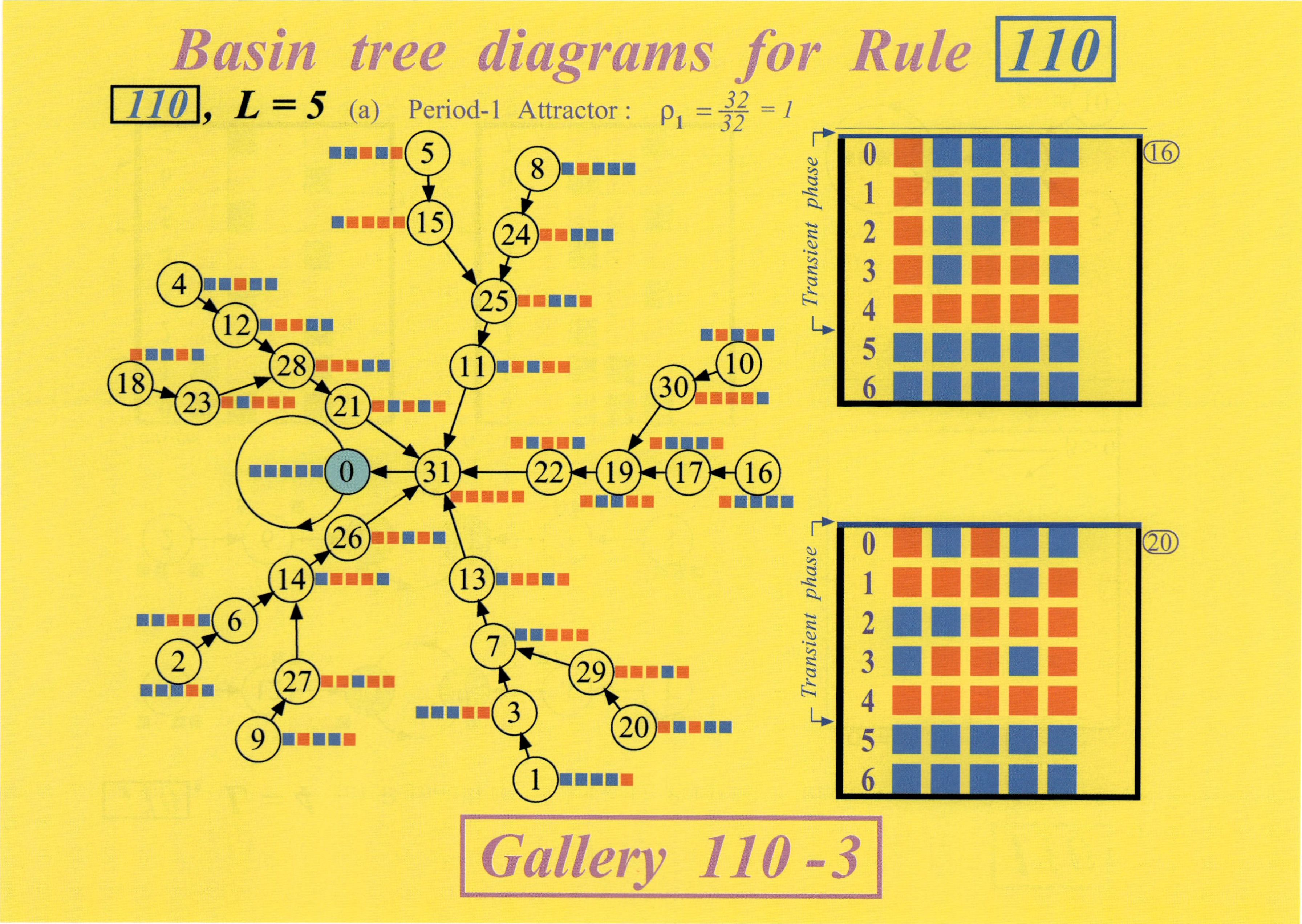
Basin tree diagrams for Rule 110
110, L = 5 (a) Period-1 Attractor : ρ1 = 32/32 = 1
Transient phase
16
Transient phase
20
Gallery 110 -3

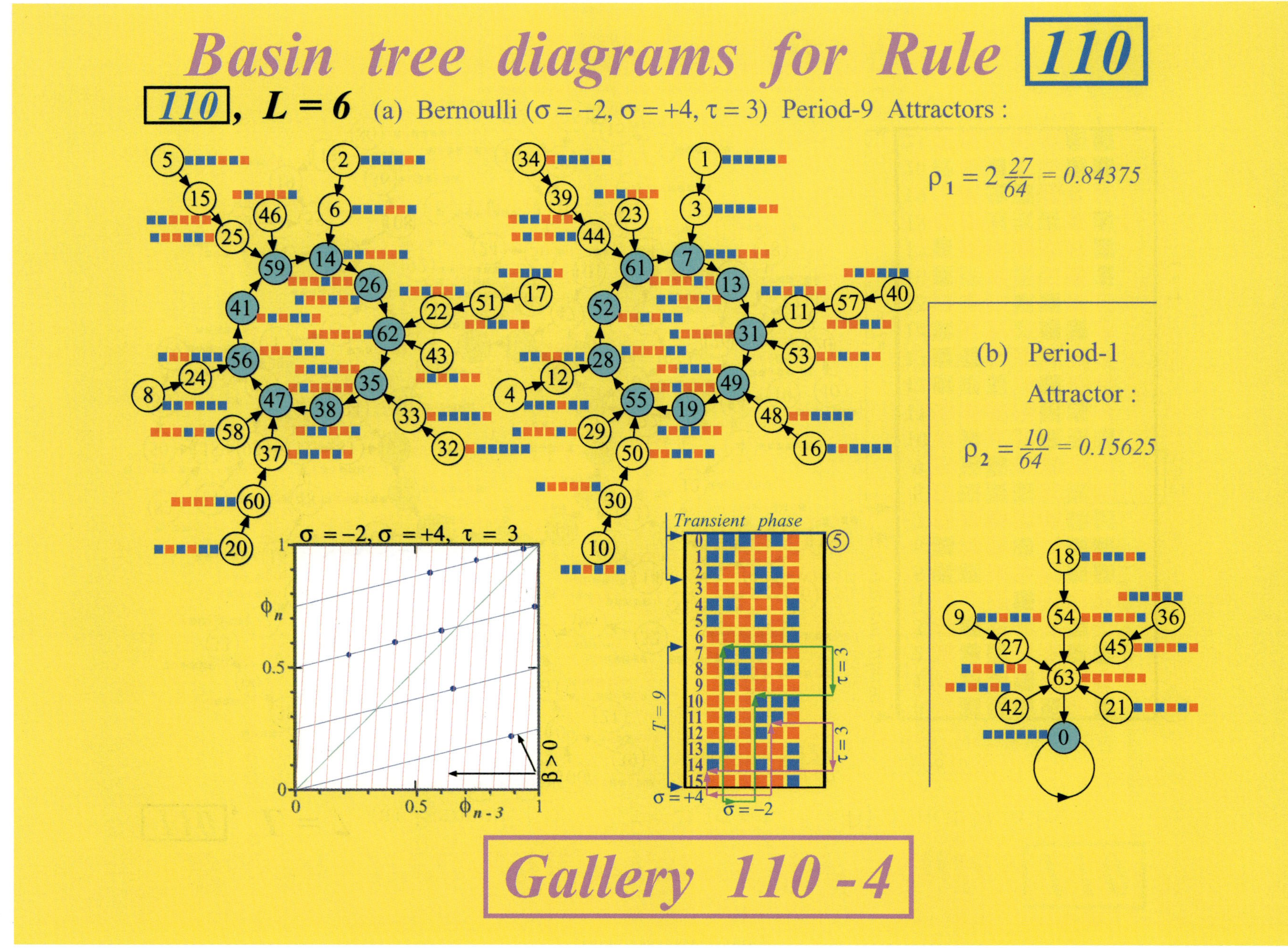
Basin tree diagrams for Rule 110
110, L = 6 (a) Bernoulli (σ = −2, σ = +4, τ = 3) Period-9 Attractors :
ρ₁ = 2 27/64 = 0.84375
(b) Period-1
Attractor :
ρ₂ = 10/64 = 0.15625
σ = −2, σ = +4, τ = 3
Transient phase
T = 9
ε = 1
ε = 1
σ = −2
σ = +4
φ_n
φ_n−3
β > 0
Gallery 110 -4

Table 11. (*Continued*)

Basin tree diagrams for Rule $\boxed{110}$

$\boxed{110}$, $\boldsymbol{L = 7}$ (a) Bernoulli $\{(\sigma = -3, \tau = 2), (\sigma = +1, \tau = 4)\}$ Period-14 Attractor :

$$\rho_1 = \frac{119}{128} = 0.9296875$$

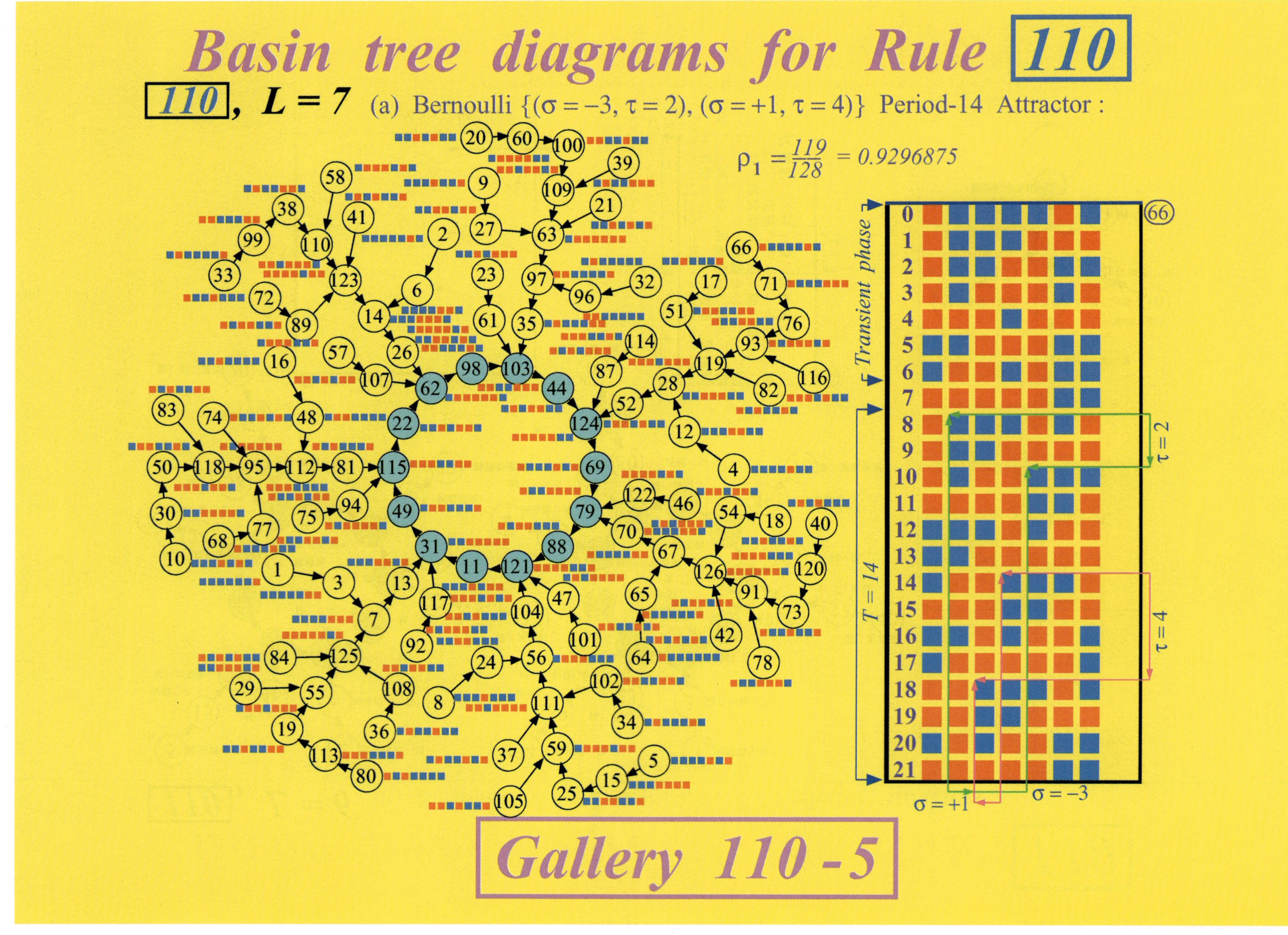

Gallery 110 - 5

Basin tree diagrams for Rule $\boxed{110}$

$\boxed{110}$, $L = 7$ (a) Bernoulli $\{(\sigma = -3, \tau = 2), (\sigma = +1, \tau = 4)\}$ Period-14 Attractor *(continued)* :

(b) Period-1 Attractor : $\rho_2 = \dfrac{9}{128} = 0.0703125$

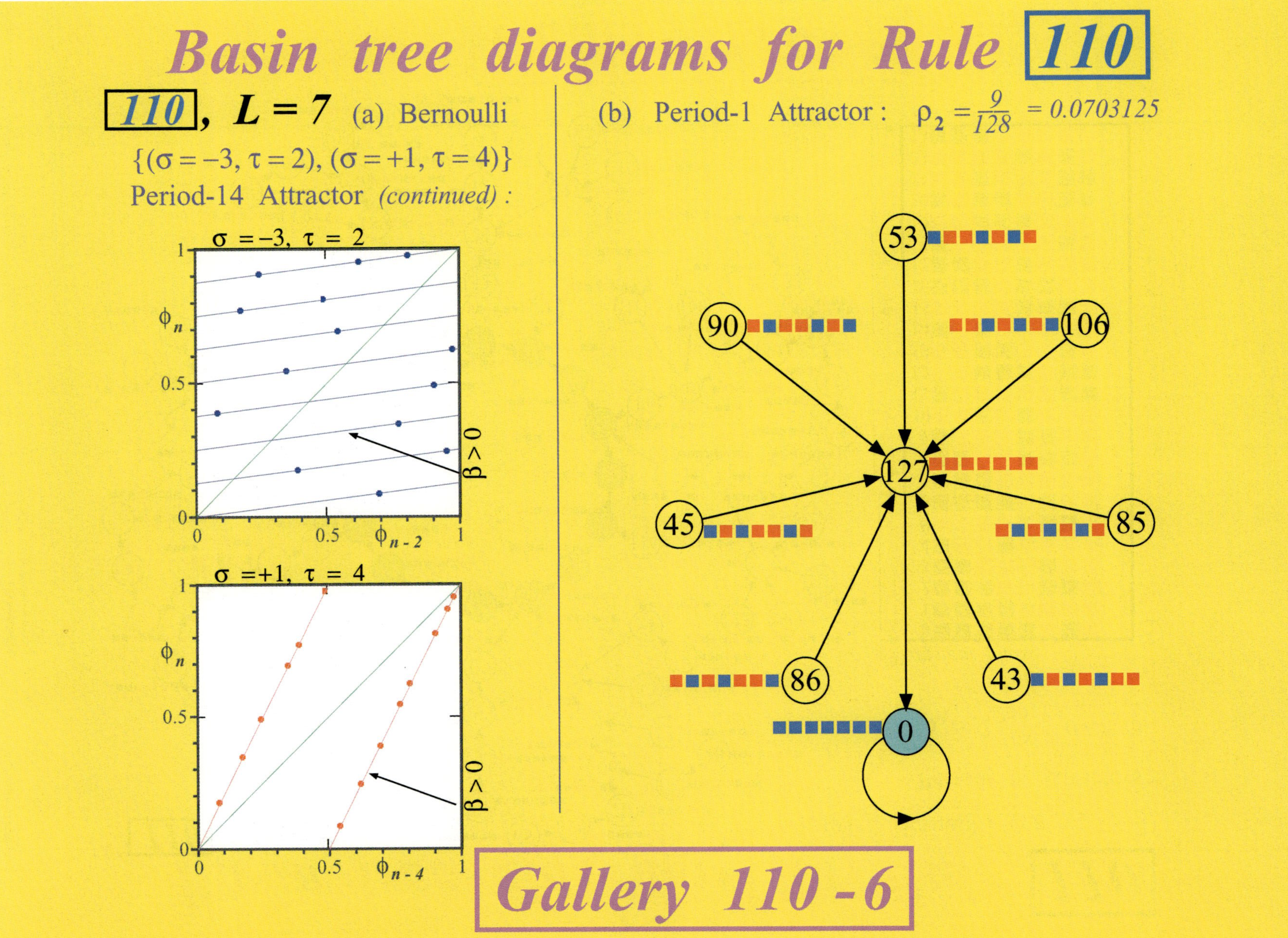

$\boxed{Gallery\ 110\ \text{-}\ 6}$

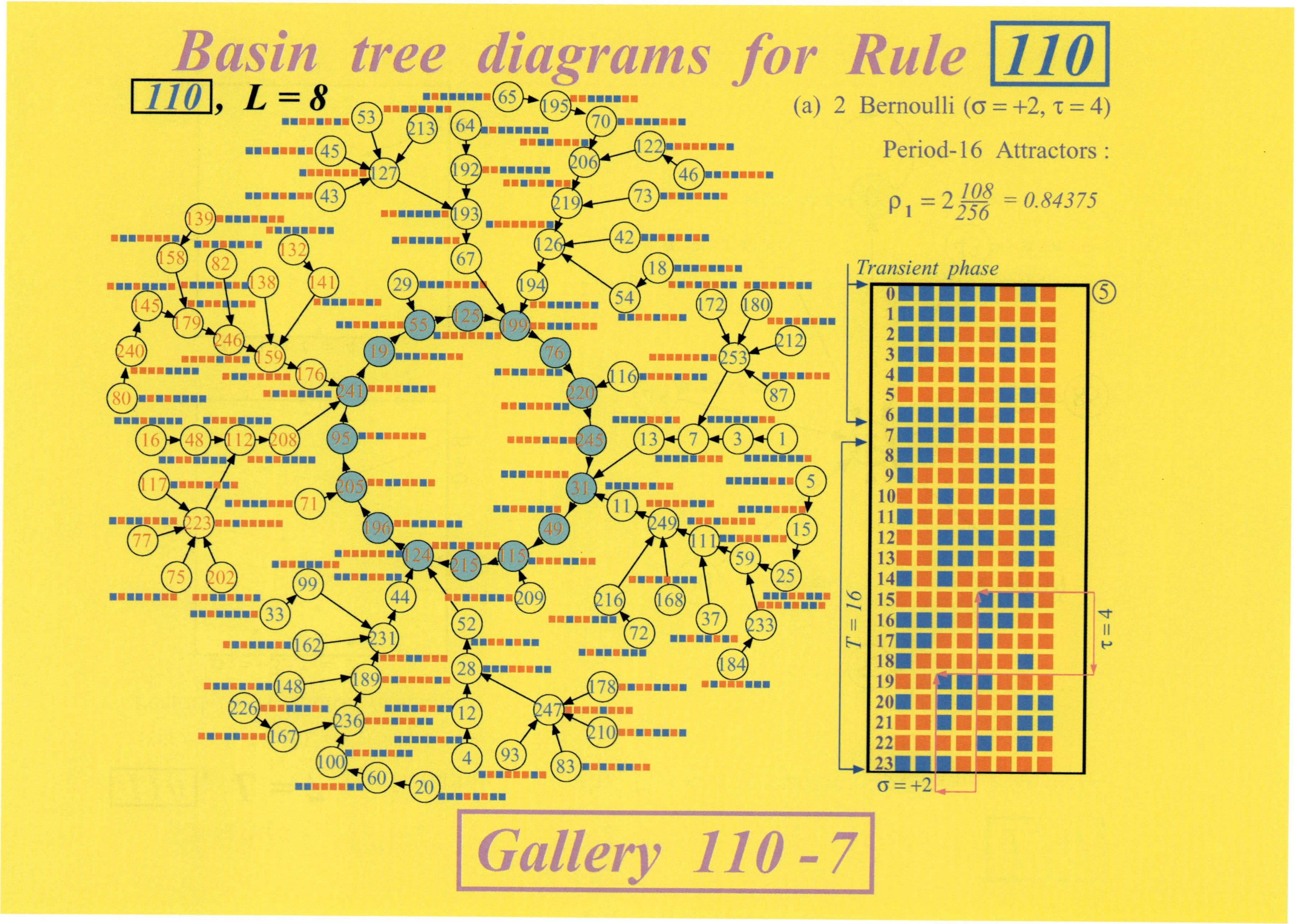

Table 11. (*Continued*)

Basin tree diagrams for Rule 110

110, $L = 8$

(a) 2 Bernoulli ($\sigma = +2$, $\tau = 4$)

Period-16 Attractors :

$$\rho_1 = 2\frac{108}{256} = 0.84375$$

Gallery 110 - 7

Basin tree diagrams for Rule 110

110, *L = 8*

(a) 2 Bernoulli ($\sigma = +2$, $\tau = 4$)

Period-16 Attractors

(*continued*) :

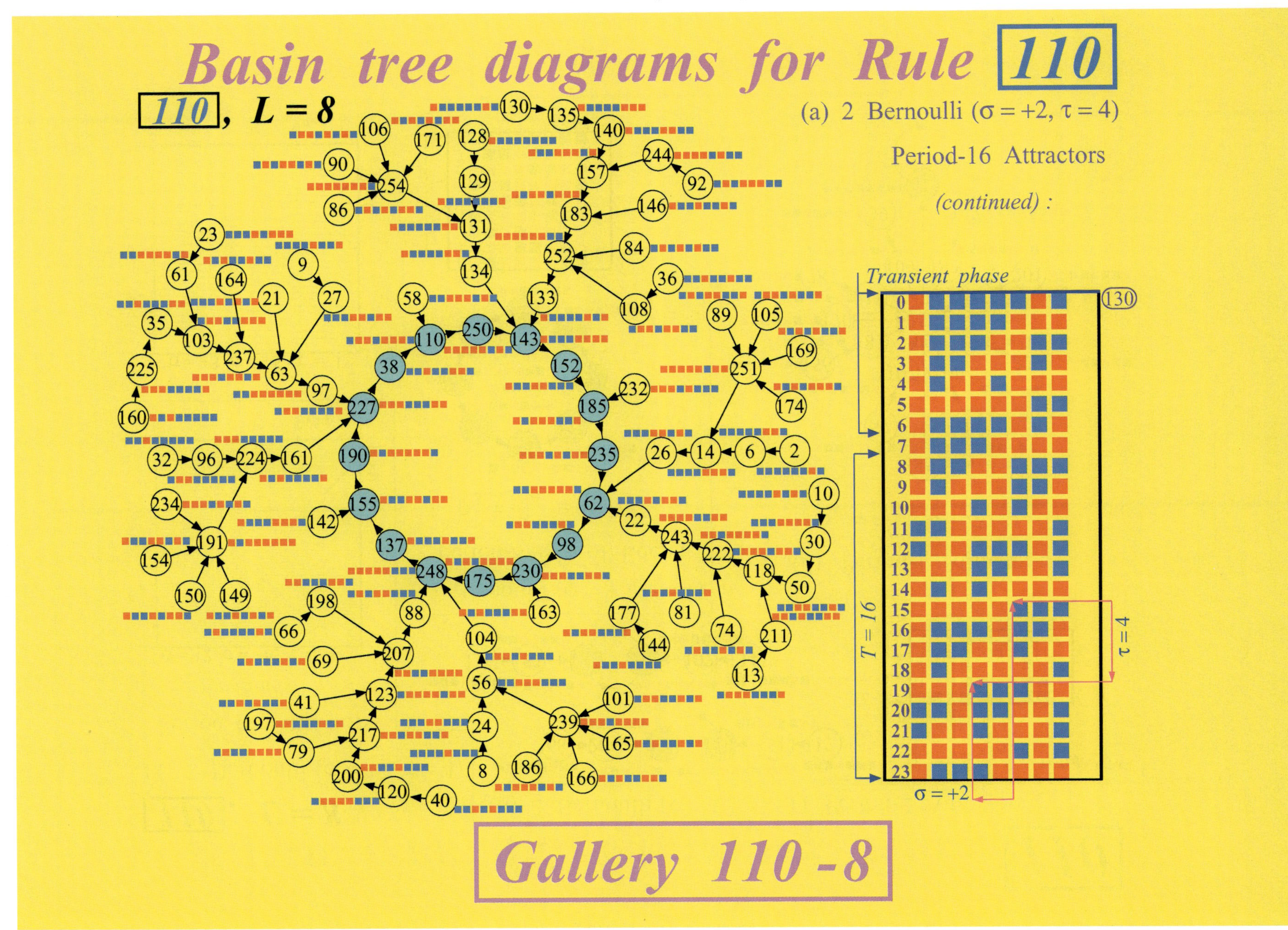

Gallery 110 - 8

Basin tree diagrams for Rule $\boxed{110}$

$\boxed{110}$, $L = 8$

(a) 2 Bernoulli ($\sigma = +2$, $\tau = 4$)

Period-16 Attractors

(continued) :

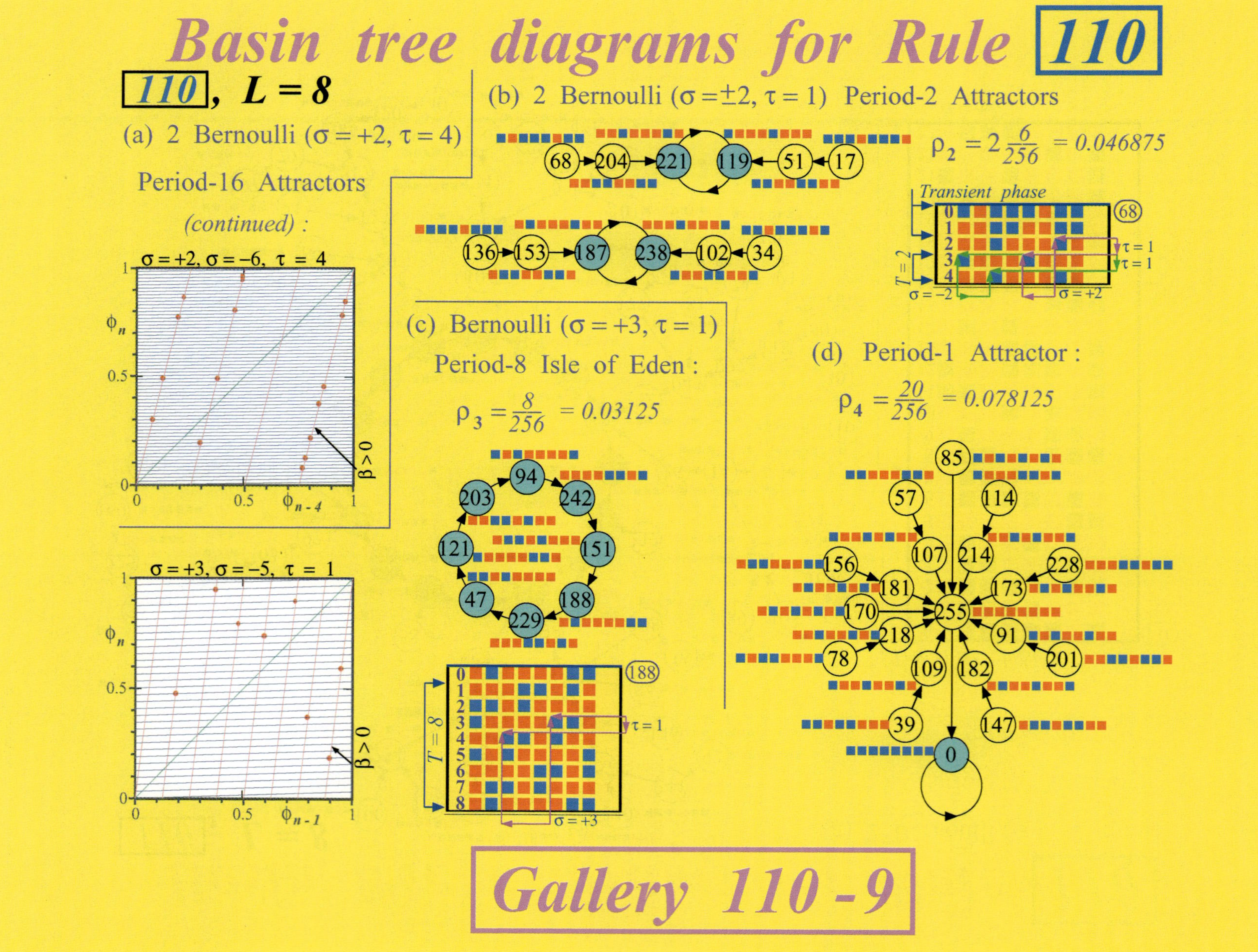

(b) 2 Bernoulli ($\sigma = \pm 2$, $\tau = 1$) Period-2 Attractors

$$\rho_2 = 2\frac{6}{256} = 0.046875$$

(c) Bernoulli ($\sigma = +3$, $\tau = 1$)

Period-8 Isle of Eden :

$$\rho_3 = \frac{8}{256} = 0.03125$$

(d) Period-1 Attractor :

$$\rho_4 = \frac{20}{256} = 0.078125$$

Gallery 110 - 9

(a) There are two *period-1 Isles of Eden* with a combined robustness coefficient $\rho_1 = 2(2/32) = 0.0625$. The dynamics on these Isles of Eden obeys a *degenerate* Bernoulli σ_τ-shift law with $\sigma = 0$, $\tau = 1$ and $\beta > 0$.

(b) There is a *period-10 Isle of Eden* with a robustness coefficient $\rho_2 = 10/32 = 0.3125$. The dynamics on this Isle of Eden obeys a Bernoulli σ_τ-shift law with $\sigma = 2$, or -3, $\tau = 2$ and $\beta > 0$. The period obeys the formula $T = \tau L = 10$, and not $\tau L/|\sigma|$ because $10/|-3|$ is not an integer.

(c) There is a *period-20 Isle of Eden* with a robustness coefficient $\rho_3 = 20/32 = 0.625$. The dynamics on this Isle of Eden obeys a Bernoulli σ_τ-shift law with $\sigma = -1$, or 4, $\tau = 4$ and $\beta > 0$. The period obeys the formula $T = \tau L/|\sigma| = 20$.

$$\boxed{\text{Gallery } 154\text{-}5, 154\text{-}6, 154\text{-}7 : L = 6, \quad n\left(\Sigma^6\right) = 64}$$

(a) There is a *period-1 attractor* $\{\textcircled{0}\}$ with a robustness coefficient $\rho_1 = 3/64 = 0.046875$. The dynamics on this attractor obeys a *degenerate* Bernoulli σ_τ-shift law with $\sigma = 0$, $\tau = 1$ and $\beta > 0$.

(b) There is a *period-1 Isle of Eden* $\{\textcircled{63}\}$ with a robustness coefficient $\rho_2 = 1/64 = 0.015625$. The dynamics on this Isle of Eden obeys a *degenerate* Bernoulli σ_τ-shift law with $\sigma = 0$, $\tau = 1$ and $\beta > 0$.

(c) There are three *period-2 attractors* with a combined robustness coefficient $\rho_3 = 3(4/64) = 0.1875$. The dynamics on these attractors obeys a Bernoulli σ_τ-shift law with $\sigma = 3$, or -3, $\tau = 1$ and $\beta > 0$. The period obeys the formula $T = \tau L/|\sigma| = 2$.

(d) There are two *period-3 Isles of Eden* with a combined robustness coefficient $\rho_4 = 2(3/64) = 0.09375$. The dynamics on these Isles of Eden obeys a Bernoulli σ_τ-shift law with $\sigma = -2$, or -4,

$\tau = 1$ and $\beta > 0$. The period obeys the formula $T = \tau L/|\sigma| = 3$.

(e) There is a *period-6 Isle of Eden* with a robustness coefficient $\rho_5 = 6/64 = 0.09375$. The dynamics of this Isle of Eden obeys a Bernoulli σ_τ-shift law with $\sigma = -1$, or $\tau = 2$ and $\beta > 0$. The period obeys the formula $T = \tau L/|\sigma| = 12$ with a minimal period $T_{\min} = 6$.

(f) There are two *period-6 Isles of Eden* with a combined robustness coefficient $\rho_6 = 2(6/64) = 0.1875$. The dynamics on these Isles of Eden obeys a Bernoulli σ_τ-shift law with $\sigma = 2$, or -4, $\tau = 2$ and $\beta > 0$. The period obeys the formula $T = \tau L/|\sigma| = 6$.

(g) There are two *period-12 Isles of Eden* with a combined robustness coefficient $\rho_7 = 2(12/64) = 0.375$. The dynamics on these Isles of Eden obeys a Bernoulli σ_τ-shift law with $\sigma = -2$, or 4, $\tau = 4$ and $\beta > 0$. The period obeys the formula $T = \tau L/|\sigma| = 12$.

$$\boxed{\text{Gallery } 154\text{-}8, 154\text{-}9, 154\text{-}10, 154\text{-}11 : L = 7, \quad n\left(\Sigma^7\right) = 128}$$

(a) There are three *period-28 Isles of Eden* with a robustness coefficient $\rho_1 = 3(28/128) = 0.65625$. The dynamics on these Isles of Eden obeys a Bernoulli σ_τ-shift law with $\sigma = -3$, or 4, $\tau = 4$ and $\beta > 0$. The period obeys the formula $T = \tau L = 28$, and not $\tau L/|\sigma|$ because $28/|-3|$ is not an integer.

(b) There are three *period-14 Isles of Eden* with a combined robustness coefficient $\rho_2 = 3(14/128) = 0.328125$. The dynamics on these Isles of Eden obeys a Bernoulli σ_τ-shift law with $\sigma = 2$, or -5, $\tau = 2$ and $\beta > 0$. The period obeys the formula $T = \tau L = 14$, and not $\tau L/|\sigma|$ because $14/|-5|$ is not an integer.

(c) There are two *period-1 Isles of Eden* with a combined robustness coefficient $\rho_3 = 2(1/128) = 0.015625$. The dynamics on these Isles of Eden obeys a *degenerate* Bernoulli σ_τ-shift law with $\sigma = 0$, $\tau = 1$ and $\beta > 0$.

Gallery 154-12, 154-13, 154-14, 154-15, 154-16, 154-17 : $L = 8$,

$$n\left(\textstyle\sum^8\right) = 256$$

(a) There is a *period-1 attractor* $\{0\}$ with a robustness coefficient $\rho_1 = 31/256 = 0.12109375$. The dynamics on this attractor obeys a *degenerate* Bernoulli σ_τ-shift law with $\sigma = 0$, $\tau = 1$ and $\beta > 0$.

(b) There is a *period-1 Isle of Eden* $\{255\}$ with a robustness coefficient $\rho_2 = 1/256 = 0.00390625$. The dynamics on this Isle of Eden obeys a *degenerate* Bernoulli σ_τ-shift law with $\sigma = 0$, $\tau = 1$ and $\beta > 0$.

(c) There are 20 *period-8 Isles of Eden* with a combined robustness coefficient $\rho_3 = 20(8/256) = 0.625$. The dynamics on these Isles of Eden obeys a Bernoulli σ_τ-shift law with $\sigma = 4$, or -4, $\tau = 4$ and $\beta > 0$. The period obeys the formula $T = \tau L/|\sigma| = 8$.

(d) There are seven *period-8 Isles of Eden* with a combined robustness coefficient $\rho_4 = 7(8/256) = 0.21875$. The dynamics on these Isles of Eden obeys a Bernoulli σ_τ-shift law with $\sigma = 2$, $\tau = 2$ and $\beta > 0$. The period obeys the formula $T = \tau L/|\sigma| = 8$.

(e) There are two *period-4 Isles of Eden* with a robustness coefficient $\rho_5 = 2(4/256) = 0.03125$. The dynamics on these Isles of Eden obeys a Bernoulli σ_τ-shift law with $\sigma = 2$, or -2, $\tau = 2$ and $\beta > 0$. The period obeys the formula $T = \tau L/|\sigma| = 8$ with a minimal period $T_{\min} = 4$.

The qualitative properties of local rule $\boxed{154}$ extracted from the above basin-tree Galleries 154-1 to 154-17 are summarized below:

Summary of Qualitative properties of local rule $\boxed{154}$ *extracted from Gallery 154 for Rule* $\boxed{154}$

L	ID Number i	Number of Period-n attractors	Number of Period-n Isles of Eden	Period n	σ_1	τ_1	β_1	σ_2	τ_2	β_2	Robustness coefficient ρ
3	1		2	1	0	1	+				$\rho_1 = 0.25$
	2		1	6	-1	2	+	2	2	+	$\rho_2 = 0.75$
4	1	1		1	0	1	+				$\rho_1 = 0.0625$
	2		1	1	0	1	+				$\rho_2 = 0.4375$
	3		2	4	2	2	+	-2	2	+	$\rho_3 = 0.5$
5	1		2	1	0	1	+				$\rho_1 = 0.0625$
	2		1	10	2	2	+	-3	2	+	$\rho_2 = 0.3125$
	3		1	20	-1	4	+	4	4	+	$\rho_3 = 0.625$
6	1	1		1	0	1	+				$\rho_1 = 0.046875$
	2		1	1	0	1	+				$\rho_2 = 0.015625$
	3	3		2	3	1	+	-3	1	+	$\rho_3 = 0.1875$
	4		2	3	-2	1	+	4	1	+	$\rho_4 = 0.09375$
	5		1	6	-1	2	+				$\rho_5 = 0.09375$
	6		2	6	-2	2	+	4	2	+	$\rho_6 = 0.1875$
	7		2	12	-2	4	+	4	4	+	$\rho_7 = 0.375$
7	1		3	28	-3	4	+	4	4	+	$\rho_1 = 0.65625$
	2		3	14	2	2	+				$\rho_2 = 0.328125$
	3		2	1	0	1	+				$\rho_3 = 0.015625$
8	1	1		1	0	1	+				$\rho_1 = 0.12109375$
	2		1	1	0	1	+				$\rho_2 = 0.00390625$
	3		20	8	4	4	+	-4	4	+	$\rho_3 = 0.625$
	4		7	8	2	2	+				$\rho_4 = 0.21875$
	5		2	4	2	2	+	-2	2	+	$\rho_5 = 0.031255$

Table 12. Basin tree diagrams for rule 154.

Basin tree diagrams for Rule 154
154, L = 3
(a) Period-1 Isles of Eden :
ρ1 = 2/8 = 0.25
0
7
(b) Bernoulli (σ = −1, σ = +2, τ = 2)
Period-6 Isle of Eden : ρ2 = 6/8 = 0.75
2
3
5
4
1
6
σ = −1, σ = +2, τ = 2
ϕn
ϕn - 2
0
0.5
1
β > 0
④
T = Lτ = 6
τ = 2
τ = 2
σ = +2
σ = −1
Gallery 154 -1

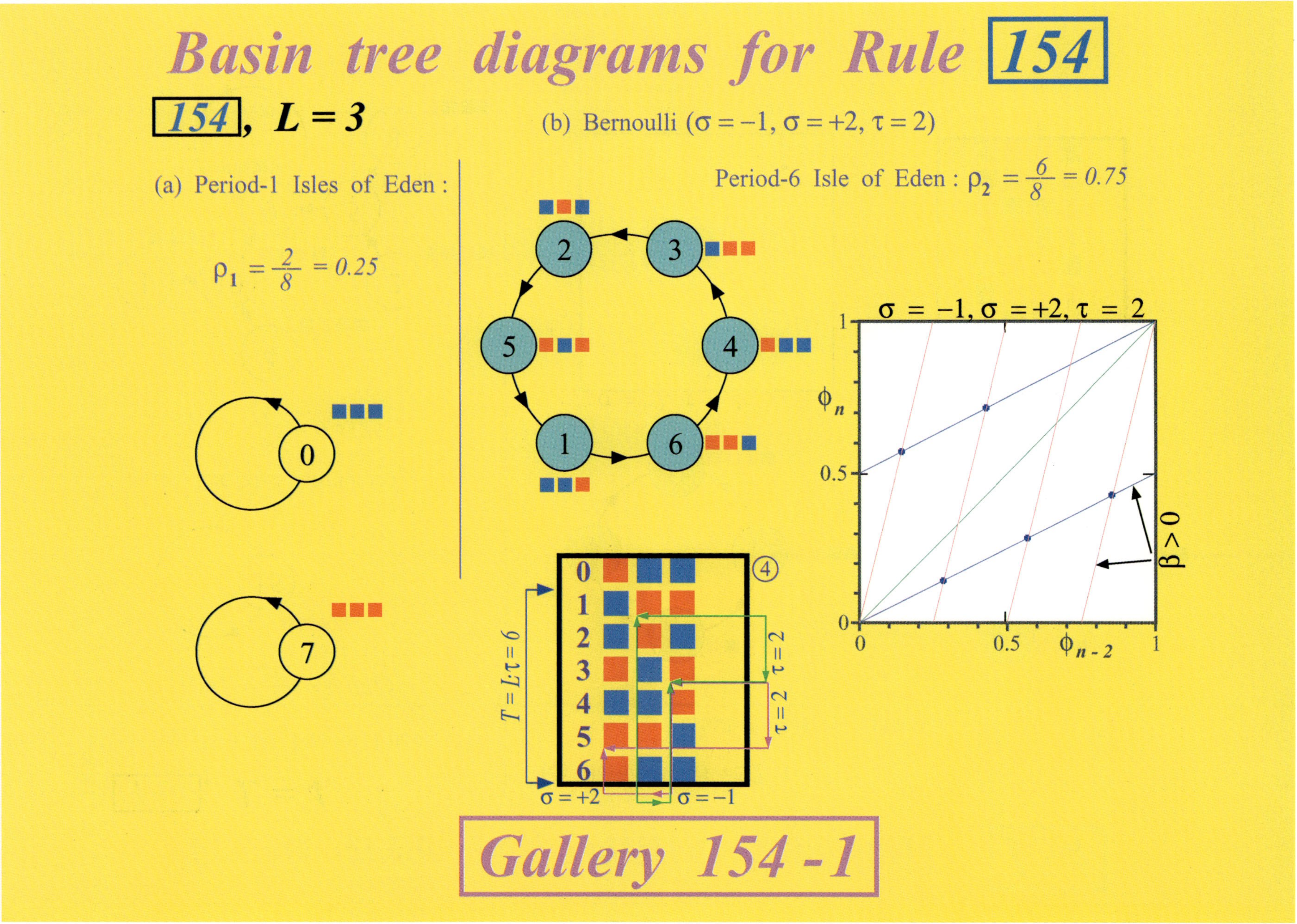

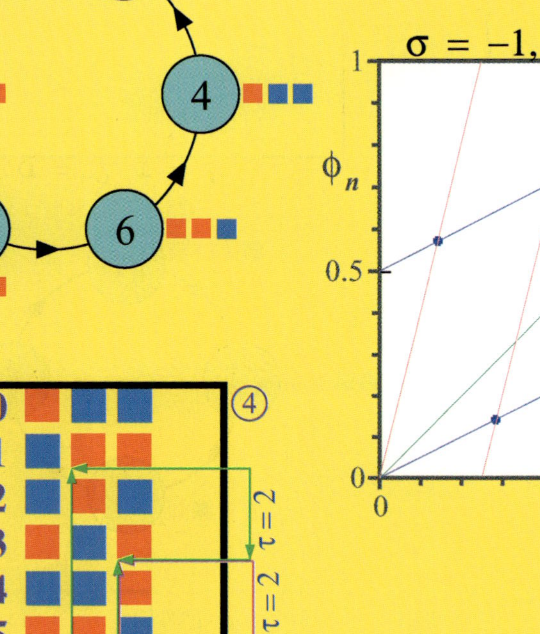

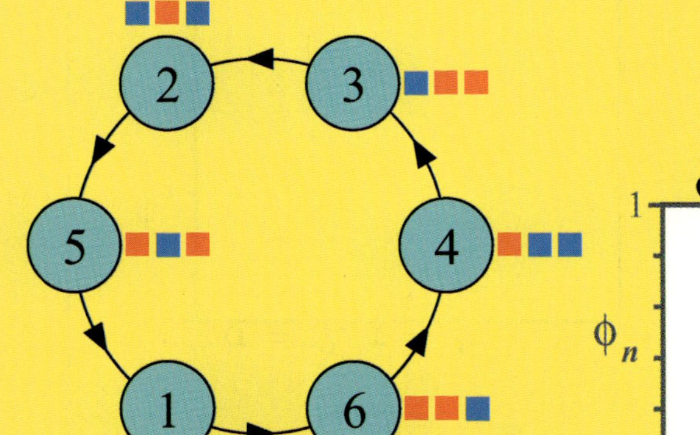

Basin tree diagrams for Rule 154

154 , L = 4

(a) Period-1 Isle of Eden :

$$\rho_1 = \frac{1}{16} = 0.0625$$

(b) Period-1 Attractor :

$$\rho_2 = \frac{7}{16} = 0.4375$$

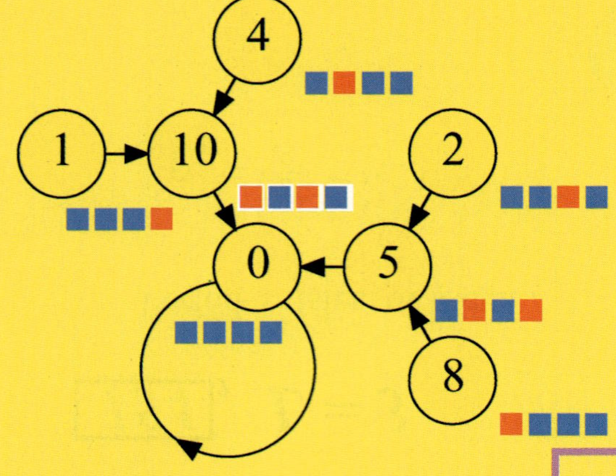

(c) Bernoulli ($\sigma = \pm 2$, $\tau = 2$)

Period-4 Isles of Eden : $\rho_3 = 2\,\frac{4}{16} = 0.5$

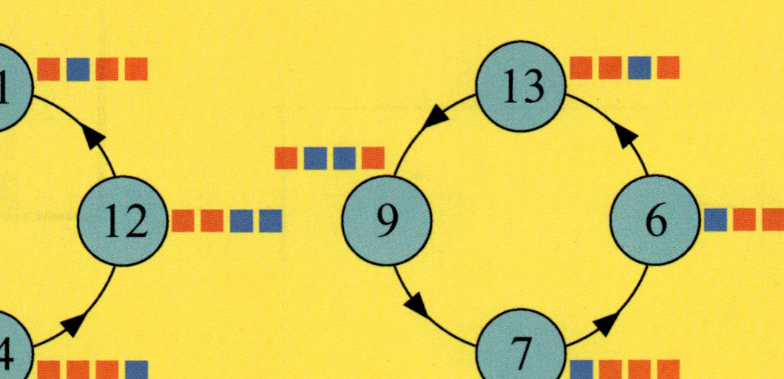

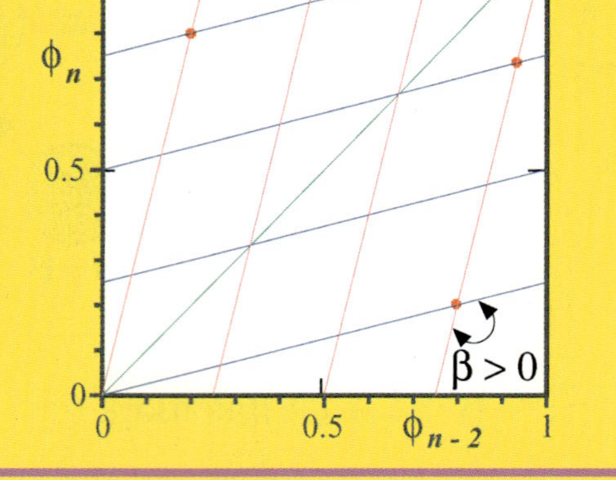

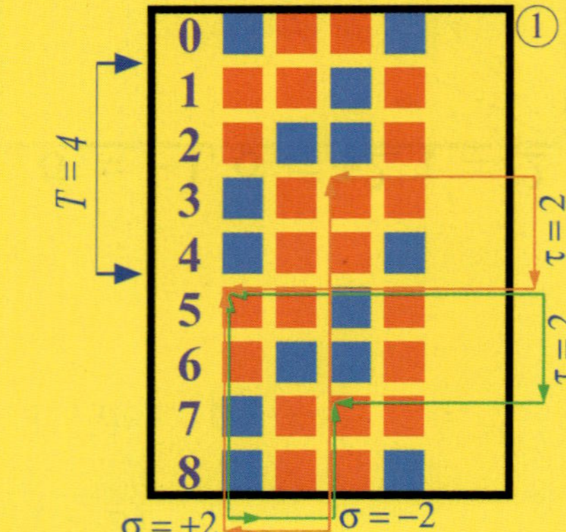

Gallery 154 - 2

Basin tree diagrams for Rule $\boxed{154}$

$\boxed{154}$, $L = 5$ (b) Bernoulli ($\sigma = 2$, $\sigma = -3$, $\tau = 2$)

(a) Period-1 Isles of Eden :   Period-10 Isle of Eden : $\rho_2 = \dfrac{10}{32} = 0.3125$

$\rho_1 = \dfrac{2}{32} = 0.0625$

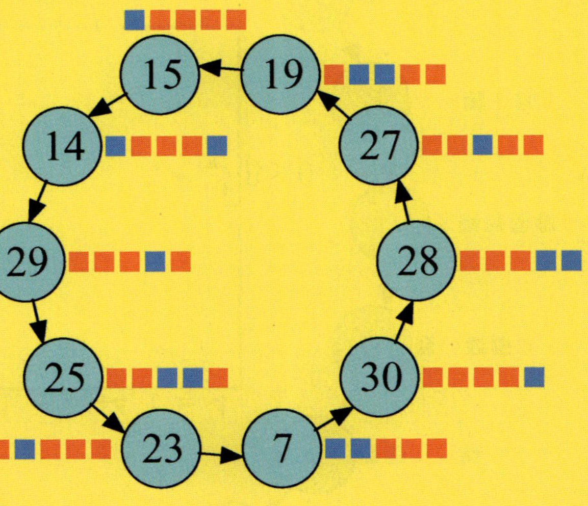
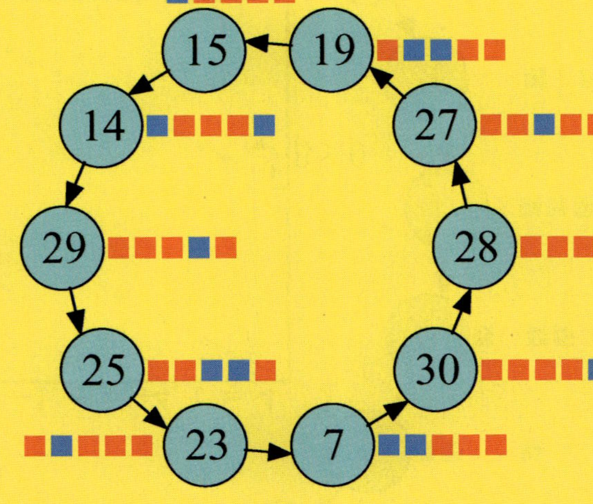

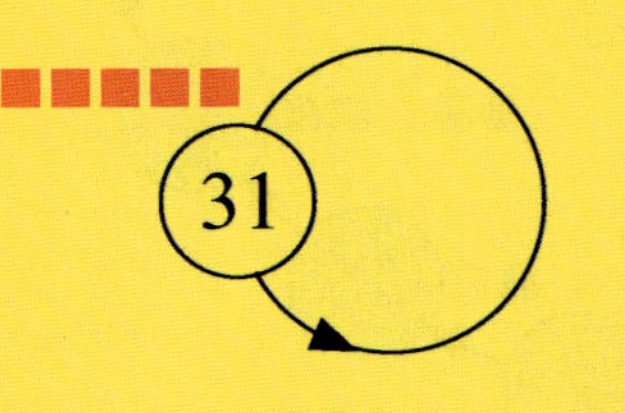

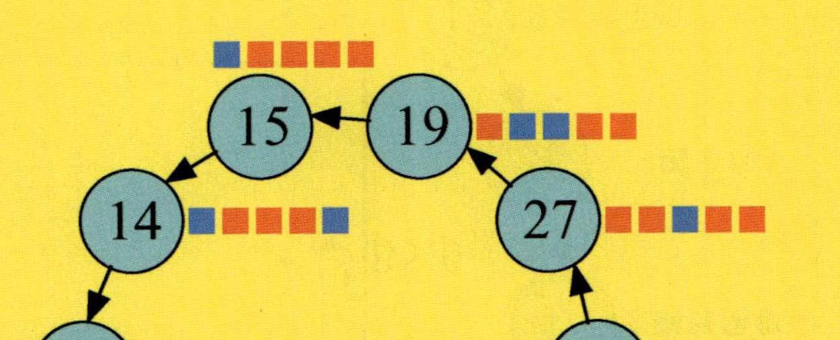

$\boxed{\textit{Gallery 154 -3}}$

Table 12. (Continued)

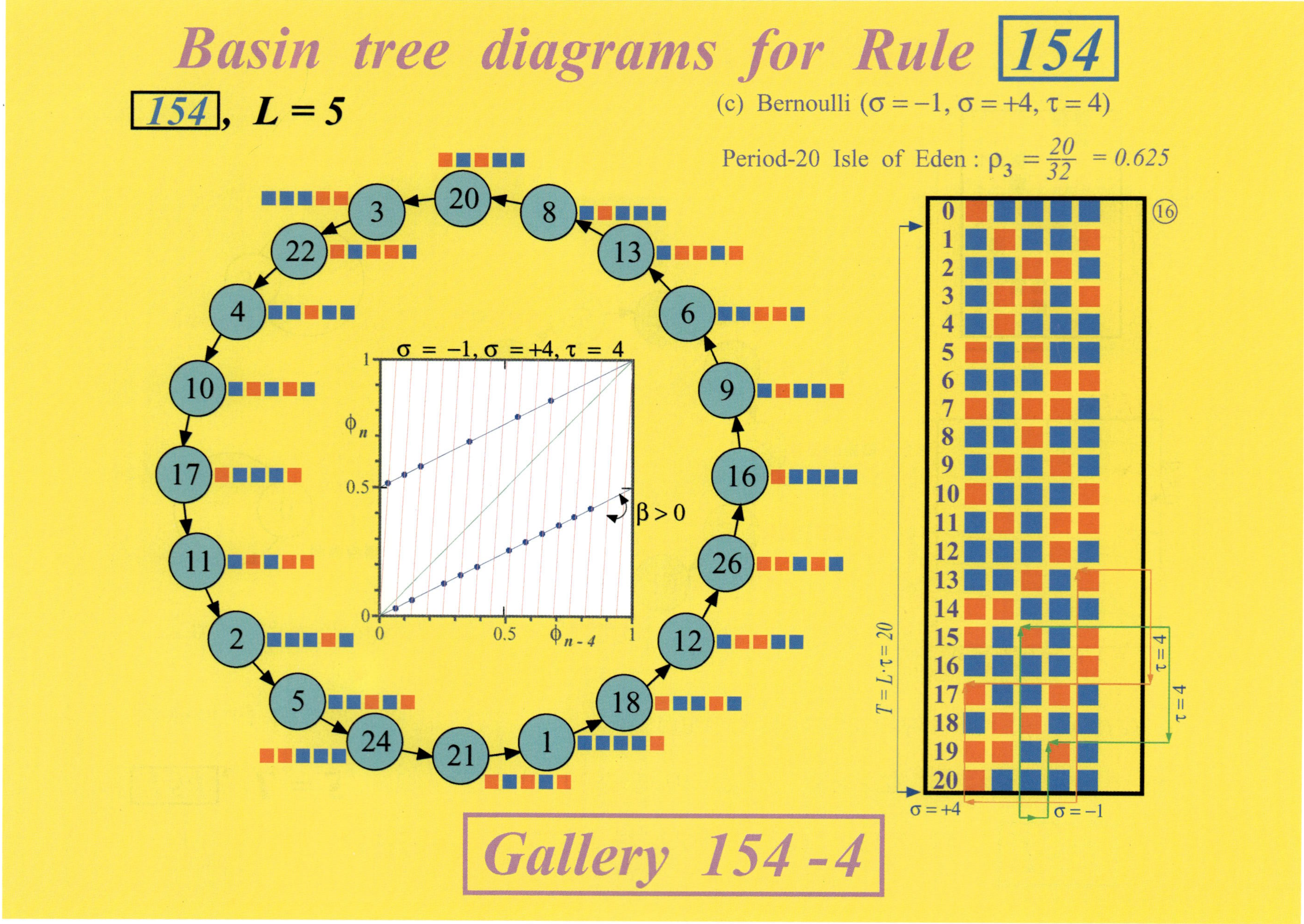

Basin tree diagrams for Rule 154
154, L = 5
(c) Bernoulli (σ = −1, σ = +4, τ = 4)
Period-20 Isle of Eden : ρ3 = 20/32 = 0.625
σ = −1, σ = +4, τ = 4
β > 0
ϕn
ϕn-4
T = L·τ = 20
τ = 4
τ = 4
σ = +4
σ = −1
Gallery 154 - 4

Basin tree diagrams for Rule 154

154, $L = 6$

(c) Bernoulli ($\sigma = \pm 3$, $\tau = 1$) Period-2 Attractors : $\rho_3 = 3\frac{4}{64} = 0.1875$

(a) Period-1 Attractor :

$$\rho_1 = \frac{3}{64} = 0.046875$$

(b) Period-1 Isle of Eden :

$$\rho_2 = \frac{1}{64} = 0.015625$$

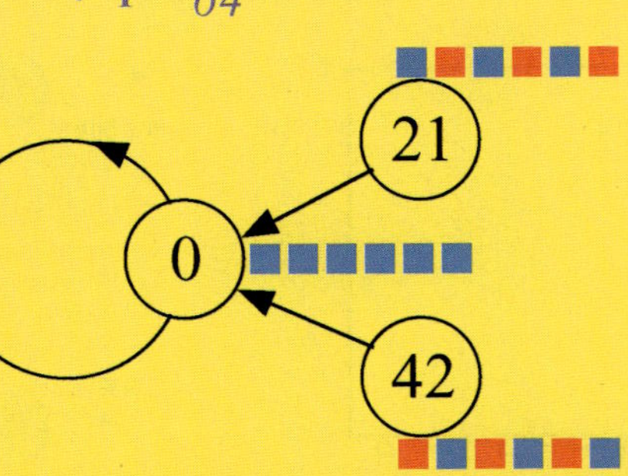

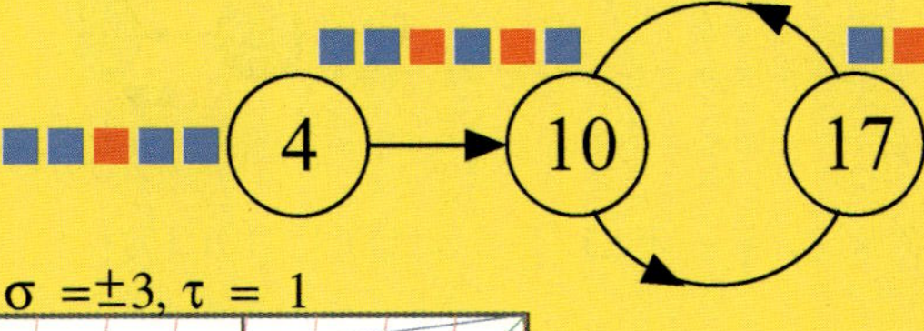

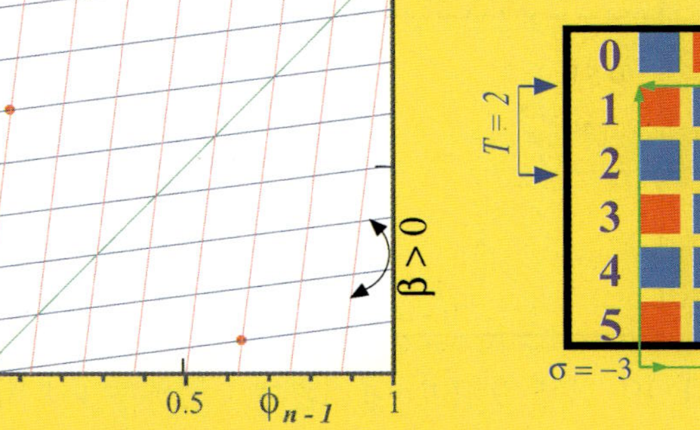

Gallery 154 - 5

Table 12. (*Continued*)

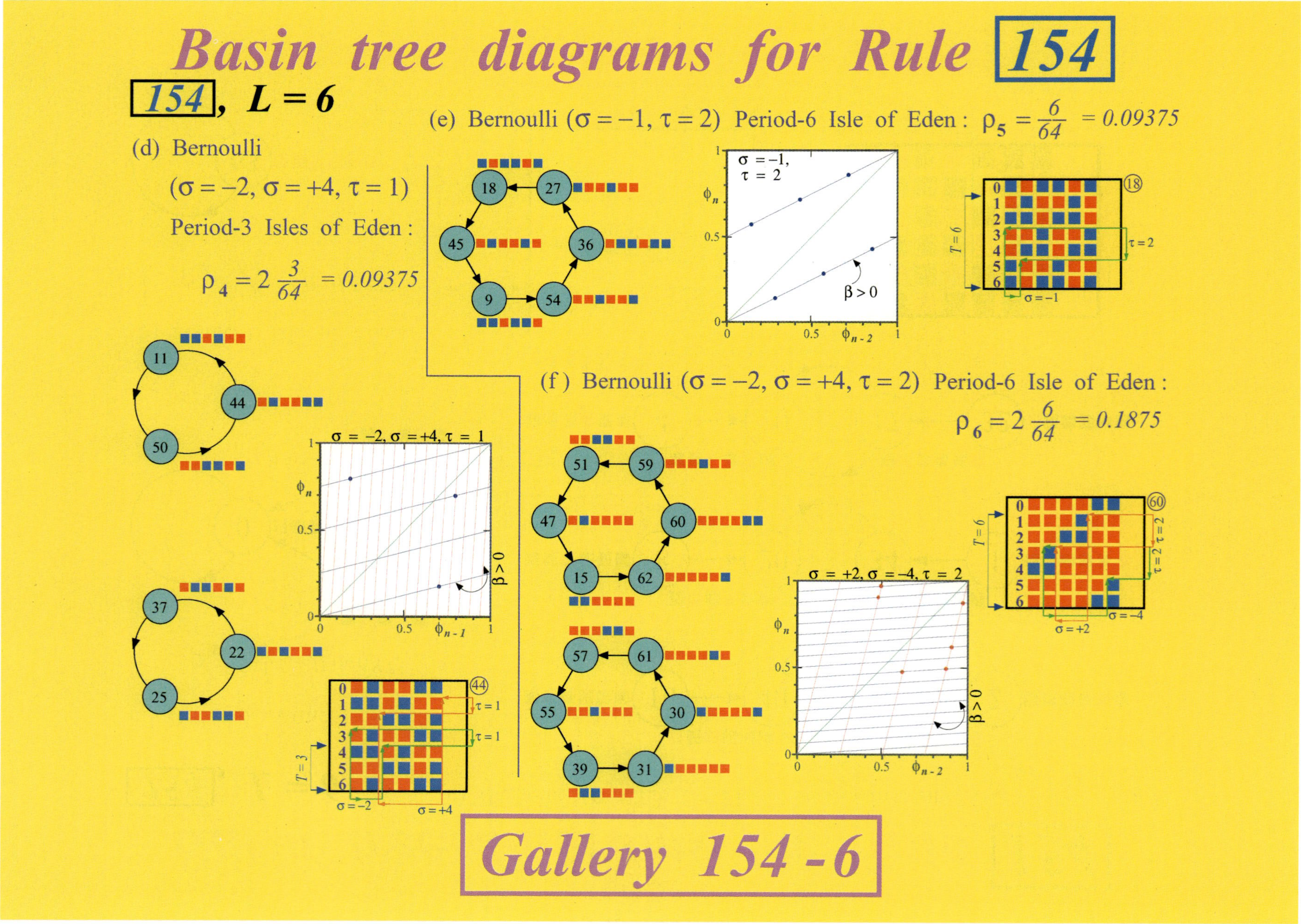

Basin tree diagrams for Rule $\boxed{154}$

$\boxed{154}$, $L = 6$

(d) Bernoulli

$(\sigma = -2,\ \sigma = +4,\ \tau = 1)$

Period-3 Isles of Eden :

$$\rho_4 = 2\,\frac{3}{64} = 0.09375$$

(e) Bernoulli $(\sigma = -1,\ \tau = 2)$ Period-6 Isle of Eden : $\rho_5 = \dfrac{6}{64} = 0.09375$

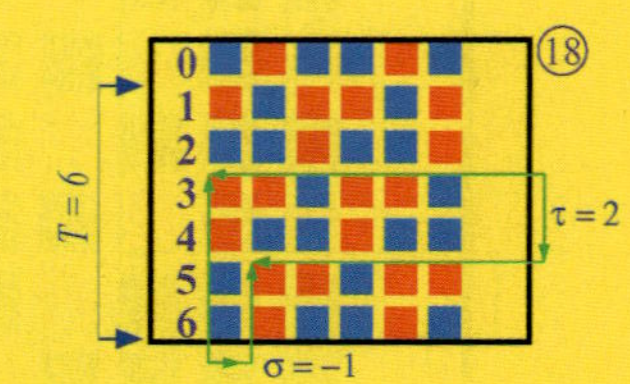

(f) Bernoulli $(\sigma = -2,\ \sigma = +4,\ \tau = 2)$ Period-6 Isle of Eden :

$$\rho_6 = 2\,\frac{6}{64} = 0.1875$$

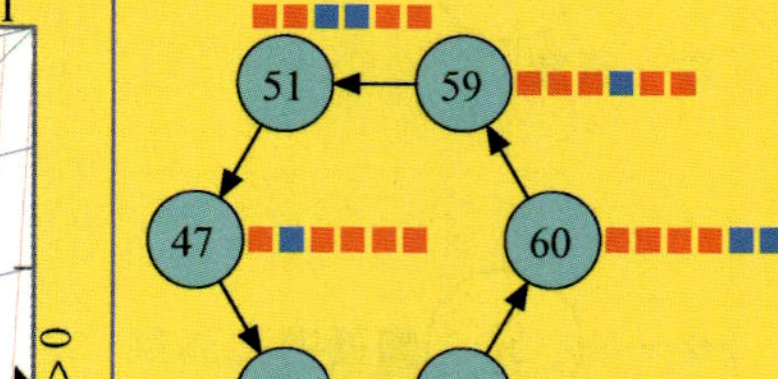
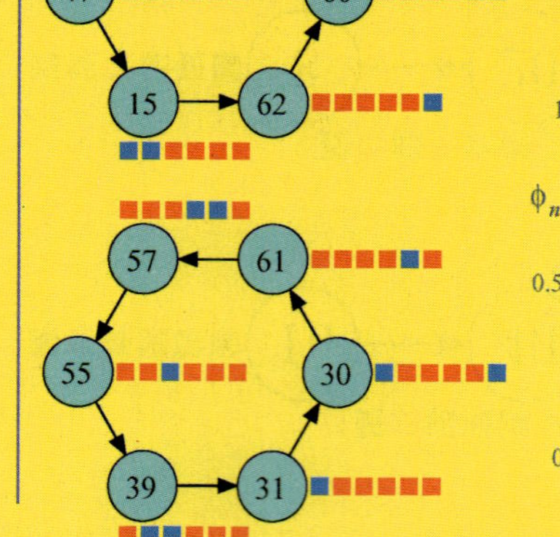
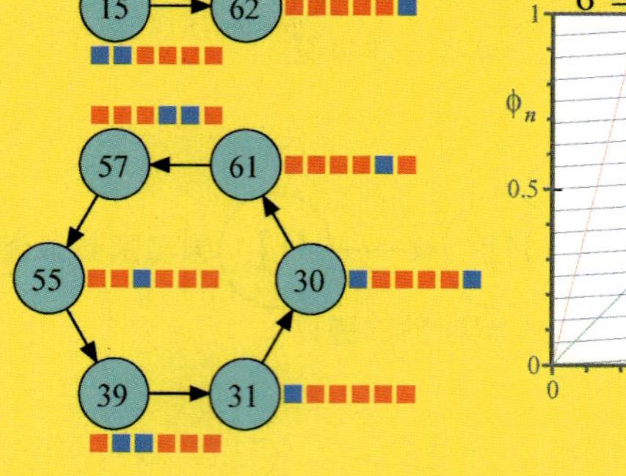

$\boxed{Gallery\ 154\text{-}6}$

Basin tree diagrams for Rule 154

154 , $L = 6$

(g) Bernoulli ($\sigma = -2$, $\sigma = +4$, $\tau = 4$)

Period-12 Isles of Eden :

$$\rho_7 = 2\,\frac{12}{64} = 0.375$$

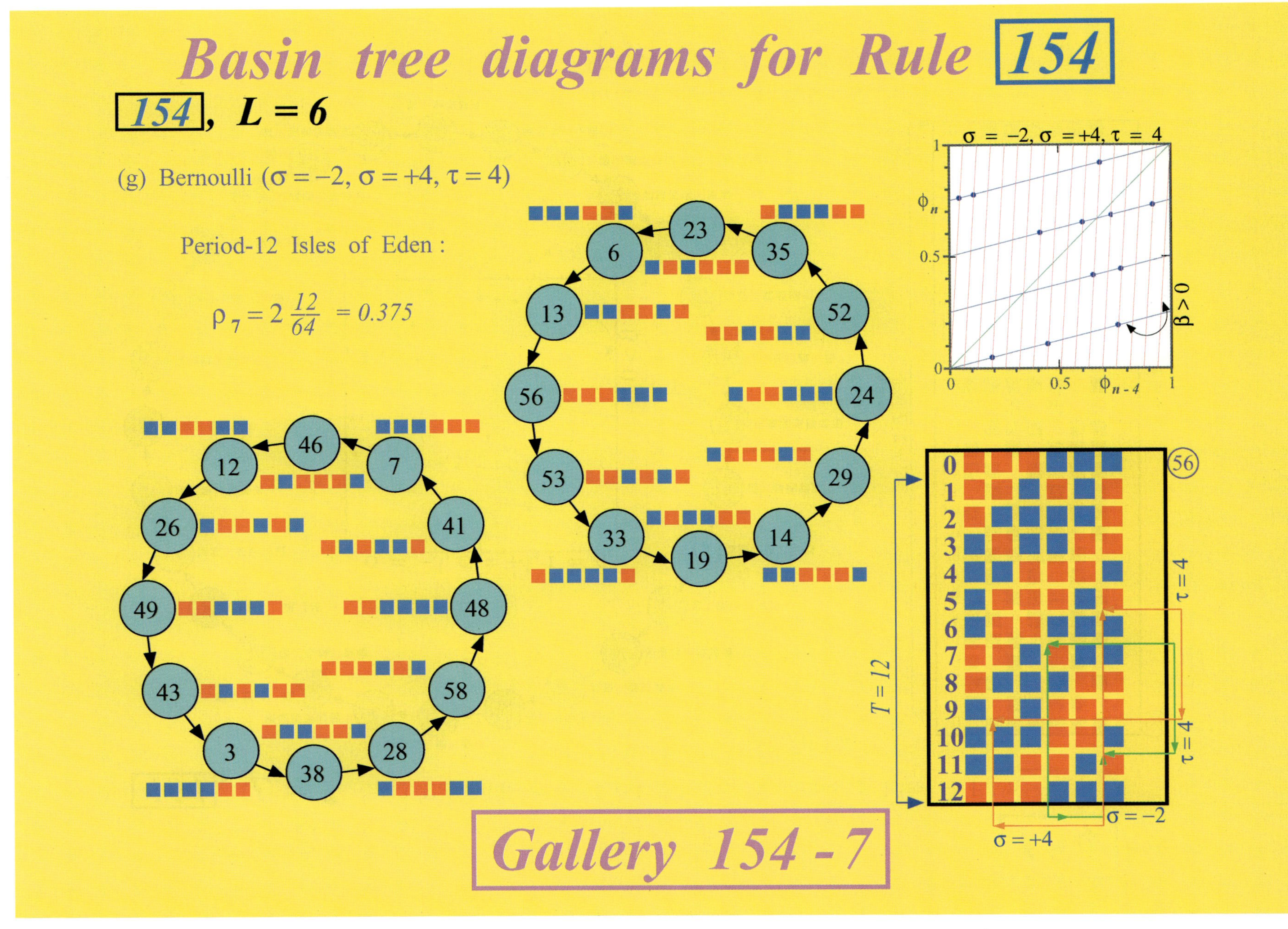

Gallery 154 - 7

Table 12.   (Continued)

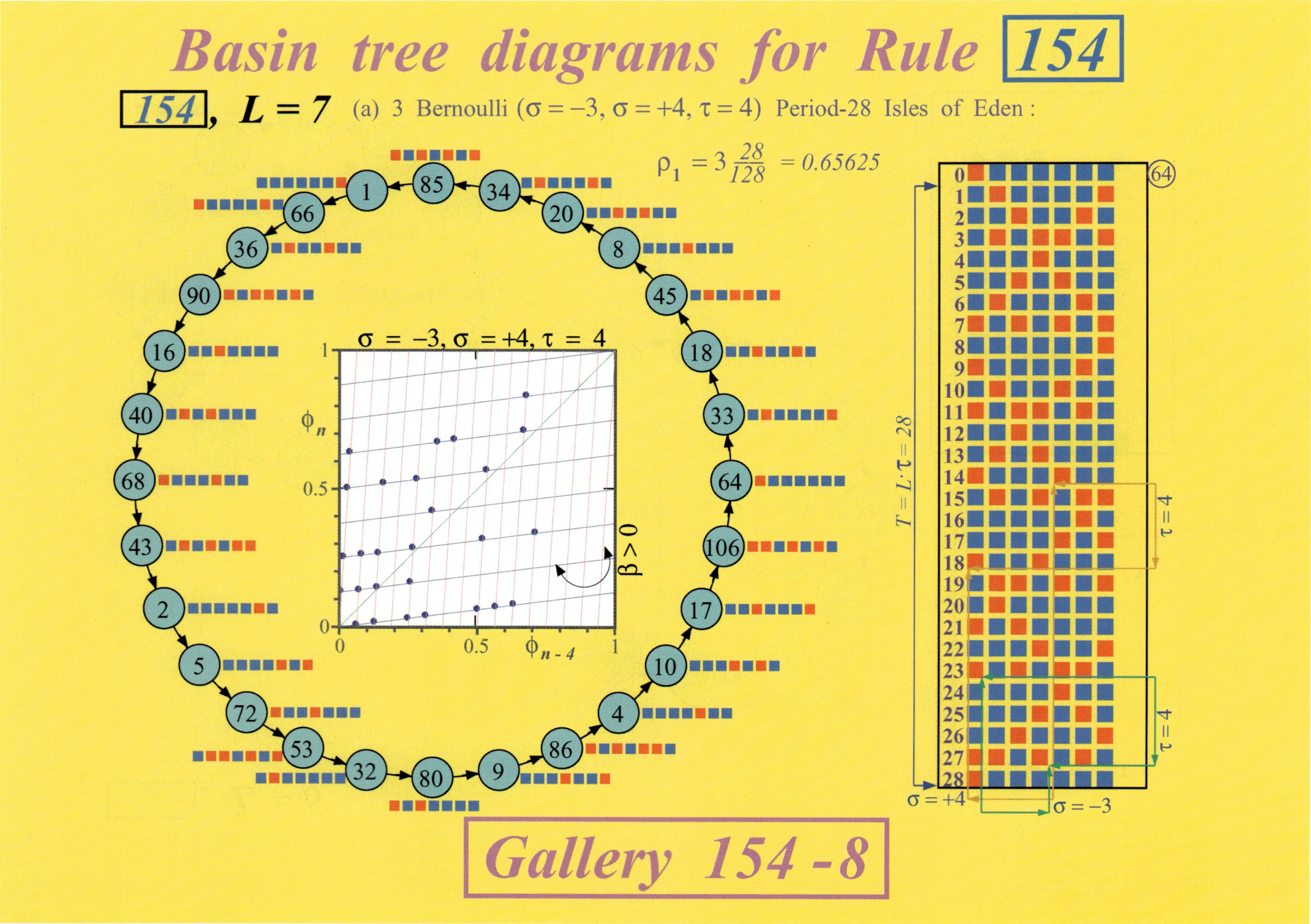

Basin tree diagrams for Rule 154

154, **L = 7**   (a) 3 Bernoulli ($\sigma = -3$, $\sigma = +4$, $\tau = 4$) Period-28 Isles of Eden :

$$\rho_1 = 3\frac{28}{128} = 0.65625$$

Gallery 154 - 8

Basin tree diagrams for Rule 154

154, **L = 7** (a) 3 Bernoulli ($\sigma = -3$, $\sigma = +4$, $\tau = 4$) Period-28 Isles of Eden :

(continued)

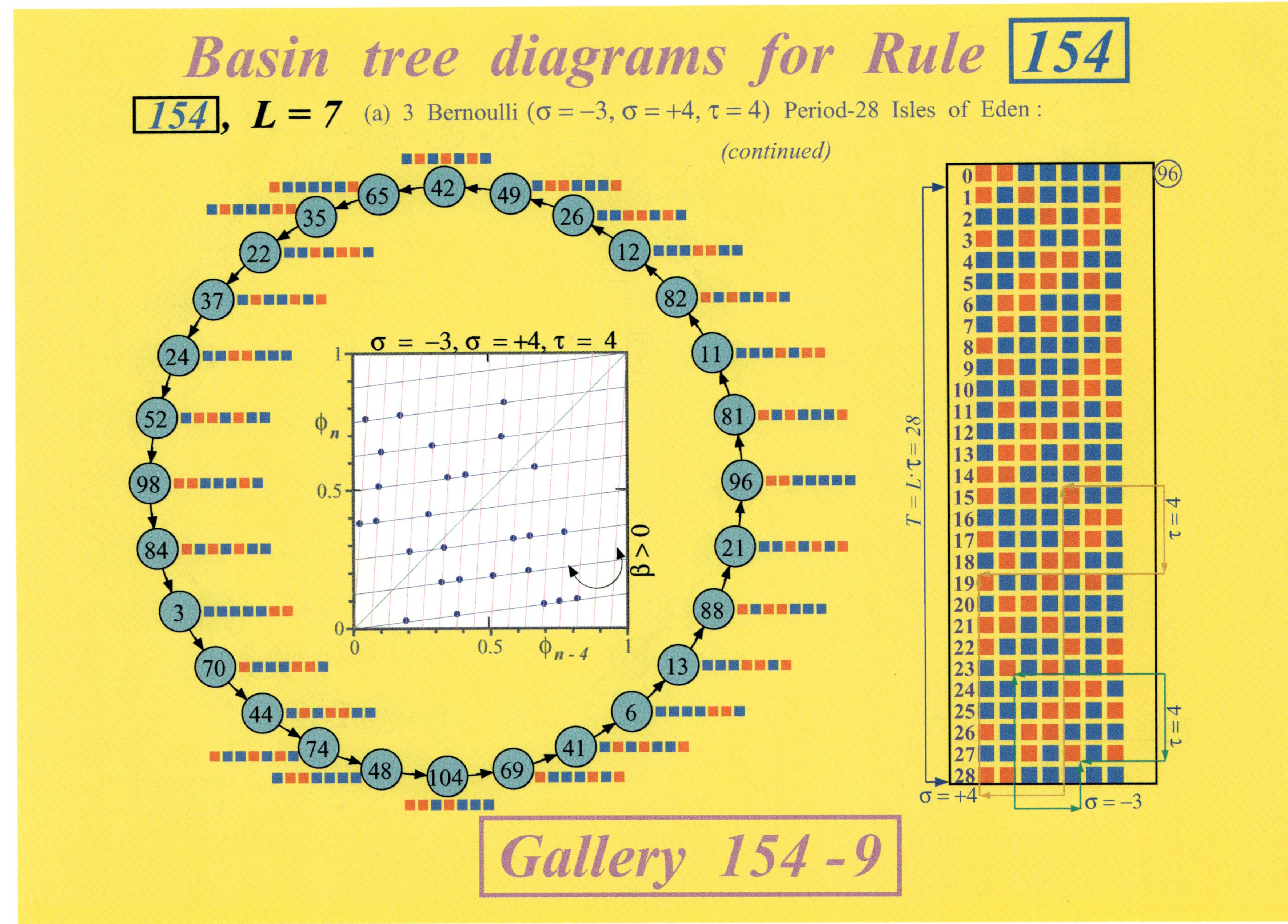

Gallery 154 - 9

Table 12. (*Continued*)

Basin tree diagrams for Rule $\boxed{154}$

$\boxed{154}$, **_L_ = 7** (a) 2 Bernoulli ($\sigma = -3$, $\sigma = +4$, $\tau = 4$) Period-28 Isles of Eden :

(continued)

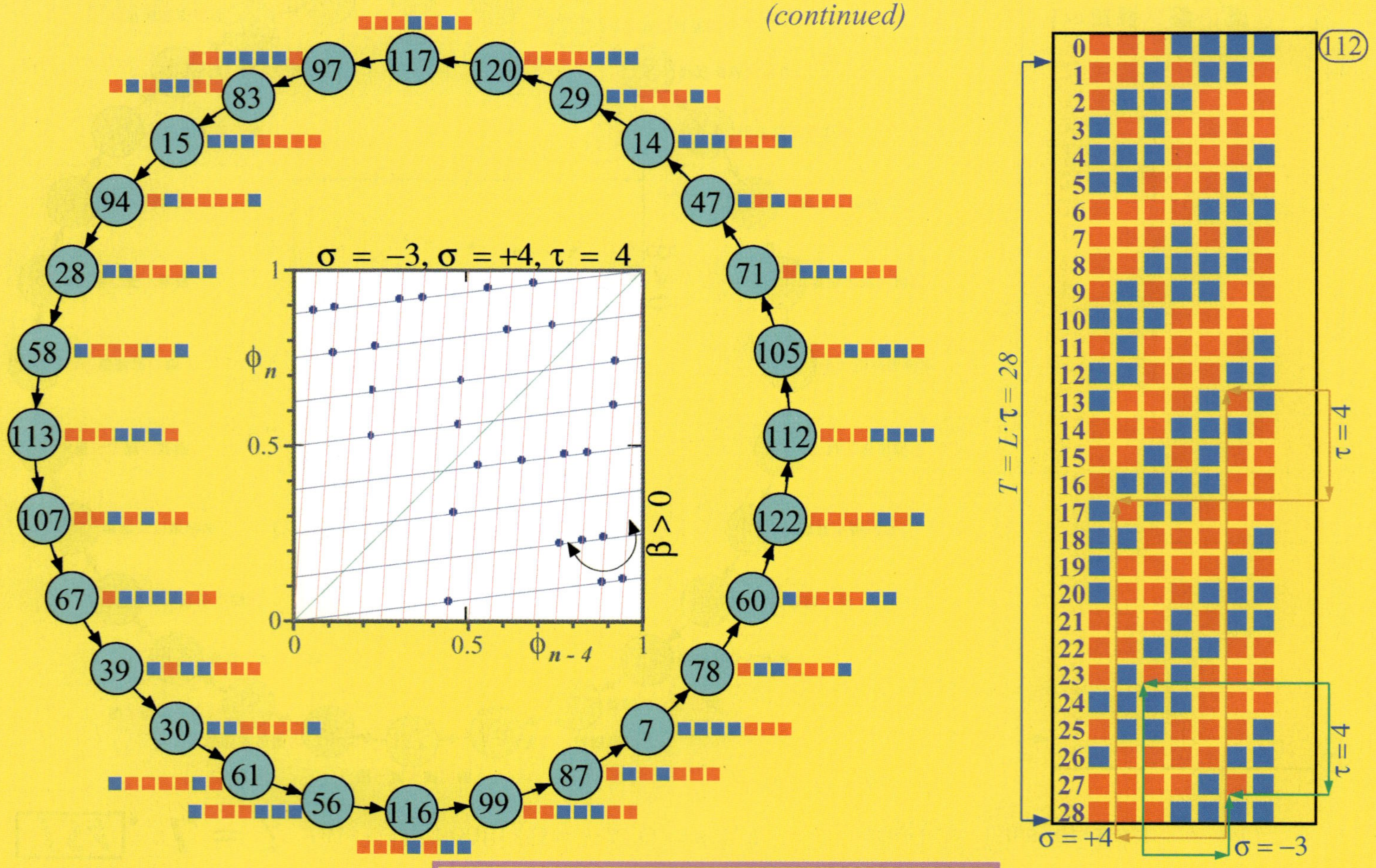

Gallery 154 -10

274

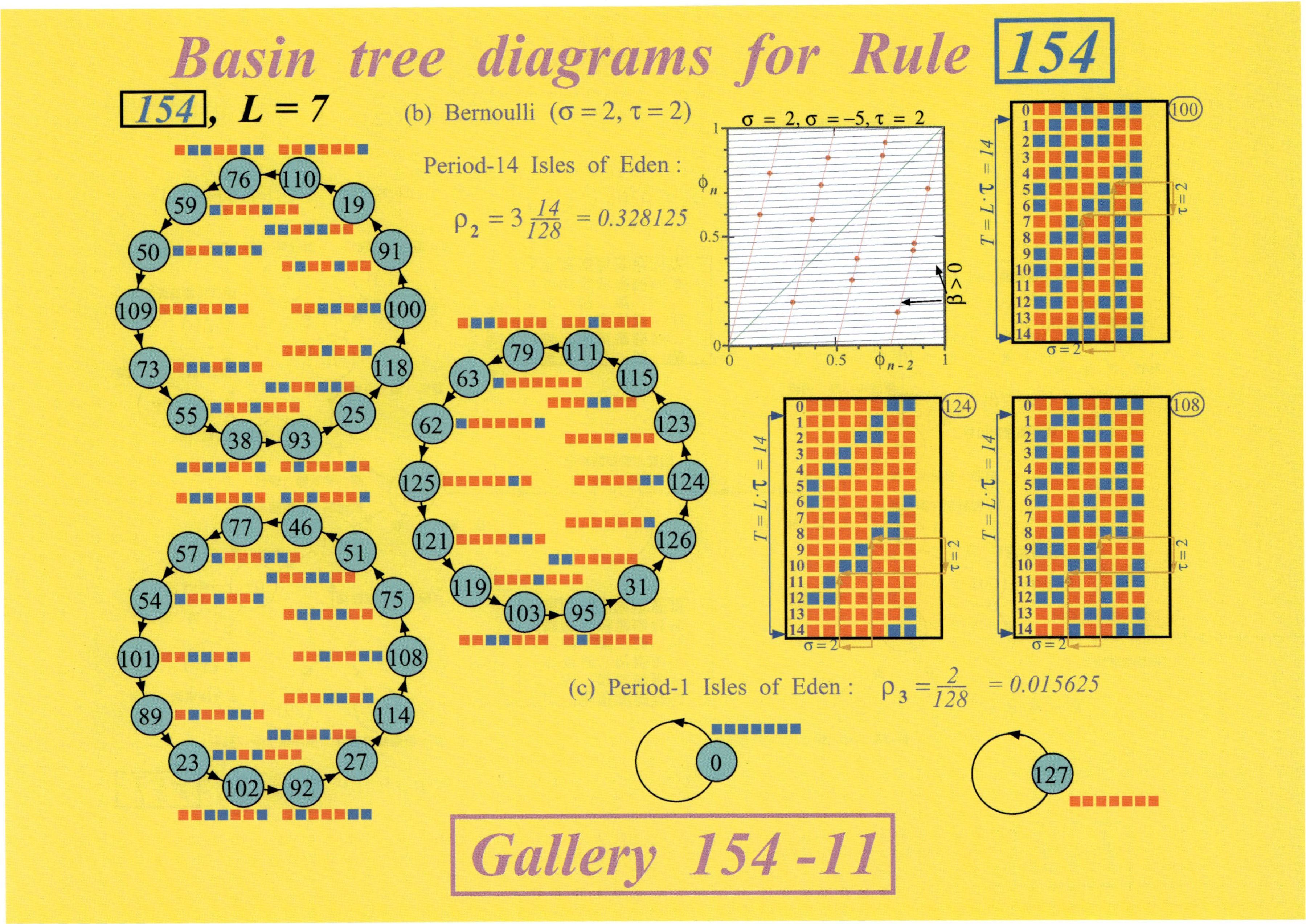

Table 12.   (*Continued*)

Basin tree diagrams for Rule 154

154, $L = 7$

(b) Bernoulli ($\sigma = 2$, $\tau = 2$)

Period-14 Isles of Eden :

$$\rho_2 = 3\frac{14}{128} = 0.328125$$

(c) Period-1 Isles of Eden : $\rho_3 = \frac{2}{128} = 0.015625$

Gallery 154 -11

275

Basin tree diagrams for Rule 154

154, $L = 8$ (a) Period-1 Attractor : $\rho_1 = \dfrac{31}{256}$ $= 0.12109375$

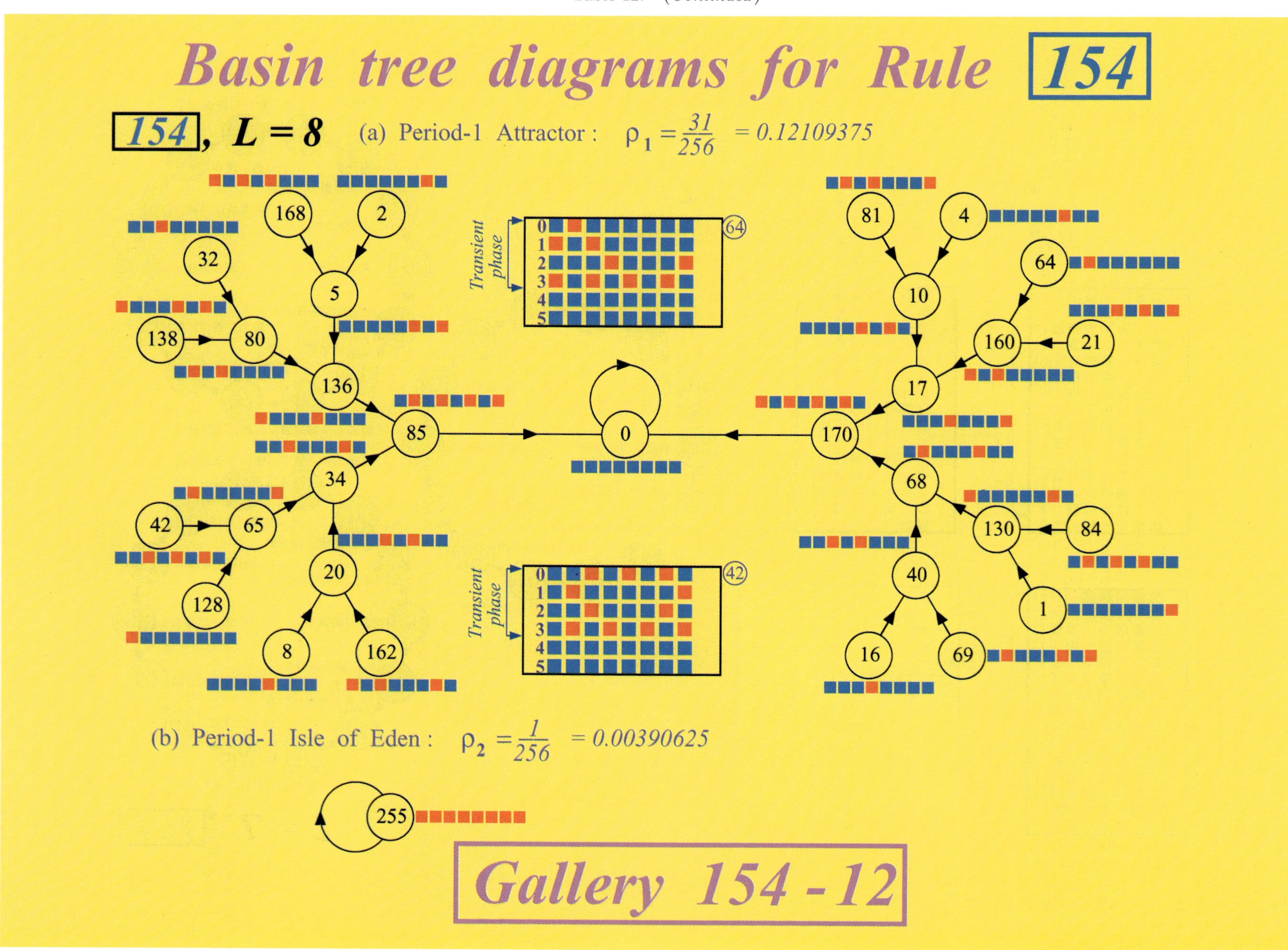

(b) Period-1 Isle of Eden : $\rho_2 = \dfrac{1}{256}$ $= 0.00390625$

Gallery 154 - 12

Table 12. (*Continued*)

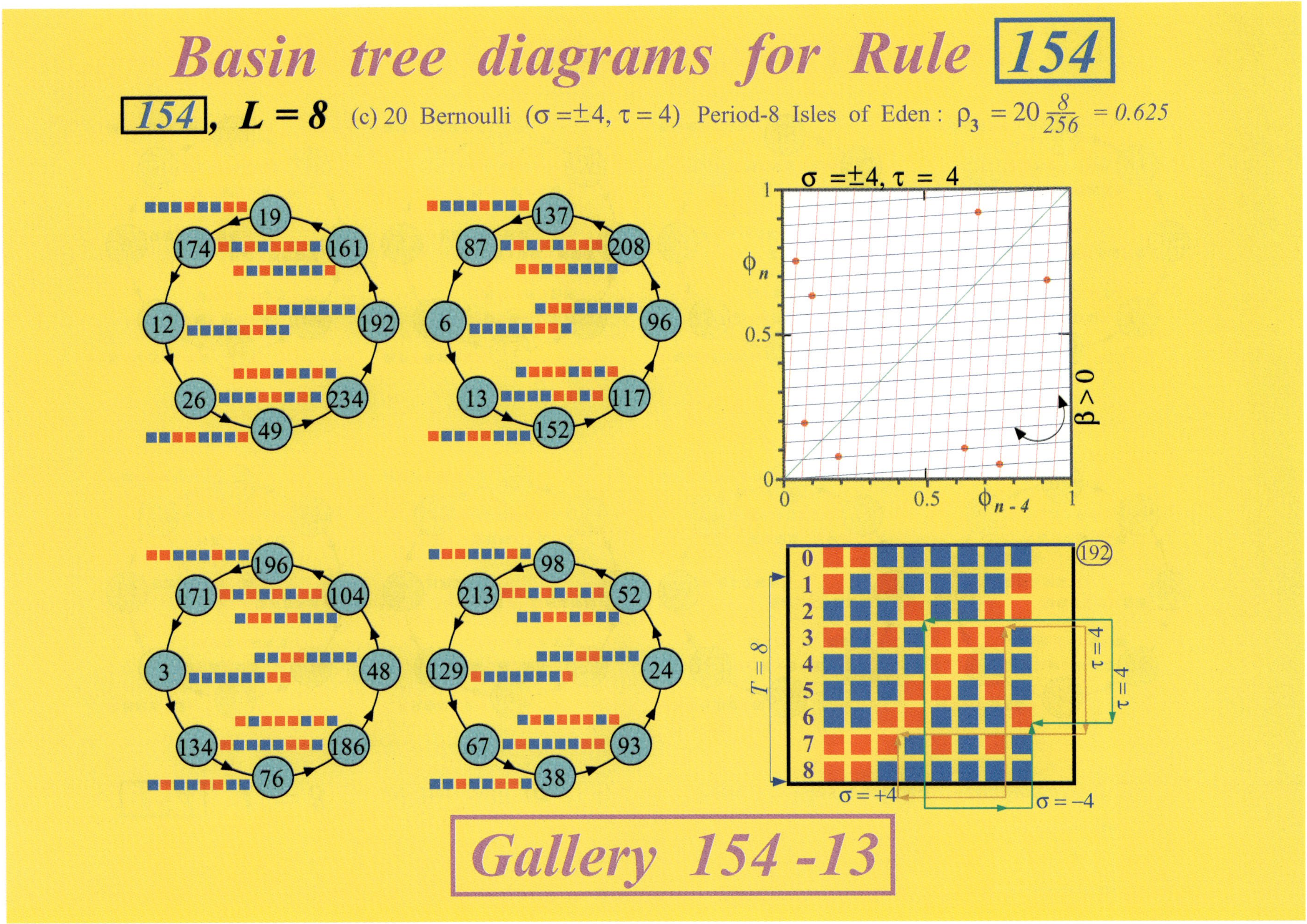
Basin tree diagrams for Rule 154
154, L = 8 (c) 20 Bernoulli (σ = ±4, τ = 4) Period-8 Isles of Eden : $\rho_3 = 20 \frac{8}{256} = 0.625$
19
174
161
12
192
26
234
49
137
87
208
6
96
13
117
152
196
171
104
3
48
134
186
76
98
213
52
129
24
67
93
38
σ = ±4, τ = 4
ϕ_n
ϕ_{n-4}
0
0.5
1
β > 0
192
T = 8
0
1
2
3
4
5
6
7
8
τ = 4
τ = 4
σ = +4
σ = −4
Gallery 154 -13

Table 12. (*Continued*)

Basin tree diagrams for Rule 154

154 , $L = 8$ (c) 20 Bernoulli ($\sigma = \pm 4$, $\tau = 4$) Period-8 Isles of Eden : *(continued)*

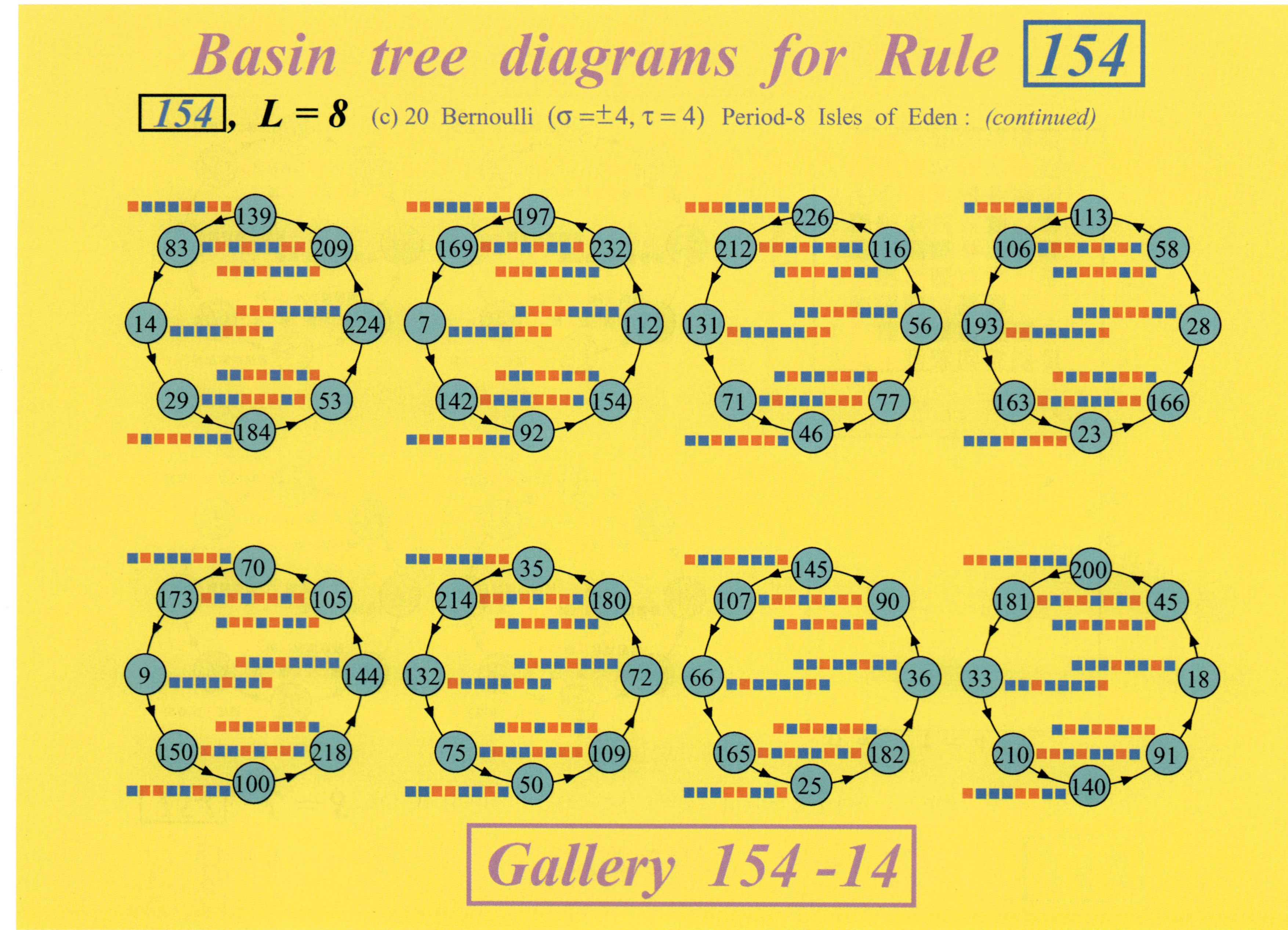

Basin tree diagrams for Rule 154

154, **L = 8** (c) 20 Bernoulli ($\sigma = \pm 4$, $\tau = 4$) Period-8 Isles of Eden :

(continued)

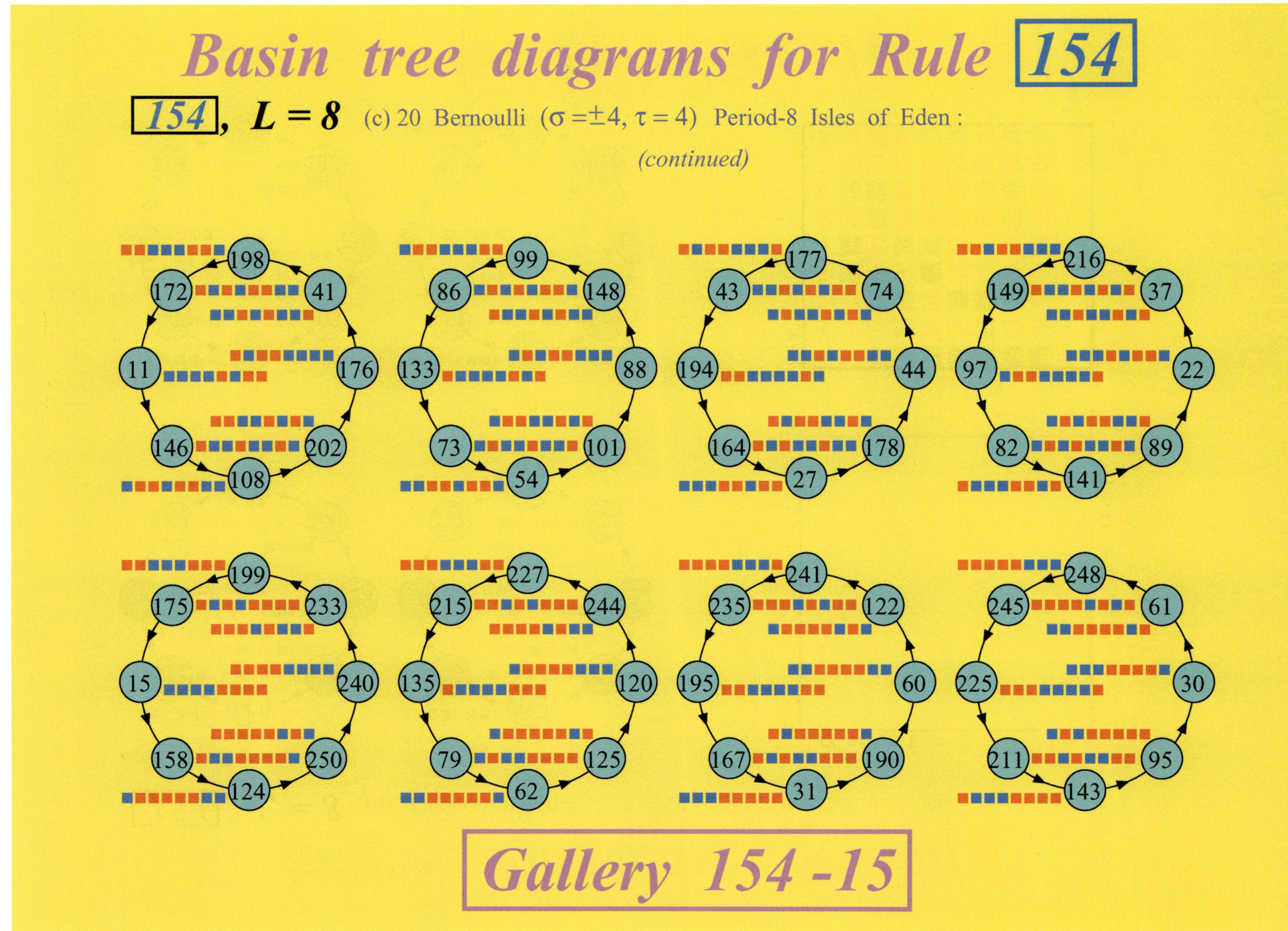

Gallery 154 -15

Basin tree diagrams for Rule $\boxed{154}$

$\boxed{154}$, $L = 8$ (d) 7 Bernoulli ($\sigma = 2, \tau = 2$)   Period-8 Isles of Eden : $\rho_4 = 7\frac{8}{256} = 0.21875$

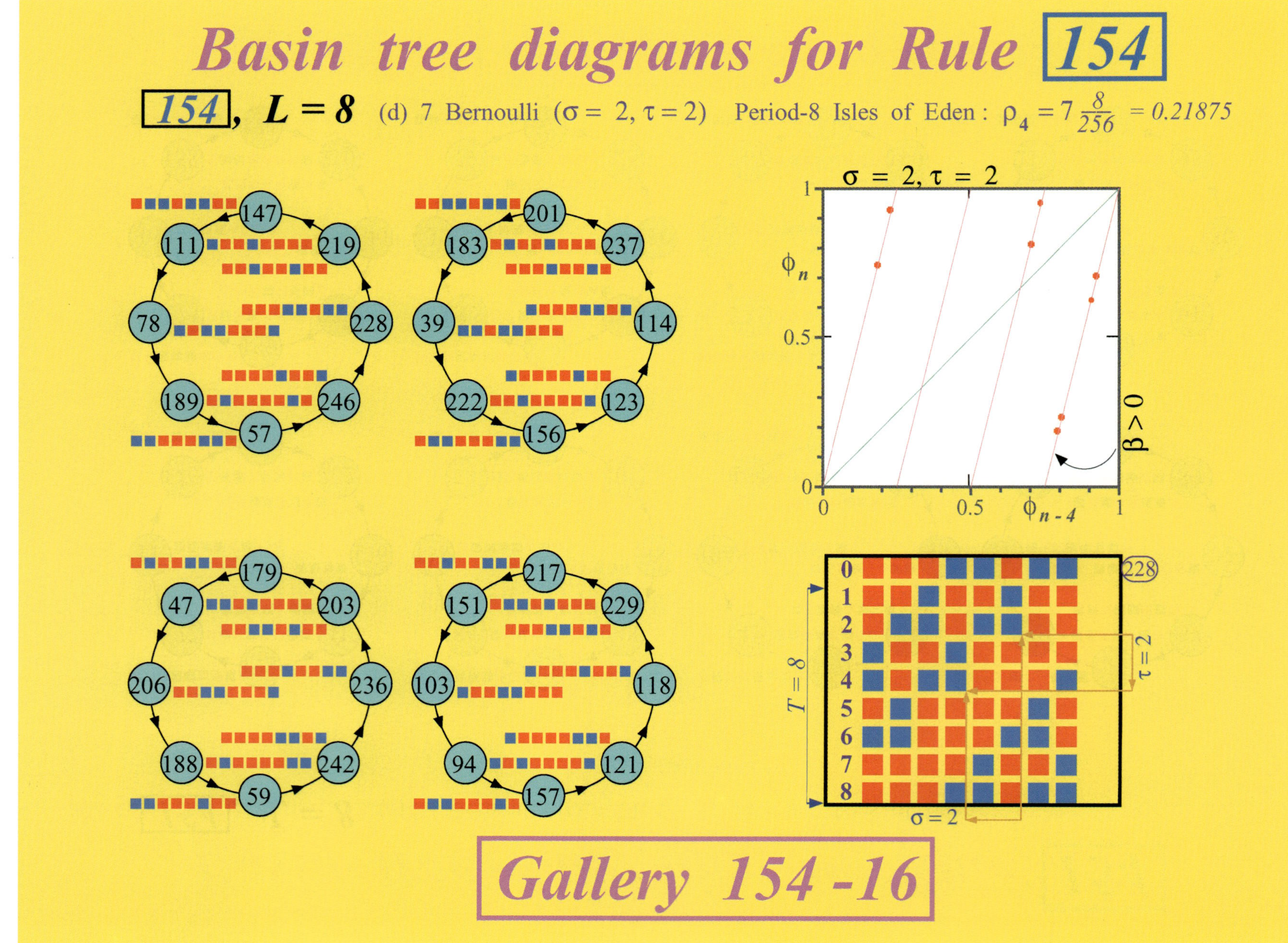

Gallery 154 -16

Basin tree diagrams for Rule 154

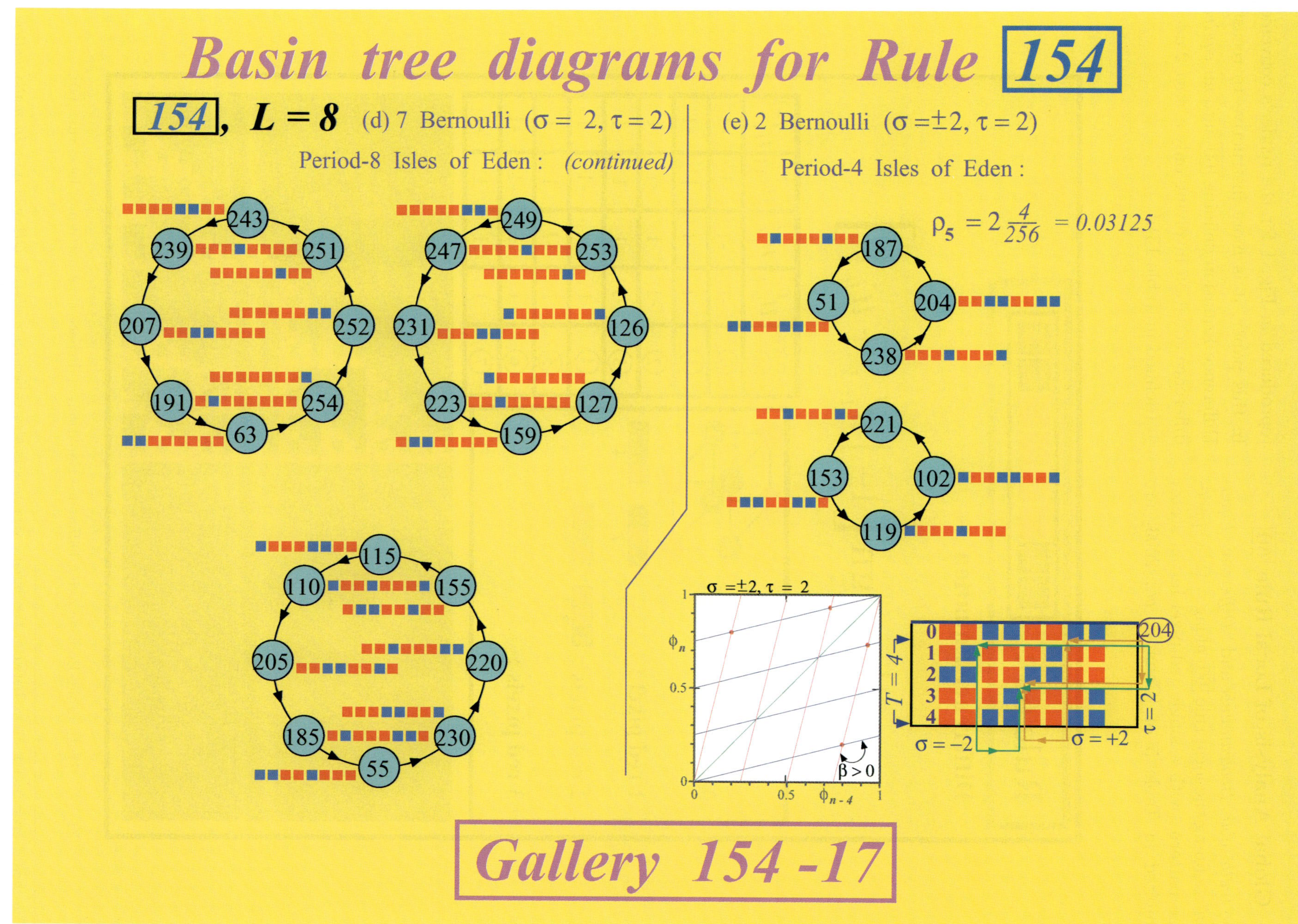

Gallery 154 -17

3. Global Analysis of Local Rule $\boxed{60}$

The *truth table, Boolean cube* and "*Difference Equation*" defining the local rule $\boxed{60}$ along with a space-time pattern (with a single red-pixel initial state) exhibited in Table 5 of [Chua *et al.*, 2003] are reproduced in Fig. 1 for the reader's convenience. In this paper, it is more instructive to recast the Difference Equation defining $\boxed{60}$ into an *equivalent* difference equation involving only a *mod 2 addition* $\oplus$ (defined in Table 13).[2]

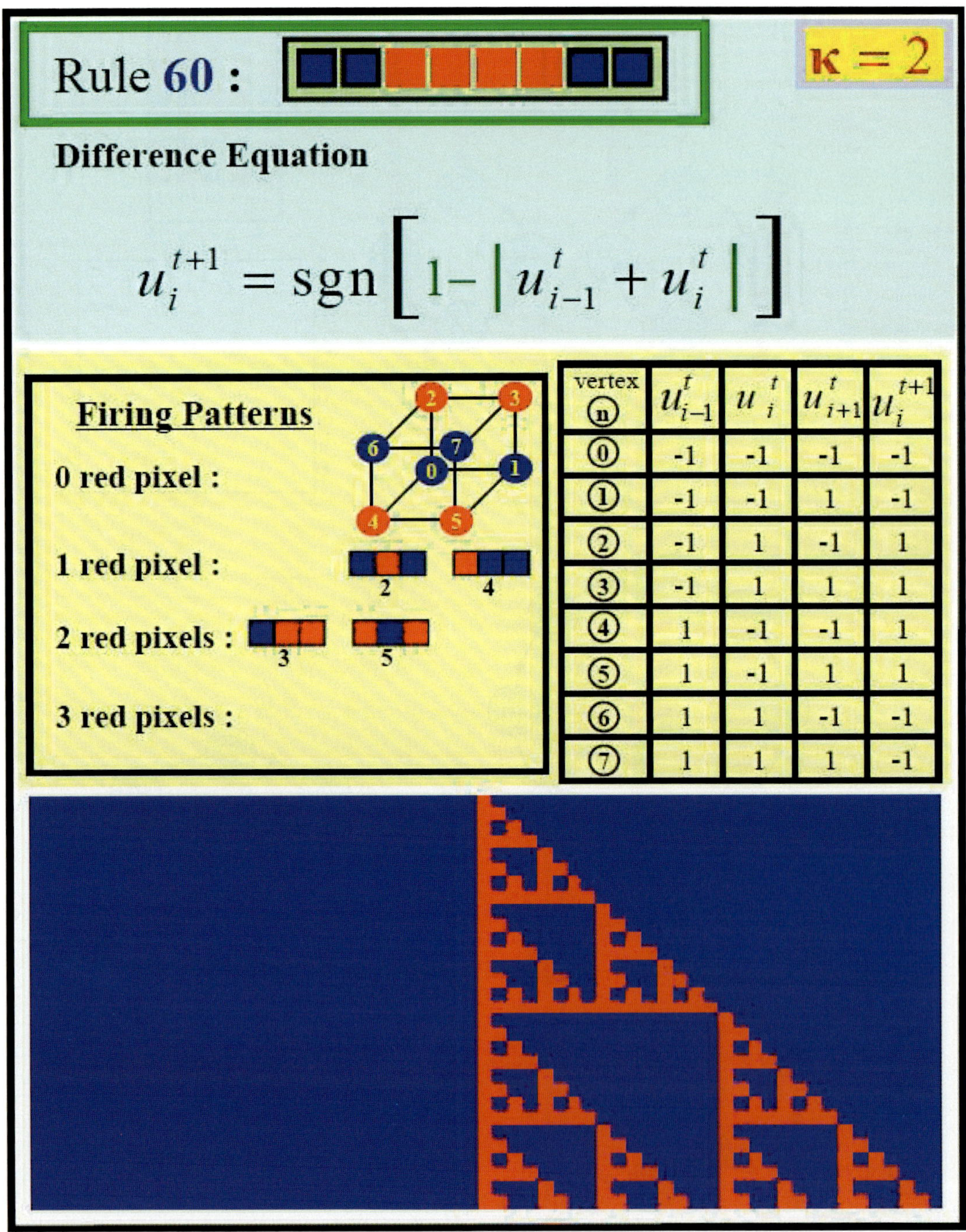

$$u_i^{t+1} = \mathrm{sgn}\left[1 - \left| u_{i-1}^t + u_i^t \right|\right]$$

vertex (n)	u_{i-1}^t	u_i^t	u_{i+1}^t	u_i^{t+1}
0	-1	-1	-1	-1
1	-1	-1	1	-1
2	-1	1	-1	1
3	-1	1	1	1
4	1	-1	-1	1
5	1	-1	1	1
6	1	1	-1	-1
7	1	1	1	-1

Fig. 1.   Truth table, Boolean cube, Difference Equation, and space-time pattern of local rule $\boxed{60}$.

---

[2]The *mod 2 operation* $\mathbf{x}_i \oplus \mathbf{x}_j$ between *two* binary variables is also called an *exclusive OR* operation in *mathematical logic*, and denoted by $\mathbf{x}_i \oplus \mathbf{x}_j \triangleq \mathbf{x}_i$ *XOR* $\mathbf{x}_j$.

Table 13. Table defining $\mathbf{x}_i \oplus \mathbf{x}_j \overset{\Delta}{=} \mathbf{x}_i \; XOR \; \mathbf{x}_j$.

$\oplus$	0	1
0	0	1
1	1	0

state transition matrix:

$$
\underbrace{\begin{bmatrix} x_0^{t+1} \\ x_1^{t+1} \\ x_2^{t+1} \\ x_3^{t+1} \\ \vdots \\ x_{L-2}^{t+1} \\ x_{L-1}^{t+1} \end{bmatrix}}_{\mathbf{x^{t+1}}} = \underbrace{\begin{bmatrix} 1 & 0 & 0 & 0 & 0 & \cdots & 1 \\ 1 & 1 & 0 & 0 & 0 & \cdots & 0 \\ 0 & 1 & 1 & 0 & 0 & \cdots & 0 \\ 0 & 0 & 1 & 1 & 0 & \cdots & 0 \\ \vdots & & & & \ddots & & \vdots \\ 0 & 0 & 0 & 0 & \cdots & 0 & 0 \\ 0 & 0 & 0 & 0 & \cdots & 1 & 0 \\ 0 & 0 & 0 & 0 & \cdots & 1 & 1 \end{bmatrix}}_{\mathbf{M}(\boxed{60})} \underbrace{\begin{bmatrix} x_0^{t} \\ x_1^{t} \\ x_2^{t} \\ x_3^{t} \\ \vdots \\ x_{L-2}^{t} \\ x_{L-1}^{t} \end{bmatrix}}_{\mathbf{x^{t}}}
$$

$$(5)$$

Substituting $u_i = 2x_i - 1$ from Eq. (4) of [Chua et al., 2005a] for u_i in the Difference Equation for $\boxed{60}$, we obtain

$$
\begin{aligned}
2\mathbf{x}_i^{t+1} - 1 &= \operatorname{sgn}[1 - |2\mathbf{x}_{i-1}^t + 2\mathbf{x}_i^t - 2|] \\
&= \operatorname{sgn}[1 - 2|\mathbf{x}_{i-1}^t + \mathbf{x}_i^t - 1|] \quad (1)
\end{aligned}
$$

Simplifying Eq. (1) using Table 13, we obtain the following equivalent Difference Equation[3]:

$$
\text{Rule } \boxed{60} \quad \boxed{\begin{aligned} \mathbf{x}_i^{t+1} &= (\mathbf{x}_{i-1}^t + \mathbf{x}_i^t) \bmod (2) \\ &= \mathbf{x}_{i-1}^t \oplus \mathbf{x}_i^t \end{aligned}} \quad (2)
$$

3.1. Rule $\boxed{60}$ has no Isles of Eden

A cursory glimpse at the basin-tree diagram of rule $\boxed{60}$ in Table 9 reveals that all bit strings converge to an attractor for $3 \leq L \leq 8$. We now prove this property is true for all L.

Theorem 3.1. *Rule $\boxed{60}$ does not have any Isle of Eden.*

Proof. It follows from Eq. (2) that an arbitrary bit string

$$
\mathbf{x^t} = (\mathbf{x_0^t} \quad \mathbf{x_1^t} \quad \mathbf{x_2^t} \quad \cdots \quad \mathbf{x_{L-1}^t}) \quad (3)
$$

at time "t" is *linearly* related (mod (2)) to its *image* (via rule $\boxed{60}$)

$$
\mathbf{x^{t+1}} = (\mathbf{x_0^{t+1}} \quad \mathbf{x_1^{t+1}} \quad \mathbf{x_2^{t+1}} \quad \cdots \quad \mathbf{x_{L-1}^{t+1}}) \quad (4)
$$

at time "$t+1$" via an $L \times L$ *circulant matrix* [Davis, 1979] $\mathbf{M}(\boxed{60})$, *henceforth called the local time-1*

The matrix algebra in Eq. (5) is implemented in mod (2) addition, as defined in Table 13. We omit the mod (2) notation to avoid clutter.

Note the *diagonal* elements of the circulant matrix $\mathbf{M}(\boxed{60})$ are all equal to *one*. Observe also the elements directly below the diagonal of $\mathbf{M}(\boxed{60})$ are all equal to *one*. All other elements are zero, except for the top rightmost element, which is equal to *one*. It follows from this special structure that the *leftmost* column of $\mathbf{M}(\boxed{60})$ is equal to the *mod 2 sum* of the remaining L-1 columns. Since the columns of $\mathbf{M}(\boxed{60})$ are not *linearly independent*, mod (2), it follows that $\mathbf{M}$ does *not* have an inverse. Since the bit string $\mathbf{x^{t+1}}$ on the left side of Eq. (2) does not have a *unique preimage*, it follows that the bit string $\mathbf{x^t}$ is *not* an *Isle of Eden* of $\boxed{60}$: Since $\mathbf{x^t}$ is an arbitrary bit string, it follows that $\boxed{60}$ cannot possess an *Isle of Eden* for any L. ∎

3.2. Period of rule $\boxed{60}$ grows with L

Since rule $\boxed{60}$ does *not* have Isles of Eden, all bit strings of $\boxed{60}$ must converge to some *period-T attractors* whose *period "T"* is bounded by

$$1 \leq T \leq T_{\max} \quad (6)$$

where $T_{\max} = 2^L$ as defined in Eq. (6) in [Chua et al., 2007]. As an example, the period T of an attractor of $\boxed{60}$ is listed in Table 14 for $3 \leq L \leq 100$. Observe that the period T for *some L* (e.g. $L = 47, 49, 53$, etc.) is not listed in Table 14 because it is so large that it exceeded the maximum simulation time allocated. A bit string belonging to one of the many period-T attractors for $3 \leq L \leq 25$ is

---

[3]Although one could derive Eq. (2) directly from the truth table, we opted for our analytic approach to illustrate how our analytical formula can be reduced to simpler equivalent formulas for some local rules.

Table 14.   Period "T" of attractors of local rule $\boxed{60}$ for $3 \le L \le 100$.

$\boxed{60}$ L	Isles of Eden	Period-T Attractors	$\boxed{60}$ L	Isles of Eden	Period-T Attractors	$\boxed{60}$ L	Isles of Eden	Period-T Attractors
3		$T = 3, ...$	36		$T = 252, ...$	69		
4		$T = 1, ...$	37			70		$T = 8190, ...$
5		$T = 15, ...$	38		$T = 19418, ...$	71		
6		$T = 6, ...$	39		$T = 4095, ...$	72		$T = 504, ...$
7		$T = 7, ...$	40		$T = 120, ...$	73		$T = 511, ...$
8		$T = 1, ...$	41		$T = 41943, ...$	74		
9		$T = 63, ...$	42		$T = 126, ...$	75		$T = 1048575, ..$
10		$T = 30, ...$	43		$T = 5461, ...$	76		$T = 38836, ...$
11		$T = 341, ...$	44		$T = 1364, ...$	77		
12		$T = 12, ...$	45		$T = 4095, ...$	78		$T = 8190, ...$
13		$T = 819, ...$	46		$T = 4094, ...$	79		
14		$T = 14, ...$	47			80		$T = 240, ...$
15		$T = 15, ...$	48		$T = 48, ...$	81		
16		$T = 1, ...$	49			82		$T = 83886, ...$
17		$T = 255, ...$	50		$T = 51150, ...$	83		
18		$T = 126, ...$	51		$T = 255, ...$	84		$T = 252, ...$
19		$T = 9709, ...$	52		$T = 3276, ...$	85		$T = 255, ...$
20		$T = 60, ...$	53			86		$T = 10922, ...$
21		$T = 63, ...$	54		$T = 27594, ...$	87		
22		$T = 682, ...$	55		$T = 1048575, ..$	88		$T = 2728, ...$
23		$T = 2047, ...$	56		$T = 56, ...$	89		$T = 2047, ...$
24		$T = 24, ...$	57		$T = 29127, ...$	90		$T = 8190, ...$
25		$T = 25575, ...$	58		$T = 950214, ...$	91		$T = 4095, ...$
26		$T = 1638, ...$	59			92		$T = 8188, ...$
27		$T = 13797, ...$	60		$T = 60, ...$	93		$T = 1023, ...$
28		$T = 28, ...$	61			94		
29		$T = 475107, ...$	62		$T = 62, ...$	95		
30		$T = 30, ...$	63		$T = 63, ...$	96		$T = 96, ...$
31		$T = 31, ...$	64		$T = 1, ...$	97		
32		$T = 1, ...$	65		$T = 4094, ...$	98		
33		$T = 1023, ...$	66		$T = 2046, ...$	99		
34		$T = 510, ...$	67			100		$T = 102300, ...$
35		$T = 4095, ...$	68		$T = 1020, ...$			

Table 15. Bit strings for generating a period-T attractor of rule $\boxed{60}$.

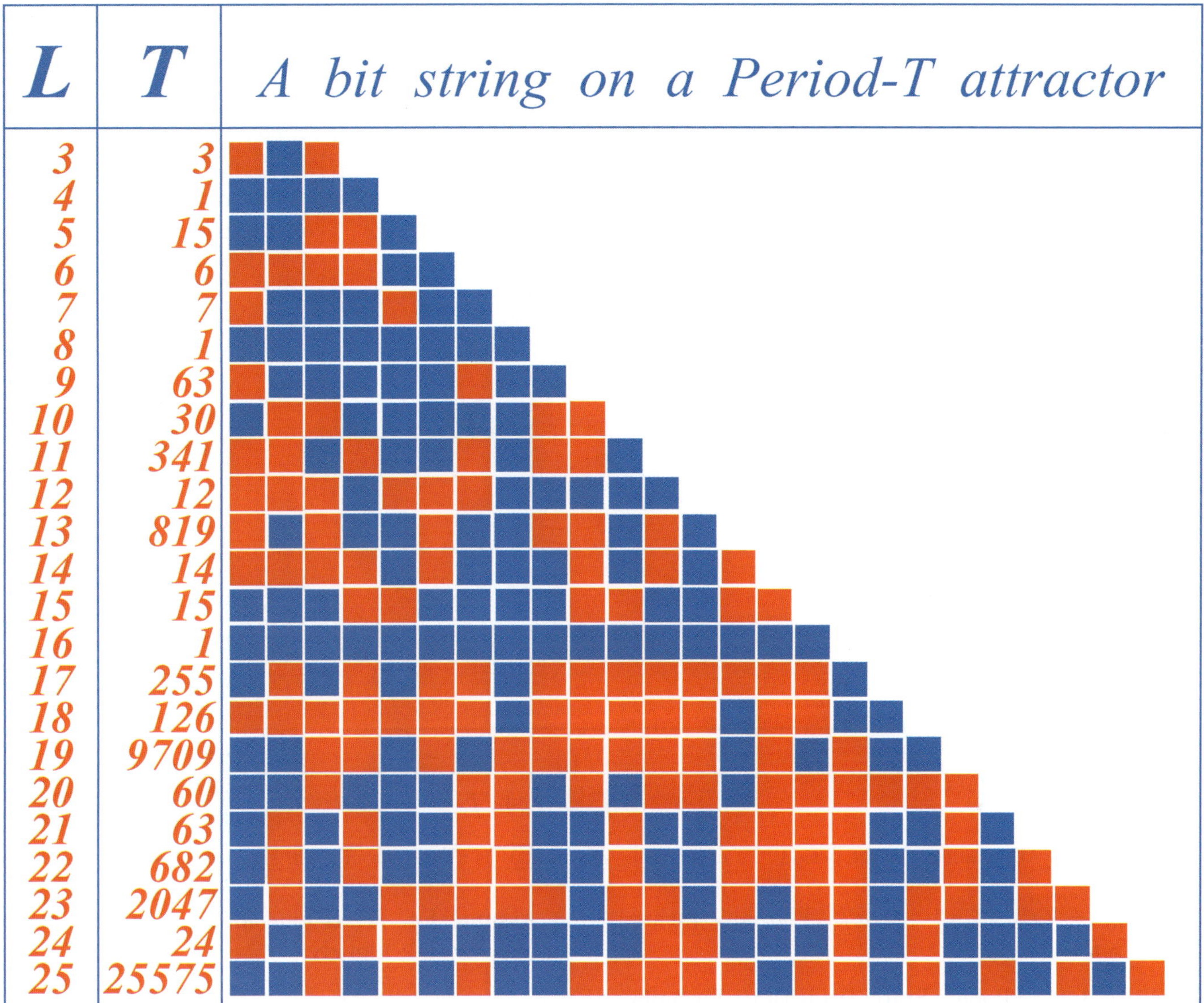

L	T	A bit string on a Period-T attractor
3	3	
4	1	
5	15	
6	6	
7	7	
8	1	
9	63	
10	30	
11	341	
12	12	
13	819	
14	14	
15	15	
16	1	
17	255	
18	126	
19	9709	
20	60	
21	63	
22	682	
23	2047	
24	24	
25	25575	

given in Table 15. For example, the bit string listed for $L = 3$ corresponds to the *period-3* attractor listed in Gallery 60-1 of Table 9. The bit string listed for $L = 5$ corresponds to node ⑥ of Gallery 60-3 of Table 9, of period-15 attractor. The bit string listed for $L = 6$ corresponds to node ㉚ in the second period-6 attractor shown on the left of Gallery 60-4 of Table 9. The bit string listed for $L = 7$ corresponds to node ㊇ in the period-7 attractor of $\boxed{60}$ shown on the top left of Gallery 60-6 of Table 9.

As examination of Table 14 shows that unlike the period-1, -2 and -3 local rules listed in Tables 7 and 8 shown in [Chua *et al.*, 2007], which have a relatively small period, and independent of L, the period T of rule $\boxed{60}$ increases at an *exponential* rate as a function of L, as

depicted in Fig. 2. Such exponential growth of T as a function of L is a signature of all *complex Bernoulli* rules in Table 11, and *hyper-Bernoulli* rules in Table 12, where both tables are given in [Chua *et al.*, 2007].

3.3. *Global state-transition formula for rule* $\boxed{60}$

The state transition formula given in Fig. 1 and Eq. (2) for rule $\boxed{60}$ is *local in time* in the sense that it generates from a bit string $\mathbf{x}^t = (x_0^t \; x_1^t \; x_2^t \; \cdots \; x_{L-1}^t)$ at time "t" the next bit string $\mathbf{x}^{t+1} = (x_0^{t+1} \; x_1^{t+1} \; x_2^{t+1} \; \cdots \; x_{L-1}^{t+1})$ at time "$t+1$". Our next theorem gives an explicit formula which is *global in time* in the sense that it generates a bit

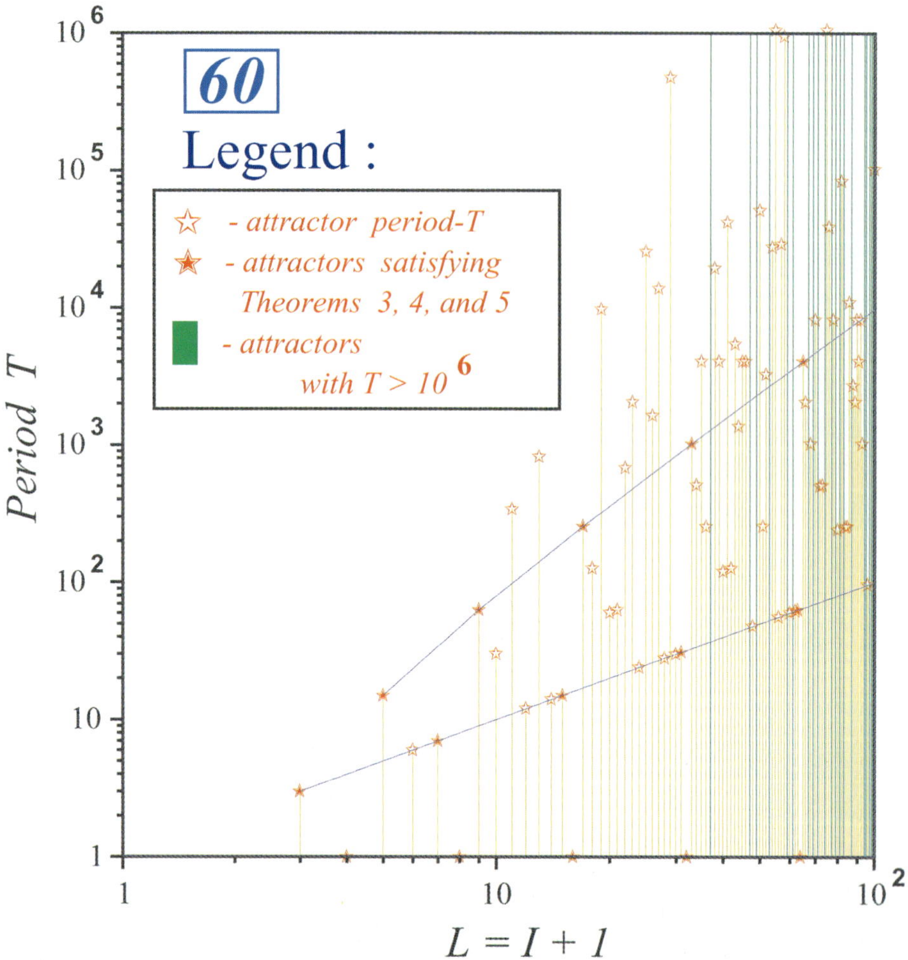

Fig. 2.   Dependence of the period "T" of attractor of rule $\boxed{60}$ as a function of L (in logarithmic scale).

string $\mathbf{x}_0^n = (x_0^n \; x_1^n \; x_2^n \; \cdots \; x_{L-1}^n)$ at any future time $n > t$.

Theorem 3.2.   Global State-Transition Formula for $\boxed{60}$.

Each pixel x_i^n at time $n > t$ is determined from "$n+1$" initial pixels $x_{i-n}^0, x_{i-n+1}^0, \ldots, x_{i+n-1}^0, x_{i+n}^0$ at $t = 0$ via the binomial formula.

$$\boxed{x_i^n = \sum_{k=0}^{n} \frac{n!}{k!(n-k)!} \bullet x_{i-n+k}^0 \quad \mathrm{mod}\,(2)} \quad (7)$$

Proof.   Apply *mathematical induction* as follows:

(a) $n = 1$

Applying $n = 1$ in Eq. (7), we obtain[4]

$$x_i^1 = x_{i-1}^0 + x_i^0 \quad \mathrm{mod}\,(2) \quad (8)$$

which is Eq. (2) for $t = 0$.

(b) Assume Eq. (7) is true for $n = m$ (induction hypothesis); namely,

$$x_i^m = \sum_{k=0}^{m} \frac{m!}{k!(m-k)!} \bullet x_{i-m+k}^0 \quad \mathrm{mod}\,(2) \quad (9)$$

We must show that incrementing "m" to "$m+1$" in Eq. (9) gives Eq. (7) with $n = m + 1$.

Substituting Eq. (9) to Eq. (2), we obtain

$$x_i^{m+1} = x_{i-1}^m + x_i^m \quad \mathrm{mod}\,(2)$$

$$= \sum_{k=0}^{m} \frac{m!}{k!(m-k)!} x_{(i-1)-m+k}^0$$

$$+ \sum_{k=0}^{m} \frac{m!}{k!(m-k)!} x_{i-m+k}^0 \quad \mathrm{mod}\,(2)$$

$$(10)$$

---

[4] Recall the factorial notation $0! \triangleq 1$.

Changing symbol "m" on the right-hand side of Eq. (10) to $m' - 1$ gives

$$\sum_{k=0}^{m'-1} \frac{(m'-1)!}{k!(m'-1-k)!} x_{i-m'+k}^0$$

$$+ \sum_{k=0}^{m'-1} \frac{(m'-1)!}{k!(m'-1-k)!} x_{i-m'+1+k}^0 \quad \mathrm{mod}\,(2) \tag{11}$$

Changing symbol k in the second summation terms in Eq. (11) to $k' - 1$ gives

$$\sum_{k=0}^{m'-1} \frac{(m'-1)!}{k!(m'-1-k)!} x_{i-m'+k}^0$$

$$+ \sum_{k'=1}^{m'} \frac{(m'-1)!}{(k'-1)!(m'-k')!} x_{i-m'+k'}^0 \tag{12}$$

Changing the dummy index k' in Eq. (12) back to k, we obtain

$$\left[\sum_{k=0}^{m'-1} \frac{(m'-1)!}{k!(m'-1-k)!} + \sum_{k=1}^{m'} \frac{(m'-1)!}{(k-1)!(m'-k)!} \right]$$

$$\times x_{i-m'+k}^0 \tag{13}$$

The terms inside the bracket can be simplified by observing for $k = 1$ to $m' - 1$, we have

$$\frac{(m'-1)!}{k!(m'-1-k)!} + \frac{(m'-1)!}{(k-1)!(m'-k)!}$$

$$= \frac{(m'-1)!}{(k-1)!(m'-1-k)!} \left[\frac{1}{k} + \frac{1}{(m'-k)} \right]$$

$$= \frac{(m'-1)!}{(k-1)!(m'-1-k)!} \left[\frac{(m'-k)+k}{k(m'-k)} \right]$$

$$= \frac{m'!}{k!(m'-k)!} \tag{14}$$

Moreover, when $k = 0$ and $k = m'$, Eq. (14) gives the same value as the first term on the left of Eq. (13), and the last term on the right of Eq. (13), respectively. Substituting back $m = m' - 1$ in Eq. (14), and making use of Eqs. (10)–(14), we obtain

$$x_i^{m+1} = \sum_{k=0}^{m+1} \frac{(m+1)!}{k!(m+1-k)!} \bullet x_{i-(m+1)+k}^0 \quad \mathrm{mod}\,(2) \tag{15}$$

which is identical to incrementing m in the induction hypothesis (9) to $m + 1$. ∎

Table 16 gives the global state-transition formula (7) of rule $\boxed{60}$ for $n = 1, 2, 3, 4$ and 5. Observe that the coefficients $\binom{n}{k}$ for each time $n \geq 1$ is

Table 16.   Global state-transition formula for rule $\boxed{60}$ for $1 \leq n \leq 5$.

n	$x_i^n = \displaystyle\sum_{k=0}^{n} \frac{n!}{k!(n-k)!} \cdot x_{i-n+k}^0 \quad \mathbf{mod(2)}$
1	$x_i^1 = x_{i-1}^0 + x_i^0 \quad \mathbf{mod(2)}$
2	$x_i^2 = x_{i-2}^0 + 2x_{i-1}^0 + x_i^0 \quad \mathbf{mod(2)}$
3	$x_i^3 = x_{i-3}^0 + 3x_{i-2}^0 + 3x_{i-1}^0 + x_i^0 \quad \mathbf{mod(2)}$
4	$x_i^4 = x_{i-4}^0 + 4x_{i-3}^0 + 6x_{i-2}^0 + 4x_{i-1}^0 + x_i^0 \quad \mathbf{mod(2)}$
5	$x_i^5 = x_{i-5}^0 + 5x_{i-4}^0 + 10x_{i-3}^0 + 10x_{i-2}^0 + 5x_{i-1}^0 + x_i^0 \quad \mathbf{mod(2)}$
$\cdots$	$\cdots$

Table 17. Table of $\binom{n}{k} \triangleq (n!/k!(n-k)!)$, $n = 1, 2, \ldots, 11$, $k = 0, 1, 2, \ldots, 11$.

n	k											
	0	1	2	3	4	5	6	7	8	9	10	11
1	1	1										
2	1	2	1									
3	1	3	3	1								
4	1	4	6	4	1							
5	1	5	10	10	5	1						
6	1	6	15	20	15	6	1					
7	1	7	21	35	35	21	7	1				
8	1	8	28	56	70	56	28	8	1			
9	1	9	36	84	126	126	84	36	9	1		
10	1	10	45	120	210	252	210	120	45	10	1	
11	1	11	55	165	330	462	462	330	165	55	11	1

identical to the binomial coefficients in the expansion of $(x + y)^n$, as listed in Table 17 for $n = 1, 2, \ldots, 11$. These binomial coefficients are repackaged in Table 18 into the form of a Pascal's triangle where each coefficient under the pyramid is obtained by adding adjacent left and right coefficients above it.

The "skewed" (consisting of a system of slanted parallel lines terminating on arrowheads) grid represents a frame of reference (i, t) for calculating the binary variable $\mathbf{x}_i^{t+1} = \mathbf{x}_{i-1}^t \oplus \mathbf{x}_i^t$.

Taking the mod (2) equivalent of each coefficient in Table 16, we obtain the more compact but equivalent expansion in Table 19 where all nonzero terms correspond to those in Table 16 with "*odd number*" coefficients. The equivalent mod (2) coefficients are repackaged in Table 20. Observe that Table 20 can be obtained from Table 18 by replacing each odd (respectively, even) coefficient in Table 18 by a *one* (respectively, a *zero*). If we fill in the blank slots to the left and to the right of each row outside

the Pascal's triangle with "zeros", and rotate the "skewed" grid by 45° counterclockwise, we would obtain the pyramidal "fractal" space-time pattern of rule $\boxed{60}$ in Table 21, which is identical to that shown in the bottom of Fig. 1, where the initial configuration consists of a single red bit at the center, as in [Wolfram, 2002].

Example 3.1. Table 22 shows the *space-time pattern* obtained from the *global* state-transition formula of rule $\boxed{60}$ in (a) where the initial configuration consists of a single red bit at the center.

The corresponding pattern obtained from the *local* state-transition formula is shown in (b). They are identical, as expected. The minor differences in the graphics and color is due to the differences in the softwares used to generate these patterns.

Example 3.2. Table 23 shows the corresponding results where the initial configuration consists of a string of *random* bits.

Table 18.   Binomial coefficients $\binom{n}{k}$ repackaged into a Pascal's triangle.

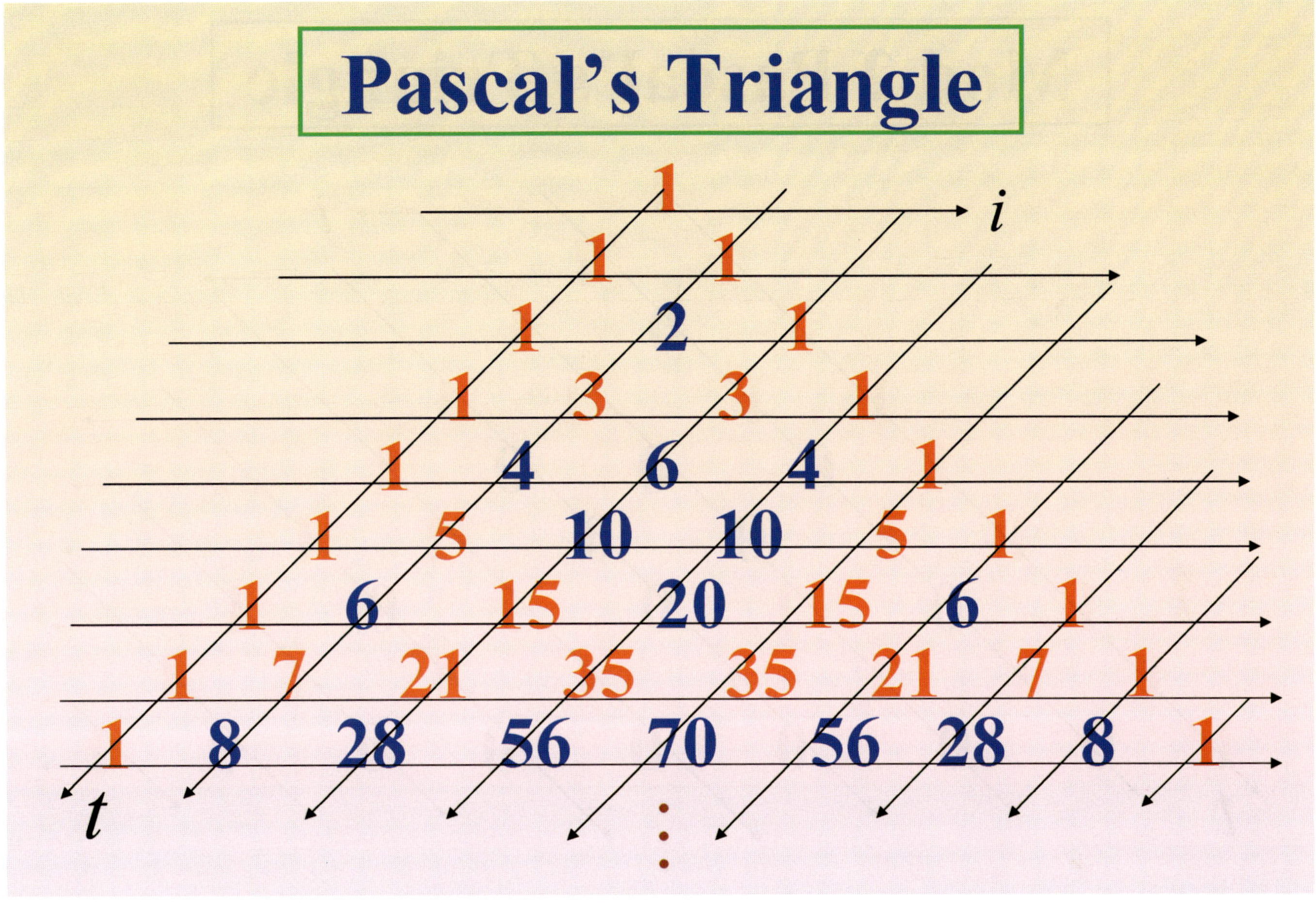

Table 19.   Compact global state-transition formula for rule $\boxed{60}$ for $1 \leq n \leq 5$.

n	$x_i^n = \sum\limits_{k=0}^{n} \dfrac{n!}{k!(n-k)!} \cdot x_{i-n+k}^0 \quad \mathbf{mod(2)}$
1	$x_i^1 = x_{i-1}^0 + x_i^0 \quad \mathbf{mod(2)}$
2	$x_i^2 = x_{i-2}^0 \qquad\qquad + x_i^0 \quad \mathbf{mod(2)}$
3	$x_i^3 = x_{i-3}^0 + x_{i-2}^0 + x_{i-1}^0 + x_i^0 \quad \mathbf{mod(2)}$
4	$x_i^4 = x_{i-4}^0 \qquad\qquad\qquad\qquad + x_i^0 \quad \mathbf{mod(2)}$
5	$x_i^5 = x_{i-5}^0 + x_{i-4}^0 \qquad\qquad\quad + x_{i-1}^0 + x_i^0 \quad \mathbf{mod(2)}$
...	...

Table 20.   Mod 2 binomial coefficients $\binom{n}{k}$ repackaged into a mod 2 Pascal's triangle.

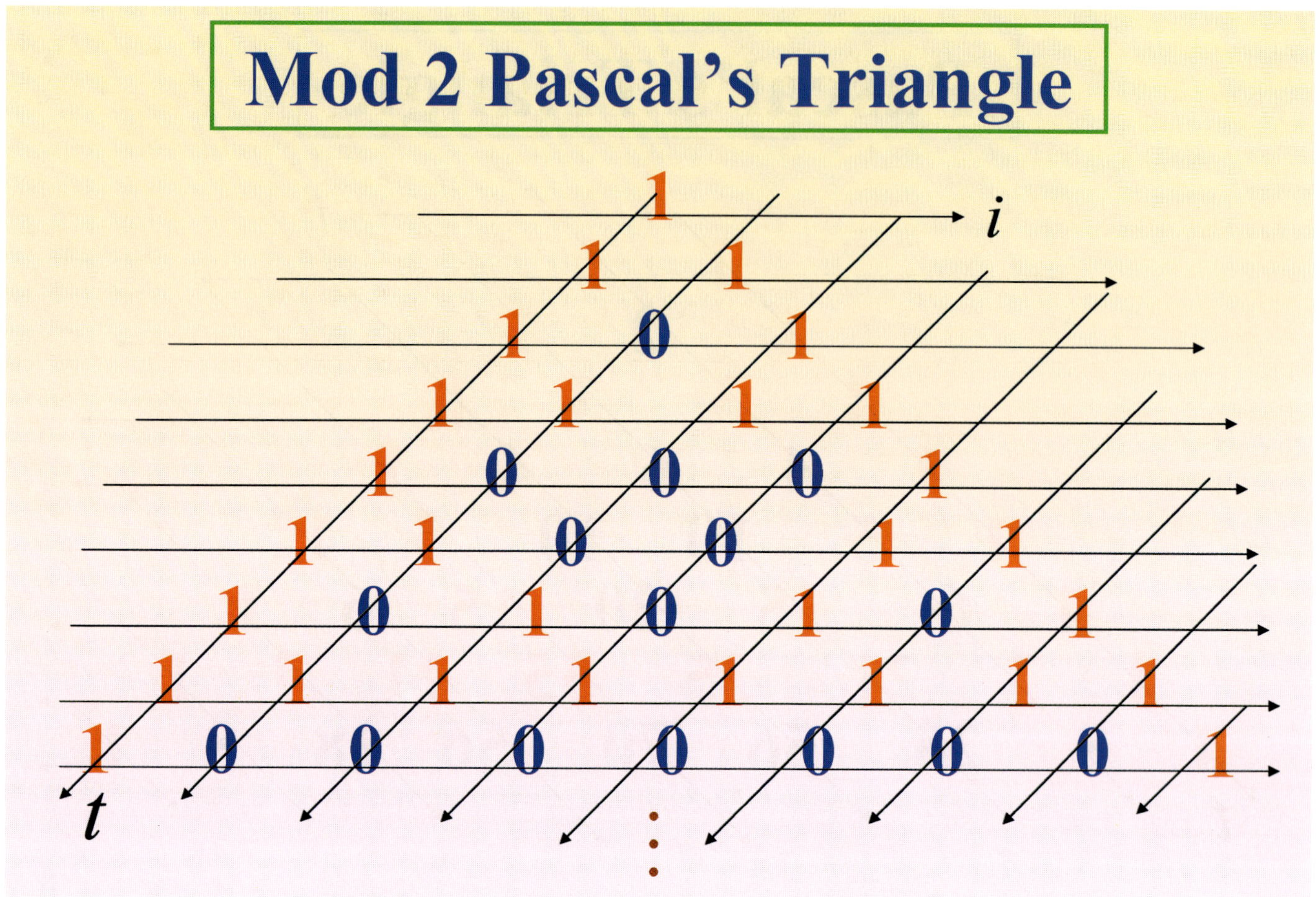

3.4. *Periodicity constraints of rule* $\boxed{60}$

Theorem 3.1 implies that all bit strings of rule $\boxed{60}$ must converge to a period-T attractor, where $T \leq T_{\max} \leq 2^L$. We will prove in this subsection that for finite length $L \triangleq I + 1$, the period T must satisfy certain constraints. Such *periodicity constraints* are useful on many occasions, such as verifying whether certain periodic orbit can exist or to generate new periodic orbits, etc. The proof of many of these results depend on the following easily verifiable identities:

Lemma 3.1. Binomial Coefficient Identities.
   If $n = 2^m$, where $m \geq 2$, then the following identities hold:

$$\text{(i)} \quad \binom{n-1}{k} = 1 \bmod (2), \quad \text{for } k = 0, 1, 2, \ldots, n-1 \tag{16}$$

$$\text{(ii)} \quad \binom{n}{k} = \begin{cases} 0 \bmod (2), & \text{for } k = 1, 2, \ldots, n-1 \\ 1 \bmod (2), & \text{for } k = 0, n \end{cases} \tag{17}$$

$$\text{(iii)} \quad \binom{n+1}{k} = \begin{cases} 0 \bmod (2), & \text{for } k = 2, 3, \ldots, n-1 \\ 1 \bmod (2), & \text{for } k = 0, 1, n, n+1 \end{cases} \tag{18}$$

where

$$\binom{n}{k} \triangleq \frac{n!}{k!(n-k)!} \tag{19}$$

Table 21. **Mod (2)** coefficients for global-transition formula for $\boxed{60}$ for $1 \le n \le 8$.

n	-8	-7	-6	-5	-4	-3	-2	-1	0	1	2	3	4	5	6	7	8
1	0	0	0	0	0	0	0	0	1	1	0	0	0	0	0	0	0
2	0	0	0	0	0	0	0	0	1	0	1	0	0	0	0	0	0
3	0	0	0	0	0	0	0	0	1	1	1	1	0	0	0	0	0
4	0	0	0	0	0	0	0	0	1	0	0	0	1	0	0	0	0
5	0	0	0	0	0	0	0	0	1	1	0	0	1	1	0	0	0
6	0	0	0	0	0	0	0	0	1	0	1	0	1	0	1	0	0
7	0	0	0	0	0	0	0	0	1	1	1	1	1	1	1	1	0
8	0	0	0	0	0	0	0	0	1	0	0	0	0	0	0	0	1

Theorem 3.3. Periodicity Condition: $L = 2^m$.

For $L = 2^m$, $m = 2, 3, 4, \ldots$, rule $\boxed{60}$ has a global period-1 attractor Γ; namely,

$$\mathbf{x}(\Gamma) = \underbrace{(0 \quad 0 \quad 0 \quad \cdots \quad 0)}_{L = 2^m} \tag{20}$$

All *bit strings not belonging to the attractor Γ converge to Γ in at most 2^m iterations.*

Proof. Let $n = 2^m$ in the global state-transition formula (7). It follows from Eqs. (17) and (19) that

$$\frac{n!}{k!(n-k)!} \; \mathrm{mod}\,(2) = \begin{cases} 0, & \text{for } k = 1, 2, \ldots, n-1 \\ 1, & \text{for } k = 0, n = 2^m \end{cases} \tag{21}$$

It follows from Eq. (21) and the global state-transition formula (7) that x_i^n contains only two nonzero terms; namely, the leftmost and the rightmost terms. Hence,

$$x_{(i \ \mathrm{mod}\,(L))}^n = x_{((i-n) \ \mathrm{mod}\,(L))}^0 + x_{(i \ \mathrm{mod}\,(L))}^0 \quad \mathrm{mod}\,(2) \tag{22}$$

where $n = 2^m$. Substituting $i = n = 2^m = L$ in Eq. (22), we obtain

$$\begin{aligned} x_{(n \ \mathrm{mod}\,(L))}^n &= x_{((n-n) \ \mathrm{mod}\,(L))}^0 + x_{(n \ \mathrm{mod}\,(L))}^0 \quad \mathrm{mod}\,(2) \\ &= x_0^0 + x_{(L \ \mathrm{mod}\,(L))}^0 \quad \mathrm{mod}\,(2) \\ &= 2x_0^0 \quad \mathrm{mod}\,(2) \\ &= 0 \end{aligned} \tag{23}$$

because $x_0^0 = x_{(L \ \mathrm{mod}\,(L))}^0$.

Since x_i^0 is arbitrary, it follows that all bit strings must converge to a period-T orbit $\Gamma_T(\boxed{60})$ at most 2^{m-1} iterations, see Eq. (7) in [Chua *et al.*, 2007]. ∎

Corollary 3.1. Corollary to Theorem 3.2.

A bit string

$$\mathbf{x}^0 = (x_0^0 \quad x_1^0 \quad x_2^0 \quad \cdots \quad x_{L-1}^0) \tag{24}$$

of length $L = I + 1$ (under periodic boundary condition) is a period-n attractor *of local rule $\boxed{60}$ if,*

Table 22.   Space-time pattern of the rule $\boxed{60}$ with red central bit initial configuration: (a) from global state-transition formula; (b) from local state-transition formula.

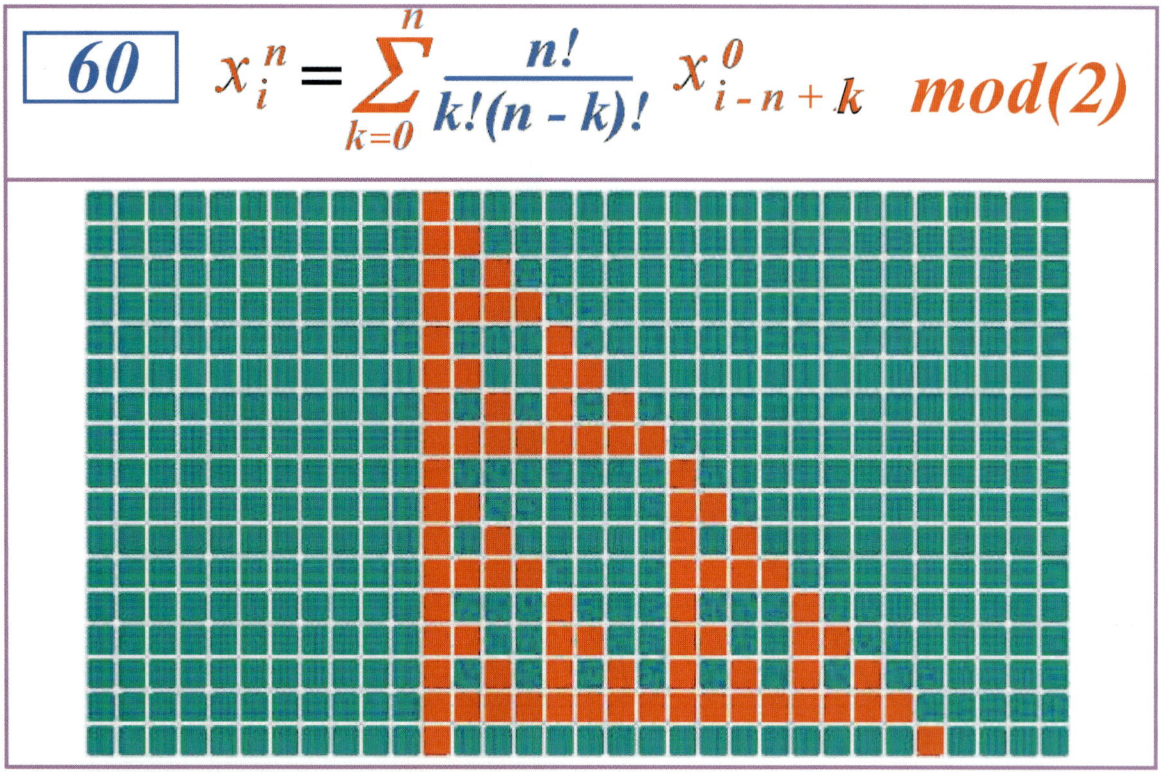

$$\boxed{60} \quad x_i^n = \sum_{k=0}^{n} \frac{n!}{k!(n-k)!}\, x_{i-n+k}^0 \ \ mod(2)$$

(a)

$$\boxed{60} \quad x_i^{n+1} = x_{i-1}^n + x_i^n \ \ mod(2)$$

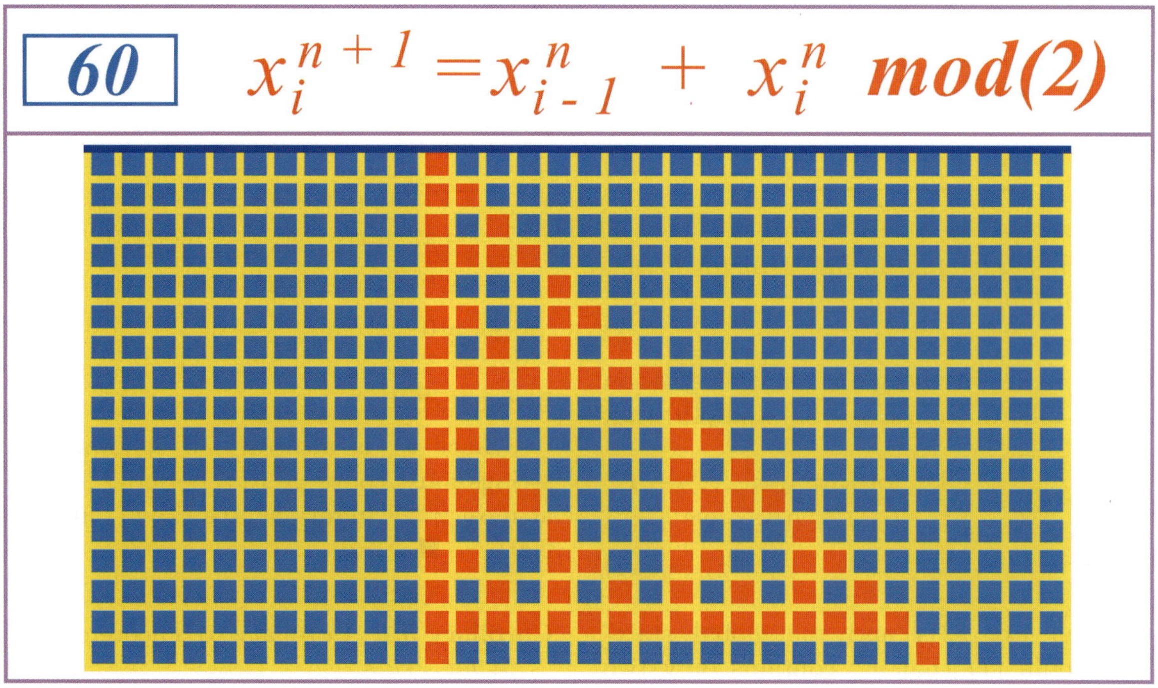

(b)

Table 23.   Space-time pattern of rule $\boxed{60}$ obtained from random initial state generated by: (a) global state-transition formula; and (b) local state-transition formula.

$$\boxed{\textbf{60}}\quad x_i^n = \sum_{k=0}^{n} \frac{n!}{k!(n-k)!}\, x_{i-n+k}^0 \quad \textbf{\textit{mod}}(2)$$

(a)

$$\boxed{\textbf{60}}\quad x_i^{n+1} = x_{i-1}^n + x_i^n \quad \textbf{\textit{mod}}(2)$$

(b)

and only *if, the* periodicity condition

$$x_{(i \bmod (L))}^n = x_i^0 = \sum_{k=0}^{n} \frac{n!}{k!(n-k)!} \times x_{((i-n+k)\bmod(L))}^0 \quad \bmod(2)$$

(25)

is satisfied for all i.

Proof. Follows directly from *Theorem 3.2* and the periodic boundary condition. ∎

The *periodicity constraint* equation (25) is applicable to *any* period-n attractor of rule $\boxed{60}$. The "**mod (L)**" operation attached to the subscript index of x^0 is just a mathematically precise algorithm for implementing the *periodic boundary conditions*. It is also mathematically equivalent to *concatenating* replicas of the L-bit string $x_0\ x_1\ x_2\ \cdots\ x_I$ *ad infinitum*; namely,

$$\cdots \underbrace{x_0\ x_1\ x_2\ \cdots\ x_I}_{L\text{ bits}}\ \underbrace{x_0\ x_1\ x_2\ \cdots\ x_I}_{L\text{ bits}}$$
$$\underbrace{x_0\ x_1\ x_2\ \cdots\ x_I}_{L\text{ bits}}\cdots$$

(26)

where $L = I + 1$.

In the special case where

$$n = 2^m - 1$$

(27)

all binomial coefficients in Eq. (25) are equal to *unity*, in view of the Binomial Coefficient Lemma;

namely,

$$\boxed{\frac{n!}{k!(n-k)!} \mod (2) = 1, \quad k = 0, 1, 2, \ldots, n} \tag{28}$$

$$\boxed{\begin{array}{l} \text{Valid if} \\ n = 2^m - 1 \end{array} \; x^n_{(i \, \mod (L))} = x^0_i = \sum_{k=0}^{2^m - 1} x^0_{((i-n+k) \, \mod (L))} \quad \mod (2) \qquad \text{for all } i.} \tag{30}$$

If we impose the additional constraint $L = n = 2^m - 1$, then we obtain the following simple method for finding period-$(2^m - 1)$ attractors:

Theorem 3.4. Periodicity Condition: $L = 2^m - 1$. *Rule* $\boxed{60}$ *has a* period-n attractor *where* $n = 2^m - 1$ and $L = 2^m - 1$ if, and only if,

$$\boxed{\begin{array}{l} \text{Valid for} \\ n = 2^m - 1 \\ L = 2^m - 1 \end{array} \; \sum_{i=0}^{L-1} x^0_i \mod (2) = 0} \tag{31}$$

Proof. Let us list all terms from Eq. (30) as follows:

$$x^0_{i \, \mod (L)} = x^0_{((i-n) \, \mod (L))} + x^0_{((i-n+1) \, \mod (L))}$$
$$+ \cdots + x^0_{((i-n+(n-1)) \, \mod (L))}$$
$$+ x^0_{((i-n+n) \, \mod (L))} \quad \mod (2) \tag{32}$$

Since "i" is an arbitrary index in Eq. (32), let it be "n". Substituting $i = n$ in Eq. (32), we obtain

$$x^0_0 = x^0_0 + x^0_1 + \cdots + x^0_{(n-1) \, \mod (L)}$$
$$+ x^0_{(n) \, \mod (L)} \quad \mod (2) \tag{33}$$

Observe next that for $n = L = 2^m - 1$, we have

$$(n) \mod (L) = 0, \quad ((n-1) \mod (L)) = L - 1,$$
$$((n-2) \mod (L)) = L - 2.$$

Observe also that $L = 2^m - 1$ implies that $L - 2$, $L - 4$, etc. are *odd* numbers. Substituting these $\mod (L)$ equivalent indices into Eq. (33), we obtain

$$x^0_0 = x^0_0 + x^0_1 + x^0_2 + \cdots + x^0_{L-2} + x^0_{L-1} + x^0_0$$
$$\mod (2) \tag{34}$$

Observe that while the first x^0_0 on the right-hand side of Eq. (34) comes from the corresponding first term of Eq. (33), the last x^0_0 of Eq. (34) comes from

Equation (28) is obtained by substituting $n + 1 = 2^m$ from Eq. (27) in place of n in Eq. (16):

$$\binom{(n+1)-1}{k} = 1 \quad \mod (2), \quad k = 0, 1, 2, \ldots, n \tag{29}$$

Substituting Eq. (29) into Eq. (25), we obtain the following simplified periodicity constraint:

the last bit $x^0_{(n) \, \mod (L)} = x^0_0$ of Eq. (33). Rearranging the terms in increasing subscript order in Eq. (34), we obtain

$$x^0_0 = x^0_0 + x^0_0 + x^0_1 + x^0_2 + \cdots$$
$$+ x^0_{L-2} + x^0_{L-1} \quad \mod (2) \tag{35}$$

Substituting $(x^0_0 + x^0_0) \mod (2) = 0$ in Eq. (35), we obtain

$$x^0_0 = x^0_1 + x^0_2 + x^0_3 + \cdots + x^0_{L-2} + x^0_{L-1}$$
$$\mod (2) \tag{36}$$

By adding the bit x^0_0 to both sides of Eq. (36), we obtain

$$\underbrace{x^0_0 + x^0_0}_{0 \, \mod (2)} = x^0_0 + x^0_1 + x^0_2 + \cdots + x^0_{L-2} + x^0_{L-1}$$
$$\mod (2) \tag{37}$$

It follows from Eq. (37) that

$$\sum_{i=0}^{L-1} x^0_i \quad \mod (2) = 0. \tag{38}$$

Our next theorem shows that in the case $L = 2^m + 1$, the period of the output pattern of rule $\boxed{60}$ is given by $n = 2^{2m} - 1$. This period is different from that of the corresponding Theorem 3.5 for rule $\boxed{90}$, where the period n is given by $2^m - 1$, for some $L = 2^m + 1$. ∎

Theorem 3.5. Periodicity Condition: $L = 2^m + 1$. *Rule* $\boxed{60}$ *has a* period-n attractor *where* $n = 2^{2m} - 1$ and $L = 2^m + 1$ if, and only if,

$$\boxed{\begin{array}{l} \text{Valid for} \\ n = 2^{2m} - 1 \\ L = 2^m + 1 \end{array} \; \sum_{i=0}^{L-1} x^0_i \mod (2) = 0} \tag{39}$$

Proof. Since $n = 2^{2m} - 1$, it follows from Eq. (16) that the following equality for the binomial coefficients holds:

$$\binom{n}{k} = 1 \mod (2), \quad \forall\, k = 0, 1, 2, \ldots, 2^{2m} - 1. \quad (40)$$

Observe also that $n = 2^{2m} - 1 = (2^m - 1)(2^m + 1) = (2^m - 1)L$.

In this case, Eq. (25) can be rewritten as follows

$$x^0_{(i \bmod (L))} = \sum_{k=0}^{n} x^0_{((i-n+k) \bmod (L))} \quad \mod (2)$$

$$= \sum_{k=0}^{2^{2m}-1} x^0_{((i-(2^m-1)(2^m+1)+k) \bmod (L))}$$

$$\mod (2) \quad (41)$$

Since the *periodic boundary condition* is mathematically equivalent to *concatenating* replicas of the L-bit string $x_0\, x_1\, x_2\, \cdots\, x_I$ *ad infinitum*, let us choose $i = n = 2^{2m} - 1$. Observe that $((2^{2m} - 1) \bmod (L)) = ((2^m - 1)(2^m + 1) \bmod (L)) = ((2^m - 1) \cdot L \bmod (L)) = 0$ and we can rewrite Eq. (41) as follows

$$x^0_0 = \sum_{k=0}^{(2^m-1)\cdot L} x^0_{((k) \bmod (L))} \quad \mod (2). \quad (42)$$

The summation terms in this equation can be rewritten as follow:

$$
\begin{aligned}
x^0_0 = x^0_0 &\quad + x^0_1 &\cdots&\quad + x^0_{L-1} \\
+ x^0_{((L+0) \bmod (L))} &\quad + x^0_{((L+1) \bmod (L))} &\cdots&\quad + x^0_{((L+L-1) \bmod (L))} \\
+ x^0_{((2L+0) \bmod (L))} &\quad + x^0_{((2L+1) \bmod (L))} &\cdots&\quad + x^0_{((2L+L-1) \bmod (L))} \\
\cdots &\quad \cdots &\cdots&\quad \cdots \\
+ x^0_{(((2^m-2)L+0) \bmod (L))} &\quad + x^0_{(((2^m-2)L+1) \bmod (L))} &\cdots&\quad + x^0_{(((2^m-2)L+L-1) \bmod (L))} \\
&\quad + x^0_{(((2^m-1)L+0) \bmod (L))} &\mathbf{mod\,(2)}.&
\end{aligned}
\quad (43)
$$

Observe that $(k \cdot L \bmod (L)) = 0$, $((k \cdot L + 1) \bmod (L)) = 1$, $((k \cdot L + 2) \bmod (L)) = 2, \ldots$, $((k \cdot L + L - 1) \bmod (L)) = L - 1$, for $k = 0, 1, 2, \ldots, (2^m - 1)$. Hence, the sum (43) can be simplified as follow:

$$
\begin{aligned}
x^0_0 = x^0_0 &\quad + x^0_1 \quad \cdots &\quad + x^0_{L-1} \\
+ x^0_0 &\quad + x^0_1 \quad \cdots &\quad + x^0_{L-1} \\
+ x^0_0 &\quad + x^0_1 \quad \cdots &\quad + x^0_{L-1} \\
\cdots &\quad \cdots \quad \cdots &\quad \cdots \\
+ x^0_0 &\quad + x^0_1 \quad \cdots &\quad + x^0_{L-1} + x^0_0 \quad \mathbf{mod\,(2)}
\end{aligned}
\quad (44)
$$

Observe that we do not include $n = L, 2L, 3L, \ldots, (2^m - 2)L$, but choose only $n = (2^m - 1)L$ for the following reason: In the case of rule $\boxed{60}$ where the index k in Eq. (7) ranges over $0, 1, \ldots, n$, the *binomial coefficients are equal exactly to one* only when $n = (2^m - 1)L = (2^m - 1)(2^m + 1) = 2^{2m} - 1$, in view of Eq. (16). In this case Eq. (44) contains a large number of binary variables x^0_0, x^0_1, $x^0_2, \ldots, x^0_{L-1}$:

$$x^0_0 = (2^m - 1)x^0_0 + (2^m - 1)x^0_1 + \cdots$$
$$+ (2^m - 1)x^0_{L-1} + x^0_0 \quad \mod (2) \quad (45)$$

Here, the first bit x^0_0 comes from the leftmost column of Eq. (44). The rightmost bit x^0_0 is the last term in Eq. (44). Since $(2^m - 1)$ is an odd number we can simplify Eq. (45) further:

$$x^0_0 = x^0_0 + x^0_1 + \cdots + x^0_{L-1} + x^0_0 \quad \mod (2) \quad (46)$$

Let us add x^0_0 to the right and the left parts of this equation, and regroup these terms from the right to obtain:

$$\underbrace{x^0_0 + x^0_0}_{=0 \bmod (2)} = \underbrace{x^0_0 + x^0_0 + x^0_0}_{=x^0_0 \bmod (2)} + x^0_1 + x^0_2 + \cdots + x^0_{L-2}$$

$$+ x^0_{L-1} \quad \mod (2) \quad (47)$$

Hence, we have

$$\sum_{i=0}^{L-1} x^0_i \quad \mod (2) = 0 \quad (48)$$

∎

Corollary 3.2. Corollary to Theorems 3.4 and 3.5. *The total number of* red *pixels in the period* $2^m - 1$ *attractors of Theorem 3.4, and period* $2^{2m} - 1$ *attractors of Theorem 3.5, must be an* even *number.*

Corollary 3.3. Corollary to Theorem 3.3–3.5. *Theorems 3.3, 3.4, and 3.5 hold also for* infinite *bit strings* $(L \to \infty)$.

Recap. Theorems 3.3–3.5 give necessary and sufficient conditions for rule $\boxed{60}$ to have the following *period-n attractors*:

$$
\begin{aligned}
&Theorem\ 3.3 \\
&\quad n = 1 \quad and \quad L = 2^m \\
&Theorem\ 3.4 \\
&\quad n = 2^m - 1 \quad and \quad L = 2^m - 1 \\
&Theorem\ 3.5 \\
&\quad n = 2^{2m} - 1 \quad and \quad L = 2^m + 1
\end{aligned}
$$

As illustrations of the applications of these *analytically* derived results, let us examine the basin tree diagrams exhibited in Tables 5–12. The locations of (L, T) which satisfy Theorems 3.3–3.5 are shown in Fig. 2 by **"fully-filled"** *stars* lying along violet lines.

Applications of Theorem 3.3

1. $m = 2, \quad L = 2^m = 4$

Gallery 60-2 shows all bit strings converge to the unique global attractor $\boxed{0}$, as predicted by Theorem 3.3. The maximal transient time is $T_{\text{tr}} = 2^m = 4$.

2. $m = 3, \quad L = 2^m = 8$

Gallery 60-8 shows all bit strings converge to the global attractor $\boxed{0}$, as predicted by Theorem 3.3. The maximal transient time is $T_{\text{tr}} = 2^m = 8$.

Applications of Theorem 3.4

$$m = 3, \quad n = 2^m - 1 = 7, \quad L = 2^m - 1 = 7$$

Gallery 60-6 and 60-7 show nine period-7 attractors as predicted. Observe the number of red pixels in all attractors is an even number, as predicted. The only other attractor is a *period-1 attractor*, $\boxed{0}$, which qualifies also as a period-7 attractor, with "0" red pixels, an even number, as predicted.

Applications of Theorem 3.5

$$m = 2, \quad n = 2^{2m} - 1 = 15, \quad L = 2^m + 1 = 5$$

Gallery 60-3 shows a single period-15 attractor. Observe that all bit strings in this attractor contain an *even* number of red pixels. The only other attractor is a *period-1 attractor*, $\boxed{0}$, which qualifies also as a period-15 attractor. In this case, there are no red bits, which is an *even* number, as predicted.

4. Global Analysis of Local Rule $\boxed{154}$ and $\boxed{45}$

The truth table, Boolean cube and "Difference Equation" defining the local rules $\boxed{154}$ and $\boxed{45}$ along with a space-time pattern (with a single red pixel initial state) exhibited in [Chua *et al.*, 2003] is, reproduced in Figs. 3 and 4, respectively, for the readers' convenience. For a more representative picture over a longer time period, we have included two space-time patterns using *two random initial configurations*, and two different $L = I + 1$, *à la* Wiener [Chua *et al.*, 2005a], in Figs. 5(a) and 5(b) for local rule $\boxed{154}$, and in Figs. 6(a) and 6(b) for local rule $\boxed{45}$.

An examination of the basin tree diagrams in Table 12 for rule $\boxed{154}$, and in Table 8 for rule $\boxed{45}$, reveals many *Isles of Eden* (of different periods) co-existing with *Attractors* (of different periods). The period T of each Isle of Eden exhibited in Table 12 for $L = 3, 4, \ldots, 8$ is listed in the left column of the first six rows of Table 24. Similarly, the period T of each *attractor* exhibited in Table 12 for $L = 3, 4, \ldots, 8$ is listed in the right column. Observe that unlike the 70 local rules covered by the first four rows of Table 4, where the period T does not depend on L, we now see T increases with L, making it increasingly time consuming to calculate them by brute force computer simulations. Consequently, for rows $9, 10, \ldots, 100$ in Table 24, we have listed only *a few* periods found by computer simulations.

We have repeated the above exercise for local rule $\boxed{45}$ and compiled the results in Table 25 for $L = 3, 4, \ldots, 28$. Observe that while the co-existence of both *Isles of Eden* and *Attractors* of different periods is also observed for rule $\boxed{45}$, the period T increases at a much higher rate than that of $\boxed{154}$. Consequently, Table 25 is restricted only to $L = 3, 4, \ldots, 28$.

The rate of increase in T as a function of L is plotted (in base-10 logarithmic scale) in Fig. 7 for rule $\boxed{154}$, and in Fig. 8 for rule $\boxed{45}$. Observe that most data points in Fig. 7 fall on parallel lines with unit slope, indicating a *scale-free* phenomenon, similar to that of Fig. 7 for rule $\boxed{150}$ and Fig. 8 for rule $\boxed{105}$ in [Chua *et al.*, 2007]. In sharp contrast, the data points in Fig. 8 for rule $\boxed{45}$ do *not* exhibit a *scale-free* property, but diverge at an exponential rate with period $T > 10^6$ for $L > 30$. In fact, as already noted in Table 28 of [Chua *et al.*, 2007], when $L = 9$, rule $\boxed{45}$ has an Isle of Eden made of 504 distinct bit strings, which is almost equal to

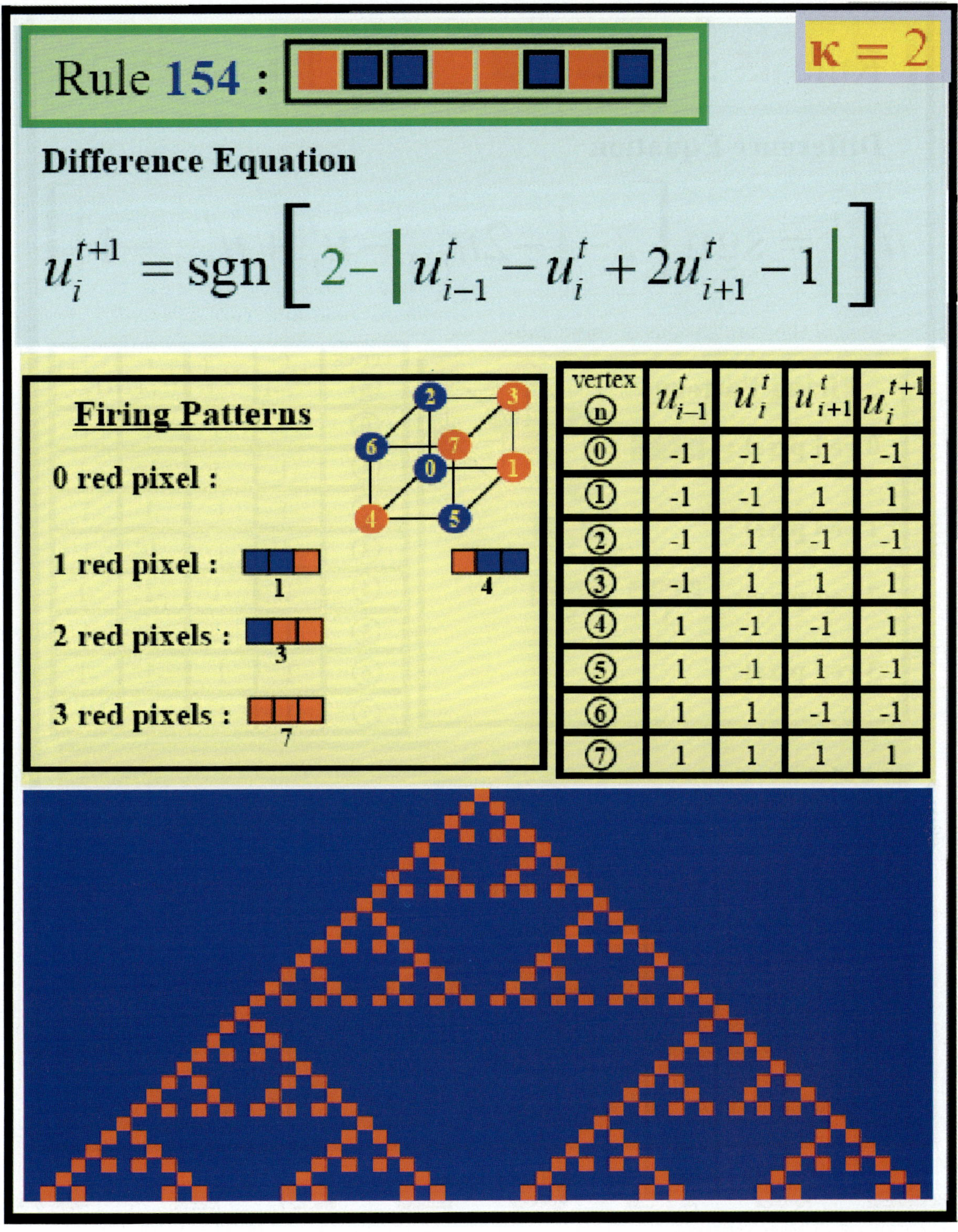

vertex n	u_{i-1}^t	u_i^t	u_{i+1}^t	u_i^{t+1}
0	-1	-1	-1	-1
1	-1	-1	1	1
2	-1	1	-1	-1
3	-1	1	1	1
4	1	-1	-1	1
5	1	-1	1	-1
6	1	1	-1	-1
7	1	1	1	1

Fig. 3. Truth table, Boolean cube, difference equation, and space-time pattern of local rule 154.

the length of the maximum period $T_{\max} = 2^9 = 512$. For $L = 11$, there are two Isles of Eden with period $T = 935$ and $T = 979$, respectively. A list of the complete bit-string sequence of the latter is exhibited in Fig. 9.

A careful scrutiny of Tables 24 and 25 reveals that *all* orbits of both local rules 154 and 45 are Isles of Eden *if, and only if,* the length $L = I + 1$ of the bit string is an *odd* integer. In other words, local rules 154 and 45 are inhabited by a *continuum* of

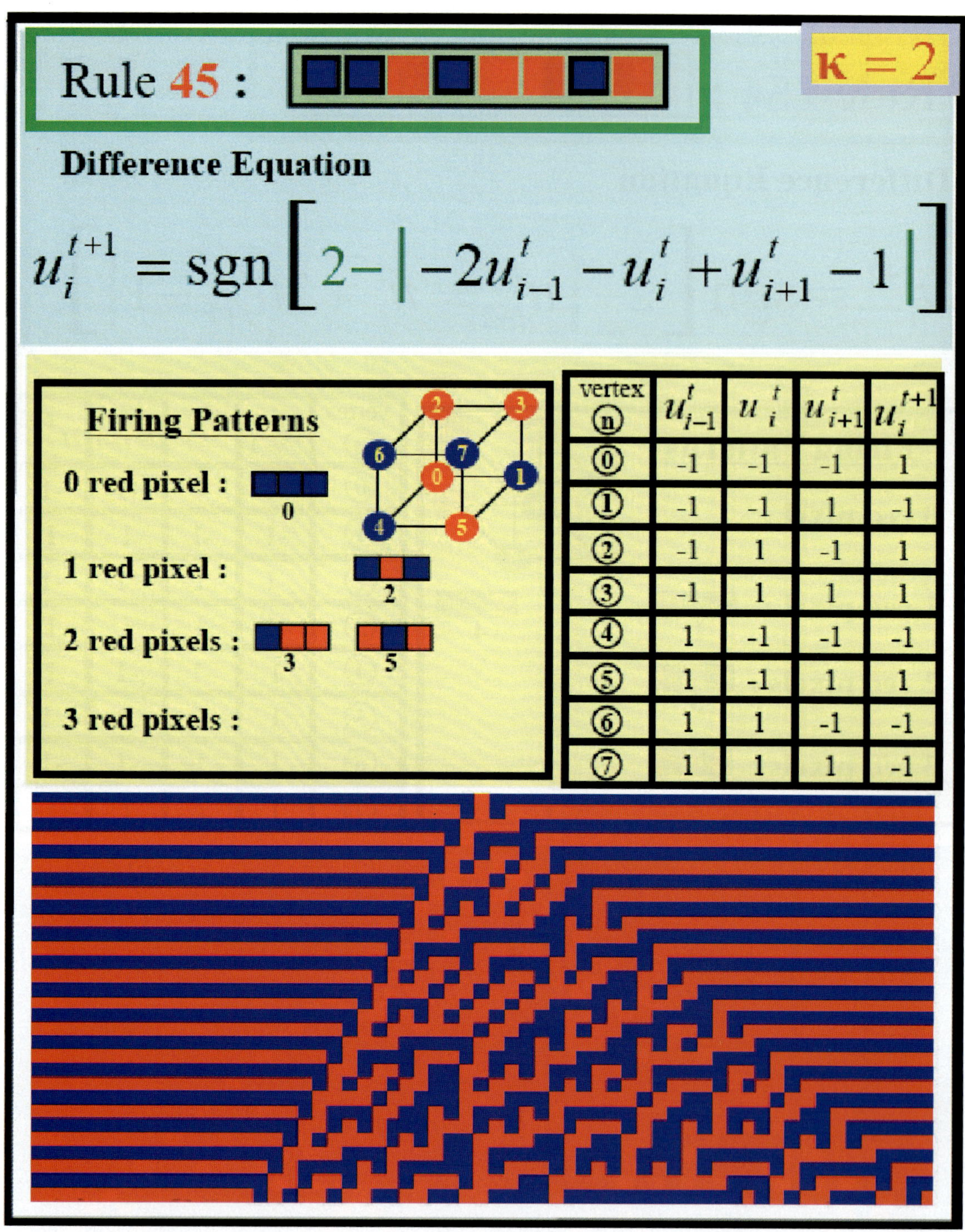

$$u_i^{t+1} = \mathrm{sgn}\left[2 - \left| -2u_{i-1}^t - u_i^t + u_{i+1}^t - 1 \right|\right]$$

vertex (n)	u_{i-1}^t	u_i^t	u_{i+1}^t	u_i^{t+1}
⓪	-1	-1	-1	1
①	-1	-1	1	-1
②	-1	1	-1	1
③	-1	1	1	1
④	1	-1	-1	-1
⑤	1	-1	1	1
⑥	1	1	-1	-1
⑦	1	1	1	-1

Fig. 4.   Truth table, Boolean cube, difference equation, and space-time pattern of local rule 45.

Isles-of-Eden *if, and only if, L* is *not divisible* by 2. In this case, every orbit is an *invariant orbit*, as in the case of rules 15, 85, 170 and 240, for all L, and in rules 150 and 105 for all L *not divisible* by 3, as proved in *Theorem 4.1* of [Chua *et al.*, 2007]. However, unlike rules 150 and 105, where all orbits are *attractors* if L is divisible by 3, Tables 24 and 25 show that for all L divisible by 2, local rules 154 and 45 are inhabited by *both* Isles of Eden and attractors; they *co-exist* so to speak.

A rigorous proof of the above observations is *nontrivial* and will be postponed to the next section. Meanwhile, we end this section with a formal statement of the main result of this paper, which will be restated and proved in Theorems 5.2 and 5.3 in Sec. 5.

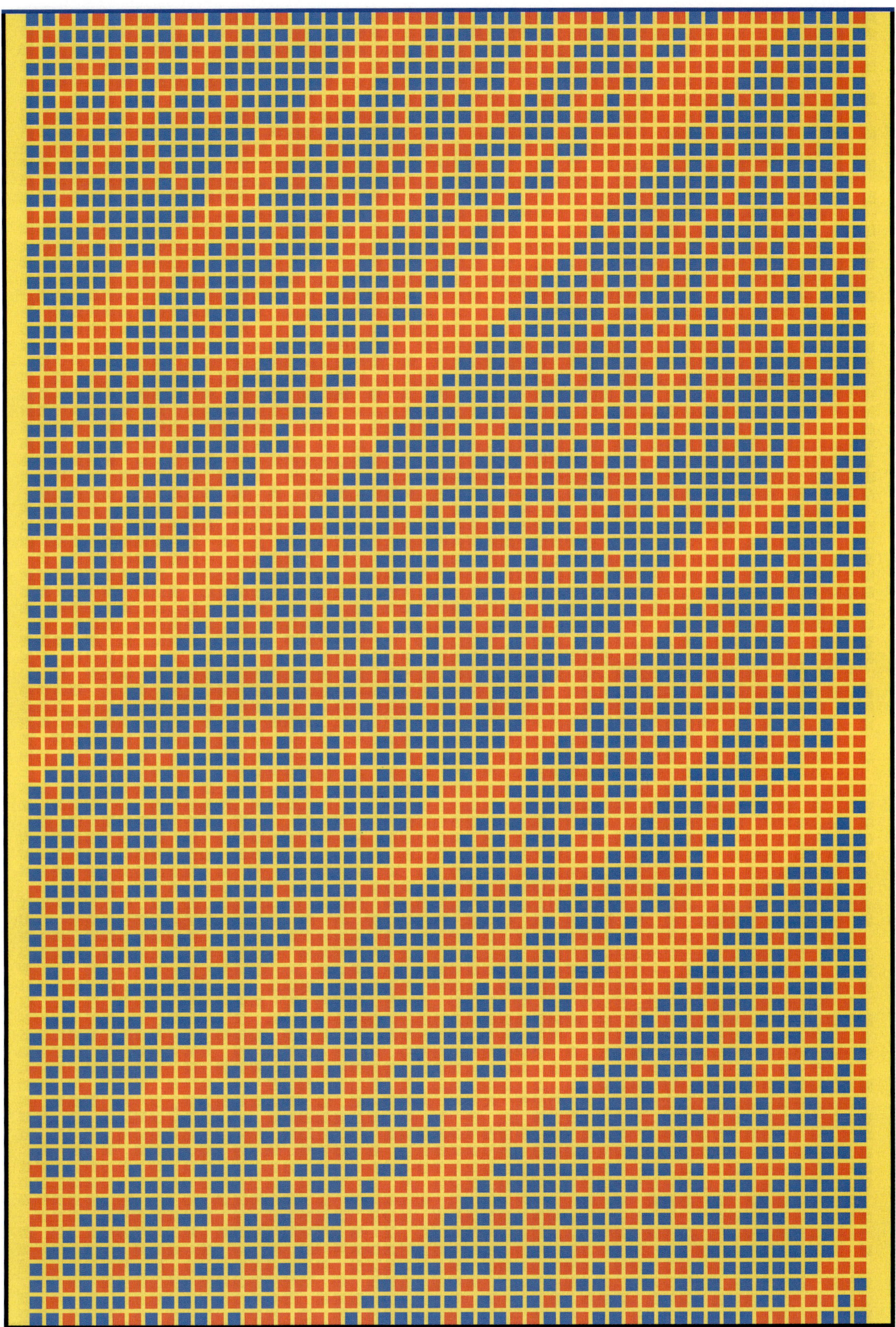

(a) rule $\boxed{154}$, $L = 51 : T = 408; \sigma = 2, \tau = 104, \beta > 0$

Fig. 5. Space-time patterns of local rule $\boxed{154}$ for two random initial configurations.

(b) rule $\boxed{154}$, $L = 52 : T = 252; \sigma = 0, \tau = 252, \beta > 0$

Fig. 5.   (*Continued*)

(a) rule $\boxed{45}$, $L = 50$

Fig. 6.   Space-time patterns of local rule $\boxed{45}$ for two random initial configurations. The *period T* and *Bernoulli parameters* σ and τ are too large to compute for $L > 30$.

(b) rule $\boxed{45}$, $L = 51$

Fig. 6.   (*Continued*)

Table 24.   A listing of *some* periods T found by computer simulations for local rule 154 for $L = 3, 4, \ldots, 100$. Those listed in the left column are for *period-T Isles of Eden*. Those listed in the right column are for *period-T attractors*.

154	Period - T	
L	Isles of Eden	Attractors
3	$T=1,\quad 6,$	
4	$T=4,$	$T=1,$
5	$T=1,\quad 10,\quad 20,$	
6	$T=1,\quad 3,\quad 6,\quad 12,$	$T=1, 2$
7	$T=14,\quad 28,$	
8	$T=8,\quad 1,\quad 4,$	$T=1,$
9	$T=36,\quad 72,\ldots$	
10	$T=20,\ldots$	$T=6,\ldots$
11	$T=44,\quad 88,\ldots$	
12	$T=12,\ldots$	$T=4,\ldots$
13	$T=52,\quad 104,\ldots$	
14	$T=28,\quad 56,\ldots$	$T=14,\ldots$
15	$T=120,\ldots$	
16	$T=16,\ldots$	$T=1,\ldots$
17	$T=68,\quad 139,\ldots$	
18	$T=36,\quad 72,\ldots$	$T=14,\ldots$
19	$T=76,\ldots$	
20	$T=20,\quad 40,\ldots$	$T=12,\ldots$
21	$T=168,\ldots$	
22	$T=44,\quad 88,\quad 176,\ldots$	$T=62,\ldots$
23	$T=92,\quad 184,\ldots$	
24	$T=24,\ldots$	$T=8,\ldots$
25	$T=100,\quad 200,\quad 400,\ldots$	
26	$T=52,\quad 104,\ldots$	$T=130,\ldots$
27	$T=216,\ldots$	
28	$T=28,\quad 56,\ldots$	$T=28,\ldots$
29	$T=232,\ldots$	
30	$T=120,\ldots$	$T=30,\ldots$
31	$T=124,\quad 248,\ldots$	
32	$T=32,\ldots$	$T=1,\ldots$
33	$T=132,\quad 264,\ldots$	
34	$T=68,\quad 136,\quad 272,\ldots$	$T=30,\ldots$
35	$T=140, 280,\ldots$	

Table 24.   (*Continued*)

154	Period - T	
L	**Isles of Eden**	**Attractors**
36	$T = 72, \ldots$	$T = 28, \ldots$
37	$T = 148, \quad 296, \ldots$	
38	$T = 152, \ldots$	$T = 1022, \ldots$
39	$T = 156, \quad 312, \ldots$	
40	$T = 40, \ldots$	$T = 24, \ldots$
41	$T = 164, \quad 328, \quad 654, \ldots$	
42	$T = 84, \quad 168, \ldots$	$T = 126, \ldots$
43	$T = 172, \quad 344, \quad 688, \ldots$	
44	$T = 88, \ldots$	$T = 124, \ldots$
45	$T = 180, \quad 360, \ldots$	
46	$T = 92, \quad 184, \ldots$	$T = 4094, \ldots$
47	$T = 188, \quad 376, \ldots$	
48	$T = 48, \ldots$	$T = 16, \ldots$
49	$T = 196, \quad 392, \quad 784, \ldots$	
50	$T = 100, \quad 200, \ldots$	$T = 2046, \ldots$
51	$T = 408, \ldots$	
52	$T = 104, \quad 208, \ldots$	$T = 252, \ldots$
53	$T = 212, \quad 424, \ldots$	
54	$T = 108, \quad 216, \quad 432, \ldots$	$T = 1022, \ldots$
55	$T = 220, \quad 440, \quad 880, \ldots$	
56	$T = 56, \ldots$	$T = 56, \ldots$
57	$T = 228, \quad 456, \quad 912, \ldots$	
58	$T = 116, \quad 232, \ldots$	$T = 32766, \ldots$
59	$T = 236, \quad 472, \ldots$	
60	$T = 120, \quad 240, \ldots$	$T = 60, \ldots$
61	$T = 488, \ldots$	
62	$T = 124, \quad 248, \quad 496, \ldots$	$T = 62, \ldots$
63	$T = 252, \quad 504, \ldots$	
64	$T = 64, \ldots$	$T = 1, \ldots$
65	$T = 260, \quad 520, \quad 1040, \ldots$	
66	$T = 264, \ldots$	$T = 62, \ldots$
67	$T = 536, \ldots$	
68	$T = 68, \quad 136, \ldots$	$T = 60, \ldots$

Table 24. (*Continued*)

154	Period - T	
L	*Isles of Eden*	*Attractors*
69	$T = 552,\quad 1104, \ldots$	
70	$T = 140,\quad 280,\quad 560, \ldots$	$T = 8190, \ldots$
71	$T = 284,\quad 568,\quad 1136, \ldots$	
72	$T = 72,\quad 144, \ldots$	$T = 56, \ldots$
73	$T = 584, \ldots$	
74	$T = 148,\quad 296, \ldots$	$T = 174762, \ldots$
75	$T = 300,\quad 600, \ldots$	
76	$T = 76,\quad 152,\quad 304, \ldots$	$T = 2044, \ldots$
77	$T = 308,\quad 616, \ldots$	
78	$T = 156,\quad 312, \ldots$	$T = 8190, \ldots$
79	$T = 316,\quad 632,\quad 1264, \ldots$	
80	$T = 80, \ldots$	$T = 48, \ldots$
81	$T = 324,\quad 648,\quad 1296, \ldots$	
82	$T = 328, \ldots$	$T = 2046, \ldots$
83	$T = 332,\quad 664, \ldots$	
84	$T = 84,\quad 168, \ldots$	$T = 252, \ldots$
85	$T = 340,\quad 680, \ldots$	
86	$T = 172,\quad 344,\quad 688, \ldots$	$T = 254, \ldots$
87	$T = 348,\quad 696,\quad 1392, \ldots$	
88	$T = 88,\quad 176, \ldots$	$T = 248, \ldots$
89	$T = 356,\quad 712,\quad 1424, \ldots$	
90	$T = 180,\quad 360,\quad 720, \ldots$	$T = 8190, \ldots$
91	$T = 364,\quad 728, \ldots$	
92	$T = 184, \ldots$	$T = 8188, \ldots$
93	$T = 744, \ldots$	
94	$T = 188,\quad 376, \ldots$	$T > 10000000, \ldots$
95	$T = 380,\quad 760,\quad 1520, \ldots$	
96	$T = 96, \ldots$	$T = 32, \ldots$
97	$T = 388,\quad 776, \ldots$	
98	$T = 196,\quad 392, \ldots$	$T > 10000000, \ldots$
99	$T = 396,\quad 792,\quad 1584, \ldots$	
100	$T = 100,\quad 200, \ldots$	$T = 4092, \ldots$

Table 25.   A listing of *some* periods T found by computer simulations for local rule 45 for $L = 3, 4, \ldots, 28$. Those listed in the left column are for *period-T Isles of Eden*. Those listed in the right column are for *period-T attractors*.

45 L	Period - T	
	Isles of Eden	Attractors
3	$T = 1,\ 2,\ 3,$	
4		$T =\ 2,$
5	$T = 2,\ 30,$	
6	$T = 1,\ 3,$	$T = 2,\ 18,\ \ldots$
7	$T = 2,\ 126,$	
8	$T = 4,\ 16,\ 24,\ 32,$	$T = 2,\ \ldots$
9	$T =\ 504,\ \ldots$	
10		$T = 2,\ 430,\ \ldots$
11	$T = 935,\ 979,\ \ldots$	
12		$T = 2,\ 18,\ \ldots$
13	$T =\ 443,\ \ldots$	
14		$T = 2,\ 2198,\ \ldots$
15	$T =\ 6820,\ \ldots$	
16		$T = 2,\ 700,\ 2816,\ \ldots$
17	$T =\ 78812,\ \ldots$	
18		$T = 2,\ 7812,\ \ldots$
19	$T = 183920,\ \ldots$	
20		$T = 2,\ 142580,\ \ldots$
21	$T = 149758,\ \ldots$	
22		$T = 2,\ 122870,\ \ldots$
23	$T = 757955,\ \ldots$	
24		$T = 2,\ 421188,\ \ldots$
25		
26		$T = 2,\ 4511,\ \ldots$
27		
28		$T = 2,\ 304913,\ \ldots$

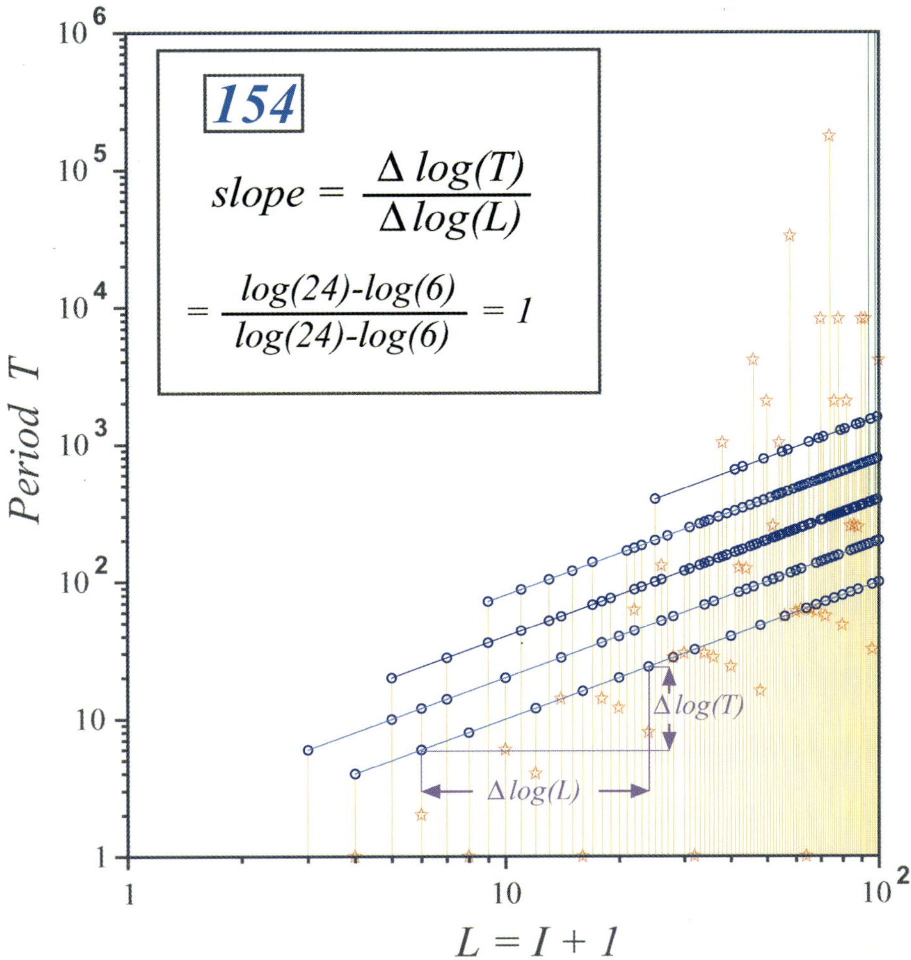

Fig. 7. Dependence of the period "T" of *Isle of Eden* (denoted by blue circles), and *attractors* (denoted by red stars), of rule $\boxed{154}$ as a function of L (in logarithmic scale).

> **Theorem 4.1.** Dense Isles-of-Eden.
> Every *orbit of local rules* $\boxed{154}$ *and* $\boxed{45}$ *is an* Isle of Eden, if, and only if, $L \overset{\Delta}{=} I + 1$ *is an* odd *integer*.

5. Dense Isles-of-Eden Property

The objective of this section is to present a rigorous proof of Theorem 4.1 from *Sec.* 4. We have developed a graph-theoretic proof which is more general than we need because the main tool of this proof, henceforth called Isles *of Eden digraph*, is applicable to *all* local rules that are inhabited by a *dense* set (i.e. a *continuum*) of *Isles* of Eden. The proof consists of constructing a *digraph* (*directed graph*) $\mathcal{G}_{IE}(\boxed{N})$ for *any* local rule $\boxed{N}$ under scrutiny for possible existence of a *dense* set of Isles of Eden. Once the digraph $\mathcal{G}_{IE}(\boxed{N})$ is constructed for any rule $\boxed{N}$, proving the existence of a *dense* set of Isles-of-Eden orbits reduces to an inspection of all oriented loops which circulate in only one direction, as in one-way traffics, and by counting the number of *edges* around each such loop.

5.1. *Notations and definitions*

Let $\sum^L$ denote the set of L-bit binary strings. Let $T_{\boxed{N}} : \sum \to \sum$ denote the mapping function of local rule $\boxed{N}$ that maps $\sum^L$ to $\sum^L$ for every $L \geq 3$. Let $T_{\boxed{N}}^r : \sum \to \sum$ denote the r-times iterated mapping function of rule $\boxed{N}$. Let $T_{\boxed{N}}(\sum^L)$ denote the image of $\sum^L$ under $T_{\boxed{N}}$:

$$T_{\boxed{N}}\left(\sum^L\right) = \bigcup_{x \in \sum^L} T_{\boxed{N}}(\mathbf{x}) \tag{49}$$

Definition 5.1. *Isle of Eden.* A bit string

$$\mathbf{x} = (x_0 \quad x_1 \quad x_2 \quad \cdots \quad x_{L-1})$$

is said to be a *period-n Isle of Eden* of a local rule $\boxed{N}$ *if, and only if,* its *preimage* under $\chi_{\boxed{N}}^n$ is *itself,*

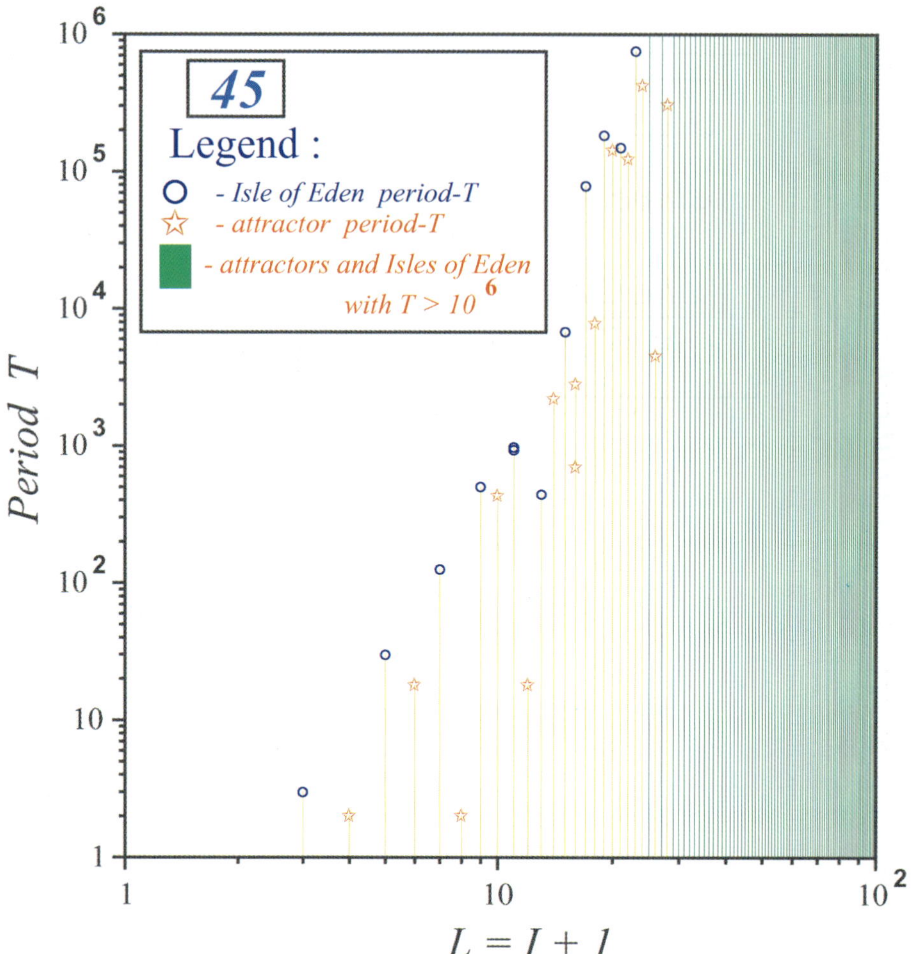

Fig. 8.   Dependence of the period "*T*" of *Isle of Eden* (denoted by blue circles), and *attractors* (denoted by red stars), of rule 45 as a function of *L* (in logarithmic scale). The thin green-color vertical lines on the right ($L > 30$) indicate that period T exceeds 10^6.

where $\chi_{\boxed{N}}^n : [0,1] \rightarrow [0,1]$ is the *time-n characteristic function of* $\boxed{N}$ [Chua *et al.*, 2005a]. More precisely, $\mathbf{x}$ is a *period-n Isle of Eden* of a local rule $\boxed{N}$ *iff*

$$(\chi_{\boxed{N}}^n)^{-1}(\mathbf{x}) = \mathbf{x}$$

5.2. *Four basic lemmas*

Lemma 5.1.   *An L-bit binary string* $\mathbf{x} = (x_0 \ x_1 \ x_2 \ \cdots \ x_{L-1})$ *is a* period-n Isle of Eden of *a local rule* $\boxed{N}$ *if, and only if,* $\mathbf{x}$ *has a unique preimage under* $\mathbf{T}_{\boxed{N}}^n$:

$$\mathbf{x}_n \stackrel{\Delta}{=} (\mathbf{T}_{\boxed{N}}^n)^{-1}(\mathbf{x}) = \mathbf{x} \qquad (50)$$

Proof.   The definition of Isle of Eden implies that $\mathbf{x}$ is an Isle of Eden *iff* the point $\Phi(\mathbf{x})$ is an Isle of Eden, where $\Phi : [0,1] \rightarrow [0,1]$ is the unit interval map defined in [Chua *et al.*, 2005a]

Necessity

$\Phi(\mathbf{x})$ being an Isle of Eden can be formally written as

$$(\chi_{\boxed{N}}^n)^{-1}(\Phi(\mathbf{x})) = \Phi(\mathbf{x}) \qquad (51)$$

Applying $\chi_{\boxed{N}}^n$ to both sides we obtain

$$\Phi(\mathbf{x}) = \chi_{\boxed{N}}^n(\Phi(\mathbf{x})) \qquad (52)$$

According to the definition of Φ [Chua *et al.*, 2005a] $\Phi \circ \mathbf{T}_{\boxed{N}} = \chi_{\boxed{N}} \circ \Phi$. Hence, we can write the right-hand side as

$$\chi_{\boxed{N}}^n(\Phi(\mathbf{x})) = \underbrace{\chi_{\boxed{N}} \circ \chi_{\boxed{N}} \circ \cdots \circ \chi_{\boxed{N}}}_{n \text{ times}} \circ \Phi(\mathbf{x})$$

$$= \Phi \circ \underbrace{\mathbf{T}_{\boxed{N}} \circ \mathbf{T}_{\boxed{N}} \circ \cdots \circ \mathbf{T}_{\boxed{N}}}_{n \text{ times}}(\mathbf{x})$$

$$= \Phi(\mathbf{T}_{\boxed{N}}^n(\mathbf{x})) \qquad (53)$$

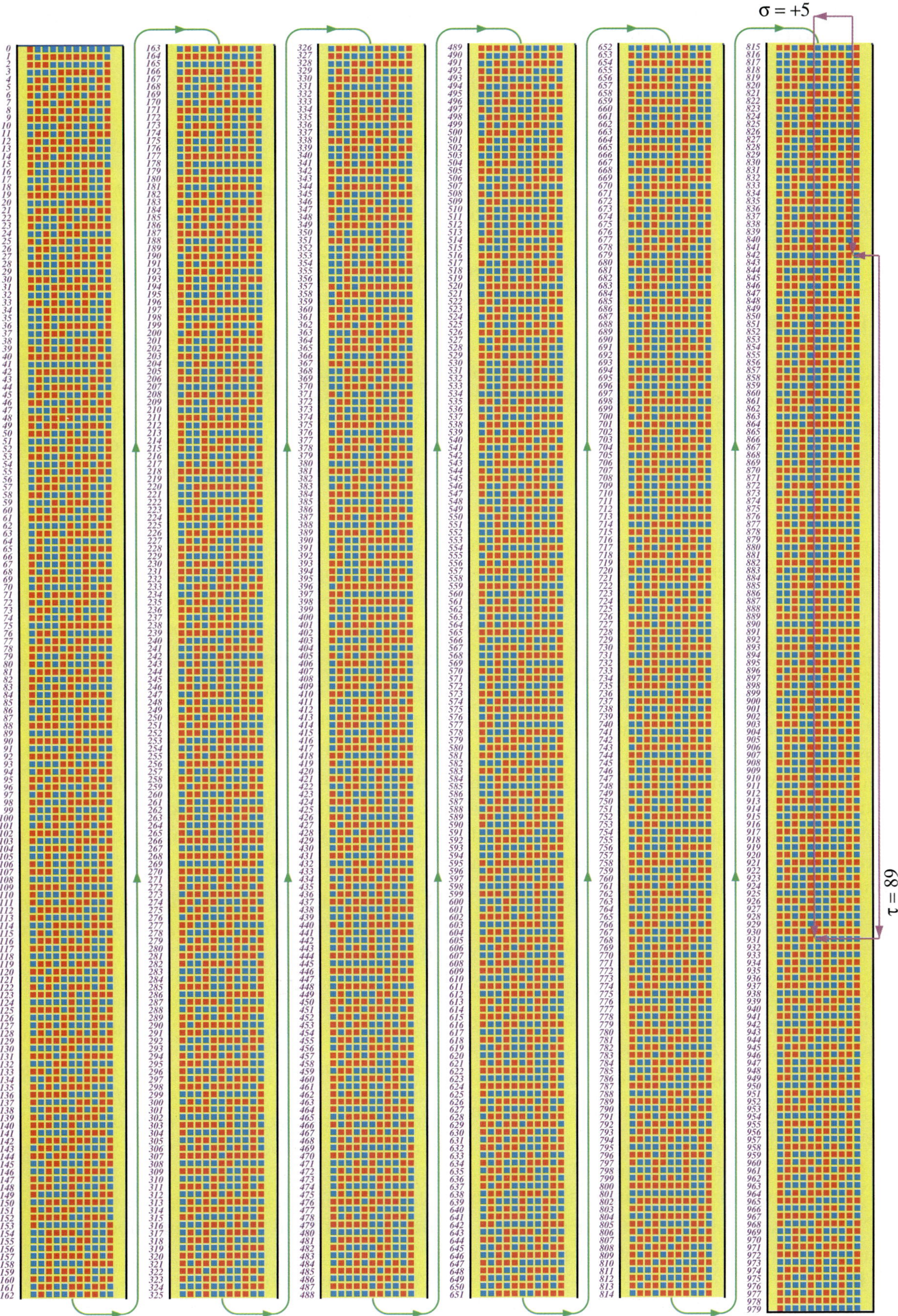

Fig. 9.   For $L = 11$, rule $\boxed{45}$ has an Isle of Eden made of 979 distinct bit strings evolving via a Bernoulli σ_τ-shift dynamics with $\sigma = 5$, $\tau = 89$, and $\beta > 0$.

By substituting Eq. (53) into Eq. (52) and applying $(\mathbf{T}_{\boxed{N}}^{n})^{-1} \circ \Phi^{-1}$ to both sides we obtain

$$(\mathbf{T}_{\boxed{N}}^{n})^{-1}(\mathbf{x}) = \mathbf{x} \qquad (54)$$

which is exactly what the lemma states.

Sufficiency

Starting from Eq. (54), applying $\Phi \circ \mathbf{T}_{\boxed{N}}^{n}$ to both sides and substituting Eq. (53) into the result we obtain Eq. (52). By applying $(\mathbf{\chi}_{\boxed{N}}^{n})^{-1}$ to both sides we obtain Eq. (51), from which it follows that $\mathbf{x}$ is an Isle of Eden. ∎

Lemma 5.2. *Every orbit of a local rule $\boxed{N}$ over $\sum^{L}$ is an Isle of Eden if, and only if, every $x \in \sum^{L}$ has a unique preimage under $T_{\boxed{N}}$.*

Proof

Necessity

Let us suppose the contrary that $\mathbf{x}_0 \in \sum^{L}$ is an Isle of Eden, but it either does not have a preimage under $T_{\boxed{N}}$, or its preimage is not unique. Using Lemma 5.1, we know that $\mathbf{x}_0$ has a unique preimage under $T_{\boxed{N}}^{n}$ (namely itself).

It is trivial that if a binary string does not have a preimage under $T_{\boxed{N}}$, then it cannot have a preimage under $T_{\boxed{N}}^{r}$ for any $r > 1$. Thus $\mathbf{x}_0$ does not have a preimage under $T_{\boxed{N}}^{n}$, which is a contradiction, so $\mathbf{x}_0$ must have a preimage under $T_{\boxed{N}}$. If the preimage under $T_{\boxed{N}}$ is not unique, then its preimage under $T_{\boxed{N}}^{n}$ cannot be unique either, thus we arrived at a contradiction once again, so $\mathbf{x}_0$ must have a preimage under $T_{\boxed{N}}$ in this case too.

Sufficiency

If every $\mathbf{x} \in \sum^{L}$ has a unique preimage $\mathbf{x}_1 \overset{\Delta}{=} T_{\boxed{N}}^{-1}(\mathbf{x})$ under $T_{\boxed{N}}$, then their preimages, being members of $\sum^{L}$, also have unique preimages under $T_{\boxed{N}}$, and this holds recursively for every preimage. Therefore every $\mathbf{x} \in \sum^{L}$ has a unique preimage $\mathbf{x}_r \overset{\Delta}{=} (T_N^{r})^{-1}(\mathbf{x})$ under T_N^{r} for every $r > 1$.

Since there are 2^{L} binary strings of length L, every $\mathbf{x} \in \sum^{L}$ has to map to itself under the inverse map $(T_N^{r(\mathbf{x})})^{-1}$ for some $r(\mathbf{x}) \leq 2^{L}$ and thus it is, by definition, a period-$r(\mathbf{x})$ Isle of Eden of local rule N. ∎

Lemma 5.3. *Every $\mathbf{x} \in \sum^{L}$ has a preimage under $T_{\boxed{N}}$ if, and only if, every $\mathbf{y} \in \sum^{L}$ has at most one*

preimage under $T_{\boxed{N}}$, or equivalently $T_{\boxed{N}} : \sum^{L} \to \sum^{L}$ is surjective if, and only if, it is injective.

Proof

Necessity

Let us suppose the contrary that every $\mathbf{x} \in \sum^{L}$ has a preimage under $T_{\boxed{N}}$ for a local rule $\boxed{N}$, and there are $\mathbf{x}_1 \in \sum^{L}$ and $\mathbf{x}_2 \in \sum^{L}$ for which $\mathbf{x}_1 \neq \mathbf{x}_2$ and $T_{\boxed{N}}(\mathbf{x}_1) = T_{\boxed{N}}(\mathbf{x}_2)$. Then the size of $T_{\boxed{N}}(\sum^{L})$ must be less than the size of $\sum^{L}$, because $\mathbf{x}_1$ and $\mathbf{x}_2$ are mapped to the same string and no $\mathbf{x} \in \sum^{L}$ can be mapped to more than one string. Thus, there is a string $\mathbf{z} \in \sum^{L}$ that is not in $T_{\boxed{N}}(\sum^{L})$, which means, by definition, that there is no $\mathbf{x} \in \sum^{L}$ that maps to $\mathbf{z}$, so $\mathbf{z}$ does not have a preimage under $T_{\boxed{N}}$, which contradicts our initial assumption.

Sufficiency

Let us suppose the contrary that every $\mathbf{y} \in \sum^{L}$ has at most one preimage under $T_{\boxed{N}}$, but there is an $\mathbf{x}_0 \in \sum^{L}$ that does not have a preimage under $T_{\boxed{N}}$. Since there is no $\mathbf{x} \in \sum^{L}$ for which $T_{\boxed{N}}(\mathbf{x}) = \mathbf{x}_0$ thus $\mathbf{x}_0 \notin T_{\boxed{N}}(\sum^{L})$, and therefore the number of elements of $T_{\boxed{N}}(\sum^{L})$ is less than the number of elements of $\sum^{L}$. It follows that there must exist a $\mathbf{y}_0 \in T_{\boxed{N}}(\sum^{L})$ and an $\mathbf{x}_1 \in \sum^{L}$ and an $\mathbf{x}_2 \in \sum^{L}$ for which $\mathbf{x}_1 \neq \mathbf{x}_2$ and $T_{\boxed{N}}(\mathbf{x}_1) = \mathbf{y}_0$ and $T_{\boxed{N}}(\mathbf{x}_2) = \mathbf{y}_0$. This, however, contradicts our initial assumption. ∎

Lemma 5.4. *Let $R^{(n)}(\mathbf{u})$ denote the vector obtained by circularly shifting the bits of an arbitrary vector $\mathbf{u}$ by n bits, i.e.*

$$R^{(n)}(\mathbf{u}) = \{u_{n \bmod L}, u_{(n+1) \bmod L}$$
$$, u_{(n+2) \bmod L}, \cdots, u_{(n+L-1) \bmod L}\},$$

Let $\mathbf{x} = (x_0\ x_1\ x_2\ \cdots\ x_{L-1})$ be an L-bit binary string and let $n \in \mathbb{Z}$. The shifting operator $R^{(n)}$ and the mapping function of a local rule $\boxed{N}$ can be exchanged, that is $T_{\boxed{N}}(R^{(n)}(\mathbf{x})) = R^{(n)}(T_{\boxed{N}}(\mathbf{x}))$.

Note: Lemma 5.4 expresses the rotation invariance property of local cellular automaton rules.

Proof. The ith bit of $R^{(n)}(\mathbf{y})$ is defined by

$$\mathbf{y}_{(n+i) \bmod L} = f_{\boxed{N}}(\mathbf{x}_{(n+i-1) \bmod L}, \mathbf{x}_{(n+i) \bmod L}$$
$$, \mathbf{x}_{(n+i+1) \bmod L}) \qquad (55)$$

for all $0 \le i \le L - 1$, where $f_{\boxed{N}} : I^3 \to I$ is the local mapping function of rule $\boxed{N}$, with $I = \{0, 1\}$. These equations are equivalent to those that define the jth bit of $\mathbf{y}$, $j = ((i + n) \bmod L)$. Thus the stated equation holds for every position.   ∎

Example 5.1. Let us consider rule $\boxed{45}$, whose truth table is given in Table 26, and let $\mathbf{x} = 0110100$. The output string $\mathbf{T}_{\boxed{45}}(\mathbf{x})$ is 0101101. If we circularly shift $\mathbf{x}$ by three characters ($n = 3$), we would obtain $R^{(3)}(\mathbf{x}) = 0100011$. The output after the shifting can be calculated by circularly shifting $\mathbf{T}_{\boxed{45}}(\mathbf{x})$ too by three characters, i.e. $T_{\boxed{45}}(R^{(3)}(\mathbf{x})) = R^{(3)}(T_{\boxed{45}}(\mathbf{x})) = 1101010$.

Example 5.2. Now let us consider rule $\boxed{90}$, whose truth table is given in Table 26, and let $\mathbf{x} = 00001111$. Thus $\mathbf{T}_{\boxed{90}}(\mathbf{x}) = 10011001$. Let us circularly shift $\mathbf{x}$ by $n = -7$ characters: $R^{(-7)}(\mathbf{x}) = 00011110$. The output after the shifting operator can be obtained by applying circular shifting by -7 characters to $\mathbf{T}_{\boxed{90}}(\mathbf{x})$ as well: $T_{\boxed{90}}(R^{(-7)}(\mathbf{x})) = R^{(-7)}(T_{\boxed{90}}(\mathbf{x})) = 00110011$.

Table 26.   Mapping rules of local rules $\boxed{45}$ and $\boxed{90}$.

Input pattern	Output bit	
	$\boxed{45}$	$\boxed{90}$
000	1	0
001	0	1
010	1	0
011	1	1
100	0	1
101	1	0
110	0	1
111	0	0

5.3.   *Locating points with multiple preimages*

In this subsection we define the *Isles of Eden Digraph* (and give a method to create it) that can generate the bits of two binary strings of any length L in parallel in every possible way so that their image under $T_{\boxed{N}}$ is equal. We use the digraph in a theorem that allows one to decide if it is possible to generate different bit strings of a length L (inputs) that have the same image under $T_{\boxed{N}}$, by simply examining the cycles of the Isles of Eden digraph.

The idea behind the construction is that every move along an edge in the digraph adds one bit to the length of the input strings, as well as to the output string, so L moves will generate strings of length L. The edges are labeled with these three bits. The nodes are labeled with the previous two bits of $\mathbf{x}$ and $\mathbf{y}$, thus making it possible to compute the output based on the present node and on the edge through which we leave the node. The label of the node where the edge leads to is composed of the second bits of the previous node and the corresponding bits of the edge. Actually, the nodes of the graph refer to a sliding double window of width 2 over $\mathbf{x}$ and $\mathbf{y}$, and each move along an edge moves the windows to the right by one bit. To compute the first move we need an initial condition of two bits which are also used at the end of the generation, due to the cyclic boundary condition. This means that to meet the required boundary condition we need to return to the starting node. Let us give a brief overview on some basic concepts from graph theory that we will use.

Definition 5.2. *Directed Graph.* A finite *directed graph* (*digraph*) G is given by an ordered pair (V, E) and two functions $i, t \colon E(G) \to V(G)$, where

- $V(G)$ is a finite set of *vertices* (also called nodes),
- $E(G)$ is a finite set of *edges*,
- $i(e)$ and $t(e)$ are the initial and terminal vertices of edge $e \in E(G)$, respectively.

Definition 5.3. *Walks and Cycles.* A *walk of length* n in a directed graph is a finite sequence $\pi = e_1 \ e_2 \ \cdots \ e_{n-1}$ of edges such that $t(e_k) = i(e_{k+1})$ for $k = 1, 2, \ldots, n - 1$. A vertex can be included many times in the sequence. The vertex $i(e_1)$ is called the *start vertex* and the vertex $t(e_{n-1})$ is called the *end vertex*. A *walk* with the same starting and ending vertex, i.e. where $i(e_1) = t(e_{n-1})$, is called a *closed walk or a cycle*.

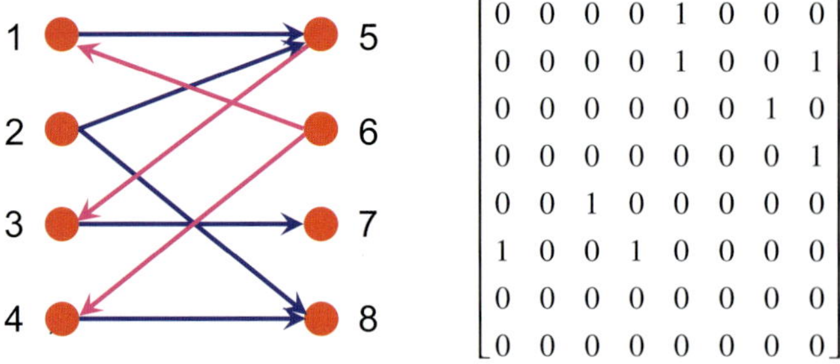

Fig. 10.   A bipartite graph and its adjacency matrix.

Definition 5.4. *Bipartite Graph.* A graph, the vertices of which can be decomposed into two disjoint sets such that no two graph vertices within the same set are adjacent is called a *bipartite* graph. A sample bipartite graph is shown in Fig. 10.

Definition 5.5. *Adjacency Matrix of a Directed Graph.* The *adjacency* matrix of a finite directed graph G on n vertices is defined to be an $n \times n$ matrix $\mathbf{A}$ whose ijth element $\mathbf{A}(i, j)$ is the number of edges from vertex i to vertex j. The powers of $\mathbf{A}$ refer to the number of walks in G, that is $\mathbf{A}^n(i, j)$ is the number of walks of length n from vertex i to vertex j.

Definition 5.6. *Isles of Eden Digraph (Double preimage locator).* Let us consider a directed graph with nodes referring to pairs of binary strings of length two. There is an edge from node $\mathbf{u}$, labeled $(\{u_{1,x}, u_{2,x}\}, \{u_{1,y}, u_{2,y}\})$, to node $\mathbf{v}$, labeled $(\{v_{1,x}, v_{2,x}\}, \{v_{1,y}, v_{2,y}\})$, if $v_{1,x} = u_{2,x}$, $v_{1,y} = u_{2,y}$ and local rule $\boxed{N}$ maps both $x^* = \{u_{1,x}, u_{2,x}, v_{2,x}\}$ and $y^* = \{u_{1,y}, u_{2,y}, v_{2,y}\}$ to the same output bit $\mathbf{z}$, and the edge is labeled $(v_{2,x}, v_{2,y}, z)$. Let us call such a graph the *Isles of Eden digraph*[5] $\mathcal{G}_{IE}(\boxed{N})$ of local rule $\boxed{N}$. Normally it is enough to include nodes for which $u_{1,x} \neq u_{1,y}$ (generator nodes) and those that can be reached from them. If all 16 possible nodes are included we call it the *full Isles of Eden digraph* (see Sec. 5.5).

It is also called a *double preimage locator* because it is only possible to generate two strings that are mapped to the same output if there is a string that has two distinct preimages, that is when there is a fork in the inverse map and therefore the map is not invertible. As an example, Fig. 11 shows the Isles of Eden digraph of rule $\boxed{154}$. Colors of nodes refer to different properties: black

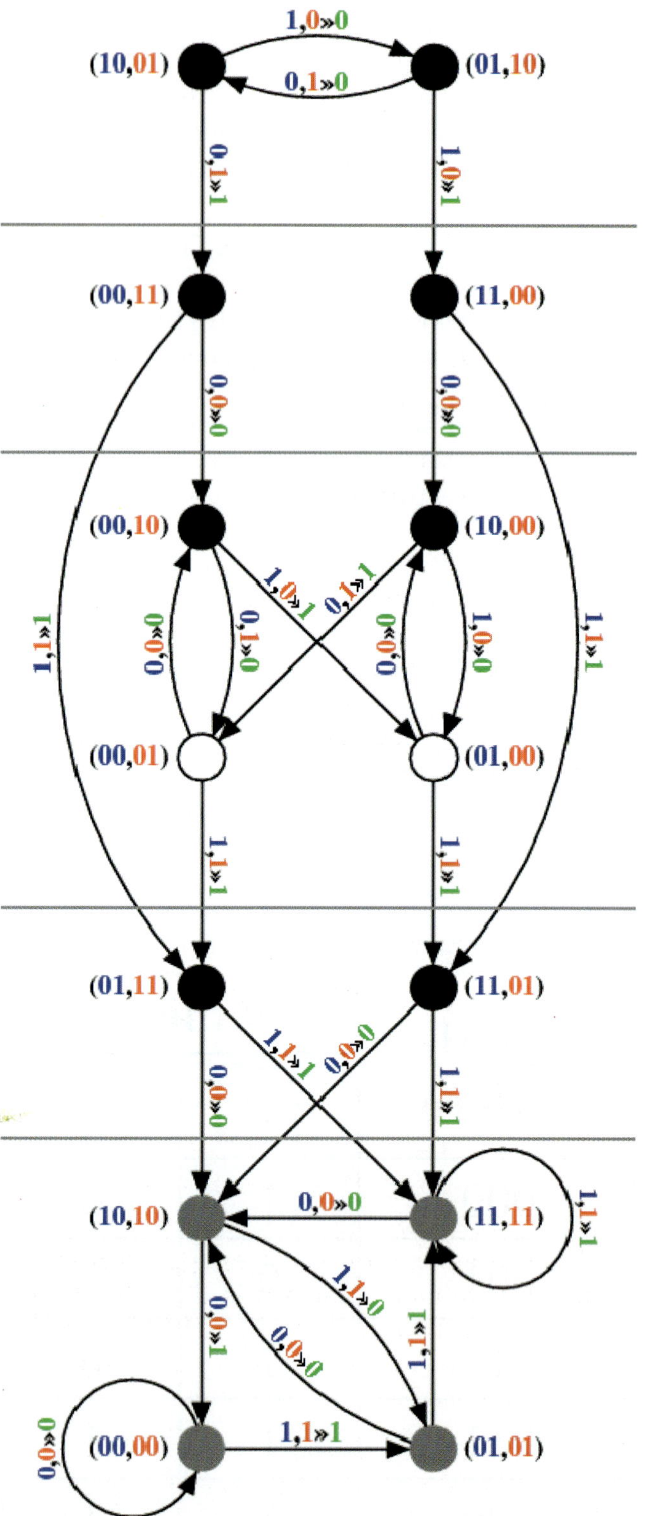

Fig. 11.   Isles of Eden digraph for local rule $\boxed{154}$. The nodes shaded in gray are *degenerate* in the sense that any path consisting solely of them can only generate equal strings. Horizontal gray lines partition the graph in a way that they cross only downward edges.

---

[5]The Isles of Eden digraph is basically the *de Bruijn* graph of strings of length L with the nodes being the Cartesian product of the set of two bit strings. For the definition of de Bruijn graph see the Appendix.

nodes are *generators*, gray nodes are *degenerate*, and every other node is white. Bits of the two generated input strings are colored blue and red, respectively, whereas the bits of their image, the output string, are green. A *walk* $e_0, e_1, \ldots, e_{n-1}$ in $G_{IE}(\boxed{N})$, e_i labeled $(e_{i,x}, e_{i,y}, e_{i,z})$, starting from a node $\mathbf{u}$, labeled $(\{u_{1,x}, u_{2,x}\}, \{u_{1,y}, u_{2,y}\})$, generates two input strings $\mathbf{x'}$ and $\mathbf{y'}$ and one output string $\mathbf{z}$:

$$
\begin{aligned}
\mathbf{x'} &= \{u_{1,x}, u_{2,x}, e_{0,x}, e_{1,x}, \ldots, e_{n-1,x}\}, \\
\mathbf{y'} &= \{u_{1,y}, u_{2,y}, e_{0,y}, e_{1,y}, \ldots, e_{n-1,y}\}, \quad (56) \\
\mathbf{z} &= \{e_{0,z}, e_{1,z}, \ldots, e_{n-1,z}\}
\end{aligned}
$$

for which local rule $\boxed{N}$ maps both $\{e_{i-2,x}, e_{i-1,x}, e_{i,x}\}$ and $\{e_{i-2,y}, e_{i-1,y}, e_{i,y}\}$ to $e_{i,z}$ for all $0 \leq i < n$, where $e_{-2,x}, e_{-1,x}, e_{-2,y}$ and $e_{-1,y}$ refer to $u_{1,x}$, $u_{2,x}$, $u_{1,y}$, and $u_{2,y}$, respectively.

The *cyclic boundary condition* of the local cellular automaton rules makes interesting those walks that are *closed* (cycles), which implies $n = L$. The cyclic boundary condition can be formalized by the following equations:

$$
\begin{aligned}
e_{-2,x} &\equiv u_{1,x} = e_{n-2,x}, \\
e_{-1,x} &\equiv u_{2,x} = e_{n-1,x}, \\
e_{-2,y} &\equiv u_{1,y} = e_{n-2,y}, \quad (57) \\
e_{-1,y} &\equiv u_{2,y} = e_{n-1,y}
\end{aligned}
$$

and thus $T_{\boxed{N}}(\mathbf{x}) = \mathbf{z}$ and $T_{\boxed{N}}(\mathbf{y}) = \mathbf{z}$, where

$$
\begin{aligned}
\mathbf{x} &= \{e_{L-1,x}, e_{0,x}, e_{1,x}, \ldots, e_{L-2,x}\}, \\
\mathbf{y} &= \{e_{L-1,y}, e_{0,y}, e_{1,y}, \ldots, e_{L-2,y}\}, \quad (58) \\
\mathbf{z} &= \{e_{0,z}, e_{1,z}, \ldots, e_{L-1,z}\}
\end{aligned}
$$

It follows that the initial node is labeled $\{e_{L-2,x}, e_{L-1,x}\}, \{e_{L-2,y}, e_{L-1,y}\}$.

We can assume $e_{L-2,x} \neq e_{L-2,y}$ because from Lemma 5.4 we know that if it is the nth bit where $\mathbf{x}$ and $\mathbf{y}$ differ, then we can take $\mathbf{x}^{(L-n)}$ and $\mathbf{y}^{(L-n)}$ instead of $\mathbf{x}$ and $\mathbf{y}$. Thus the first generated output bit is $\mathbf{z}_0$ and the last is $\mathbf{z}_{L-1}$.

Definition 5.7. *Degenerate Node and Cycle.* A node $\mathbf{u}$ of the Isles of Eden digraph, labeled $(\mathbf{u}_x, \mathbf{u}_y)$, is called *degenerate if, and only if,* $\mathbf{u}_x = \mathbf{u}_y$. A cycle in the Isles of Eden digraph is called *degenerate* if all the nodes in it are degenerate.

5.4. *Constructing the Isles of Eden digraph*

We give a constructive method to create the Isles of Eden digraph for a given local rule $\boxed{N}$.

Algorithm for constructing the Isles of Eden digraph

1. Add a node for every possible pair of two-bit binary strings $(x_{L-2}, x_{L-1}) - (y_{L-2}, y_{L-1})$ so that $x_{L-2} \neq y_{L-2}$, and label them accordingly. (There are eight such nodes.) These are the *generator nodes*, marked with *black* circles in the digraph.

2. Now take a node, labeled $(x_{i-1}, x_i) - (y_{i-1}, y_i)$, and examine it to see if an additional pair of bits (x_{i+1} and y_{i+1}) can be added so that (x_{i-1}, x_i, x_{i+1}) and (y_{i-1}, y_i, y_{i+1}) generates the same output. (We suppose that all previous output bits are equal, thus in this case $\mathbf{x}$ and $\mathbf{y}$ will also generate the same output, and to calculate the upcoming output bit the three input bits are available.) For every possible pair add an edge with a label containing x_{i+1} and y_{i+1} as the first two bits, and the generated bit $N(x_{i-1}, x_i, x_{i+1}) = N(y_{i-1}, y_i, y_{i+1})$ as the third bit. (Nodes with no appropriate continuation will have no outgoing edges, and thus will be *dead ends*.)

3. If there is a node labeled $(x_i, x_{i+1}) - (y_i, y_{i+1})$, then connect the given edge to it, if not, then create a new node for each such edge, and connect the edge to this new node.

4. If all nodes have been examined for possible continuations, then quit; otherwise pick a node and go to step 2.

Example 5.3. *Constructing the Isles of Eden Digraph for Rule* $\boxed{45}$. Initial nodes to be added are $(00, 10)$, $(00, 11)$, $(01, 10)$, $(01, 11)$, and their mirrors: $(10, 00)$, $(11, 00)$, $(10, 01)$ and $(11, 01)$.

Let us first examine node $(00, 10)$. According to the construction the label of a node contains the first two bits of the three bit strings for the next mapping. Thus 00 refers to the first two rows of Table 27, and can generate a 1 and a 0 with continuations of 0 and 1, respectively. The 2 bits 10 refers to rows 5 and 6 and can generate a 0 and a 1 with continuations of 0 and 1, respectively. Therefore, we have to add two edges from this node corresponding to the two possible outputs.

One will be labeled $(0, 1, 1)$, which means we concatenate a 0 to 00 and a 1 to 10, obtaining 000 and 101, respectively. Both generate an output 1 that is the third digit on the label of the edge. This edge will point to the node $(00, 01)$ that

Table 27.   Mapping rules of local rule $\boxed{45}$.

Input pattern	Output bit
000	1
001	0
010	1
011	1
100	0
101	1
110	0
111	0

Table 28.   Mapping rules of local rule $\boxed{154}$.

Input pattern	Output bit
000	0
001	1
010	0
011	1
100	1
101	0
110	0
111	1

is obtained by taking the last two digits of 000 and 101, respectively.

The other outgoing edge will be labeled $(1, 0, 0)$. The first digit is concatenated to 00, giving 001; the second digit is concatenated to 10, giving 100. Both strings generate 0, the third digit of the label. The edge will point to node $(01, 00)$ that refers to the last two digits of 001 and 100, respectively.

The nodes $(00, 01)$ and $(01, 00)$ have to be added to the graph, because they are not present yet. This process has to be done for the newly created nodes, and repeated until every node added to the graph was examined.

The nodes $(01, 11)$ and its mirror $(11, 01)$ are interesting, because they have no possible continuation to generate the same output: rows 3 and 4 (having the prefix 01) both generate 1, whereas rows 7 and 8 (having the prefix 11) both generate 0.

Example 5.4.   *Constructing the Isles of Eden Digraph for Rule* $\boxed{154}$. This example is a bit more complicated since it includes *degenerate* nodes as well. We have already used the Isles of Eden Digraph for Rule $\boxed{154}$ as the sample digraph chosen for illustrating Definition 5.6, shown in Fig. 11.

Initial nodes to be added are $(00, 10)$, $(00, 11)$, $(01, 10)$, $(01, 11)$, and their mirrors: $(10, 00)$, $(11, 00)$, $(10, 01)$ and $(11, 01)$.

Let us first examine node $(10, 01)$. According to our construction algorithm, the label of a node contains the first two bits of the three bit strings for the next mapping. Thus 10 refers to the rows 5 and 6 of Table 28, and can generate a 1 and a 0 with continuations of 0 and 1, respectively. The 2 bits 01 refers to rows 3 and 4 and can generate a 0 and a 1 with continuations of 0 and 1, respectively. Therefore we have to add two edges from this node corresponding to the two possible outputs.

One will be labeled $(0, 1, 1)$, which means we concatenate a 0 to 10 and a 1 to 01, thereby obtaining 100 and 011, respectively. Both generate an output 1 that is the third digit on the label of the edge. This edge will point to the node $(00, 11)$ that is obtained by taking the last two digits of 100 and 011, respectively.

The other outgoing edge will be labeled $(1, 0, 0)$. The first digit is concatenated to 10, giving 101; the second digit is concatenated to 01, giving 010. Both strings generate 0, the third digit of the label. The edge will point to node $(01, 10)$ corresponding to the last two digits of 001 and 100, respectively.

The nodes $(00, 11)$ and $(01, 10)$ must be added to the digraph, because they are not present yet. This process must be applied to the newly created nodes, and repeated until every node added to the graph was examined.

Remarks. Note that both digraphs are laid out and drawn to have the mirrored nodes next to each other, except for the degenerate (self-symmetric) ones. Also note that the statement of Lemma 5.4 means that a closed walk on a cycle can be started from any of its nodes, and the corresponding input and output strings can be transformed to each other by circular shifting of the same length.

5.5. *The full Isles of Eden digraph*

The number of nodes in the Isles of Eden digraph is between 8 and 16. The lower limit is due to the fact that eight nodes are always added in the first step of the algorithm. The upper bound is given by the number of possible pairs of bit-strings of length 2 that is equal to $(2^2)^2 = 16$. The reason for not including some nodes for certain local rules is that they have no incoming edges and thus they cannot be part of a cycle in the Isles of Eden digraph. Nevertheless, we may include all 16 nodes in any Isles of Eden digraph, since the number of nondegenerate cycles will not change. This digraph is called the *full Isles of Eden digraph* $G_{IE}^F(\boxed{N})$ of a given local rule $\boxed{N}$. It can be constructed using the following algorithm:

Algorithm for constructing the full Isles of Eden digraph

1. Add a node for every possible pair of 2-bit binary strings and label them accordingly. (There are 16 such nodes.)
2. Now take a node, labeled $(x_{i-1}, x_i) - (y_{i-1}, y_i)$, and examine it to see if an additional pair of bits (x_{i+1} and y_{i+1}) can be added so that $(x_{i-1}, x_i, x_{i+1}) - (y_{i-1}, y_i, y_{i+1})$ generate the same output. (We suppose that all previous output bits are equal, thus in this case **x** and **y** will also generate the same output, and to calculate the upcoming output bit the three input bits are available.) For every possible pair add an edge from this node to the node labeled $(x_i, x_{i+1}) - (y_i, y_{i+1})$, with a label containing x_{i+1} and y_{i+1} as the first two bits, and the generated bit $N(x_{i-1}, x_i, x_{i+1}) = N(y_{i-1}, y_i, y_{i+1})$ as the third bit. (Nodes with no appropriate continuation will have no outgoing edges, and thus will be dead ends.)
3. If all nodes have been examined for possible continuations, then quit, otherwise pick a node and go to step 2.

Since the full Isles of Eden digraph contains all 16 possible nodes, it also contains all four *degenerate* nodes. Note that the degenerate subgraph always contains the same edges, since an edge from one degenerate node to a second (not necessarily different) degenerate node refers to the same three-bit pattern for both **x** and **y**, and therefore to the same output bit. These edges are invariant for the full Isles of Eden digraphs of all 256 local rules. There are eight of them, two per each degenerate node.

5.6. *Nondegenerate cycles and Isles of Eden*

As we have seen, in order to generate an output with correct cyclic boundary conditions we need to return to the starting node. It follows that for every input pair $(\mathbf{x}, \mathbf{y}) \in \sum^L$ that can generate the same output string $\mathbf{z} \in \sum^L$ and for which $\mathbf{x} \neq \mathbf{y}$, there is a cycle of length L in the Isles of Eden digraph, and every cycle of length L containing at least one nondegenerate node in the graph represents such an input pair $(\mathbf{x}, \mathbf{y}) \in \sum^L$. Thus the length of the input strings equals the length of the cycles, and all lengths for which there is no cycle in the graph will be a contradiction. The following theorem is a joint result of this observation and Lemma 5.2:

Theorem 5.1. *Every orbit of a local rule* $\boxed{N}$ *over* $\sum^L$ *is an Isle of Eden if, and only if, its associated Isles-of-Eden digraph* $G_{IE}(\boxed{N})$ *has no nondegenerate cycle of length* L.

Proof

Necessity

Let us suppose the contrary that every orbit of a local rule $\boxed{N}$ over $\sum^L$ is an Isle of Eden, but there is a nondegenerate cycle of length L in $G_{IE}(\boxed{N})$. The input strings $\mathbf{x} \in \sum^L$ and $\mathbf{y} \in \sum^L$ generated by this cycle have the same image $z \in \sum^L$ under $T_{\boxed{N}}$, but they are different because it has a node labeled $(\mathbf{u}_x, \mathbf{u}_y)$ with $\mathbf{u}_x$ and $\mathbf{u}_y$ being different substrings of **x** and **y**, respectively, starting from the same index. Therefore **z** has two different preimages under $T_{\boxed{N}}$. This contradicts the fact that every $\mathbf{z} \in \sum^L$ has a unique preimage under $T_{\boxed{N}}$ that follows from Lemma 5.2, since we have assumed that every orbit of a local rule $\boxed{N}$ over $\sum^L$ is an Isle of Eden.

Sufficiency

Since there is no nondegenerate cycle of length L in $\mathcal{G}_{IE}(\boxed{N})$, thus there exist no two different strings that are mapped to the same $\mathbf{z} \in \sum^L$. Therefore every $\mathbf{z} \in \sum^L$ has at most one preimage under $T_{\boxed{N}}$, and using Lemmas 5.3 and 5.2 respectively, it follows that every orbit of a local rule $\boxed{N}$ over $\sum^L$ is an Isle of Eden. ∎

Finding cycles in directed graphs is generally a difficult task, but by topologically ordering the nodes it can be made much easier, if most edges point toward the same direction. We organized the Isles of Eden digraphs in Figs. 11 and 12 so that most of the edges point downwards. It is trivial that a cycle has to contain edges of the other directions

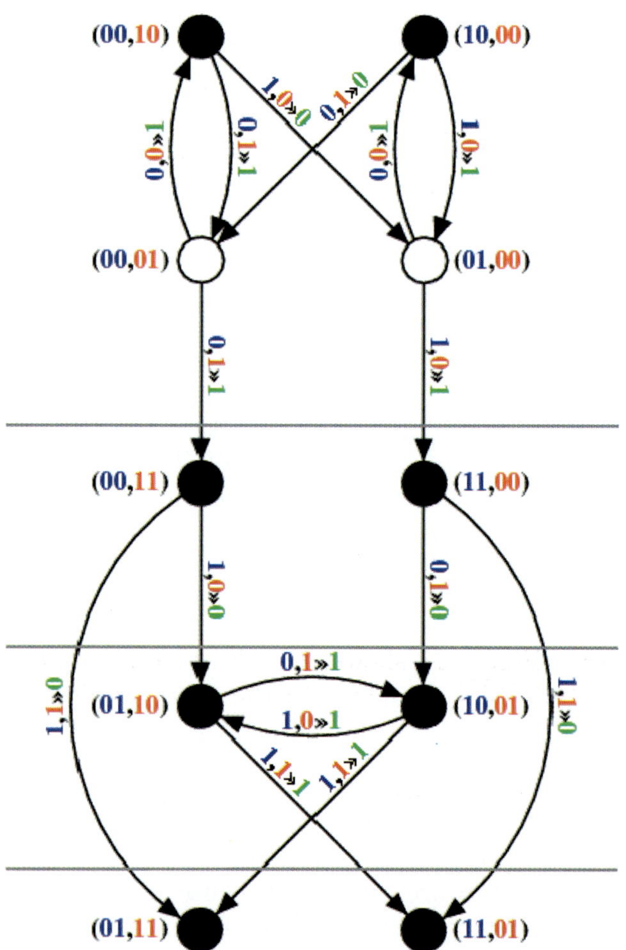

Fig. 12.   Isles of Eden digraph for local rule $\boxed{45}$. Cycles in it refer to different inputs that map to the same output. Horizontal gray lines partition the digraph in a way that they cross only downward edges.

(horizontal or upward) too. Some cuts are drawn on the digraphs that cross only downward edges. Observe that no cycles can contain nodes from the two sides of any such cut. Therefore these cuts partition the graph into subgraphs, and we only need to examine these subgraphs for cycles. We will use this observation to find *all* nondegenerate cycles of an Isles of Eden digraph.

It is enough to look for simple cycles (all nodes are different except the starting and ending nodes), because other cycles can be decomposed into single cycles, therefore their lengths are sums of the length of the simple cycles they are composed of.

5.7. *Effect of global equivalence transformations on Isles of Eden digraphs*

Based on the three global equivalence transformations $T^\dagger, \overline{T}$ and T^* [Chua *et al.*, 2007], as well as the alternating transformation $\widetilde{T}$ [Chua *et al.*, 2007], corresponding transformations can be established for the Isles of Eden digraphs of globally equivalent, or quasi-equivalent rules; namely,

> 1. Left-right transformation:
> $$\mathcal{G}^\dagger_{IE}(\boxed{N}) \triangleq \mathcal{G}_{IE}(T^\dagger(\boxed{N}))$$
> 2. Global complementation:
> $$\overline{\mathcal{G}}_{IE}(\boxed{N}) \triangleq \mathcal{G}_{IE}(\overline{T}(\boxed{N}))$$
> 3. Left-right complementation:
> $$\mathcal{G}^*_{IE}(\boxed{N}) \triangleq \mathcal{G}_{IE}(T^*(\boxed{N}))$$
> 4. Alternating transformation:
> $$\widetilde{\mathcal{G}}_{IE}(\boxed{N}) \triangleq \mathcal{G}_{IE}(\widetilde{T}(\boxed{N}))$$

The *global complementation* on $\mathcal{G}_{IE}(\boxed{N})$ can be easily characterized for any local rule $\boxed{N}$. The corresponding Isles-of-Eden digraph $\overline{\mathcal{G}}_{IE}(\boxed{N})$ can be generated from $\mathcal{G}_{IE}(\boxed{N})$ by inverting all bits in the labels of the nodes and the edges. Naturally, $\mathcal{G}^*_{IE}(\boxed{N})$ can be generated from $\mathcal{G}^\dagger_{IE}(\boxed{N})$ in the same way. The *alternating* transformation is even more simple: to compute $\widetilde{\mathcal{G}}_{IE}(\boxed{N})$ from $\mathcal{G}_{IE}(\boxed{N})$, only the output bits have to be inverted.

Characterization of the *left-right transformation* is more complicated. Twenty out of the 64 possible edges are not affected by it, whereas changes regarding the rest of the edges depend on the given rule.[6]

---

[6]Eight out of the 20 are the invariant edges in the *degenerate* subgraph. The remaining 12 edges start from the 12 nondegenerate nodes of $\mathcal{G}_{IE}(\boxed{N})$.

5.8. *Dense Isles of Eden from rule* $\boxed{45}$ *and rule* $\boxed{154}$

Theorem 5.2. *Every orbit of rule* $\boxed{45}$ *over* $\sum^L$ *is an* Isle of Eden *if, and only if,* L *is an* odd *number.*

Proof. Figure 12 shows the Isles of Eden digraph for Rule $\boxed{45}$. We will see that for any positive even number there is a cycle in $G_{IE}\big(\boxed{45}\big)$ with this length. From this, using Theorem 5.1, it follows that if L is an even number, then there is a cycle of length L in the graph and thus there is an orbit of Rule $\boxed{45}$ over $\sum^L$ that is not an Isle of Eden. Then we will prove that all cycles in the graph are of even length, therefore if L is an odd number, then there are no cycles of length L, and using Theorem 5.1 we can conclude that every orbit of Rule $\boxed{45}$ over $\sum^L$ is an Isle of Eden.

Necessity

For every length $L = 2n$ there is a cycle[7] in $G_{IE}\big(\boxed{45}\big)$, e.g. starting from $(01, 10)$ and taking the cycle of length 2 through $(10, 01)$ back to $(01, 10)$ n times. These cycles refer to the strings $(01)^n$ and $(10)^n$ and both map to $(11)^n$.

Sufficiency

The upper four nodes form a *bipartite* subgraph, within which there are only cycles of even length. If the starting state is not in the upper four nodes, then the only possibility for a cycle is between node $(01, 10)$ and $(10, 01)$, because the rest of the subgraphs are not even connected. This is also a bipartite subgraph, so this allows only for cycles of even length too. ∎

5.8.1. *Another Proof for Theorem 5.2*

If we determine the adjacency matrix $A_{\boxed{45}}$ of the graph in Fig. 12, the number of cycles of length n can be easily determined. There are no degenerate nodes in $G_{IE}\big(\boxed{45}\big)$, so all cycles are nondegenerate. Since the number of cycles of length n starting at node i is $A_{\boxed{45}}^n(i,j)$, thus the total number of cycles of length n is $\mathrm{Tr}(A_{\boxed{45}}^n)$, the trace of $A_{\boxed{45}}^n$. The trace $\mathrm{Tr}(A_{\boxed{45}}^n)$ is equal to the sum of the eigenvalues of $A_{\boxed{45}}^n$, which allows us to compute it for any n without actually computing $A_{\boxed{45}}^n$. The adjacency matrix

for rule $\boxed{45}$ is defined as follow:

$$A_{\boxed{45}} = \begin{bmatrix} 0 & 1 & 1 & 0 & 0 & 0 & 0 & 0 & 0 & 0 \\ 1 & 0 & 0 & 1 & 0 & 0 & 0 & 0 & 0 & 0 \\ 0 & 0 & 0 & 0 & 1 & 1 & 0 & 0 & 0 & 0 \\ 0 & 0 & 0 & 0 & 0 & 0 & 1 & 0 & 1 & 0 \\ 0 & 0 & 0 & 0 & 0 & 0 & 0 & 1 & 0 & 1 \\ 0 & 0 & 0 & 0 & 0 & 0 & 0 & 0 & 0 & 0 \\ 1 & 0 & 0 & 1 & 0 & 0 & 0 & 0 & 0 & 0 \\ 0 & 0 & 0 & 0 & 1 & 1 & 0 & 0 & 0 & 0 \\ 0 & 0 & 0 & 0 & 0 & 0 & 0 & 1 & 0 & 1 \\ 0 & 0 & 0 & 0 & 0 & 0 & 0 & 0 & 0 & 0 \end{bmatrix}$$

The eigenvalues of $A_{\boxed{45}}$ can be computed from its characteristic polynomial:

$$x^6(x^4 - 3x^2 + 2) = 0 \tag{59}$$

The nonzero eigenvalues are $\lambda_{1,2} = \pm\sqrt{2}$ and $\lambda_{3,4} = \pm 1$. Note that the eigenvalues of $A_{\boxed{45}}$ are real and sign symmetric, and hence their sum is zero. Since the eigenvalues of $A_{\boxed{45}}^n$ are $(\lambda_i)^n$, they are also sign symmetric for odd n, whereas they are real positive numbers for even n.

Therefore

$$\begin{aligned} \mathrm{Tr}(A_{\boxed{45}}) &= \sum (\lambda_i)^n > 0, \quad \text{for } n = 2k \\ \mathrm{Tr}(A_{\boxed{45}}) &= \sum (\lambda_i)^n = 0, \quad \text{for } n = 2k+1 \end{aligned} \tag{60}$$

where $k \in \mathbb{N}$. This is what we wanted to prove.

5.8.2. *Isles-of-Eden density criterion for rule* $\boxed{154}$

Theorem 5.3. *Every orbit of rule* $\boxed{154}$ *over* $\sum^L$ *is an* Isle of Eden *if, and only if,* L *is an* odd *number.*

Proof. Figure 11 shows the Isles of Eden digraph for Rule $\boxed{154}$. Similarly to the proof of Theorem 5.2 we only have to show that for any positive even number there is a cycle in the Isles of Eden digraph of this length touching a nondegenerate node, and that all cycles in the graph touching a nondegenerate node are of even length.

Necessity

For every length $L = 2n$ there is a cycle in the Isles-of-Eden digraph $G_{IE}\big(\boxed{154}\big)$, e.g. starting from $(01, 10)$ and taking the cycle of length 2 through $(10, 01)$ back to $(01, 10)$ n times. None of these

---

[7]See Table 29 for some illustrative examples.

Table 29.   Examples of cycles of even length for rule $\boxed{45}$.

L	Route	Input strings	Output string
4	(01,10)- (10,01)- (01,10)- (10,01)- (01,10)-	0 1 0 1 1 0 1 0	1111
4	(00,01)- (00,10)- (01,00)- (10,00)- (00,01)-	0 1 0 0 0 0 0 1	0101
6	(01,10)- (10,01)- (01,10)- (10,01)- (01,10)- (10,01)- (01,10)-	0 1 0 1 0 1 1 0 1 0 1 0	111111
6	(00,10)- (00,01)- (00,10)- (00,01)- (00,10)- (00,01)- (00,10)-	0 0 0 0 0 0 1 0 1 0 1 0	111111
8	(01,10)- (10,01)- (01,10)- (10,01)- (01,10)- (10,01)- (01,10)- (10,01)- (01,10)-	0 1 0 1 0 1 0 1 1 0 1 0 1 0 1 0	11111111

nodes are degenerate. These cycles refer to the strings $(01)^n$ and $(10)^n$ and both map to $(00)^n$.

Sufficiency

Let us take the indicated subgraphs of $\mathcal{G}_{IE}(\boxed{154})$ in a top-down order. The upper two nodes form a bipartite subgraph, within which there are only cycles of even length. The next subgraph (nodes $(00,11)$ and $(11,00)$) is not connected. The next subgraph of four nodes is also a bipartite subgraph, thus it contains only cycles of even length. The next subgraph (nodes $(01,11)$ and $(11,01)$) is not connected again. Although the bottommost subgraph contains cycles of odd length, these cycles include only degenerate nodes. If the starting node is in this subgraph, then the two input strings will be equal since the pairs of these four nodes contain equal strings, and also, all edges have equal digits for the two input strings. Therefore, it does not contain any cycles corresponding to different input strings.   ∎

5.8.3.   *Another Proof for Theorem 5.3*

Since the Isles-of-Eden digraph $\mathcal{G}_{IE}(\boxed{154})$ of $\boxed{154}$ contains degenerate nodes, the number of total cycles is higher than the number of nondegenerate cycles. Thus, we cannot use the trace of its adjacency matrix $A_{\boxed{154}}$. But since degenerate nodes in $\mathcal{G}_{IE}(\boxed{154})$ do not have outgoing edges going to a nondegenerate node, therefore no nondegenerate cycles can include degenerate nodes, and we only need to consider the adjacency matrix of the nondegenerate subgraph[8] of $\mathcal{G}_{IE}(\boxed{154})$. Let $\hat{A}_{\boxed{154}}$ denote the adjacency matrix of the nondegenerate subgraph of $A_{\boxed{154}}$. Computing $\hat{A}_{\boxed{154}}$ gives

$$\hat{A}_{\boxed{154}} = \begin{bmatrix} 0 & 1 & 0 & 0 & 0 & 1 & 0 & 0 & 0 & 0 \\ 1 & 0 & 0 & 1 & 0 & 0 & 0 & 0 & 0 & 0 \\ 0 & 1 & 0 & 0 & 0 & 1 & 0 & 0 & 0 & 0 \\ 0 & 0 & 0 & 0 & 0 & 0 & 1 & 0 & 0 & 1 \\ 0 & 0 & 0 & 0 & 0 & 0 & 0 & 1 & 1 & 0 \\ 0 & 0 & 0 & 0 & 0 & 0 & 0 & 0 & 0 & 0 \\ 1 & 0 & 0 & 1 & 0 & 0 & 0 & 0 & 0 & 0 \\ 0 & 0 & 1 & 0 & 1 & 0 & 0 & 0 & 0 & 0 \\ 0 & 0 & 0 & 0 & 0 & 0 & 1 & 0 & 0 & 1 \\ 0 & 0 & 0 & 0 & 0 & 0 & 0 & 0 & 0 & 0 \end{bmatrix}$$

Although the nondegenerate subgraph of $\mathcal{G}_{IE}(\boxed{154})$ is not isomorphic to $\mathcal{G}_{IE}(\boxed{45})$, but their characteristic polynomials are the same. Therefore their spectra are equal, and $\hat{A}_{\boxed{154}}$ has the same nonzero eigenvalues as $A_{\boxed{45}}$. Therefore the rest of the proof is the same as for $A_{\boxed{45}}$.   ∎

5.9.   *Dense Isles of Eden from rule* $\boxed{105}$ *and rule* $\boxed{150}$

As a further application of Theorem 5.1, in addition to the new results in Theorems 5.2 and 5.3, we will now prove Theorem 6 from [Chua *et al.*, 2007]:

Theorem 5.4.   *Every bit string of rules* $\boxed{150}$ *and* $\boxed{105}$ *is an Isle of Eden if, and only if, $L/3$ is not an integer.*

Proof. At first we construct the Isles of Eden digraphs for rules $\boxed{150}$ and $\boxed{105}$. These are shown in Fig. 13.

Since rules $\boxed{105}$ and $\boxed{150}$ are the alternating transforms of each other, their digraphs $\mathcal{G}_{IE}(\boxed{105})$ and $\mathcal{G}_{IE}(\boxed{150})$ are very similar. Observe that only the green output bits are inverted. Since the output bit does not affect the topology of the digraph, their adjacency matrices are equal:

$$A_{\boxed{105}} = A_{\boxed{150}}$$
$$= \begin{bmatrix} 0 & 0 & 1 & 0 & 1 & 0 & 0 & 0 & 0 & 0 & 0 & 0 \\ 1 & 0 & 0 & 1 & 0 & 0 & 0 & 0 & 0 & 0 & 0 & 0 \\ 0 & 1 & 0 & 0 & 0 & 1 & 0 & 0 & 0 & 0 & 0 & 0 \\ 0 & 0 & 0 & 0 & 0 & 0 & 0 & 1 & 0 & 1 & 0 & 0 \\ 0 & 0 & 0 & 0 & 0 & 0 & 1 & 0 & 0 & 0 & 1 & 0 \\ 0 & 0 & 0 & 0 & 0 & 0 & 0 & 0 & 1 & 0 & 0 & 1 \\ 1 & 0 & 0 & 1 & 0 & 0 & 0 & 0 & 0 & 0 & 0 & 0 \\ 0 & 1 & 0 & 0 & 0 & 1 & 0 & 0 & 0 & 0 & 0 & 0 \\ 0 & 0 & 1 & 0 & 1 & 0 & 0 & 0 & 0 & 0 & 0 & 0 \\ 0 & 0 & 0 & 0 & 0 & 0 & 1 & 0 & 0 & 0 & 1 & 0 \\ 0 & 0 & 0 & 0 & 0 & 0 & 0 & 0 & 1 & 0 & 0 & 1 \\ 0 & 0 & 0 & 0 & 0 & 0 & 0 & 1 & 0 & 1 & 0 & 0 \end{bmatrix}$$

The characteristic polynomial of the matrix is

$$x^9(x^3 - 8) = 0 \tag{61}$$

The nonzero eigenvalues are $\lambda_1 = 2$ and $\lambda_{2,3} = -1 \pm i\sqrt{3}$. Note that the nonzero eigenvalues are the complex cubic roots of 8, that is, vectors of length 2 and argument 0, $2\pi/3$, and $4\pi/3$, respectively. Therefore the following statements hold for the

---

[8]The same reasoning also works if the degenerate subgraph is connected to the nondegenerate subgraph only by outgoing edges.

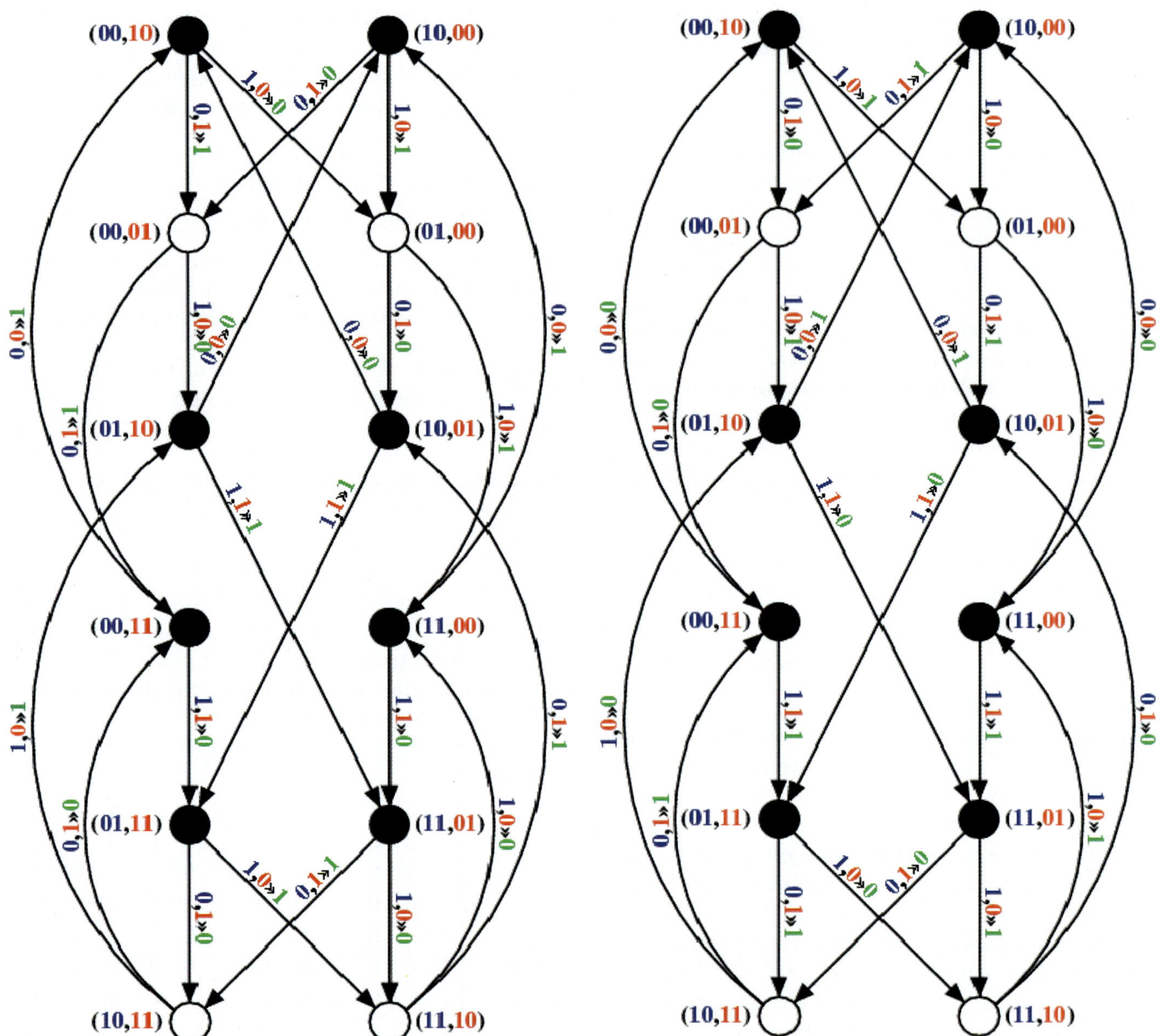

Fig. 13.   Isles of Eden digraph of local rules $\boxed{150}$ (left) and $\boxed{105}$ (right).

complex eigenvalues and $k \in \mathbb{N}$:

$$2(\lambda_2)^{3k+1} = (\lambda_3)^{3k+2} = 2^{3k+1}\lambda_2 \qquad (62)$$

and

$$2(\lambda_3)^{3k+1} = (\lambda_2)^{3k+2} = 2^{3k+1}\lambda_3 \qquad (63)$$

The sum of the eigenvalues is zero, and due to Eqs. (62) and (63) the sum of powers of the eigenvalues is also zero for every integer not divisible by 3, whereas they are real positive numbers for $n = 3k$, because $(\lambda_i)^{3k} = 8^k$, for $i = 1, 2, 3$.

Therefore

$$\mathrm{Tr}(A_{\boxed{105}}) = \sum(\lambda_i)^n > 0, \quad \text{for } n = 3k$$

$$\mathrm{Tr}(A_{\boxed{105}}) = \sum(\lambda_i)^n = 0, \quad \text{for } n = 3k+1 \qquad (64)$$

$$\text{and} \quad n = 3k+2$$

where $k \in \mathbb{N}$. Since $A_{\boxed{105}} = A_{\boxed{150}}$, it follows that all cycles in $\mathcal{G}_{IE}\left(\boxed{105}\right)$ and $\mathcal{G}_{IE}\left(\boxed{150}\right)$ are of lengths divisible by 3. The statement of the theorem now follows directly from Theorem 5.1.   ∎

5.10. *Gallery of Isles of Eden digraphs of eight representative local rules*

To illustrate that the Isles of Eden digraph is applicable to *any* local rule $\boxed{N}$, we have constructed the Isles of Eden digraphs of eight representative local rules $\boxed{90}$, $\boxed{60}$, $\boxed{170}$, $\boxed{102}$, $\boxed{15}$, $\boxed{62}$, $\boxed{30}$, and $\boxed{110}$, and exhibited them in Figs. 14–21 respectively.

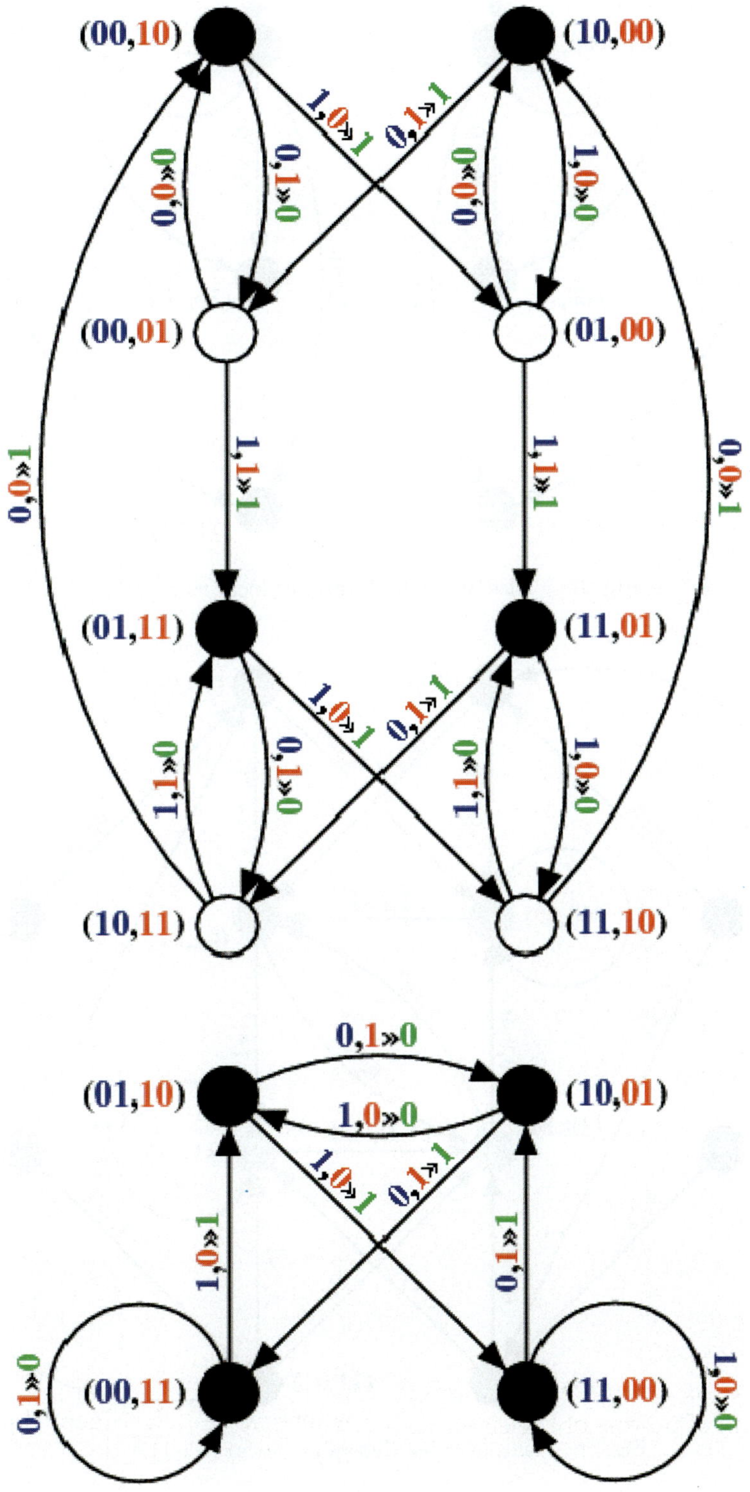

Fig. 14. Isles of Eden digraph of local rule $\boxed{90}$.

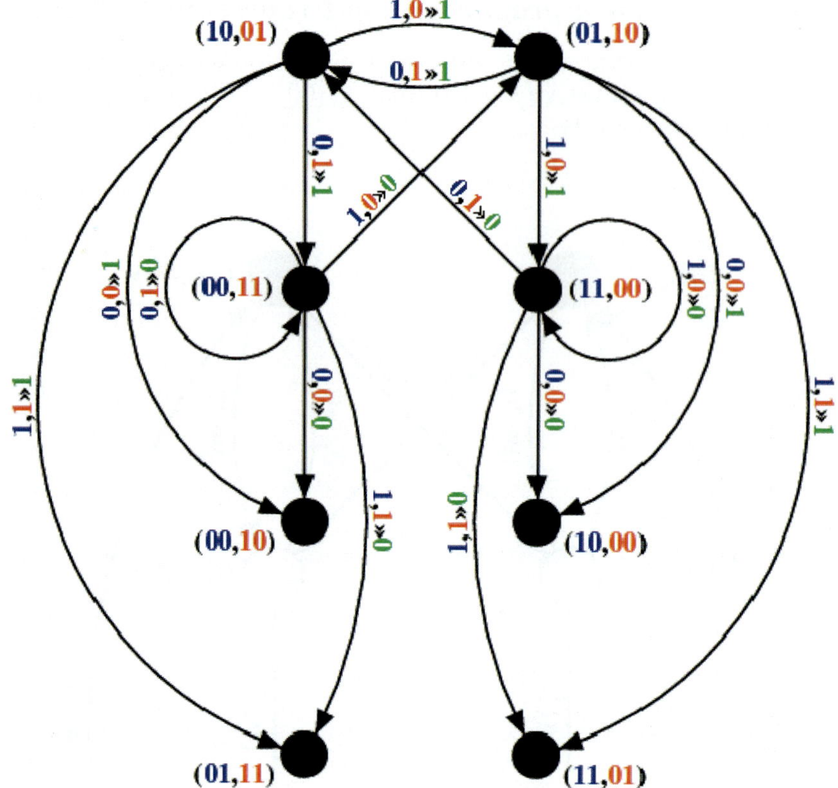

Fig. 15. Isles of Eden digraph of local rule 60.

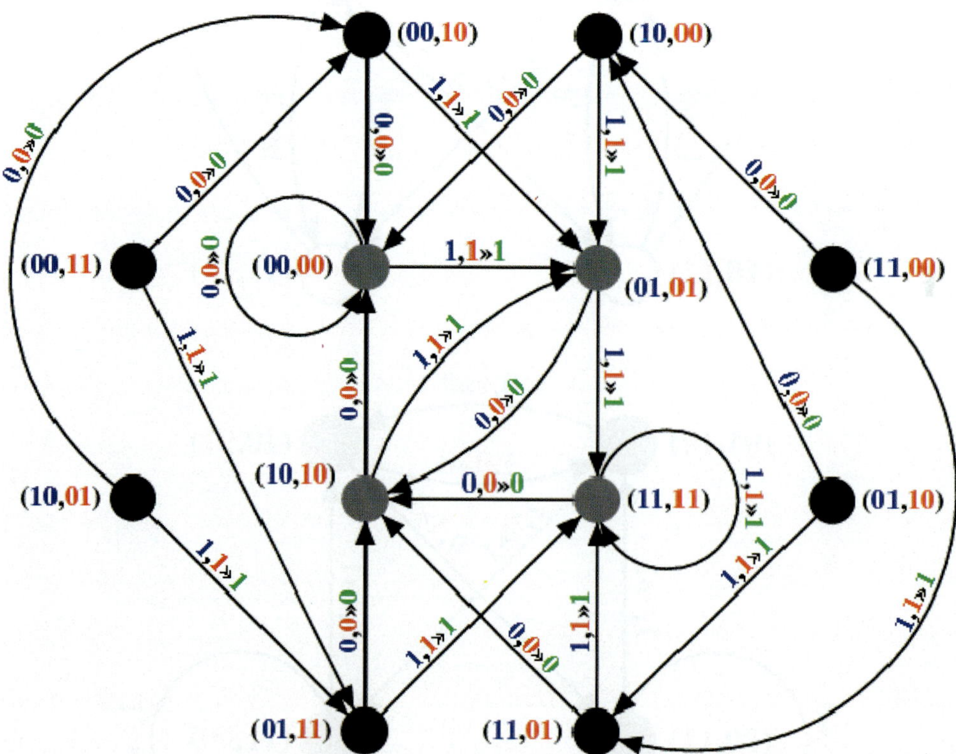

Fig. 16. Isles of Eden digraph of local rule 170.

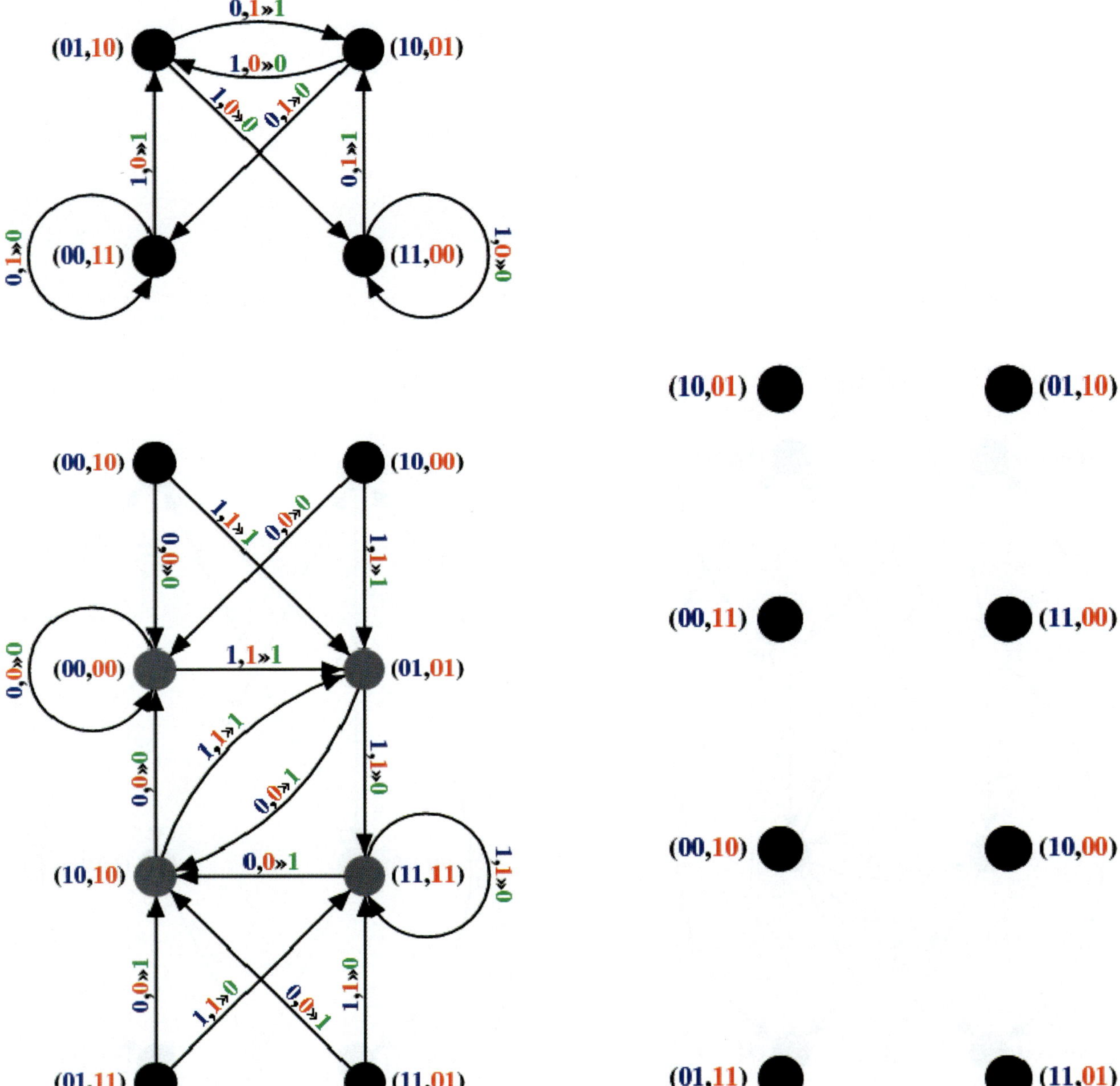

Fig. 17.   Isles of Eden digraph of local rule $\boxed{102}$.

Fig. 18.   Isles of Eden digraph of local rule $\boxed{15}$.

Fig. 19.   Isles of Eden digraph of local rule ⑥②.

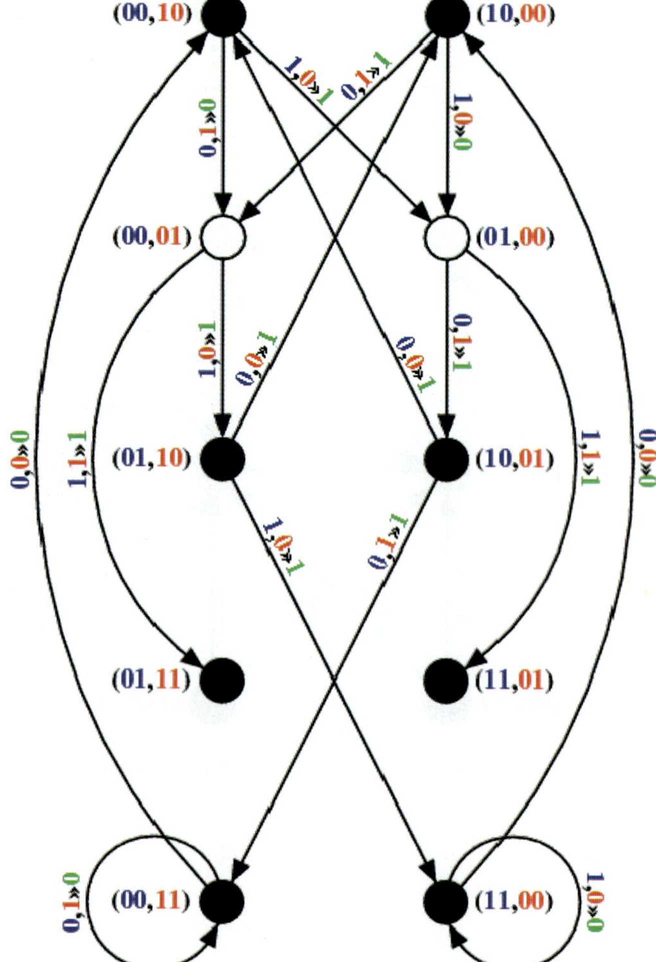

Fig. 20.   Isles of Eden digraph of local rule ③⓪.

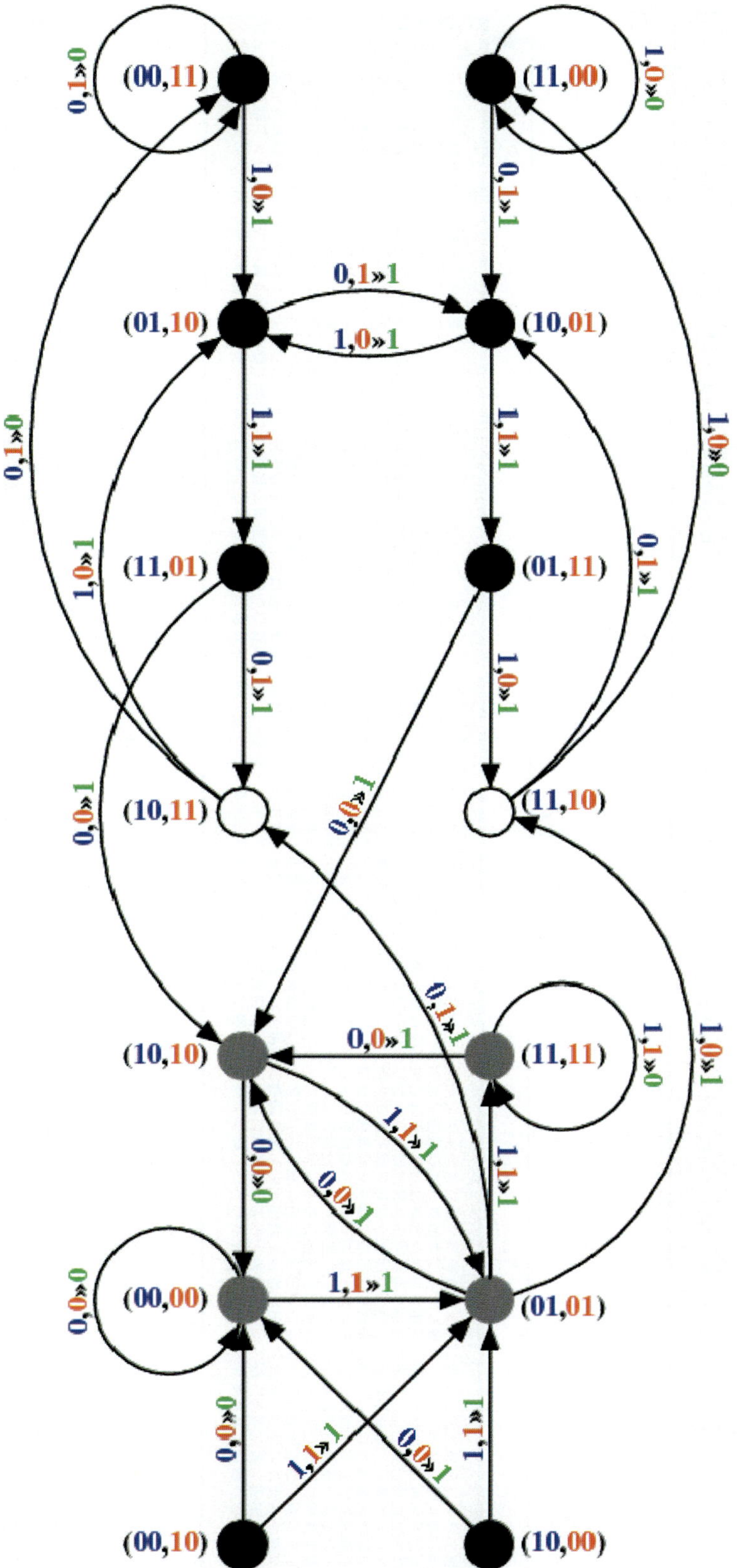

Fig. 21.   Isles of Eden digraph of local rule 110.

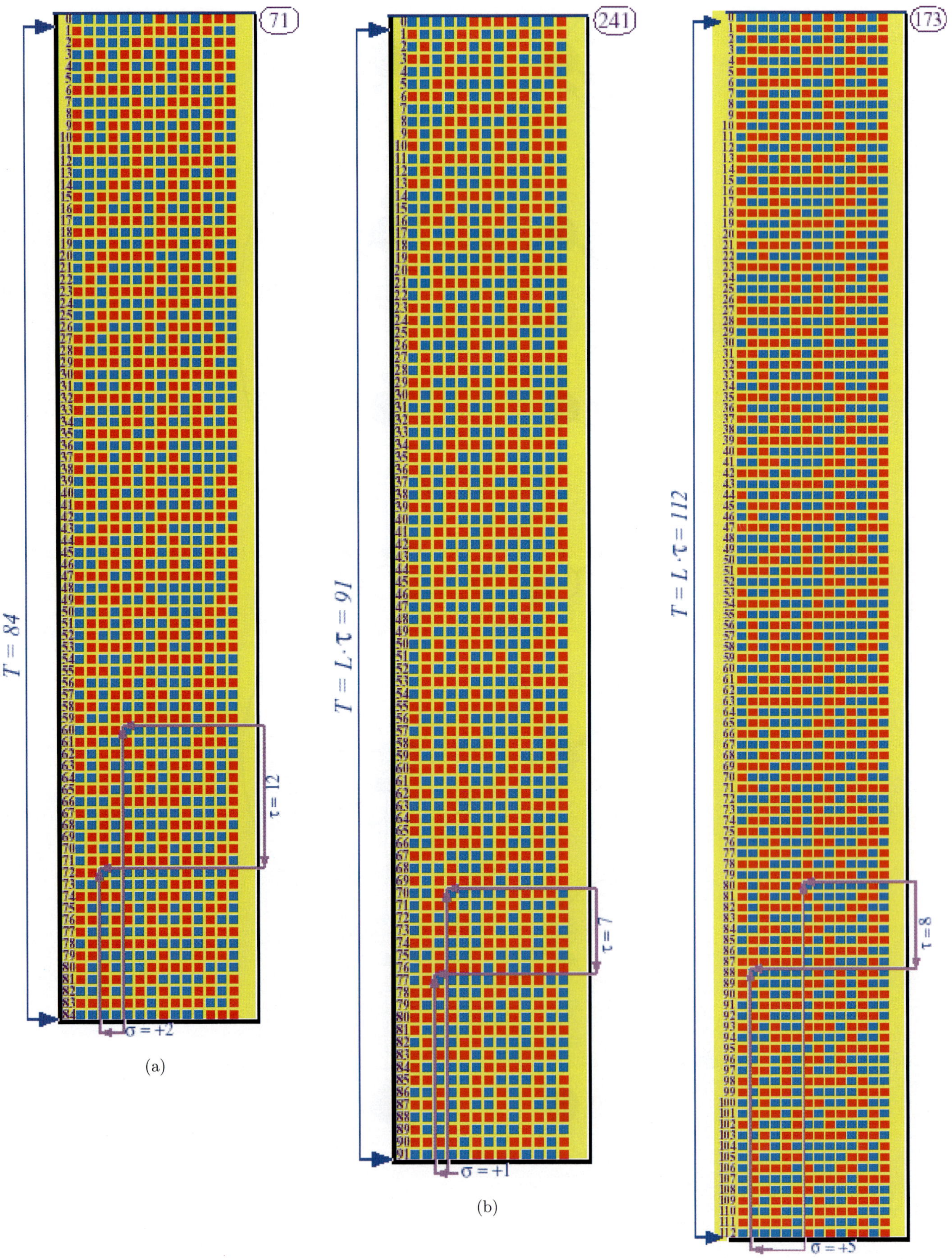

Fig. 22.   Three isolated Isles of Eden of local rule $\boxed{30}$.

The *dense Isles of Eden Criterion* can be used to derive the conditions, if any, for a local rule to be endowed with *dense* Isles of Eden. For example, since the preceding *Isles of Eden digraphs* of local rules $\boxed{170}$ and $\boxed{15}$ do *not* have any *non-degenerate* cycles (i.e., cycles which do not contain *gray* nodes) for all L, they are endowed with *dense* Isles of Eden for all $L \geq 3$. It follows that local rules $\boxed{170}$ and $\boxed{15}$, as well as their globally-equivalent rules $\boxed{240}$ and $\boxed{85}$, have *no transient regimes*. Similarly, since the *Isles of Eden digraphs* of rules $\boxed{150}$ and $\boxed{105}$ do *not* have any *non-degenerate* cycles for all L not divisible by 3, it follows that rules $\boxed{150}$ and $\boxed{105}$ are endowed by *dense* Isles of Eden for all L which is not a multiple of 3.

6. Concluding Remarks

We have proved that local rule $\boxed{60}$ is devoid of any *Isles of Eden* for all $L < \infty$. We have derived a *global state transition formula* for local rule $\boxed{60}$ which resembles that derived in [Chua *et al.*, 2007] for rule $\boxed{90}$. We have proved that local rules $\boxed{154}$ and $\boxed{45}$ are inhabited exclusively by *Isles of Eden* for all L not divisible 2 (i.e. when L is an *odd* integer), just like the *dense Isles of Eden* inhabiting rules $\boxed{150}$ and $\boxed{105}$ when L is *not* divisible by 3 [Chua *et al.*, 2007]. However, unlike rules $\boxed{150}$ and $\boxed{105}$ where all orbits are *exclusive* attractors when L is divisible by 3, we found both Isles of Eden and attractors to be *coexisting* in rules $\boxed{154}$ and $\boxed{45}$, when L is divisible by 2 (i.e. when L is an *even* integer). These properties are truly remarkable because they are not easily recognizable from their space-time patterns. For example, there is no *telltale* characteristics in Figs. 5(a) and 5(b) to suggest that the pattern in Fig. 5(a) is an *Isle of Eden*, but that in Fig. 5(b) is an *attractor*. Likewise, although the space-time patterns for rule $\boxed{45}$ in Figs. 6(a) and 6(b) look quite similar to each other, yet the

Fig. 23.   The island of *Stromboli*. This picture was taken by Roberto Rinaldi and featured in the exquisitely beautiful book "The Eolian Islands". Permission granted by the publisher Società Editrice "Affinità Elettive".

former [Fig. 6(a)] is an *attractor* while the latter [Fig. 6(b)] is an *Isle of Eden*. Our main theorem in Sec. 4 predicts that *all* orbits of Fig. 5(a) (with $L = 51$), and all orbits of Fig. 6(b) (with $L = 51$), are Isles of Eden by virtue of the fact that $L = 51$ is an *odd* integer.

Finally, we remark that the *Isle-of-Eden digraph* is a powerful tool not only for identifying local rules inhabited by *dense* Isles of Eden, but also for *excluding* them. For example, the presence of a *self loop* at any *nondegenerate* node (black or white-colored node in the Isles-of-Eden digraph) would exclude the presence of *dense* Isles of Eden. Unfortunately, this digraph cannot be used to detect the presence of *isolated* Isles of Eden, such as those exhibited in Table 24 for rule $\boxed{154}$, and in Table 25 for rule $\boxed{45}$, when L is an *even* integer, and those shown in Fig. 22 of local $\boxed{30}$ for $L = 13, 14$. In general, it would be extremely lucky for any one to discover an *isolated Isle of Eden* of any local rule with $L > 50$, specially the *hyper* Bernoulli shift rules in Table 12 of [Chua *et al.*, 2007]. This is why every *isolated Isle of Eden* with very large periods is a *gem*. We end this paper with a picturesque metaphor depicted in Fig. 23 of such a gem.

Appendix

Definition. *De Bruijn graph* [Wikipedia]. An n-dimensional *de Bruijn graph* of m symbols is a directed graph representing overlaps between sequences of symbols. It has m^n vertices, consisting of all possible length-n sequences of the given symbols; the same symbol may appear multiple times in a sequence. Given m symbols $s_1, s_1, \ldots, s_m$ the set of vertices is:

$$V = \left\{ \underbrace{(s_1, \ldots, s_1, s_1)}_{n \text{ times}}, \underbrace{(s_1, \ldots, s_1, s_2)}_{n \text{ times}}, \ldots, \underbrace{(s_m, \ldots, s_m, s_m)}_{n \text{ times}} \right\} \tag{A.1}$$

If one of the vertices can be expressed by shifting all symbols by one place to the left and adding a new symbol at the end of another vertex, then the latter has a directed edge to the former vertex. Thus the set of (directed) edges is given by

$$E = \{((v_1, v_2, \ldots, v_n), (w_1, w_2, \ldots, w_n)) : v_2 = w_1, v_3 = w_2, \ldots, v_n = w_{n-1}\} \tag{A.2}$$

As an example, Fig. 24 shows a de Bruijn graph of length 2.

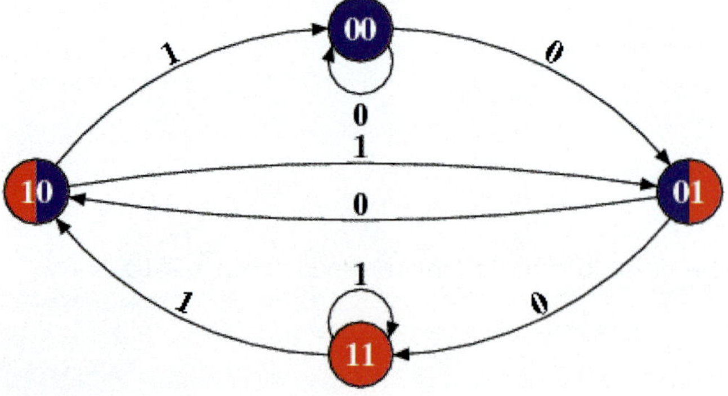

Fig. 24.   De Bruijn graph of strings of length two.

ERRATA FOR VOLUME I

1. *Page x* (front matter)

   Section 3.1: Paritioning should read Partitioning, and 89 should be changed to 88

2. *Page 6*

   Line 9 (column 1): Change u_{u-1} to u_{i-1}

3. *Page 76*

   There is a typo in Eq. (9):
   change "fof" to "for"

4. *Page 86*

   Equations (22) and (23) should read:

$$w(\sigma) = (\sigma - 3)(\sigma + 1) = \sigma^2 - 2\sigma - 3 \tag{22}$$

$$\dot{x}_i = g(x_i) + (u_{i-1} + 2u_i - 3u_{i+1})^2 - 2(u_{i-1} + 2u_i - 3u_{i+1}) - 3 \tag{23}$$

5. *Page 91*

   Top figure (a): Change the color of the leftmost "red" from *red* to *blue*.

6. *Page 141*

   The equation for rule 63 should read:

$$u_i^{t+1} = \text{sgn}[-u_{i-1}^t - u_i^t + 1]$$

7. *Page 151*

   The equation for rule 101 should read:

$$u_i^{t+1} = \text{sgn}[2 - |u_{i-1}^t - u_i^t - 2u_{i+1}^t - 1|]$$

8. *Page 233*

   Change the color of the last 3 bits on the right of (12) from "blue, red, red" to "red, blue, blue".

ERRATA FOR VOLUME II

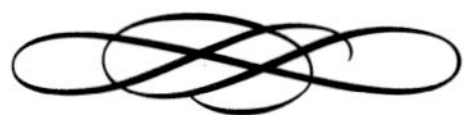

1. *Page vii* (front matter)

   Line 5 below Leon O. Chua: Change vol. 1 to Vol. 1.
   Line 7 below Leon O. Chua: Delete repeated pp.

2. *Page 454*

   Line 2 and Line 6 below Sec. 4.1: Change the number 93 to 67.

3. *Page 457*

   Line 14 above Sec. 4.2: Change the number 69 to 67.

4. *Page 461*

   Caption of Fig. 15: Change last sentence to "Each pattern has 26 rows and 68 columns."

5. *Page 475*

   First line above Sec. 5.2: Change $\{2, 3, 5\}$ to $\{1, 2, 3, 5\}$.

REFERENCES

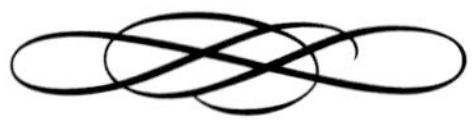

Adamatzky, A. [2009] "Book review on a nonlinear dynamics perspective of Wolfram's new kind of science," *J. Cellular Automata* **4**(3), 247–249.

Chang, J. & Muthuswamy, B. [2007] "Extracting optimal CNN templates for linearly-separable one-dimensional cellular automata," *Int. J. Bifurcation and Chaos* **17**, 749–779.

Chua, L. O. & Roska, T. [2002] *Cellular Neural Networks and Visual Computing* (Cambridge University Press, Cambridge).

Chua, L. O., Yoon, S. & Dogaru, R. [2002] "A nonlinear dynamics perspective of Wolfram's new kind of science. Part I: Threshold of complexity," *Int. J. Bifurcation and Chaos* **12**, 2655–2766.

Chua, L. O., Sbitnev, V. I. & Yoon, S. [2003] "A nonlinear dynamics perspective of Wolfram's new kind of science. Part II: Universal neuron," *Int. J. Bifurcation and Chaos* **13**, 2377–2491.

Chua, L. O., Sbitnev, V. I. & Yoon, S. [2004] "A nonlinear dynamics perspective of Wolfram's new kind of science. Part III: Predicting the unpredictable," *Int. J. Bifurcation and Chaos* **14**, 3689–3820.

Chua, L. O., Sbitnev, V. I. & Yoon, S. [2005a] "A nonlinear dynamics perspective of Wolfram's new kind of science. Part IV: From Bernoulli shift to $1/f$ spectrum," *Int. J. Bifurcation and Chaos* **15**, 1045–1183.

Chua, L. O., Sbitnev, V. I. & Yoon, S. [2005b] "A nonlinear dynamics perspective of Wolfram's new kind of science. Part V: Fractals everywhere," *Int. J. Bifurcation and Chaos* **15**, 3701–3849.

Chua, L. O. [2006] *A Nonlinear Dynamics Perspective of Wolfram's New Kind of Science*, Vol. I (World Scientific, Singapore).

Chua, L. O., Sbitnev, V. I. & Yoon, S. [2006] "A nonlinear dynamics perspective of Wolfram's new kind of science. Part VI: From time-reversible attractors to the arrow of time," *Int. J. Bifurcation and Chaos* **16**, 1097–1373.

Chua, L. O. [2007] *A Nonlinear Dynamics Perspective of Wolfram's New Kind of Science*, Vol. II (World Scientific, Singapore).

Chua, L. O., Guan, J., Sbitnev, V. I. & Shin, J. [2007a] "A nonlinear dynamics perspective of Wolfram's new kind of science. Part VII: Isles of Eden," *Int. J. Bifurcation and Chaos* **17**, 2839–3012.

Chua, L. O., Karacs, K., Sbitnev, V. I., Guan, J. & Shin, J. [2007b] "A nonlinear dynamics perspective of Wolfram's new kind of science. Part VIII: More Isles of Eden," *Int. J. Bifurcation and Chaos* **17**, 3741–3894.

Davis, P. J. [1979] *Circulant Matrices* (Wiley-Interscience, NY).

Garay, B. M. & Hofbauer, J. [2003] "Robust permanence for ecological differential equations, minimax, and discretizations," *SIAM J. Math. Anal.* **34**, 1007–1039.

Walker, C. [1971] "Behavior of a class of complex systems: The effect of system size on properties of terminal cycles," *J. Cybern.* **1**, 55–67.

Walker, C. C. & Aadryan, A. A. [1971] "Amount of computation preceding externally detectable steady state behavior in a class of complex systems," *J. Bio-Med. Comput.* **2**, 85–94.

Wikipedia [website] http://en.wikipedia.org/wiki/De_Bruijn_graph.

Wolfram, S. [2002] *A New Kind of Science* (Wolfram Media, Inc., Champaign, IL).

Wuensche, A. & Lesser, M. [1992] *The Global Dynamics of Cellular Automata* (Addison-Wesley Publishing Company, Reading, MA).

INDEX

Local Rules

Book Review

A Nonlinear Dynamics Perspective of Wolfram's New Kind of Science
(in 2 Volumes)

BY LEON O CHUA (UNIVERSITY OF CALIFORNIA AT BERKELEY, USA)
WORLD SCIENTIFIC SERIES ON NONLINEAR SCIENCE, SERIES A — VOL. 57

VOL. 1, 396 PP. PUB. DATE: JUNE 2006 ISBN: 981-256-977-4
VOL. 2, 580 PP. PUB. DATE: JUNE 2007 ISBN: 981-256-642-2

Reviewed by Andrew Adamatzky

"Our over-riding goal is to introduce cellular automata from the perspective of nonlinear dynamics for the lay readers unfamiliar with cellular automata", writes Prof. Leon Chua. The book is a colourful presentation indeed, which will please everyone with fresh ideas and attractive illustrations. The text is not about cellular automata, it is about a tiny but fundamentally complex class of one-dimensional binary-state threecell neighbourhood cellular automata.

The journey starts with Boolean-cube representation of the cell state transition rules. Every rule is shown by a cube such that every state of threecell neighbourhood is uniquely represented by a vertex of the cube. Vertices take values, 0 or 1, of cell-state transition function over corresponding states of the neighbourhood. A concept of linear separability of functions and indices of complexity are introduced then.

Complexity index of a transition function is a minimal number of parallel planes necessary to separate vertices of the Boolean-cube representing the function into the clusters of the same values. All functions are classified by three possible values of complexity index. It is illustrated many functions with complexity index one exhibit a repetitive behaviour while those with index two support mobile self-localizations, gliders, and non-trivially interacting propagating patterns. Few functions having complexity index three, it is claimed, have complex — based on the degree of unpredictability — behaviour.

Introducing a universal neuron is a culmination of the first volume. A universal neuron is a scalar nonlinear differential equation which generates each of 256 rules. The equation has eight parameters, which can be interpreted as synaptic weights hence the name. The parameters of the neuron-equation relate to complexity as follows. One needs just four parameters to represent a cell-state transition function with complexity index one, six parameters to represent a function with index two, and all eight parameters are necessary for functions with index three.[1] An abstract interpretation of universal neuron parameters and cell-state transition functions in terms of genotype which is provided and illustrated in the book, could be useful in future studies on evolution of cellular automata.

First volume ends with chapter "Predicting the Unpredictable", where 256 rules are partitioned into 88 global equivalence classes. Two rules are globally equivalent if they have identical nonlinear dynamics for all initial input patterns. Complexity index one is typical for 38 classes, index two for 41 classes, and just nine classes[2] have highest index of complexity.

Second volume proceeds along, in Chua's words, "*. . . a paradigm shift in research in Cellular Automata, which has hitherto been either empirical ⟨. . .⟩ or highly abstract. Our approach is both analytical and constructive, made possible by our discovery of an explicit unified formula for ⟨. . .⟩ characteristic functions, which was derived from an associated nonlinear differential equation, or a non-linear difference equation.*"

There we enjoy complete characterisation of behaviour of studied cellular automata in terms of attractors and invariant orbits. Main findings include exact classification of invertible and non-invertible rules (with period one to three), selection of Bernoulli rules. Also complete table of rules is provided which can be used to predict automaton global behaviour from any initial configuration.

Out of 256 rules, 112 non-periodic rules remarkably obey an explicit generalized Bernoulli shift formula, thereby allowing precise prediction of the global (time-asymptotic) dynamics. This fundamental result may have a substantial impact on future research in cellular automata. Indeed, the remaining 18 equivalence classes (including rules 30, 90, 110, etc.) also exhibit a more complex form of Bernoulli shift reminiscent but topologically different from the 112 Bernoulli rules reported so far.

It is also shown that the four universal rules exhibit $1/f$ power spectrum,[3] which is widely accepted as a signature of dynamical complexity in many disciplines, including humanities and arts.

Then author invites us to share his findings on fractal geometry of the characteristics function, explicit formulas for generation of characteristics functions from binary bit-strings, geometrical and analytic properties of characteristics functions. We also become acquainted with identification and classification of non-constructible configurations and fixed points. At this point the author introduces "isle of Eden", a configuration, whose only predecessor is the configuration itself, and which is a fixed point without transients in global evolution of cellular automaton.[4]

Rest of the volume deals with time-reversibility and invertibility of cellstate transition rules. These are studied by analysing, sometimes with the help of generalized Bernoulli maps, attractors of the rules' global dynamics. Dynamics of each attractor of a time-reversible rule is mirrored, in space and time, by its bilateral twin rule. Over half of the rules, 170 out of 256, are time-reversible in Chua's framework; other 86 rules are irreversible in the sense that attractors mirror each other only in space not time.

A test for time-reversibility of attractor is designed, and considered in relation with an idea that having attractor and its mirror we can move between time periods and thus mimic cosmological phenomena in cellular automata.

The book appeals to a wide audience. Apart of hard-core cellular automatists, those studying in nonlinear sciences, electronic engineering, mathematics and logics, complexity and emergent phenomena, and possibly even chemistry and biology will certainly discover exciting concepts, analogies and research tools in this refreshing text. Anyone from freshmen to elderly academics will find parts interesting to them. The volumes are somewhat special and exciting because they posses a unique "Chua brand" and show gradual development of ideas and concepts in an educational and entertaining yet mathematically rigorous manner.

---

[1]At this point one can be sceptical about representation potential of generative complexity at the global dynamics level, Leon Chua however provides certain demonstrative examples to strengthen the idea.

[2]Exact structure of the highest-complexity classes, in Wolfram coding, is $\{27, 83, 39, 53\}$, $\{29, 71\}$, $\{46, 116, 139, 209\}$, $\{58, 114, 163, 177\}$, $\{78, 92, 121, 197\}$, $\{105\}$, $\{150\}$, $\{172, 228, 202, 216\}$, $\{184, 226\}$.

[3]Such particular spectrum in cellular automata may be a result of glider interaction and reproduction.

[4]As Leon Chua poetically said, time really stood still on an isle-of-Eden, as in a black hole.

Rule Number	Left — Right Transformation	Global Complementation	Left — Right Complementation	Rule Number	Left — Right Transformation	Global Complementation	Left — Right Complementation
N	$T^{\dagger}[N]$	$\overline{T}[N]$	$T^{*}[N]$	N	$T^{\dagger}[N]$	$\overline{T}[N]$	$T^{*}[N]$
128	128	254	254	160	160	250	250
129	129	126	126	161	161	122	122
130	144	190	246	162	176	186	242
131	145	62	118	163	177	58	114
132	132	222	222	164	164	218	218
133	133	94	94	165	165	90	90
134	148	158	214	166	180	154	210
135	149	30	86	167	181	26	82
136	192	238	252	168	224	234	248
137	193	110	124	169	225	106	120
138	208	174	244	170	240	170	240
139	209	46	116	171	241	42	112
140	196	206	220	172	228	202	216
141	197	78	92	173	229	74	88
142	212	142	212	174	244	138	208
143	213	14	84	175	245	10	80
144	130	246	190	176	162	242	186
145	131	118	62	177	163	114	58
146	146	182	182	178	178	178	178
147	147	54	54	179	179	50	50
148	134	214	158	180	166	210	154
149	135	86	30	181	167	82	26
150	150	150	150	182	182	146	146
151	151	22	22	183	183	18	18
152	194	230	188	184	226	226	184
153	195	102	60	185	227	98	56
154	210	166	180	186	242	162	176
155	211	38	52	187	243	34	48
156	198	198	156	188	230	194	152
157	199	70	28	189	231	66	24
158	214	134	148	190	246	130	144
159	215	6	20	191	247	2	16